NON-STANDARD PARAMETRIC STATISTICAL INFERENCE

Non-Standard Parametric Statistical Inference

RUSSELL CHENG

OXFORD
UNIVERSITY PRESS

OXFORD
UNIVERSITY PRESS

Great Clarendon Street, Oxford, OX2 6DP,
United Kingdom

Oxford University Press is a department of the University of Oxford.
It furthers the University's objective of excellence in research, scholarship,
and education by publishing worldwide. Oxford is a registered trade mark of
Oxford University Press in the UK and in certain other countries

First Edition published in 2017

Impression: 1

Published in the United States of America by Oxford University Press
198 Madison Avenue, New York, NY 10016, United States of America

British Library Cataloguing in Publication Data
Data available

Library of Congress Control Number: 2017934662

ISBN 978–0–19–850504–4

Printed and bound by
CPI Group (UK) Ltd, Croydon, CR0 4YY

Acknowledgements

I am grateful to a number of colleagues and past students without whose interest and encouragement this book would never have come into being. Firstly, I must thank Dr Louise Traylor, a former research student whom I had the pleasure of supervising many years ago, who displayed great enthusiasm in tackling many non-standard statistical problems, and who wrote an exceptional doctoral thesis that was the initial catalyst in stirring my interest in the field. Had I been more single-minded, she would have made an ideal co-author, but she had long flown the academic nest by the time this book became a serious possibility. Two colleagues dating back to that time, Phil Brown, then Professor of Statistics at Liverpool, now Emeritus Professor at the University of Kent, and Tony Lawrance, then Professor of Statistics at Birmingham University, now Emeritus Professor at the University of Warwick, provided early encouragement. Tony was instrumental in having the early work presented as a read paper to the Royal Statistical Society by Louise and myself. I am also happy to acknowledge the contribution of Steve Wenbin Liu, Professor of Management Science and Computational Mathematics at the University of Kent. We collaborated in studying a particularly challenging aspect of non-standard behaviour when I was at Kent myself, and he provided much insightful input on the more subtle aspects of the subject. Steve was at one time an intended co-author, before I moved away from Kent. The work on finite mixture models in the final two chapters of the book is based on joint work with Dr Christine Currie, Associate Professor at the University of Southampton that is still ongoing. Most recently, Barry Nelson, Walter P. Murphy Professor at Northwestern University, whom I have known since his graduate student days, has provided the final encouraging push, reading early chapters and giving kind advice on how to complete the job. I must also thank Mr Keith Mansfield and latterly Mr Dan Taber, the commissioning editors at Oxford University Press, and the project manager Ms Indu Srinivasan of Integra Software Services, the partner production company, for the kind and careful way in which they have shepherded me though the entire publication process. Finally, I owe a great debt of gratitude to my wife, Ann, who has provided regular encouragement in my pursuit of an academic career since graduate student days.

Book Cover Visual: This represents a typical scatterplot of bootstrap estimated parameter points, colour-coded according to their distance from the parameter point estimated from the original data. Such scatterplots are used and discussed throughout the book.

Contents

1 Introduction

This book discusses the fitting of parametric probability distributions and regression functions, both nonlinear and linear, to data samples. Emphasis is placed on two aspects of this: (i) how to recognize and handle situations where the fitting process is non-standard in ways that will be described; parameter estimates can then behave unusually and not in accordance with standard theory, and (ii) the use of bootstrap resampling methods for analysing such problems and studying the actual behaviour of estimated quantities.

The statistical models, whether probability distributions or regression models, are discussed mainly from a frequentist likelihood viewpoint, though cases will also be considered from a Bayesian viewpoint.

Likelihood-based methods are appealing because of their wide application and because they are based on a well-established, powerful, and elegant theory. The likelihood function plays a central role; its behaviour determines the distributional properties of the statistics of interest. The most commonly encountered situation is where the likelihood can be assumed to satisfy certain regularity conditions. The theory is then considered to be 'standard' and at its most satisfying, with, asymptotically at least, test statistics and estimators that are normally or chi-squared distributed, and efficient in a well-defined sense. Such is the position of this theory and such is its wide applicability, that there is a tendency, at least in workers less experienced in statistical theory, to assume that regularity conditions will always hold, without checking that this is indeed the case. However, there are problems which can appear quite innocuous, where the theory does break down, sometimes in a spectacular way. With the growth of computing power, increasingly more complicated models are being proposed and considered, where often regularity conditions will not hold, or at least where the context of their application needs to be more carefully defined.

In this book, we investigate likelihood-based methods of parameter estimation and model fitting for non-standard situations, and investigate different ways in which regularity conditions can break down. Most statisticians would agree on what constitutes a regular problem. A number of variants of regular conditions have appeared in the literature, but the essentials are much the same. In that such conditions can be set out quite

Non-Standard Parametric Statistical Inference. Russell Cheng.

precisely, the range of non-regular problems looks intimidatingly large as, by definition, it encompasses all problems that do not satisfy the narrow specification of regularity. However, though we do not claim to be comprehensive in our treatment, it is possible, even if we cannot be entirely systematic, to make a structured study by setting out the main conditions that a problem must satisfy if it is to be regarded as standard, and then categorizing the most common ways in which each such condition might break down. We adopt a fairly geometric approach, as most of the departures from regularity can be most straightforwardly described in Euclidean geometrical terms. This approach seems particularly appropriate because it allows us to meet the main aim of the book, which is to give someone familiar with basic statistical methodology some additional insight into situations where such methodology might fail, and to provide methods that are easy to understand which overcome these difficulties. To make the technicalities as transparent as possible, we have not therefore attempted a presentation that is always fully rigorous mathematically, but to include sufficient theoretical framework to enable the essential character of any non-regularity to be properly identified, thereby allowing the adequacy of proposed practical solutions to be properly assessed. This should prevent incorrect or improper analyses being used and allow a sensible, secure, and practical formulation for any given problem. Using this approach we are able to discuss specific practical procedures in a broader context, showing how a range of commonly encountered and well-known examples, which have to date been discussed in isolation, can be handled within a broader theoretical framework.

In most of our examples, the properties of estimators are illustrated using parametric bootstrap resampling techniques. Parametric resampling is arguably less well regarded than nonparametric resampling. For example, Chernick (2008, Section 6.2.3) writes 'a parametric form of bootstrapping is equivalent to maximum likelihood' but that 'the existing theory on maximum likelihood estimation is adequate and the bootstrap adds little or nothing to the theory. Consequently, it is uncommon to see the parametric bootstrap used in real problems'. However, Chernick does then immediately point out that it can be used (i) as a check on the robustness/validity of the (asymptotic) parametric method and that (ii) it is often useful where the estimator of interest has a distribution that is difficult to derive or has an asymptotic distribution that does not provide a good small sample approximation.

These two aspects are actually extremely useful in practical situations, especially when combined with the fact that the parametric bootstrap is very easy, almost trivial, to implement, including where situations are non-standard. Indeed with regard to point (ii), it allows construction of estimators or statistics not previously considered in the literature that are tailored to a particular problem without worrying about whether their distributions are tractable to derive or not, as the parametric bootstrap allows one to obtain these numerically. We give an example of such a statistic in Section 4.6, which can be used to make a formal lack-of-fit test when fitting a regression function to data without the need to have replication of observations at explanatory variable points.

There is a significant literature demonstrating that bootstrapping performs at least as well as standard asymptotic theory in gauging the behaviour of estimators like those provided by maximum likelihood, and indeed for small samples bootstrapping usually performs rather better. However, perhaps the most attractive feature of bootstrapping is that the results are very amenable to graphical representation, providing easy-to-understand corroboration of the sampling behaviour of estimators. Moreover, relatively little mathematical theory needs to be invoked in applying the methodology. Of course, it is more satisfactory if mathematical theory can be invoked, but it is not usually necessary to be able to carry out bootstrapping in a meaningful way.

1.1 Terminology

We begin by discussing the view of statistical modelling adopted in this book and the terminology that will be used in discussing this. Our basic assumption is that we have a data set made up of n observations of a random variable Y drawn from a parametric probability distribution. In most of our discussion, we will assume that the observations are independently drawn from the same parametric continuous distribution, with cumulative distribution function (CDF), $F_Y(\cdot, \boldsymbol{\theta})$, depending on a vector of d unknown parameters $\boldsymbol{\theta} = (\theta_1, \theta_2, \ldots, \theta_d)^T$. Where the context is clear, we write $F_Y(\cdot, \boldsymbol{\theta})$ simply as $F(\cdot, \boldsymbol{\theta})$. The corresponding probability density function (PDF) will be denoted by $f_Y(\cdot, \boldsymbol{\theta})$ or $f(\cdot, \boldsymbol{\theta})$. In this situation, where the observations are all identically and independently distributed, we will denote the entire sample by $\mathbf{Y} = (Y_1, Y_2, \ldots, Y_n)^T$.

An actual set of observation values will be denoted by $\mathbf{y} = (y_1, y_2, \ldots, y_n)^T$. It is sometimes convenient to regard the sample, with no loss of generality, as being ordered with $y_1 < \ldots < y_n$, with no additional notation used when it is clear from the context that the sample is ordered. But in situations where there may be doubt, we will for emphasis write an ordered sample as $y_{(1)} < \ldots < y_{(n)}$, with the subscripts in round brackets.

The form of the probability distribution from which Y is drawn constitutes the *statistical model* of the data, and *fitting* the statistical model is simply the process of using the data to estimate the unknown parameter values so that the particular probability distribution obtained by setting the parameters to their estimated values provides a good (usually in a clearly defined statistical sense) representation of the data set. The best-known example of a statistical model is probably when the Y_i are all normally distributed with mean μ and variance σ^2, which we write in the conventional way, as $Y_i \sim N(\mu, \sigma^2)$, using the symbol '$\sim$' to mean 'is distributed as'.

We will also discuss *regression* models of the form

$$Y = \eta(x, \boldsymbol{\varphi}) + \varepsilon, \tag{1.1}$$

where $\eta(x, \boldsymbol{\varphi})$, the regression function, is non-random and depends on a vector, $\boldsymbol{\varphi}$, of unknown parameters, whilst ε is a random error, with distribution $F_\varepsilon(\cdot, \boldsymbol{\psi})$, depending also on a vector, $\boldsymbol{\psi}$, of unknown parameters. We allow the possibility that $\boldsymbol{\varphi}$ and $\boldsymbol{\psi}$ may have common components, and denote by $\boldsymbol{\theta} = (\boldsymbol{\varphi}, \boldsymbol{\psi})$ the column vector of all unknown parameters. Strictly speaking, it would be more correct to write $\boldsymbol{\theta} = (\boldsymbol{\varphi}^{\mathrm{T}}, \boldsymbol{\psi}^{\mathrm{T}})^{\mathrm{T}}$, but we shall not use this level of exactness because of its clumsiness. In specific examples, it will be made clear if η and F_ε both depend on a particular component θ_i. The fact that (1.1) is a statistical model can be emphasized by writing (1.1) in the form

$$Y \sim F_Y(\boldsymbol{\theta}, x), \tag{1.2}$$

where F_Y denotes the probability distribution of Y. This change from (1.1) to (1.2) might appear insignificant, but focusing on the distributional form of Y enables the likelihood, the main statistical quantity of interest in this book, to be written down explicitly.

There are two main ways that the unknown parameters can be viewed. The frequentist view is that, though unknown, the parameter values have a fixed 'true' value $\boldsymbol{\theta}^0$. Our objective in this case is to estimate $\boldsymbol{\theta}^0$ by using the data to obtain $\hat{\boldsymbol{\theta}}$, an estimated value of $\boldsymbol{\theta}^0$, which is therefore dependent on $\mathbf{y}$, i.e. $\hat{\boldsymbol{\theta}} = \hat{\boldsymbol{\theta}}(\mathbf{y})$, that will in some clear statistical sense be close to $\boldsymbol{\theta}^0$. For example, the estimator $\hat{\boldsymbol{\theta}}$, considered as a random variable depending on $\mathbf{Y}$, that is $\hat{\boldsymbol{\theta}} = \hat{\boldsymbol{\theta}}(\mathbf{Y})$, should be consistent in the sense that its distribution becomes progressively more concentrated about $\boldsymbol{\theta}^0$ as the sample size n increases. The Bayesian view is that $\boldsymbol{\theta}$ is random with a prespecified prior distribution $\pi(\boldsymbol{\theta})$ and in this case the objective is to find the posterior distribution, which reflects how the data $\mathbf{y}$ influences the prior. A commonly held Bayesian view has been that consistency is not often relevant in the Bayesian context, however, it seems now more recognized to be of importance. The version of consistency that we use when considering Bayesian methods is similar to the frequentist case. An unknown true fixed value $\boldsymbol{\theta}^0$ is assumed, only now with the prior regarded as reflecting the probabilistic uncertainty in this value. Consistency then means that the posterior distribution should become increasingly concentrated, in a well-defined statistical sense, about this true value $\boldsymbol{\theta}^0$ as the sample size $n \to \infty$. The Bayesian approach is used in Chapter 17 where finite mixture distributions are discussed.

In the next section, we outline the maximum likelihood method, the most important frequentist method for estimating $\boldsymbol{\theta}$, under conditions that are typical of what are regarded as being regular, where the theory of the method is then *standard* and at its most elegant, powerful, and complete.

1.2 Maximum Likelihood Estimation

Maximum likelihood (ML) is undoubtedly the most powerful general method available for estimating the unknown parameters $\boldsymbol{\theta}$, and it is the method, and its variations, that are used in this book. Practically any text on theoretical statistics will cover the method, however, good references for our purposes are Stuart and Ord (1991, Chap. 18), Cox and Hinkley (1974), and Young and Smith (2005).

For ease of reference, and to make the book more self-contained, we summarize, in the next chapter, the main features of the method under what can be viewed as standard conditions. The reader already conversant with the theory need only skim through the chapter, as much as anything to familiarize themselves with the notation.

We consider a PDF $f(y, \boldsymbol{\theta})$ where the d-component parameter can take values in a d-dimensional parameter space by $\boldsymbol{\Theta}$. Thus, the set of all possible PDFs $\{f(y, \boldsymbol{\theta}); \boldsymbol{\theta} \in \boldsymbol{\Theta}\}$

forms a *family* of distributions all with the same functional form $f(\cdot, \boldsymbol{\theta})$. Typically, $\boldsymbol{\Theta} = \{\boldsymbol{\theta};\ \theta_i \in R_i,\ i = 1, 2, \ldots, d\ \}$, where the R_i are prescribed, possibly non-finite intervals. The discussion in this section concerns the case where d is fixed and known.

Suppose that $\mathbf{y} = \{y_1, y_2, \ldots, y_n\}$ is a random sample drawn from the continuous distribution with PDF $f(y; \boldsymbol{\theta}^0)$, where $\boldsymbol{\theta}^0 \in \boldsymbol{\Theta}$ is the true but unknown value of $\boldsymbol{\theta}$. Then, for any given $\boldsymbol{\theta} \in \boldsymbol{\Theta}$, the *likelihood* is defined as

$$lik(\boldsymbol{\theta}, \mathbf{y}) = \prod_{i=1}^{n} f(y_i; x_i, \boldsymbol{\theta}) \tag{1.3}$$

and the *log-likelihood* as

$$L(\boldsymbol{\theta}, \mathbf{y}) = \sum_{i=1}^{n} \ln f(y_i; x_i, \boldsymbol{\theta}). \tag{1.4}$$

For simplicity, we shall often write $L(\boldsymbol{\theta}, \mathbf{y})$ as $L(\boldsymbol{\theta})$.

The *maximum likelihood estimator* (MLE) of $\boldsymbol{\theta}^0$, $\hat{\boldsymbol{\theta}}$, is that value of $\boldsymbol{\theta}$ which maximizes the log-likelihood, that is

$$\hat{\boldsymbol{\theta}} = \arg\max_{\boldsymbol{\theta} \in \boldsymbol{\Theta}} L(\boldsymbol{\theta}, \mathbf{y}). \tag{1.5}$$

The MLE is illustrated in Figure 1.1. This only depicts the one-parameter case, but is intended to show the key features of the log-likelihood responsible for the very specific and important asymptotic form of the distribution of the MLE. This is that, *in standard conditions,* $\boldsymbol{\theta}^0$ is an *interior* point of $\boldsymbol{\Theta}$, and moreover the log-likelihood is well approximated by a concave quadratic function (i.e. one with a negative definite quadratic form) in an open ball $B = B(\boldsymbol{\theta}^0, \rho) = \{\boldsymbol{\theta} : |\boldsymbol{\theta} - \boldsymbol{\theta}^0| < \rho\}$ centred on $\boldsymbol{\theta}^0$ and of radius ρ, for some $\rho > 0$, with the ball lying entirely in $\boldsymbol{\Theta}$, because $\boldsymbol{\theta}^0$ is an interior point. Then, providing only that the MLE $\hat{\boldsymbol{\theta}}$ is consistent, we can assume $\hat{\boldsymbol{\theta}}$ will lie in B once n is large enough. The assumption $B \subset \boldsymbol{\Theta}$ then allows $\hat{\boldsymbol{\theta}}$ to vary freely in B as $\mathbf{y}$ varies, and this unrestricted behaviour results in the asymptotic normality of $\hat{\boldsymbol{\theta}}$. (For simplicity we will always use just the modulus $|\cdot|$ to denote the Euclidean norm whatever the dimension of $\boldsymbol{\theta}$.)

In many of the non-standard situations to be considered, $\boldsymbol{\theta}^0$ can be on the boundary. The following definitions are therefore occasionally useful for our discussion:

$$\bar{\boldsymbol{\Theta}} = \text{Closure}(\boldsymbol{\Theta}),\ \partial\boldsymbol{\Theta} = \text{Boundary}(\boldsymbol{\Theta}), \text{Int}\boldsymbol{\Theta} = \bar{\boldsymbol{\Theta}} \setminus \partial\boldsymbol{\Theta}. \tag{1.6}$$

If $\boldsymbol{\Theta}$ is open, then $\text{Int}\boldsymbol{\Theta} = \boldsymbol{\Theta}$.

Also shown in Figure 1.1 are a height, a horizontal width, and a gradient. Any of the three can be used to test the hypothesis that the true parameter value is $\boldsymbol{\theta} = \boldsymbol{\theta}^0$, where $\boldsymbol{\theta}^0$ is a given value. Such a *hypothesis* test will be discussed in Section 3.3. When the log-likelihood is exactly quadratic, the three quantities are equivalent in that each can be calculated from either of the other two.

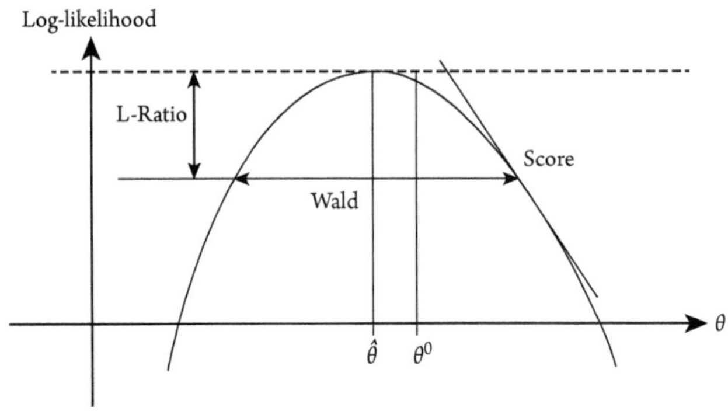

Figure 1.1 Typical form of the log-likelihood when it depends on just one scalar parameter θ. Also shown are three quantities each of which can be used to test the hypothesis that the unknown value of the parameter is a specified value θ^0.

In Chapter 3, we will set out, more formally than the preceding discussion, a set of conditions from which follows, relatively easily, the asymptotic normality of the MLE $\hat{\boldsymbol{\theta}}$ and related distributional results. In a nutshell, the proofs all rely on considering a Taylor series expansion of either the log-likelihood or its derivative, appealing to a multivariate central theorem to show that the first-order sums are asymptotically normally distributed, and the weak law of large numbers to show that the second-order derivatives, which essentially define the second-order coefficients in the expansions, can be treated as constants in probability under the regularity assumptions. Moreover, and this is why the result is so practical, these second-order derivatives can be evaluated with $\boldsymbol{\theta} = \hat{\boldsymbol{\theta}}$, so that they and therefore the asymptotic distribution can be treated as being effectively known. Asymptotic theory therefore offers a simple result, establishing conditions which guarantee that L becomes increasingly better approximated by a quadratic as the sample size $n \to \infty$.

1.3 Bootstrapping

Though it can be used in principle whatever the size of n, there is one problem that bedevils asymptotic theory, namely, that the results are only approximate for finite n. Thus when used, it has to be in the expectation that n is 'sufficiently large' for the approximation to be good enough for practical purposes. The problem is that it is not usually easy to gauge how good the approximation is for any given n; and this is the main concern if asymptotic theory is applied in an uncritical way.

An alternative to asymptotic theory is the *bootstrap* (BS). It is well recognized that bootstrap analysis often has better convergence properties than first-order asymptotic theory, so that where it can be used it would be preferable even if, like asymptotic theory, it cannot be considered to be a probabilistically exact method. Here and in the rest of the book, we use the term 'probabilistically exact' to describe any distributional result obtained for a statistic of interest where the form of the distribution is known *exactly* whatever the sample size. Thus, though asymptotic results are asymptotically exact, typically normal or chi-squared, this is only achieved in the limit. For finite n, asymptotic results are not exact in general.

We have used *parametric bootstrapping*, described by Davison and Hinkley (1997, Section 2.2), for example, who actually call it parametric simulation, to analyse examples in this book. The method is discussed in more detail in Chapter 4, which describes the basic parametric bootstrap, but also a number of uses of bootstrapping that may not be so familiar. For example, we show that bootstrapping provides a simple way of calculating confidence bands for entire CDFs and for entire regression functions, and for calculating critical values in goodness-of-fit (GoF) tests in situations not covered by theoretical methods.

In general, like asymptotic theory, parametric bootstrapping does not provide results that are probabilistically exact, though there are notable exceptions—for example, in GoF tests of location-scale models, as discussed in Chapter 4.

However, a big advantage of bootstrapping is that it can provide qualitative visual information through distributional plots and scatterplots, and if necessary (it has to be said at added computational cost), it enables sensitivity analyses to be carried out that assess probabilistic exactness.

Moreover, the samples generated by bootstrapping have exactly the same sample size n as the original data, so that the variability due to sample size that is assessed by the bootstrap analysis matches that of the actual sample.

In many real-world situations, use of bootstrapping provides a simple approach that allows discussion of statistical issues with practitioners who are not specialist statisticians, and who wish to concentrate on a problem as it arises in their own subject area without having to take time to absorb and evaluate complex statistical techniques.

There is one important final point to make. Being computationally intensive, an important practical question concerning bootstrapping is whether it is capabable of handling what are now called large-scale problems, see Efron (2010), for example, where sample size or dimensionality is large. We shall not be discussing the question in this book, because our focus is on how to recognize and handle non-standardness in statistical model structure, and this is not directly related to problem size. Examination of size issues is an undertaking that would have changed the purpose of this book unacceptably. However, the obvious approach to countering the effect of problem size is to use parallel computation. Bootstrap analysis is characterized by subjecting all the observations in a sample to exactly the same calculations, and so is particularly amenable to parallelization. Orders-of-magnitude improvement in the overall speed with which large data samples can be analysed is therefore possible if parallelization is employed. New generations of *general purpose graphical processor units* (GPGPUs) are now being regularly developed, so that parallel processing is increasingly accessible and affordable even to individual researchers and practitioners. The development of efficient parallel statistical computational and bootstrapping algorithms is a fertile area of research interest to take advantage of such developments.

1.4 Book Layout and Objectives

The book layout is very simple. The next chapter provides immediate motivation by giving elementary examples to illustrate the types of problem that will be discussed in the rest of the book.

Chapters 3 and 4, which then follow, respectively summarize standard asymptotic theory and parametric bootstrapping. As already mentioned, these are the two main approaches used in the book to statistically analyse fitted models.

The material in Chapter 3 on standard asymptotic theory is well known, so may be familiar to readers as the basis of much statistical inference. In analysing non-standard problems, we have tried to use standard methods where we can, avoiding possibly less tractable methods. Our summary therefore has an emphasis that is most apposite for the discussion of the non-standard problems considered in this book, with the material presented intended simply to provide a convenient reference.

Chapter 4 is a much fuller version of the discussion of the basic parametric bootstrap outlined in the previous section.

The book, from Chapter 5 onwards, discusses in detail each of the non-standard problems introduced in Chapter 2.

The book's objectives are also easily stated. As pointed out at the outset of this chapter, the two main objectives of this book are (i) to identify situations where model fitting is non-standard, and to show how to fit parametric statistical models to data samples in such situations, and (ii) to show how to use parametric bootstrapping methods for analysing the behaviour of estimated quantities in such situations. Within these two objectives, simple model building is an underlying overall theme.

With these objectives in mind, we end the chapter with some comments on the way the material has been set out in this book.

The general mathematical level of the book is suitable for a graduate readership. Many of the problems discussed are of current interest both theoretically and practically. The book should therefore provide an introduction to graduate students wishing to do research in the area of non-standard parametric statistical estimation. Chapters 10, 11, and 13 still draw significantly on Traylor (1994), the doctoral dissertation of Louise Traylor, a former research student and co-author of Cheng and Traylor (1995), the early paper read to the Royal Statistical Society on the subject. I am very happy to acknowledge her influential role, even if so long ago, in the genesis of this book.

The presentation aims to be reasonably rigorous, but not in a fully formal way so as not to be overly mathematical and dry. Only sufficient mathematical detail is included to make clear the nature of each non-standard problem discussed.

Our emphasis has been on presenting practical solutions that are easy to understand, and that are mathematically straightforward to implement. Focus is placed on likelihood-based estimation methods. These are asymptotically efficient in standard situations, but when modified and used in non-standard situations, their efficiency properties are then not always quite so clear. We would, however, expect the solution methods proposed to be sufficiently effective to be useful practically. Not covered in this book, but requiring further work, is a more detailed examination of both the efficiency and power of some of our proposed methods, particularly in the model building discussed towards the end of this book. A simple but time-consuming way of doing this would be to carry out extensive bootstrap simulation studies. We have not attempted such evaluations in a systematic way in this book, because of the effort involved.

The book is intended to be of interest to practitioners who carry out statistical modelling of problems that do not fall within the framework of standard statistical methodology. Such problems are often very individual, and may require solutions where the calculations are non-standard. This is illustrated in the book by the numerical examples used, which are almost all based on real data sets appearing in the literature or that have arisen in consultancy work. The calculations in these examples were mainly carried out using Excel spreadsheets with VBA macros implementing the algorithms involved. The majority of statisticians tend to use **R** nowadays, but spreadsheet use is still the more common currency in the wider world, and using these provides an easy means of communication with clients who, while quite mathematically orientated and statistically literate, are not specialists.

The results presented graphically in the book reflect what might be used with clients, especially those not wishing to invest time in learning or using a statistical package. Such visual presentation of results is particularly effective in meetings involving senior company executives who often are unfamiliar with application packages, but are used to seeing dynamic spreadsheet demonstrations.

The macro-driven spreadsheets used for all the worked examples of this book are available online at <http://www.personal.soton.ac.uk/rchc>.

2 Non-Standard Problems: Some Examples

It would be satisfying to formulate non-standard behaviour in a way that parallels the generality obtained in the study of standard behaviour as achieved in the theory of exponential families. However, given the current state of research on non-standard problems, we are some way off this ideal. In considering solutions, we have not attempted a unified approach, but instead have simply tackled each problem type on its own merits.

In this section, we provide motivation for the rest of the book by briefly giving concrete examples of the main non-standard problem types that will be discussed. As the examples are simply intended to illustrate how non-standard behaviour can occur, we focus on explicit examples that we can examine and resolve on their own, avoiding being drawn into problems involving aspects like the relationship between different problem types.

However, we can give our illustrations some structure by categorizing problem types in an informal systematic manner based on the geometric form taken by the log-likelihood $L(\boldsymbol{\theta}, \mathbf{y})$ in the parameter region $\boldsymbol{\Theta}$ of eqn (1.6) and on the location of the true parameter $\boldsymbol{\theta}^0$ in $\boldsymbol{\Theta}$.

As already mentioned, the standard situation is when $\boldsymbol{\theta}^0$ is an interior point, with $L(\boldsymbol{\theta}, \mathbf{y})$ well approximated by a concave quadratic function of $\boldsymbol{\theta}$ in the neighbourhood of $\boldsymbol{\theta}^0$ and with $L(\boldsymbol{\theta}, \mathbf{y})$ having a global maximum, $\hat{\boldsymbol{\theta}}$, in this neighbourhood that tends in an unrestricted way to $\boldsymbol{\theta}^0$ as the sample size $n \to \infty$. In the standard situation, the rate of convergence is $|\boldsymbol{\theta} - \boldsymbol{\theta}^0| = O_p(n^{-1/2})$, where the subscript p indicates that the convergence rate $O(.)$ holds in probability.

If $\boldsymbol{\theta}^0$ or $L(\boldsymbol{\theta}, \mathbf{y})$ does not fully meet the assumptions of the standard situation, then non-standard behaviour can occur. We outline a selection of ways that this can happen.

2.1 True Parameter Value on a Fixed Boundary

This is the case where the ML estimator $\hat{\boldsymbol{\theta}}$ defined in eqn (1.5) cannot vary freely about $\boldsymbol{\theta}^0$ because $\boldsymbol{\theta}^0$ actually lies on the boundary. A simple example is where

$Y_i \sim N(\mu, \sigma^2)$ $i = 1, 2, \ldots, n$, with μ and σ^2 unknown and to be estimated. If μ is unrestricted with $-\infty < \mu < \infty$, then the ML estimator $\hat{\mu} = \bar{Y} = \sum_{i=1}^{n} Y_i/n$ is unrestricted, and the situation is completely standard. However, suppose that the problem requires that $0 \leq \mu$ and the true value μ_0 is on the boundary, i.e. $\mu_0 = 0$. Then the MLE has the form

$$\hat{\mu} = \begin{cases} 0 \text{ if } \bar{Y} \leq 0 \\ \bar{Y} \text{ if } \bar{Y} > 0 \end{cases},$$

so that the distribution of $\hat{\mu}$ is non-standard, where $\hat{\mu}$ is equal to zero with probability 0.5 and has the half-normal distribution otherwise. A typical repercussion concerns hypothesis testing, which in effect becomes one-sided in this situation. We discuss the problem briefly in Section 6.2.

2.2 Infinite Likelihood: Weibull Example

Strictly speaking, this example does not involve $\hat{\boldsymbol{\theta}}$ being restricted by the boundary of $\boldsymbol{\Theta}$. However, it does involve $\hat{\boldsymbol{\theta}}$ being restricted by the data sample itself. Consider the following situation.

Suppose Y is a continuously distributed random variable whose PDF is positive only for $y \geq a$, with a unknown, so that it has to be estimated from a sample $\mathbf{y} = (y_1, y_2, \ldots, y_n)$. We write the ordered sample as $y_{(1)} < y_{(2)} < \ldots < y_{(n)}$. The condition $a \leq y$ must apply to each observation, so that $\hat{a}$, the estimator of the threshold a, must satisfy $\hat{a} \leq y_{(1)}$, the smallest observation. We therefore have a boundary condition that is dependent on the sample $\mathbf{y}$. Clearly, one must allow the estimator $\hat{a}$ to tend to $y_{(1)}$, and indeed to have $\hat{a} = y_{(1)}$ if this gives a distribution that is non-degenerate. However, allowing $a \to y_{(1)}$ can result in the log-likelihood becoming infinite with not all parameters consistently estimable using ML. This turns out to be a significant practical problem, as the following example shows.

Consider the three-parameter Weibull distribution with density

$$f_W(y; \boldsymbol{\theta}) = \frac{c}{b}\left(\frac{y-a}{b}\right)^{c-1} \exp\left(-\left(\frac{y-a}{b}\right)^c\right), \tag{2.1}$$

where all three parameters a, b, and c are unknown, a case examined by Smith and Naylor (1987). The two-parameter version, where $a = 0$, is well known in life testing and modelling. The parameter a allows an unknown threshold or shifted origin to be incorporated. A good introduction to both versions is given by Johnson, Kotz, and Balakrishnan (1994), who also provide an extensive bibliography for the distribution. A more up-to-date general reference for the Weibull distribution is Rinne (2008).

The 'natural' parameter space for the PDF (2.1) is

$$\Theta = \{\boldsymbol{\theta} : -\infty < a < \infty,\ 0 < c < \infty,\ 0 < b < \infty\}, \tag{2.2}$$

with (2.1) defining a non-degenerate model for each finite value $\boldsymbol{\theta} \in \Theta$.

Suppose $\mathbf{y}$ is a sample drawn from the Weibull distribution with PDF (2.1), where a, b, and c are unknown. We consider their estimation by the ML method.

The log-likelihood is

$$L(\boldsymbol{\theta}, \mathbf{y}) = n \ln c - nc \ln b + (c-1) \sum_{i=1}^{n} \ln(y_{(i)} - a) - \sum_{i=1}^{n} \{(y_{(i)} - a)/b\}^c.$$

As already pointed out, to be meaningful, a can only be selected subject to $a \leq y_{(1)}$.

Consider the log-likelihood as $a \to y_{(1)}$. For any $c < 1, L(\boldsymbol{\theta}, \mathbf{y})$ is dominated by the first term $(c-1)\ln(y_{(1)} - a)$ in the first summation, with $(c-1)\ln(y_{(1)} - a) \to +\infty$ as $a \to y_{(1)}$. Thus $L(\boldsymbol{\theta}, \mathbf{y})$ has no finite maximum. This in itself would not be a problem if we could obtain sensible estimates for b and c by first maximizing $L(\boldsymbol{\theta}, \mathbf{y})$ with respect to $\hat{b}$ and $\hat{c}$ for fixed a, and then letting $a \to y_{(1)}$. This does not work. As pointed out by Huzurbazar (1948), the ML estimators $\hat{b}$ and $\hat{c}$ are not consistent.

Let $(\hat{b}(a), \hat{c}(a)) = \arg\max_{b,c} L(a, b, c, \mathbf{y})$. To find $\hat{b}(a)$, $\hat{c}(a)$, we first solve $\partial L(\boldsymbol{\theta}, \mathbf{y})/\partial b = 0$. This gives $b = \left\{\sum_{i=1}^{n} (y_{(i)} - a)^c / n\right\}^{1/c} = \hat{b}(a, c)$, say. Substituting this into $L(\boldsymbol{\theta}, \mathbf{y})$, we can write $L(\boldsymbol{\theta}, \mathbf{y}) = L\{a, \hat{b}(a, c), c\}$. We then find that $\partial L\{a, \hat{b}(a, c), c\}/\partial c = 0$ has solution $\hat{c}(a) \simeq -n/\ln(y_{(1)} - a)$ as $a \to y_{(1)}$. Use of this expression in $\hat{b}(a, c)$ gives $\hat{b}(a) \simeq \left[\left\{\frac{\exp(-n)+n-1}{n}\right\}^{1/n}\right]^{-\ln(y_{(1)}-a)}$ as $a \to y_{(1)}$. Thus $\hat{b}(a) \to 0, \hat{c}(a) \to 0$ as $a \to y_{(1)}$, and so neither is consistent.

This is a celebrated problem with a large literature, involving many distributions where there is a threshold. For example, Hill (1963) expressed an early concern in the case of the three-parameter lognormal distribution, which he reiterated in Hill (1995). Other cases are listed in Chapter 8, where a number of possible solutions are discussed.

2.3 Embedded Model Problem

We now give an example involving a significant boundary problem that occurs when the parametrization used for a family of models appears perfectly adequate for all $\boldsymbol{\theta} \in \mathbf{Int}(\boldsymbol{\Theta})$ but does not properly identify all legitimate non-degenerate distributions that correspond to boundary points. The situation is of most concern when certain individual boundary points $\boldsymbol{\theta}^0$ do not correspond to a unique distribution, but to *a whole subset of distinct non-degenerate distributions*. In this situation, simply letting $\boldsymbol{\theta} \to \boldsymbol{\theta}^0$ does not yield a meaningful result at all, so it might appear that a limiting distribution corresponding to $\boldsymbol{\theta}^0$ does not exist or is degenerate, even though non-degenerate limits are possible which might be the best fit to the data. We call such a hidden model, or family of models, an *embedded model*, and will define it formally in Section 5.2. Often the difficulty is compounded by $\mathbf{Int}(\boldsymbol{\Theta})$ not being compact, with $\boldsymbol{\theta}^0$ not being a finite point, so that the embedding is not apparent. The first problem therefore is to recognize when an embedded model may be present. We stress that embeddedness is not an inherent or invariant property. The problem simply arises because of the form of the particular parametrization used in specifying the model as a whole. A reparametrization is always possible that will turn an embedded model into a regular special case, moreover, with $\boldsymbol{\theta}^0$ finite. What can be disconcerting is that embedding can occur using what appears to be a 'natural' model parametrization that is perfectly satisfactory for all $\boldsymbol{\theta} \in Int\boldsymbol{\Theta}$, but which does not then extend satisfactorily to include distributions corresponding to $\boldsymbol{\theta} \in \partial\boldsymbol{\Theta}$.

We give an example involving again the three-parameter Weibull distribution with PDF (2.1). The problem is quite different from the infinite likelihood problem previously discussed, and only concerns the behaviour of (2.1) as $\boldsymbol{\theta}$ approaches the boundary of $\boldsymbol{\Theta}$. Usually, one would not bother to check the boundary exhaustively, as this can be an unrewardingly tedious exercise, but if there is any possibility of embedding, then the only certain way of ensuring that all cases are identified is to check the entire boundary of $\boldsymbol{\Theta}$.

We illustrate in our Weibull case by first examining what might initially appear to be the entire boundary of the Weibull parameter space given in eqn (2.2).

The boundary model where any one given component is allowed to tend to one or other of the limits of its range, but with the other two components held fixed and finite, can readily be identified as being degenerate in an obvious way. If $b \to 0$, the distribution converges to just a probability atom of size unity located at a. If we let $c \to 0$, we again obtain a probability atom at a but of size $(1 - e^{-1})$, the rest of the distribution being improperly distributed, with all values of $Y > a$ becoming equally likely in the sense of $f(y_1) \to 0$ and $f(y_1)/f(y_2) \sim 1$ for any fixed y_1, y_2. Similarly, we have improper distributions if $c \to \infty$ or $b \to \infty$. Finally, the distribution becomes degenerate if $a \to +\infty$ (respectively $-\infty$) when, with probability one, $Y = +\infty$ (respectively $-\infty$).

The above obvious cases might lead one to suppose that all boundary models are degenerate, with none giving a good fit to non-degenerate data. However, this is not so.

If we set

$$b = \mu - a, c = b/\sigma, \text{with } \mu, \sigma \text{ fixed,} \tag{2.3}$$

and let $a \to -\infty$, then $b \to \infty$ and $c \to \infty$, with $a + b = \mu$ and $b/c = \sigma$ remaining finite. We then have convergence to the extreme value distribution

$$f_W(y, a, b, c) \to f_{EXT}(y, \mu, \sigma) = \sigma^{-1} \exp\{(y - \mu)/\sigma] - \exp[(y - \mu)/\sigma]\}. \tag{2.4}$$

We can plot the log-likelihood in a way similar to Figure 1.1 by selecting a and plotting the *profile* log-likelihood $L^*(a) = \sup_{b,c} L(\boldsymbol{\theta})$. In this case, unlike Figure 1.1, if $a \to -\infty$ $L(a)$ does not tend to $-\infty$, but instead $L^*(a) \to L_{EV}(\hat{\mu}, \hat{\sigma})$, a finite value corresponding to the maximized log-likelihood of the extreme value distribution with PDF (2.4). Thus, if the boundary model is the best fit, then $L^*(a)$ will be monotone increasing as $a \to -\infty$, once a is negative enough, with a horizontal asymptotic value of $L_{EV}(\hat{\mu}, \hat{\sigma})$.

We call such a boundary model an *embedded model.* Embedded models abound in the literature, but are usually not acknowledged as such, sometimes not even recognized. They are formally defined and examined in Chapters 5, 6, and 7.

2.4 Indeterminate Parameters

Another problem that can occur with boundary models that is highlighted in this book concerns the converse situation, where instead of a single point representing a whole family of distributions, we have *one* distribution represented by a whole subset of boundary points. A much-discussed example is the two-component finite mixture model, which we take here in the form given by Chen, Ponomareva, and Tamer (2014):

$$g(y;\delta,\theta) = \delta f(y;\theta) + (1-\delta)f(y;\theta^0), \tag{2.5}$$

with $f(\cdot;\theta)$ a given continuous PDF, where θ^0 is assumed known with $-a < \theta^0 < a$, where $0 < a < \infty$ with a also known, and where θ and δ satisfy $-a < \theta < a, 0 \le \delta \le 1$, but are otherwise unknown so that they have to be estimated.

A typical use of the model is where the known component $f(y;\theta^0)$ is of special interest, but there is uncertainty that the component, on its own, models the data sufficiently accurately. The other component $f(y;\theta)$ is therefore added, but weighted by δ as a precaution. One would want therefore to statistically test if this addition is needed. Figure 2.1 depicts the parameter space for this model, this being the rectangle

$$\boldsymbol{\Theta} = \{(\theta,\delta)| -a \le \theta \le a,\ 0 \le \delta \le 1\}$$

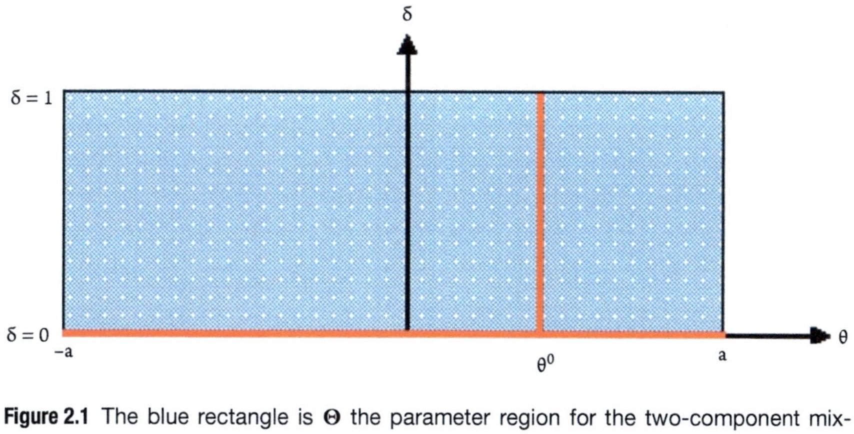

Figure 2.1 The blue rectangle is $\boldsymbol{\Theta}$ the parameter region for the two-component mixture model of eqn (2.5). The two red lines taken together are the subset of points corresponding to the model $f(y;\theta^0)$.

shown in light blue in the figure. The two-component mixture model with PDF given by eqn (2.5) is defined for all $\theta \in \mathbf{\Theta}$. The special case $f(y;\theta^0)$ is obtained if (θ,δ) lies on either of the lines $\delta = 0$ or $\theta = \theta^0$, as depicted in the figure. Thus, $f(y;\theta^0)$ corresponds to a whole subset of points and not simply to just one point of $\mathbf{\Theta}$.

A characteristic feature of a model like (2.5) is that setting a certain parameter to a particular value entirely removes another parameter from the formula. In (2.5), δ vanishes if $\theta = \theta^0$, and likewise θ vanishes if $\delta = 0$. Not surprisingly, if we try to estimate both parameters simultaneously using ML estimation, say, the resulting estimator $\hat{\boldsymbol{\theta}}$ is unstable if $f(y;\theta^0)$ is the true model. This has inference ramifications. For example, the statistical hypothesis test of whether $f(y;\theta^0)$ is the true model is no longer 'standard'. Davies (1977, 1987, 2002) characterizes this as 'hypothesis testing when a nuisance parameter [in this case δ] is present only under the alternative'. Garel (2005) refers to the problem as one of 'non-identifiability of parameters'. In this book, we follow the terminology of 'indeterminacy' used by Cheng and Traylor (1995).

We discuss the model (2.5) fully in Chapter 14, simply noting here, as shown by Feng and McCulloch (1996) and Cheng and Liu (2001), that the MLE $\hat{\boldsymbol{\theta}}$, though unstable when $f(y;\theta^0)$ is the true model, nevertheless will identify this as the correct model. Parametric bootstrapping can therefore be used to estimate the distribution of statistics used in hypothesis testing in this case.

In Chapter 17, we extend discussion of indeterminacy to general continuous finite mixture models with density

$$h(y,\boldsymbol{\mu},\boldsymbol{\sigma},\mathbf{w}) = \sum_{i=1}^{k} w_i f_i\{(y-\mu_i)/\sigma_i\}/\sigma_i, \tag{2.6}$$

where components are of location-scale type (for example, the normal) with component density functions f_i for $1 \le i \le k$, where $\boldsymbol{\mu} = (\mu_1,\mu_2,\ldots,\mu_k)$, $\boldsymbol{\sigma} = (\sigma_1,\sigma_2,\ldots,\sigma_k)$, and $w = (w_1,w_2,\ldots,w_k)$ are the parameter vectors for the component means, standard deviations (SD), and weights, respectively, with $\sigma_i, w_i \ge 0$ for all i and $\sum_{i=1}^{k} w_i = 1$.

The interesting case is where k^0, the true value of k, is unknown. If one attempts to fit the model of eqn (2.6) with $k > k^0$, the true value of the weights w_i of the excess $k - k^0$ components is then zero, making the corresponding μ_i and σ_i indeterminate. The problem, referred to as *overfitting*, is thus one of parameter indeterminacy. The estimators of μ_i and σ_i of overfitted components do not have a meaningful interpretation in this situation. Standard tests of the significance of parameter estimates do not hold, and can often be quite misleading.

A further practical consequence is that the infinite likelihood problem now occurs, not through threshold issues, but simply because of the flexibility in what a finite mixture can fit to. As only k^0 components are needed to give a PDF that fits the sample, this means that any of the remaining $k - k^0$ components, the ith, say, can be positioned with its mean $\mu_i = y_j$ for some observation y_j. The contribution of that component to the PDF is then $w_i f(0)/\sigma_i$ if $\sigma_i > 0$. But this tends to ∞ if w_i remains fixed and $\sigma_i \to 0$. Thus there are *always* paths in the parameter space where $\sigma_i \to 0$ for which the likelihood becomes infinite. The overall fit is obviously not satisfactory, as the point y_j receives a positive discrete atom of weight w_i.

This unboundedness can cause a real practical difficulty, as the likelihood function may genuinely have several, sometimes *many*, optima corresponding to alternative fits (see Titterington et al. 1985). Each such local optimum represents a possible reasonable fit, capable of some practical explanation. The method of fitting needs to offer some ranking of the alternatives in order to give an indication of which are the best fits. However, for ML, the unbounded likelihood is a potential serious impediment to this, as clearly there is no way to distinguish between different optima if they are all infinite.

Finite mixtures are the subject of a large literature and, as already mentioned, will be discussd in Chapter 17. Fitting a finite mixture model can be very much viewed as an excercise in model building, this being an underlying theme of the book.

2.5 Model Building

Parametric nested models are families whose models can be placed in a hierarchy. At the top is a full model. This has submodels that are special cases obtained by setting selected parameters to specific values. These submodels in their turn have their own submodels as special cases. The process continues downwards, ending up with all the (sub)models placed in a *nested hierarchical lattice*. The finite mixture model is an important example, where there is just one model, the k-component model, at the kth level for $k = 1, 2, .., K$, with $k = K$ being the full model. In Chapter 15, we discuss the fitting of a nested nonlinear regression model to a data sample, where the objective is to find a model with the fewest parameters that will nevertheless be a good fit to the data. A step-by-step approach is considered, starting at the bottom of the hierarchy, fitting and testing the fit of models progressively further up the hierarchy until a satisfactory fit is obtained. The fitting is non-standard if an embedded model is possible that is a good fit or if indeterminacy can occur. Unlike linear regression model fitting, there are restrictions on the way successive models can be selected, fitted, and tested, if the process is to remain standard. Such problems are considered generally in Chapter 15, and the discussion illustrated by specific examples, two involving the fitting of (different) nested models to two real data sets.

2.6 Multiform Families

Hierarchical models typically involve what might be termed *multiform* model families.

We have seen that embedded models can occur at the boundary of $\boldsymbol{\Theta}$ having a different functional form to those in the interior of $\boldsymbol{\Theta}$. There are situations where $\boldsymbol{\Theta}$ is the union of a number of disjoint subregions $\boldsymbol{\Theta}_j, j = 1, 2, \ldots, k$, so that

$$\boldsymbol{\Theta} = \bigcup_{j=1}^{k} \boldsymbol{\Theta}_j,$$

with the distribution *multiform* in the sense of taking different functional forms in each $\mathbf{\Theta}_j$. A unified model-fitting scenario would need to allow the best fit to be located in any one of the subregions. Boundaries between different regions may be subregions in their own right, with their own distinct models, so the possibility that such a boundary model is the best fit has to be allowed for. The same then applies to boundaries of boundaries, and so on, so that a hierarchy of nested models may be involved, with parameter vectors $\boldsymbol{\theta}_i$ of different dimensions.

The well-known families of Pearson distributions and Johnson distributions are examples where $\mathbf{\Theta}$ comprises different subregions, where some of the boundary models are functionally distinct from the models of $\text{Int}(\mathbf{\Theta}_j)$ and moreover, are of embedded form. Another example is the family of stable law distributions. These three families are discussed in Chapter 9.

Our next two examples are when the true parameter $\boldsymbol{\theta}^0$ is an interior point $\mathbf{\Theta}$, but non-standard behaviour arises because $L(\boldsymbol{\theta}, \mathbf{y})$ is not a quadratic near $\boldsymbol{\theta}^0$.

2.7 Oversmooth Log-likelihood

It is possible that the true parameter $\boldsymbol{\theta}^0$ is an interior point and $L(\boldsymbol{\theta}, \mathbf{y})$ has a consistent concave maximum which is not a quadratic but instead is, for example, quartic, at least in some of the parameters, so that it is functionally smoother with regard to these parameters. In this situation, the ML estimator may still converge probabilistically to $\boldsymbol{\theta}^0$, but at a rate slower than in the standard case.

The problem is best illustrated with an example. The family of distributions where the PDF takes the form

$$f_{SN}(y) = 2\phi(y)\Phi(\lambda y), \quad -\infty < y < \infty,$$

where $\phi(\cdot)$ and $\Phi(\cdot)$are the PDF and CDF of the standard $N(0, 1)$ distribution, and λ is a real finite valued parameter, is called the family of *skew-normal* (SN) distributions. A special case in this family is the standard $N(0, 1)$ distribution, obtained when $\lambda = 0$. If location and scale parameters are included so that $y = (z - \mu)/\sigma$ with μ and σ also to be estimated, then the problem becomes non-standard if the true value is $\lambda = 0$. Azzalini (1985) shows that the log-likelihood is unusually flat in the neighbourhood of the true parameter value, so that standard asymptotic results do not apply. This type of situation has been investigated by Rotnitzky et al. (2000). We discuss the skew-normal distribution and this problem in Chapter 12.

2.8 Rough Log-likelihood

As in the preceding section, the true parameter $\boldsymbol{\theta}^0$ is an interior point, but it may be that $L(\boldsymbol{\theta}, \mathbf{y})$ is not a continuous function of all the parameters. We again illustrate with an example.

Consider the change-point hazard rate model, where Y is the time to failure of a certain component, but the failure characteristics are different depending on whether the components survive until time $y = \tau$ or not, with the change-point τ unknown but assumed the same for all components. Nguyen et al. (1984) discuss the case where Y has distribution with PDF

$$\begin{aligned} f(y) &= a\exp(-ay) & 0 \leq y \leq \tau \\ &= b\exp(-a\tau - b(y-\tau)) & y > \tau, \end{aligned} \tag{2.7}$$

so that there is a discontinuity at $y = \tau$. All three parameters a, b, τ have to be estimated.

For a given sample $\mathbf{y}$, the likelihood function is highly discontinuous because, as τ varies, there is a discontinuity at $\tau = y_i$ for *every* i. We consider this problem in Chapter 11.

2.9 Box-Cox Transform

Chapter 10 discusses the well-known Box-Cox model where observations Y_i, $i = 1, 2, \ldots, n$, come from an asymmetric distribution, but it is thought there is a relatively simple nonlinear transformation of Y that will make it normally distributed. One of the models proposed by Box and Cox (1964) assumes that the shifted power transformation

$$\begin{aligned} y(\mu, \lambda) &= ((y+\mu)^{\lambda} - 1)/\lambda & \text{if } \lambda \neq 0 \\ &= \ln(y+\mu) & \text{if } \lambda = 0 \end{aligned}$$

is normally distributed. Thus, for suitably chosen λ, $y(\lambda)$ can simply be analysed as a normal variable.

There are actually three points of note here.

Firstly, if it is actually the distribution of Y that is of interest, with the transformation employed to make estimation of the paramters more straightforward, then we would still need in the end to consider the distribution of

$$Y = (1 + \lambda Z)^{1/\lambda} - \mu,$$

where $Z \sim N(\theta, \sigma^2)$. This is not a particularly tractable distribution to manipulate.

Secondly, the normality assumption is only an approximation, as the range of variation of $y(\mu, \lambda)$ is restricted if $\lambda \neq 0$, with, for example, $y(\mu, \lambda) > -\lambda^{-1}$ if $\lambda > 0$. What is actually being assumed is that $y(\mu, \lambda)$ has a *truncated* normal distribution, but that the parameters μ, λ, θ, and σ can be found so that truncation only occurs out in one of the tails of the normal, and therefore the truncation can be ignored.

Thirdly, inclusion of the shift μ makes the transformation a threshold model like the three-parameter Weibull model (2.1), discussed in Section 2.2, with μ playing the role of $-a$. The non-standard infinite likelihood problem can therefore occur in this case.

All these issues are discussed more fully in Chapter 10.

2.10 Two Additional Topics

The main emphasis of the book is on non-standard problems, examples of which were given in the previous sections. There are two chapters that concern problems which might not be regarded as all that non-standard, though they do involve an element of non-standardness in some way. One concerns generalizations of a given model, so that, in its enhanced form, it is more able to model tail behaviour than is otherwise possible. The other discusses a non-standard use of bootstrapping for model selection in the linear model. We give some introductory details of the chapters here.

2.10.1 Randomized Parameter Models

In Chapter 13, we discuss a way to generalize a given 'base' model so that there is added flexibility in modelling tail behaviour. Randomized parameter models are where a parameter in the base distribution is considered to be random, adding an extra variability which depends on an extra parameter not present in the base distribution. A useful version is where Y is a continuous random variable from a two-parameter base model with PDF $g(y; \lambda, \mu)$, depending on the parameters λ and μ. The parameter λ is then replaced by λZ, where Z is a continuous random variable with PDF $h(z; \alpha)$, depending on a parameter α. We shall call the $h(\cdot)$ the *mixing distribution*. For fixed z, $g(y; \lambda z, \mu)$ is simply the conditional PDF of Y given $Z = z$.

We can just as well regard the sampling of Z as an internal mechanism for generating Y, with Y having the unconditional three-parameter *randomized parameter distribution* (also known as a *compound distribution*) with PDF given by

$$f(y; \lambda, \mu, \alpha) = \int_z g(y; \lambda z, \mu)\, h(z; \alpha)\, dz. \tag{2.8}$$

Chapter 13 describes a number of examples of densities of the form (2.8), focusing on how inclusion of a mixing random variable yields unconditional distributions with broader tails than the original base distribution. A familiar instance is where $g(\cdot)$ is the normal PDF and $h(\cdot)$ is a gamma PDF, where, with appropriately chosen parameters λ, μ, and α, Y has the Student's t-distribution with ν degrees of freedom, a distribution that has broader tails, but with the original normal distribution recovered by letting $\nu \to \infty$.

Randomized parameter distributions play a prominent role in Bayesian analysis in situations where the assumed prior distributions of parameters contain additional parameters that have their own prior distributions, known as hyper-priors. We give such an instance in Chapter 17.

Depending on the form of $f(\cdot)$ in eqn (2.8), it may be easy to generate a variate directly from this distribution. Alternatively, a Y variate can be obtained first by sampling a Z value from the PDF $h(z; \alpha)$ and then generating Y by sampling from the conditional density $g(y; \lambda Z, \mu)$.

2.10.2 Bootstrapping Linear Models

The linear model is regarded as being completely standard. However, a common use of the linear model is in exploratory situations where a response Y may depend on a large number of explanatory variables X_i, $i = 1, 2, \ldots, P$, but it is unclear which X_i's that Y will depend on strongly, and which X_i's might be omitted from consideration because they have little influence on the behaviour of Y. Methods that build up a model by selecting X_i's to include in a stepwise manner are often proposed but, when P is large and the design is non-orthogonal, there can be uncertainty over whether a model formed in this way really is satisfactory. In Chapter 16, we discuss how bootstrapping can be used, not only in a non-standard way to efficiently identify important explanatory variables, but also to carry out sensitivity analysis that checks the robustness of a selected model.

3 Standard Asymptotic Theory

The basic theoretical aspects of standard asymptotic theory are well known. A thorough account is given by Cox and Hinkley (1974), who make clear the interplay between the key ideas and quantities underlying the theory. A more recent similar, briefer, but still thorough account is given by Young and Smith (2005). However, to establish the notation used in this book, and for completeness and ease of reference, we summarize in this chapter the main aspects of the theory that we will draw on.

As indicated in Chapter 1, maximum likelihood (ML) is the main estimation method used in this book. In practical applications, ML is implemented using numerical methods, and this will be illustrated in many examples in the book. We will therefore also summarize some general numerical aspects of ML estimation in this chapter.

Somewhat more formally than in the discussion of Chapter 1, a set of conditions for ensuring asymptotic normal behaviour of the MLE $\hat{\boldsymbol{\theta}}$ is the following:

(i) $f(y, \boldsymbol{\theta})$ is continuous in $\boldsymbol{\theta} \in \boldsymbol{\Theta}$ for almost all y.

(ii) $f(y, \boldsymbol{\theta}) \to 0$ if $|\boldsymbol{\theta}| \to \infty$.

(iii) Any two $F(y, \boldsymbol{\theta})$'s corresponding to two distinct $\boldsymbol{\theta}$'s will differ in at least one y value.

(iv) $E[|f(Y, \boldsymbol{\theta}^0)|]$ is finite.

(v) $E[f(Y, \boldsymbol{\theta})]$ satisfies a technical, but weak, smoothness assumption for $\boldsymbol{\theta}$ varying in the neighbourhood of $\boldsymbol{\theta}^0$.

(vi) $\boldsymbol{\Theta}$ is a compact region, with $\boldsymbol{\theta}^0$ an interior point.

(vii) The first three derivatives of $L(\boldsymbol{\theta}, y)$ with respect to $\boldsymbol{\theta}$ exist.

(viii) The third derivative satisfies a technical boundedness condition in the neighborhood of $\boldsymbol{\theta}^0$, typically taken as

$$n^{-1}\left|\frac{\partial^3 L(\boldsymbol{\theta}, \mathbf{y})}{\partial\theta_i \partial\theta_j \partial\theta_k}\right| < M(\mathbf{y}), \text{ all } i, j, k,$$

Non-Standard Parametric Statistical Inference. Russell Cheng.

with $M(\cdot)$ a function independent of $\boldsymbol{\theta}$, and for which

$$E[M(\mathbf{y})] < \infty.$$

(ix) $f(y, \boldsymbol{\theta})$ is regular in the sense defined, for example, by Cox and Hinkley (1974), so that differentiation of $E[f(Y, \boldsymbol{\theta})]$ with respect to $\boldsymbol{\theta}$, where this expectation involves integration, can be carried out by reversing the order of the integration and differentiation.

We will not discuss these assumptions in any detail, but note that they do not explicitly involve the quadratic nature of the log-likelihood. In fact, (viii) is the only assumption that is explicit about the form of the log-likelihood. Indeed, it is generally recognized that this assumption is somewhat artificial, in that a condition on the third derivative is not absolutely necessary. However, it is commonly used, as it relatively easy to show the asymptotic normal form of the distribution of $\widehat{\boldsymbol{\theta}}$ under this assumption. Establishing asymptotic normality when (viii) is replaced by a more 'natural' assumption seems to require much more delicate, less hyaline arguments, see Le Cam (1970), for example, and this will not be pursued further here.

We consider first the issue of consistency under these conditions. The MLE $\widehat{\boldsymbol{\theta}} = \widehat{\boldsymbol{\theta}}(\mathbf{Y})$ is said to be *consistent in probability* or *weakly consistent* if $|\widehat{\boldsymbol{\theta}}(\mathbf{Y}) - \boldsymbol{\theta}^0| \to 0$ in probability, i.e. $\lim_{n\to\infty} \widehat{\boldsymbol{\theta}} \stackrel{p}{=} \boldsymbol{\theta}^0$. We have *strong consistency* if $|\widehat{\boldsymbol{\theta}}(\mathbf{Y}) - \boldsymbol{\theta}^0| \to 0$ with probability 1, i.e. if $\lim_{n\to\infty} \widehat{\boldsymbol{\theta}} \stackrel{as}{=} \boldsymbol{\theta}^0$, where the '*as*' means 'almost surely'.

Wald (1949) actually proves that $\widehat{\boldsymbol{\theta}}$ is strongly consistent under the assumptions (i)–(v), but, as pointed out by Wolfowitz (1949), the proof is relatively easy to modify to show that $\widehat{\boldsymbol{\theta}}$ is weakly consistent as well. Differentiability of the log-likelihood is not needed for $\widehat{\boldsymbol{\theta}}$ to be consistent.

3.1 Basic Theory

We now consider more fully the mathematical form that asymptotic normality takes, the basis on which the practical methodology is built.

We begin with some definitions.

The derivative of the log-likelihood

$$\mathbf{u}(\boldsymbol{\theta}, \mathbf{y}) = \frac{\partial L(\boldsymbol{\theta}, \mathbf{y})}{\partial \boldsymbol{\theta}} = \left[\frac{\partial L(\boldsymbol{\theta}, \mathbf{y})}{\partial \theta_1}, \frac{\partial L(\boldsymbol{\theta}, \mathbf{y})}{\partial \theta_2}, \ldots, \frac{\partial L(\boldsymbol{\theta}, \mathbf{y})}{\partial \theta_d}\right]^T \tag{3.1}$$

is called the *score* or *score function*. Here the score is treated as a function of $\boldsymbol{\theta}$, but that is also dependent on the observations $\mathbf{y}$. The equation

$$\mathbf{u}(\boldsymbol{\theta}, \mathbf{y}) = \frac{\partial L(\boldsymbol{\theta}, \mathbf{y})}{\partial \boldsymbol{\theta}} = \mathbf{0} \tag{3.2}$$

is called the *likelihood equation*. We now treat the score as a random variable, emphasizing this by writing it as $\mathbf{u}(\boldsymbol{\theta}, \mathbf{Y})$. Its expected value is

$$
\begin{aligned}
E[\mathbf{u}(\boldsymbol{\theta},\mathbf{Y})] &= \int_{\mathcal{Y}} \frac{\partial L(\boldsymbol{\theta},\mathbf{y})}{\partial \boldsymbol{\theta}} f(\mathbf{y},\boldsymbol{\theta}) d\mathbf{y} \\
&= \int_{\mathcal{Y}} \frac{1}{f(\mathbf{y},\boldsymbol{\theta})} \frac{\partial f(\mathbf{y},\boldsymbol{\theta})}{\partial \boldsymbol{\theta}} f(\mathbf{y},\boldsymbol{\theta}) d\mathbf{y} \\
&= \int_{\mathcal{Y}} \frac{\partial f(\mathbf{y},\boldsymbol{\theta})}{\partial \boldsymbol{\theta}} d\mathbf{y},
\end{aligned}
$$

where $\mathcal{Y}$ is the support of $f(\cdot,\boldsymbol{\theta})$. Under the regularity assumption, (ix), integration and differentiation can be interchanged, so that

$$
\int_{\mathcal{Y}} \frac{\partial f(\mathbf{y},\boldsymbol{\theta})}{\partial \boldsymbol{\theta}} d\mathbf{y} = \frac{\partial}{\partial \boldsymbol{\theta}} \int_{\mathcal{Y}} f(\mathbf{y},\boldsymbol{\theta}) d\mathbf{y} = \mathbf{0}.
$$

Thus, in the regular case we have

$$
E[\mathbf{u}(\boldsymbol{\theta},\mathbf{Y})] = \mathbf{0}. \tag{3.3}
$$

In words, the equation shows that the expected value of the score is zero, in a regular family satisfying (ix). If we divide the likelihood equation (3.2) by n, so that it is the average

$$
n^{-1}\mathbf{u}(\boldsymbol{\theta},\mathbf{y}) = n^{-1} \sum_{j=1}^{n} \frac{\partial \ln f(y_j,\boldsymbol{\theta})}{\partial \boldsymbol{\theta}} = \mathbf{0},
$$

this emphasizes that it is just the sample analogue of (3.3).

We can also apply condition (ix) to the second derivative. This shows that the covariance between the components of $\mathbf{u}(\boldsymbol{\theta},\mathbf{Y})$ is the d by d matrix that has entries satisfying

$$
E\left[\frac{\partial L(\boldsymbol{\theta},\mathbf{Y})}{\partial \theta_i} \frac{\partial L(\boldsymbol{\theta},\mathbf{Y})}{\partial \theta_j}\right] = E\left[-\frac{\partial^2 L(\boldsymbol{\theta},\mathbf{Y})}{\partial \theta_i \partial \theta_j}\right]. \tag{3.4}
$$

This right-hand matrix is called the *Fisher information matrix,* and will be denoted by $\mathrm{I}(\boldsymbol{\theta})$. The negative of the so-called Hessian matrix of second partial derivatives appearing in the right-hand side of the equation is known as the *observed information matrix,* and will be denoted by $\mathrm{J}(\boldsymbol{\theta})$. We also write $\mathrm{i}(\boldsymbol{\theta})$ and $\mathrm{j}(\boldsymbol{\theta})$ for the Fisher and observed information matrix of a *single* observation, so that for a random sample,

$$
\mathrm{I}(\boldsymbol{\theta}) = n\mathrm{i}(\boldsymbol{\theta}) \text{ and } \mathrm{J}(\boldsymbol{\theta}) = n\mathrm{j}(\boldsymbol{\theta}). \tag{3.5}
$$

Having just one observation is enough to allow its log-likelihood to be clearly specified, with the general log-likelihood of a random sample of size n obtainable as the summation of n terms all with the same functional form. We shall throughout the book frequently give log-likelihoods in one-observation form for simplicity.

We now consider the asymptotic distribution of the maximum likelihood estimator $\widehat{\boldsymbol{\theta}}$, defined in (1.5). We use, for simplicity, the notation $\partial L(\boldsymbol{\theta}_0, \mathbf{y})/\partial\boldsymbol{\theta}$ to mean $\partial L(\boldsymbol{\theta}, \mathbf{y})/\partial\boldsymbol{\theta}\,|_{\boldsymbol{\theta}=\boldsymbol{\theta}_0}$, where calculation of the derivative is followed by taking its value at $\boldsymbol{\theta} = \boldsymbol{\theta}_0$. The same notation applies for higher derivatives. A Taylor series expansion of $\mathbf{u}$ gives

$$\frac{\partial L(\widehat{\boldsymbol{\theta}}, \mathbf{y})}{\partial\boldsymbol{\theta}} = \frac{\partial L(\boldsymbol{\theta}^0, \mathbf{y})}{\partial\boldsymbol{\theta}} + \frac{\partial^2 L(\boldsymbol{\theta}', \mathbf{y})}{\partial\boldsymbol{\theta}^2}(\widehat{\boldsymbol{\theta}} - \boldsymbol{\theta}^0),$$

where $\boldsymbol{\theta}'$ lies between $\widehat{\boldsymbol{\theta}}$ and $\boldsymbol{\theta}^0$. Setting this derivative to zero yields

$$\sqrt{n}(\widehat{\boldsymbol{\theta}} - \boldsymbol{\theta}^0) = -n\left[\frac{\partial^2 L(\boldsymbol{\theta}', \mathbf{y})}{\partial\boldsymbol{\theta}^2}\right]^{-1}\left[n^{-1/2}\frac{\partial L(\boldsymbol{\theta}^0, \mathbf{y})}{\partial\boldsymbol{\theta}}\right].$$

As $\boldsymbol{\theta}'$ lies between $\widehat{\boldsymbol{\theta}}$ and $\boldsymbol{\theta}^0$, and $\widehat{\boldsymbol{\theta}} \xrightarrow{p} \boldsymbol{\theta}^0$, one might expect $-n[\partial^2 L(\boldsymbol{\theta}', \mathbf{y})/\partial\boldsymbol{\theta}^2]^{-1} \xrightarrow{p} \mathrm{i}^{-1}(\boldsymbol{\theta}^0)$, the inverse of the information matrix for a single observation, and this can be shown under the full set of regularity assumptions given previously. Moreover, applying the central limit theorem to $n^{-1/2}\partial L(\boldsymbol{\theta}^0, \mathbf{y})/\partial\boldsymbol{\theta}$ shows that this and $\sqrt{n}(\widehat{\boldsymbol{\theta}} - \boldsymbol{\theta}^0)$ are both asymptotically normally distributed, with mean zero following from (3.3). Using (3.4) yields the asymptotic covariance matrix of the latter as

$$\begin{aligned} E[n(\widehat{\boldsymbol{\theta}} - \boldsymbol{\theta}^0)(\widehat{\boldsymbol{\theta}} - \boldsymbol{\theta}^0)^{\mathrm{T}}] &= \mathrm{i}^{-1}(\boldsymbol{\theta}^0)E\left[n^{-1}\left(\frac{\partial L(\boldsymbol{\theta}^0, \mathbf{y})}{\partial\boldsymbol{\theta}}\right)\left(\frac{\partial L(\boldsymbol{\theta}^0, \mathbf{y})}{\partial\boldsymbol{\theta}}\right)^{\mathrm{T}}\right]\mathrm{i}^{-1}(\boldsymbol{\theta}^0) \\ &= \mathrm{i}^{-1}(\boldsymbol{\theta}^0)\mathrm{i}(\boldsymbol{\theta}^0)\mathrm{i}^{-1}(\boldsymbol{\theta}^0) \\ &= \mathrm{i}^{-1}(\boldsymbol{\theta}^0). \end{aligned}$$

It will be convenient to use the less precise notation that

$$\widehat{\boldsymbol{\theta}} \sim N(\boldsymbol{\theta}^0, V^0),$$

where

$$V^0 = [\mathrm{I}(\boldsymbol{\theta}^0)]^{-1}, \tag{3.6}$$

so that for any particular component,

$$\hat{\theta}_i \sim N(\theta_i^0, V_{ii}^0).$$

3.2 Applications of Asymptotic Theory

The asymptotic theory enables confidence intervals (CI) be to constructed in the usual way by inverting probability statements about a range of values taken by $\hat{\theta}_i$, so that, for example, a two-sided $100(1-\alpha)\%$ confidence interval for $\hat{\theta}_i$ is

$$(\hat{\theta}_i + z_{\alpha/2}\sqrt{V^0_{ii}},\ \hat{\theta}_i + z_{(1-\alpha/2)}\sqrt{V^0_{ii}}), \tag{3.7}$$

where z_α is the 100α percentage point of the standard normal distribution. The unknown variance $V^0 = [\mathrm{I}(\boldsymbol{\theta}^0)]^{-1}$ can be replaced by $\widehat{V} = [\mathrm{I}(\widehat{\boldsymbol{\theta}})]^{-1}$. It is generally preferable to use $\widehat{V} = [\mathrm{J}(\widehat{\boldsymbol{\theta}})]^{-1}$, with the observed information $\mathrm{J}(\widehat{\boldsymbol{\theta}})$ replacing $\mathrm{I}(\widehat{\boldsymbol{\theta}})$. Not only does this save having to calculate $\mathrm{I}(\widehat{\boldsymbol{\theta}})$ explicitly, but there is wide evidence that use of $\mathrm{J}(\widehat{\boldsymbol{\theta}})$ gives results that are more robust. See Efron and Hinkley (1978), for example. ML estimation can be extended to a function $\mu(t,\boldsymbol{\theta})$, $a \le t \le b$, that is dependent also on the parameter values, so that CIs for the function can also be calculated when the parameters are estimated. We use the following well-known invariance property of MLEs. The result is quite general and we can show it following Dudewicz and Mishra (1988, Theorem 7.2.15(ii)). We need only suppose that $\mu = \mu(t,\boldsymbol{\theta})$ is a well-defined function of t and $\boldsymbol{\theta}$. Then define

$$\hat{L}(\mu) = \max_{\boldsymbol{\theta}:\mu(t,\boldsymbol{\theta})=\mu} L(\boldsymbol{\theta}),$$

obtained by maximizing $L(\boldsymbol{\theta})$ over all $\boldsymbol{\theta} \in \boldsymbol{\Theta}$ for which $\mu(t,\boldsymbol{\theta}) = \mu$. Thus, $\hat{L}(\mu)$ is the profile log-likelihood with profiling parameter μ. Let $\hat{\mu} = g(t,\widehat{\boldsymbol{\theta}})$. Clearly, $\widehat{\boldsymbol{\theta}} \in \{\boldsymbol{\theta} : \mu(t,\boldsymbol{\theta}) = \hat{\mu}\}$. Then, as $L(\widehat{\boldsymbol{\theta}}) = \max_{\boldsymbol{\theta}\in\boldsymbol{\Theta}} L(\boldsymbol{\theta})$, we must have

$$\hat{L}(\hat{\mu}) = \max_{\boldsymbol{\theta}:\mu(t,\boldsymbol{\theta})=\mu} L(\boldsymbol{\theta}) = L(\widehat{\boldsymbol{\theta}}).$$

Thus, while it may not be unique,

$$\hat{\mu} = \mu(t,\widehat{\boldsymbol{\theta}})$$

is an MLE of $\mu(t,\boldsymbol{\theta})$, obtained simply by evaluating $\mu(t,\boldsymbol{\theta})$ at $\boldsymbol{\theta} = \widehat{\boldsymbol{\theta}}$, treating t as a constant.

A first-order correct two-sided $(1-\alpha)100\,\%$ confidence interval for $\mu(t,\boldsymbol{\theta})$ is

$$\mu(t,\widehat{\boldsymbol{\theta}}) \pm z_{(1-\alpha/2)}\sqrt{\left.\frac{\partial\mu(t,\boldsymbol{\theta})}{\partial\boldsymbol{\theta}}\right|^T_{\widehat{\boldsymbol{\theta}}} V(\widehat{\boldsymbol{\theta}}) \left.\frac{\partial\mu(t,\boldsymbol{\theta})}{\partial\boldsymbol{\theta}}\right|_{\widehat{\boldsymbol{\theta}}}}, \tag{3.8}$$

where the term under the square root is the variance of the first-order Taylor series approximation of $\mu(t,\boldsymbol{\theta})$. This is known as the δ–method CI.

A numerical example illustrating the calculation of confidence intervals is given in Section 3.8.

3.3 Hypothesis Testing in Nested Models

A note of clarification. The material covered in this section is well known. However, though the results are clear-cut and relatively simple to apply, their derivation is not often given in much detail. For example, Cox and Hinkley (1974) and Young and Smith (2005) only outline their derivation. For completeness, we give fuller details. This makes following through every step of the algebra quite demanding. The reader not wishing such a full immersion need only note the final results, as this is all that is really required in the rest of the book.

Suppose that $\boldsymbol{\theta} = (\boldsymbol{\psi}, \boldsymbol{\lambda})$, where all the vectors are assumed column vectors, but for simplicity we have not used full vector notation. In this section, we consider only the simplest situation, where the true value of $\boldsymbol{\theta}$, $\boldsymbol{\theta} = \boldsymbol{\theta}_0 = (\boldsymbol{\psi}_0, \boldsymbol{\lambda}_0)$, say, is an interior point of $\boldsymbol{\Theta}$, where we wish to test the *null* hypothesis $H_0 : \boldsymbol{\psi} = \boldsymbol{\psi}_0$. Such a test is based on examining a test-statistic whose distribution is different depending on whether H_0 is true or not.

A note of clarification is in order concerning the parameter dimensions involved here. Let the dimensions of $\boldsymbol{\theta}$, $\boldsymbol{\psi}$, and $\boldsymbol{\lambda}$ be d, d_0, and d_1, respectively. The test is therefore one where we are comparing a full model with parameter $\boldsymbol{\theta}$, comprising d unknown components, with a submodel where, under the null hypothesis H_0, the d_0 components that comprise $\boldsymbol{\psi}$ are supposed known, so that only $\boldsymbol{\lambda}$ is still unknown, with just d_1 components. Such a submodel is said to be a *nested* submodel, in that its unknown parameters all appear in the original full model. The critical dimension as far as the test is concerned is d_0, the *difference*, $d - d_1$, between the parameter dimensions of the full and the submodel; that is, the dimension of the parameter $\boldsymbol{\psi}$, supposed known.

The full picture turns out to be quite involved, as the distribution of the test-statistic is very dependent on the form of H_0. For example, suppose each component of θ is known to lie in a given finite interval, so that $\theta_i \in [a_i, b_i]$. Then the distribution of the test-statistic changes depending which components of ψ are interior to their given interval and which components are end points, with $\psi_{0i} = a_i$, say, under the null. Similar uncertainty can occur with the components not specified in the null. The most detailed study of this issue is that of Self and Liang (1987), who examine nine cases of special interest. We return to this issue in Section 6.2, when we discuss fitting probability distributions.

Here we consider just the simplest situation, where $\boldsymbol{\Theta}$ is an open set, so that $\boldsymbol{\theta}_0$ is an interior point. Moreover, asymptotic theory covers the case where the log-likelihood function behaves essentially as a quadratic function of the parameter values, with a maximum point that converges to the true parameter value. The situation is as illustrated in Figure 1.1 given in Chapter 1, where it is pointed out that there are three geometric features, namely, the height of the maximum above the observed log-likelihood value, the distance of the observed parameter point from the true maximum point, and the slope of the tangent plane of the log-likelihood at the observed parameter value, any one of which characterizes the quadratic probabilistic behaviour of the log-likelhood in the neighbourhood of the maximum point. There are three well-known test-statistics corresponding to these features that can be used to test H_0: the likelihood ratio (LR) test, the Wald test, and the score test. All three are asymptotically equivalent, stemming from the simple geometric relationship between them illustrated in Figure 1.1.

The LR test is normally expressed in terms of the difference between the logarithms of the square of the likelihood maximized freely with respect to $\boldsymbol{\theta}$ and maximized only with respect to $\boldsymbol{\lambda}$, in this latter case with $\boldsymbol{\psi} = \boldsymbol{\psi}_0$ so that H_0 is satisfied. The LR test is based on the following key asymptotic result due to Wilks (1938), often called Wilks's Theorem, see Young and Smith (2005, §8.5), which is as follows. We write $\widehat{\boldsymbol{\lambda}}_0$ for the MLE of $\boldsymbol{\lambda}$ obtained with $\boldsymbol{\psi}$ fixed at $\boldsymbol{\psi} = \boldsymbol{\psi}_0$. The test-statistic is

$$T_{LR} = 2(L(\widehat{\boldsymbol{\theta}}) - L(\boldsymbol{\psi}_0, \widehat{\boldsymbol{\lambda}}_0)). \tag{3.9}$$

Using the asymptotic forms of the two log-likelihoods, we have, as $n \to \infty$,

$$T_{LR} = T_a + o_p(1),$$

where

$$T_a = \begin{bmatrix} \widehat{\boldsymbol{\psi}} - \boldsymbol{\psi}_0 \\ \widehat{\boldsymbol{\lambda}} - \boldsymbol{\lambda} \end{bmatrix}^{\mathrm{T}} \mathrm{i}(\boldsymbol{\theta}) \begin{bmatrix} \widehat{\boldsymbol{\psi}} - \boldsymbol{\psi}_0 \\ \widehat{\boldsymbol{\lambda}} - \boldsymbol{\lambda} \end{bmatrix} - \begin{bmatrix} \mathbf{0} \\ \widehat{\boldsymbol{\lambda}}_0 - \boldsymbol{\lambda} \end{bmatrix}^{\mathrm{T}} \mathrm{i}(\boldsymbol{\theta}) \begin{bmatrix} \mathbf{0} \\ \widehat{\boldsymbol{\lambda}}_0 - \boldsymbol{\lambda} \end{bmatrix}, \tag{3.10}$$

with $\boldsymbol{\theta} = (\boldsymbol{\psi}_0, \boldsymbol{\lambda})$ under the null.

Observe that the LR test, using either (3.9) or (3.10), requires calculation of both $\widehat{\boldsymbol{\theta}} = (\widehat{\boldsymbol{\psi}}, \widehat{\boldsymbol{\lambda}})$ and $\hat{\boldsymbol{\lambda}}_0$.

The second test, T_W, is commonly called *Wald's test,* after Abraham Wald who first proposed the test. See Wald (1943). T_W is based only on the parameter estimate $\widehat{\boldsymbol{\psi}}$, and takes the form

$$T_W = (\widehat{\boldsymbol{\psi}} - \boldsymbol{\psi}_0)^T n\mathrm{i}(\widehat{\boldsymbol{\psi}}|\widehat{\boldsymbol{\lambda}})(\widehat{\boldsymbol{\psi}} - \boldsymbol{\psi}_0), \tag{3.11}$$

where $\mathrm{i}(\boldsymbol{\psi}|\boldsymbol{\lambda})$ is the inverse of the covariance matrix of the asymptotic normal distribution of $\widehat{\boldsymbol{\psi}}$.

The asymptotic equivalence of (3.10) and (3.11) follows by showing that $\widehat{\boldsymbol{\lambda}}_0$ can be expressed in terms of $\widehat{\boldsymbol{\theta}}$. To first order the likelihood equation for $\widehat{\boldsymbol{\theta}}$, in partitioned form, is

$$\begin{bmatrix} \frac{\partial L(\widehat{\boldsymbol{\theta}})}{\partial \boldsymbol{\psi}} \\ \frac{\partial L(\widehat{\boldsymbol{\theta}})}{\partial \boldsymbol{\lambda}} \end{bmatrix} = \begin{bmatrix} \frac{\partial L(\boldsymbol{\theta})}{\partial \boldsymbol{\psi}} \\ \frac{\partial L(\boldsymbol{\theta})}{\partial \boldsymbol{\lambda}} \end{bmatrix} + \frac{1}{2} \begin{bmatrix} \frac{\partial^2 L(\boldsymbol{\theta}')}{\partial \boldsymbol{\psi}^2} & \frac{\partial^2 L(\boldsymbol{\theta}')}{\partial \boldsymbol{\psi} \partial \boldsymbol{\lambda}} \\ \frac{\partial^2 L(\boldsymbol{\theta}')}{\partial \boldsymbol{\lambda} \partial \boldsymbol{\psi}} & \frac{\partial^2 L(\boldsymbol{\theta}')}{\partial \boldsymbol{\lambda}^2} \end{bmatrix} \begin{bmatrix} \widehat{\boldsymbol{\psi}} - \boldsymbol{\psi} \\ \widehat{\boldsymbol{\lambda}} - \boldsymbol{\lambda} \end{bmatrix} = 0,$$

where $\boldsymbol{\theta}'$ lies between $\widehat{\boldsymbol{\theta}}$ and $\boldsymbol{\theta}^0$. The bottom half can be written as

$$\frac{\partial L(\boldsymbol{\theta})}{\partial \boldsymbol{\lambda}} - \frac{n}{2}\mathrm{j}_{\lambda\psi}\left[\widehat{\boldsymbol{\psi}} - \boldsymbol{\psi}\right] - \frac{n}{2}\mathrm{j}_{\lambda\lambda}\left[\widehat{\boldsymbol{\lambda}} - \boldsymbol{\lambda}\right] = 0, \tag{3.12}$$

where, to simplify the notation, we have written the second derivatives in terms of the observed information and omitted the parameter argument. As already indicated, there is some flexibility in the value of $\boldsymbol{\theta}$ used for evaluating $\mathrm{j}(\boldsymbol{\theta})$, especially in the context of the hypothesis test, and a random value, so long as it is consistent, like the MLE $\widehat{\boldsymbol{\theta}}$, is satisfactory.

Likewise, for $\widehat{\boldsymbol{\lambda}}_0$ we have

$$\frac{\partial L(\boldsymbol{\psi}_0, \widehat{\boldsymbol{\lambda}}_0)}{\partial \boldsymbol{\lambda}} = \frac{\partial L(\boldsymbol{\psi}_0, \boldsymbol{\lambda})}{\partial \boldsymbol{\lambda}} - \frac{n}{2}\mathrm{j}_{\lambda\lambda}(\widehat{\boldsymbol{\lambda}}_0 - \boldsymbol{\lambda}) = \mathbf{0},$$

which we write as

$$\boldsymbol{\lambda} = \widehat{\boldsymbol{\lambda}}_0 - 2n^{-1}\mathrm{j}_{\lambda\lambda}^{-1}\frac{\partial L(\boldsymbol{\psi}_0, \boldsymbol{\lambda})}{\partial \boldsymbol{\lambda}}.$$

Substituting this into (3.12) gives

$$\frac{\partial L(\boldsymbol{\theta})}{\partial \boldsymbol{\lambda}} - \frac{n}{2}\mathrm{j}_{\lambda\psi}\left[\widehat{\boldsymbol{\psi}} - \boldsymbol{\psi}\right] - \frac{n}{2}\mathrm{j}_{\lambda\lambda}\left[\widehat{\boldsymbol{\lambda}} - \widehat{\boldsymbol{\lambda}}_0 + 2n^{-1}\mathrm{j}_{\lambda\lambda}^{-1}\frac{\partial L(\boldsymbol{\psi}_0, \boldsymbol{\lambda})}{\partial \boldsymbol{\lambda}}\right] = \mathbf{0},$$

where $\boldsymbol{\theta} = (\boldsymbol{\psi}_0, \boldsymbol{\lambda})$ under the null hypothesis. The derivatives cancel, so that the equation reduces to

$$-n\mathrm{j}_{\psi\lambda}\left[\widehat{\boldsymbol{\psi}} - \boldsymbol{\psi}\right] - n\mathrm{j}_{\lambda\lambda}\left[\widehat{\boldsymbol{\lambda}} - \widehat{\boldsymbol{\lambda}}_0\right] = \mathbf{0}.$$

This shows how to obtain $\widehat{\boldsymbol{\lambda}}_0$ from $\widehat{\boldsymbol{\lambda}}$:

$$\widehat{\boldsymbol{\lambda}}_0 - \boldsymbol{\lambda} = \widehat{\boldsymbol{\lambda}} - \boldsymbol{\lambda} + \mathrm{j}_{\lambda\lambda}^{-1}\mathrm{j}_{\psi\lambda}\left[\widehat{\boldsymbol{\psi}} - \boldsymbol{\psi}\right].$$

If we replace j by i in the right-hand side of this equation and replace $\widehat{\boldsymbol{\lambda}}_0 - \boldsymbol{\lambda}$ in the expression (3.10) for T_a, we find, after some simplification, that T_a reduces to

$$T_W = \left[\widehat{\boldsymbol{\psi}} - \boldsymbol{\psi}\right]^{\mathrm{T}} n[\mathrm{i}_{\psi\psi} - \mathrm{i}_{\psi\lambda}\mathrm{i}_{\lambda\lambda}^{-1}\mathrm{i}_{\psi\lambda}^{\mathrm{T}}]\left[\widehat{\boldsymbol{\psi}} - \boldsymbol{\psi}\right].$$

The form of the matrix defining the quadratic form shows that it is indeed precisely $[\mathrm{i}^{\psi\psi}]^{-1}$, the inverse of the covariance matrix of the asymptotic distribution of $\widehat{\boldsymbol{\psi}}$. This matrix can be evaluated at $\widehat{\boldsymbol{\theta}}$, so that T_W does not involve $\widehat{\boldsymbol{\lambda}}_0$, but it does require calculating all the components of the full unconditional information matrix. Alternatively, the matrix can be evaluated at $(\boldsymbol{\psi}_0, \widehat{\boldsymbol{\lambda}}_0)$, i.e. under the null; this is usually easier to obtain than the full unconditional matrix.

The form of T_W in (3.11) shows that it has the chi-squared distribution with d_0 degrees of freedom, where d_0 is the dimension of $\boldsymbol{\psi}_0$, that is

$$T_W \sim \chi^2_{d_0}. \tag{3.13}$$

The asymptotic equivalence of T_{LR} and T_W means that T_{LR} also has the same chi-squared distribution.

A further alternative is to base the test on the score $\mathbf{u}(\boldsymbol{\psi}_0, \boldsymbol{\lambda})$. Different variants, all asymptotically equivalent, are possible, but the simplest is

$$T_S = \left(\frac{\partial L(\boldsymbol{\psi}_0, \widehat{\boldsymbol{\lambda}}_0)}{\partial \boldsymbol{\psi}}\right)^{\mathrm{T}} n^{-1} \mathrm{i}^{\boldsymbol{\psi}\boldsymbol{\psi}}(\boldsymbol{\psi}_0, \widehat{\boldsymbol{\lambda}}_0) \frac{\partial L(\boldsymbol{\psi}_0, \widehat{\boldsymbol{\lambda}}_0)}{\partial \boldsymbol{\psi}}, \tag{3.14}$$

which does not involve the full MLE $\widehat{\boldsymbol{\theta}}$, but only the MLE $\widehat{\boldsymbol{\lambda}}_0$ obtained under H_0, where $\boldsymbol{\psi}$ is fixed at $\boldsymbol{\psi} = \boldsymbol{\psi}_0$. Like T_{LR} and T_W, T_S has the same chi-squared asymptotic distribution (3.13) under the null hypothesis.

To summarize, the standard hypothesis test of $H_0 : \boldsymbol{\psi} = \boldsymbol{\psi}_0$, at test level α, assuming $\boldsymbol{\psi}_0$ is the component of $\boldsymbol{\psi}$ corresponding to an internal point and that the alternative is $H_1 : \boldsymbol{\psi} \neq \boldsymbol{\psi}_0$, is therefore to reject H_0 if

$$T > \chi^2_{d_0}(1 - \alpha), \tag{3.15}$$

where T is the value of one of T_{LR}, T_W, or T_S calculated from the sample, and $\chi^2_{d_0}(\alpha)$ is the α quantile of the $\chi^2_{d_0}$ distribution.

Though all three statistics have the same asymptotic distribution, their behaviour, in specific problems, when n is finite, can be rather different. The LR test seems to perform best. This is perhaps not altogether surprising, given that it uses both the MLE, $(\boldsymbol{\psi}_0, \widehat{\boldsymbol{\lambda}}_0)$, obtained under the null, and the MLE, $(\hat{\boldsymbol{\psi}}, \hat{\boldsymbol{\lambda}})$, obtained under unrestricted conditions. In contrast, the forms of T_W and T_S rely on $L(\boldsymbol{\theta})$ being well approximated by a quadratic function, and their values can be very different, as well as being different from T_{LR}. This has been commented on by many authors, see, for example, Young and Smith (2005, §8.6.3), Barndorff-Nielsen and Cox (1994).

3.3.1 Non-nested Models

Comparing models that are not nested takes us away from the comfort zone of standard theory, and so is out of the purview of this chapter. For example, consider the probability distribution with PDF

$$f(x) = (1 - a)\phi(x) + a\phi(x - \theta), \tag{3.16}$$

where $0 \leq a \leq 1$, and $\phi(x) = (2\pi)^{-1/2} \exp(-x^2/2)$, the standard normal density. If $\theta = 0$, then we have immediately that $f(x) = \phi(x)$, so that the parameter a becomes undefined. Thus $\phi(x)$ is not a nested submodel of $f(x)$ in the sense of the previous section. The distribution (3.16) is a very simple example of a finite mixture model, but nevertheless is subject to parameter *indeterminacy*, where a parameter disappears from the model when another parameter takes a special value. In (3.16), a becomes indeterminate if $\theta = 0$, and θ becomes indeterminate if $a = 0$. This example and indeterminacy in general will be discussed in Chapter 14.

A way round the difficulty is not to try to directly compare fitted models when they are not nested, but simply to make an individual assessment of how well an estimated model fits the sample. If the assessment can be quantified in terms of a p-value, then different estimated model fits can be compared by comparing p-values. Goodness-of-fit (GoF) tests are a well-known way of doing this when assessing fitted probability distributions. The main problem is that powerful goodness-of-fit statistics tend not to have null distributions that are readily calculated.

For regression problems, if the data sample has replicated observations at each observed x value, then either an F test can be carried out, or a modified LR test. Details are given in Ritz and Streibig (2008), for example.

When observations are not replicated, test statistics can still be developed for assessing individual regression fits. To avoid confusion with GoF statistics, we will alter the perspective and call them lack-of-fit (LoF) statistics. The same problem occurs as with GoF statistics of determining the distribution of the test statistic under the null hypothesis that the functional form of the regression function is correct. As with GoF statistics, the null distribution of a LoF statistic is easily obtained by parametric bootstrapping, and in Section 4.6 we give an example.

3.4 Profile Log-likelihood

Suppose $\boldsymbol{\theta}$ is a d-dimensional vector with $d \geq 2$, but there is one component, ψ, say, of major interest, with the other parameters $\boldsymbol{\lambda} = \lambda_2, \ldots, \lambda_d$ regarded as *nuisance* parameters. The profile log-likelihood is a useful construction for transforming the likelihood $L(\psi, \boldsymbol{\lambda}, \mathbf{Y})$ so that the focus is on ψ. The *profile log-likelihood*, $L_\psi(\psi, \mathbf{Y})$, with profiling parameter ψ, is defined as

$$\begin{aligned} L_\psi(\psi, \mathbf{Y}) &= \sup_{\boldsymbol{\lambda}} L(\psi, \boldsymbol{\lambda}, \mathbf{Y}) \\ &= L(\psi, \widehat{\boldsymbol{\lambda}}_\psi, \mathbf{Y}), \end{aligned}$$

where $\widehat{\boldsymbol{\lambda}}_\psi$ is the value of $\boldsymbol{\lambda}$ at which the supremum of L is obtained, taken over all possible values that $\boldsymbol{\lambda}$ can take with ψ fixed. When interest is just on ψ, the profile log-likelihood can be used instead of the full log-likelihood and, assuming $L_\psi(\psi, \mathbf{Y})$ is mathematically tractable, the maximum likelihood estimator $\widehat{\psi}$ can be analysed, using L_ψ as if it were the log-likelihood. Thus, Patefield (1977) shows that the inverse of the observed profile information is equal to the full observed information evaluated at $(\psi, \widehat{\boldsymbol{\lambda}}_\psi)$, that is,

$$\mathrm{j}_p^{-1}(\psi) = \mathrm{j}^{\psi\psi}(\psi, \widehat{\boldsymbol{\lambda}}_\psi).$$

Additional properties of this kind, involving the likelihood ratio and score statistics, are given by Young and Smith (2005, §8.6.2). The profile log-likelihood is perhaps even more useful simply as a diagnostic tool. In non-standard problems, unusual behaviour can often be revealed by visually examining how the log-likelihood varies over the parameter space Θ. Focusing on just one parameter, using its profile log-likelihood, is a very convenient way of doing this, and this has been done in a number of examples throughout the book.

3.5 Orthogonalization

Parameter orthogonality is discussed in detail by Cox and Reid (1987). Let $\boldsymbol{\theta} = (\boldsymbol{\psi}, \boldsymbol{\lambda})$ be divided into two subvectors with dimensions d_1 and d_2, with $\boldsymbol{\psi} = (\theta_1, \theta_2, \ldots, \theta_{d_1})$ and $\boldsymbol{\lambda} = (\theta_{d_1+1}, \theta_{d_1+2}, \ldots, \theta_{d_1+d_2})$. Here the notation is similar to that of Section 3.3, as splitting $\boldsymbol{\theta}$ into two components $\boldsymbol{\psi}$ and $\boldsymbol{\lambda}$ is again involved. However, the context is different, so that the components are being treated differently in this section compared with Section 3.3.

The subvectors $\boldsymbol{\psi}$ and $\boldsymbol{\lambda}$ are said to be (mutually) *globally orthogonal* if

$$i_{\theta_i\theta_j} = n^{-1} E\left[\frac{\partial^2 L}{\partial\theta_i \partial\theta_j}\right] = 0$$

for $i = 1, 2, \ldots, d_1$, $j = d_1 + 1, d_1 + 2, \ldots, d_1 + d_2$ for all points $\boldsymbol{\theta} \in \Theta$. If the condition holds at just one point $\boldsymbol{\theta}^0$, then $\boldsymbol{\psi}$ and $\boldsymbol{\lambda}$ are said to be *locally orthogonal at* $\boldsymbol{\theta}^0$. Consequences of orthogonality pointed out in Cox and Reid (1987) include: (i) the MLEs $\widehat{\boldsymbol{\psi}}$ and $\widehat{\boldsymbol{\lambda}}$ are asymptotically independent, (ii) the asymptotic variance of $\widehat{\boldsymbol{\psi}}$ is unchanged whether $\boldsymbol{\lambda}$ is known or not, and (iii) $\widehat{\boldsymbol{\psi}}_{\boldsymbol{\lambda}} = \widehat{\boldsymbol{\psi}}(\boldsymbol{\lambda})$ only varies by an amount $\left|\widehat{\boldsymbol{\psi}}_{\boldsymbol{\lambda}} - \widehat{\boldsymbol{\psi}}\right| = O_p(\frac{1}{n})$ when $\boldsymbol{\lambda}$ varies by an amount $\left|\boldsymbol{\lambda} - \widehat{\boldsymbol{\lambda}}\right| = O_p(\frac{1}{\sqrt{n}})$. Informally speaking, $\widehat{\boldsymbol{\psi}}_{\boldsymbol{\lambda}}$ only varies slowly with $\boldsymbol{\lambda}$.

We will not make much use of orthogonality, but to illustrate the calculations involved, we outline how ψ, *when it is scalar,* can be made orthogonal to $\boldsymbol{\lambda}$. This specific case is straightforward, at least in principle. The process can be made sequential by adjusting each λ_i in turn to ensure that $E[\partial^2 L(\psi, \boldsymbol{\lambda})/\partial\psi\,\partial\lambda_i] = 0$, so that λ_i is orthogonal to ψ. When ψ is not scalar, such a sequential process of making each λ_i orthogonal to ψ is not usually possible. The method is described by Cox and Reid (1987). For ease of reference, we adopt their notation, so that the initial likelihood is in terms of ψ, the single parameter of interest, with the other $d-1$ parameters denoted by $\boldsymbol{\phi}$ rather than $\boldsymbol{\lambda}$. We then redefine $\boldsymbol{\phi} = \boldsymbol{\phi}(\psi, \boldsymbol{\lambda})$, with each component $\phi_i(\psi, \boldsymbol{\lambda})$ depending on ψ, and $\boldsymbol{\lambda}$ a d-dimensional vector, so that $L(\psi, \boldsymbol{\phi}) = L^0(\psi, \boldsymbol{\phi}(\psi, \boldsymbol{\lambda}))$. The form of $\boldsymbol{\phi}(\psi, \boldsymbol{\lambda})$ is obtained by differentiating L^0 as a function $L^0(\psi, \boldsymbol{\phi})$. Twice using the chain rule (for details, see Cox and Reid, 1987), we have

$$\frac{\partial^2 L}{\partial\lambda_i \partial\psi} = \sum_{j=1}^{d-1}\left(\frac{\partial^2 L^0}{\partial\phi_j \partial\psi}\frac{\partial\phi_j}{\partial\lambda_i} + \sum_{k=1}^{d-1}\frac{\partial^2 L^0}{\partial\phi_j \partial\phi_k}\frac{\partial\phi_k}{\partial\lambda_i}\frac{\partial\phi_j}{\partial\psi}\right) + \sum_{j=1}^{d-1}\frac{\partial L^0}{\partial\phi_j}\frac{\partial^2\phi_j}{\partial\psi\partial\lambda_i}.$$

Taking expectations, the right-hand summation is zero, as $E[\partial L^0/\partial\boldsymbol{\phi}] = 0$, whilst the second derivatives of L^0 can to first order in probability be treated as constants taking their expected values $\mathrm{i}^0_{\psi\phi_j}$ and $\mathrm{i}^0_{\phi_j\phi_k}$, written in terms of the parametrization $(\psi, \boldsymbol{\phi})$. To obtain $E[\partial^2 L/\partial\lambda_i \partial\psi] = 0$, we must have

$$E\left[\frac{\partial^2 L}{\partial\lambda_i\partial\psi}\right] = \sum_{j=1}^{d-1}\frac{\partial\phi_j}{\partial\lambda_i}\left(\mathrm{i}^0_{\psi\phi_j} + \sum_{k=1}^{d-1}\mathrm{i}^0_{\phi_j\phi_k}\frac{\partial\phi_k}{\partial\psi}\right) = 0,\ i = 1,2,\ldots,d-1.$$

The $d - 1$ equations are all satisfied if

$$\mathrm{i}^0_{\psi\phi_j} + \sum_{k=1}^{d-1}\mathrm{i}^0_{\phi_j\phi_k}\frac{\partial\phi_k}{\partial\psi} = 0, \quad j = 1,2,\ldots,d-1. \tag{3.17}$$

In matrix notation,

$$\frac{\partial\boldsymbol{\phi}}{\partial\psi} = -(\mathrm{i}^0_{\boldsymbol{\phi\phi}})^{-1}\mathrm{i}^0_{\psi\boldsymbol{\phi}}. \tag{3.18}$$

The parameters are $\psi, \phi_1, \phi_2, \ldots \phi_{d-1}$ in this set of component equations, solving which involves $d - 1$ integrations. To make each integration definite requires introduction of an arbitrary 'constant of integration', and if we write these as $\boldsymbol{\lambda} = (\lambda_1, \lambda_2 \ldots, \lambda_{d-1})$, each ϕ_i can be regarded as a function $\phi_i = \phi_i(\psi, \boldsymbol{\lambda})$ with ψ and $\boldsymbol{\lambda} = (\lambda_1, \lambda_2 \ldots, \lambda_{d-1})$ comprising the parameters of the reparametrization. The end result is a system of explicit equations relating the ψ, ϕ, and $\boldsymbol{\lambda}$, from which in principle explicit expressions for $\phi_i = \phi_i(\psi, \boldsymbol{\lambda})$ can be obtained.

In practice, even in this case, when ψ is only scalar, the calculations can rapidly become difficult. We give explicit examples in Chapter 6 where the calculations are tractable.

3.6 Exponential Models

It has to be said that in practice the full set of regularity conditions discussed at the start of this chapter tend often to be assumed rather than fully checked when fitting a particular model. Indeed, it is the assumption that regularity conditions hold when they are not fully met that gives rise to a number of the problems discussed in this book.

Strictly, this section is not needed to follow the rest of the book, so could be skipped. However, it is included to show that it is possible to form a very large family of models covered by the same regularity conditions. One can then fit a model known to belong to the family, confident in the knowledge that conditions are standard and that asymptotic results will be valid. Even so, as will be discussed later, some care is needed to obtain such a family.

Our conclusion from this section therefore is that a comprehensive theory for the non-standard case is not a realistic proposition at this juncture. This therefore lends some support to the approach taken in this book of examining a limited number of non-standard situations separately, connecting together different non-standard situations only if their analyses take us in that direction.

The *exponential family of models* is the best-known family where a unified set of regular conditions can be obtained. Though it can be regarded as one family, it covers a broad range of well-known distributions.

As pointed out in (http://en.wikipedia.org/wiki/Exponential_family/, accessed 7/3/2015), the normal, exponential, lognormal, gamma, chi-squared, beta, Dirichlet, Bernoulli, Poisson, geometric, inverse Gaussian, von Mises, and von Mises-Fisher distributions are all members of the family. However, some distributions are exponential only if certain of its parameters, normally regarded as variable, are held fixed; for example, the Weibull is not a member unless the power parameter is kept fixed. Mixture models, whether finite mixtures or infinite mixtures, as in Student's t-distribution, are not members.

Other distributions not members are those whose support is parametrically variable. Thus, the gamma distribution cannot be extended to include a variable threshold and remain a member. Important distributions not belonging to the exponential family include the F-distribution, Cauchy distribution, hypergeometric distribution, and logistic distributions.

We consider the existence and properties of MLEs in the exponential family. An authoritative account is given by Barndorff-Nielsen (1978). A clear introduction is given by Brown (1986), this reference being accessible online under Project Euclid. Our account is based on the more recent formulation and description given in Jørgensen and Labouriau (2012), and couched in measure-theoretic terms to make discussion succinct. We stress, however, that this is the only section where measure theory is even mentioned, with the rest of the book essentially practical.

The definition of an exponential family is formally based on an overall measure space $(\mathcal{Y}, \mathcal{A}, \nu)$, as given by Kingman and Taylor (1966), for example, where $\mathcal{Y}$ is a metric space, $\mathcal{A}$ the $\sigma-$ field of subsets of $\mathcal{Y}$, and ν is a $\sigma-$ finite measure on $\mathcal{A}$.

The *canonical form* for an exponential family $\mathcal{P}$ is defined as the family of probability measures on $(\mathcal{Y}, \mathcal{A})$ whose members $P_{\boldsymbol{\theta}}, \boldsymbol{\theta} \in \boldsymbol{\Theta}$, have density function with respect to ν that take the form

$$\frac{dP_{\boldsymbol{\theta}}}{d\nu} = a(\boldsymbol{\theta})b(y)\exp[\boldsymbol{\theta}^{\mathrm{T}}\mathbf{t}(y)], \tag{3.19}$$

where $\boldsymbol{\theta}$ and $\mathbf{t}(y)$ are k-dimensional vectors, with $\boldsymbol{\theta} \in \boldsymbol{\Theta} \subseteq R^k$, $a(\cdot)$ is a strictly positive real function, and $b(\cdot)$ is a positive real function. The parameter $\boldsymbol{\theta}$ is called the canonical parameter, and $\boldsymbol{\Theta}$ is the domain of the canonical parameter. The dimension k is called the order of the family.

This definition can be simplified by taking μ as the measure whose density with respect to ν is b, that is, $(d\mu/d\nu)(y) = b(y)$. The probability measures then have density with respect to μ of the form

$$\frac{dP_{\boldsymbol{\theta}}}{d\mu} = a(\boldsymbol{\theta})\exp[\boldsymbol{\theta}^{\mathrm{T}}\mathbf{t}(y)]. \tag{3.20}$$

The quantity $a(\boldsymbol{\theta})$ is the normalizing factor satisfying

$$1/a(\boldsymbol{\theta}) = c(\boldsymbol{\theta}) = \int_{\mathcal{Y}} \exp[\boldsymbol{\theta}^{\mathrm{T}}\mathbf{t}(y)]\mu(dy),$$

so that

$$\mathcal{K}(\boldsymbol{\theta}) = \ln(c(\boldsymbol{\theta}))$$

is the cumulant generating function.

Non-degenerate distributions are obtained for those $\boldsymbol{\theta}$ where $c(\boldsymbol{\theta}) < \infty$, and in consequence

$$\boldsymbol{\Theta} = \{\boldsymbol{\theta} \in R^k : c(\boldsymbol{\theta}) < \infty\} \tag{3.21}$$

is defined as the *natural parameter space.* With $\boldsymbol{\Theta}$ defined in this way, the family is said to be *full.* It may be that, in the context of a given problem, the parameter space is only a proper subset of $\boldsymbol{\Theta}$, in which case the family would not then be full.

These initial definitions in measure-theoretic terms are as given by Brown (1986) and Jørgensen and Labouriau (2012), and this simplifies development of the theory. For simplicity, we refer to the latter reference as J&L in this section. In practice, it is probably most straightforward to keep the representation in the original form (3.19), taking ν simply as Lebesgue measure, with actual calculations using Riemannian differentiation and integration in the continuous case, and elementary summations to deal with probability mass atoms in the discrete case.

The representations (3.19) and (3.20) are said to be minimal if no other representation of $\mathcal{P}$ is possible with order $m < k$. If, in addition, $\boldsymbol{\Theta}$ is an open region, then $\mathcal{P}$ is said to be *regular.*

Consider now estimation of $\boldsymbol{\theta}$ from a sample $y_1, y_2, \ldots, y_n$ drawn from the family $\mathcal{P}$. The log-likelihood is

$$\mathcal{L}(\boldsymbol{\theta}) = \sum_{i=1}^{n} b(y_i) + \boldsymbol{\theta}^{\mathrm{T}} \sum_{i=1}^{n} \mathbf{t}(y_i) - n\mathcal{K}(\boldsymbol{\theta}).$$

We define $\boldsymbol{\tau}(\boldsymbol{\theta})$ by

$$\boldsymbol{\tau}(\boldsymbol{\theta}) = E_{\boldsymbol{\theta}}[\mathbf{t}(Y)].$$

When $\mathcal{P}$ is regular, it can be parametrized by $\boldsymbol{\tau}$, and the family is then said to be *parametrized by the mean*. The results in J&L particularly relevant to our discussion cover the conditions under which the MLE exists and is unique, and are as follows:

J&L Theorem 1.14

The region $\boldsymbol{\Theta}$ as defined in (3.21) is convex and $\mathcal{K}(\boldsymbol{\theta})$ is strictly convex in $\boldsymbol{\Theta}$.

From these convexity properties, it is clear that $\mathcal{L}$ has a maximum if and only if

$$\frac{\partial \mathcal{K}(\boldsymbol{\theta})}{\partial \boldsymbol{\theta}} = \sum_{i=1}^{n} \mathbf{t}(y_i)/n (= \mathbf{t}, \text{ say}) \tag{3.22}$$

has a solution; that is, if and only if $\mathbf{t} \in \boldsymbol{\tau}(\boldsymbol{\Theta})$. Such a solution will be unique. Geometrically, it is also intuitively clear that if $\mathcal{K}(\boldsymbol{\theta})$ increases sufficiently sharply as $\boldsymbol{\theta}$

approaches any point on the boundary $\partial\boldsymbol{\Theta}$ of $\boldsymbol{\Theta}$, then (3.22) will have an (interior) solution. Such behaviour is characterized by the property of *steepness*. $\mathcal{K}$ is said to be *steep* if for any given $\boldsymbol{\theta} \in \text{int}\,\boldsymbol{\Theta}$, and $\tilde{\boldsymbol{\theta}} \in \partial\boldsymbol{\Theta}$

$$(\tilde{\boldsymbol{\theta}} - \boldsymbol{\theta})^{\mathrm{T}} \frac{\partial \mathcal{K}}{\partial \theta}[\alpha\boldsymbol{\theta} + (1-\alpha)\tilde{\boldsymbol{\theta}}] \to \infty \text{ as } \alpha \to 0.$$

It turns out that regularity and steepness are closely connected in exponential families. We have

J&L Theorem 1.12

If $\mathcal{P}$ is regular then $\mathcal{P}$ is steep. The converse does not necessarily hold.

J&L Theorem 1.18

If $\mathcal{P}$ is regular (with minimal representation), then the MLE exists if and only if $\mathbf{t} = \sum_{i=1}^{n} \mathrm{t}(y_i)/n \in \boldsymbol{\tau}(\boldsymbol{\Theta})$, and is given by

$$\widehat{\boldsymbol{\theta}} = \boldsymbol{\tau}^{-1}(t). \tag{3.23}$$

If the family is steep, then the same result holds, moreover, with $\boldsymbol{\theta} \in \text{int}\,\boldsymbol{\Theta}$. In view of J&L Theorem 1.12, we can draw this conclusion always when $\mathcal{P}$ is regular.

A point of interest is that the representation (3.19), which we can write as $\exp(\boldsymbol{\theta}^T\mathbf{t} - \mathcal{K}(\boldsymbol{\theta}))$, is not unique. Brown (1986, Proposition 1.6) shows that if we make an *affine transformation*

$$\mathbf{Z} = \mathbf{M}^{\mathrm{T}}\mathbf{t} + \mathbf{z}_0$$
$$\phi = (\mathbf{M}^{\mathrm{T}})^{-1}\boldsymbol{\theta} + \phi_0,$$

where $\mathbf{M}$ is a given non-singular $k \times k$ matrix and $\mathbf{z}_0, \boldsymbol{\phi}_0$ are given k-dimensional vectors, then $\exp(\boldsymbol{\theta}^{\mathrm{T}}\mathbf{t} - \mathcal{K}(\boldsymbol{\theta}))$ becomes

$$\exp(\boldsymbol{\phi}^{\mathrm{T}}\mathbf{z} - \mathcal{K}_1(\boldsymbol{\phi})), \tag{3.24}$$

where

$$\mathcal{K}_1(\boldsymbol{\phi}) = \mathcal{K}(\mathbf{M}^{\mathrm{T}}(\boldsymbol{\phi} - \boldsymbol{\phi}_0)) - \boldsymbol{\phi}^{\mathrm{T}}\mathbf{z}_0 + \boldsymbol{\phi}_0^{\mathrm{T}}\mathbf{z}_0$$

is an exponential family with parameter space $\boldsymbol{\Phi} = (\mathbf{M}^{\mathrm{T}})^{-1}\boldsymbol{\Theta} + \boldsymbol{\phi}_0$. This version is equivalent to the original. Thus, any such parametrization can be used to study the probability distributions forming the model without affecting the statistical inferences drawn.

An important point of note is that the equivalence of these representations under affine transformation depends strongly on the fact that $\boldsymbol{\Theta}$ is open, so that the true parameter value $\boldsymbol{\theta}^0$ is an internal point. This makes the importance of steepness geometrically clear in ensuring that the true parameter point will be internal. This is not the case when we come to consider certain non-standard situations where the natural parameter space is *closed* with $\boldsymbol{\Theta} = \bar{\boldsymbol{\Theta}} = \text{Int}\boldsymbol{\Theta} \cup \partial\boldsymbol{\Theta}$ and where $\boldsymbol{\theta}^0 \in \partial\boldsymbol{\Theta}$ is possible. In this situation, the model structure can be different depending on whether the true $\boldsymbol{\theta}^0 \in \textbf{Int}\boldsymbol{\Theta}$ or $\boldsymbol{\theta}^0 \in \partial\boldsymbol{\Theta}$. The form of the parametrization can then affect the statistical inferences that can be drawn. We will take up this point again when we consider non-standard behaviour more fully in Chapter 5.

The satisfactory behaviour of regular exponential families extends to asymptotic normality theory. For example, J&L examine the situation where $\boldsymbol{\theta} = \boldsymbol{\theta}(\boldsymbol{\beta})$ is a function of a vector $\boldsymbol{\beta}$ with a dimension m, $1 \leq m \leq k$, with $\boldsymbol{\beta} \in B$ an open convex m-dimensional region, giving simple conditions (J&L, §1.7, Conditions G1-G4) when $\hat{\boldsymbol{\beta}}$ is asymptotically normal.

The previous outline, albeit very brief, nevertheless shows that exponential families are covered by an elegant unified theory. In particular, regular exponential families offer a wide range of models for use in model fitting, where ML estimation can be carried out without any concerns that regularity conditions will not be met.

Inverse Gaussian Example

An interesting application of these results is to the inverse Gaussian family, $\mathcal{P}$, whose PDF takes the form

$$f(y; \mu, \lambda) = \left(\frac{\lambda}{2\pi}\right)^{1/2} x^{-3/2} \exp\left[-\frac{\lambda(x-\mu)^2}{2\mu^2 x}\right], \; x > 0, \tag{3.25}$$

with $\boldsymbol{\Theta} = \{\boldsymbol{\theta} : \mu, \lambda > 0\}$. If we reparametrize to canonical form, with $\psi = \lambda/\mu^2$ to eliminate μ, then

$$f(y; \psi, \lambda) = \left(\frac{1}{2\pi}\right)^{1/2} x^{-3/2}\sqrt{\lambda} \exp\sqrt{\psi\lambda} \exp(-\psi x/2 - \lambda/2x), \quad x > 0.$$

This shows that $\mathcal{P}$ is not full in this case, as the boundary case, where $\psi = 0$, is the valid PDF

$$f(y; 0, \lambda) = \left(\frac{1}{2\pi}\right)^{1/2} x^{-3/2}\sqrt{\lambda} \exp\left(-\frac{\lambda}{2x}\right), \; x > 0, \tag{3.26}$$

the Lévy stable law distribution with index $1/2$. Thus the full family $\tilde{\mathcal{P}}$ is not regular. However, it can be shown to be steep, so that the MLE exists and is at an interior point of $\boldsymbol{\Theta}$, and cannot therefore be obtained at $\psi = 0$. Thus, the stable law distribution is never in probability the best fit.

In this example, we can corroborate the exponential family theory directly by finding the MLEs explicitly from the likelihood equations

$$\frac{\partial\mathcal{L}}{\partial\psi} = \frac{\partial\mathcal{L}}{\partial\lambda} = 0$$

to get

$$\frac{1}{\hat{\psi}} = A_n^2(H_n^{-1} - A_n^{-1}), \frac{1}{\hat{\lambda}} = (H_n^{-1} - A_n^{-1}),$$

where $H_n = n/\sum_{i=1}^{n}(1/y_i)$ and $A_n = (\sum_{i=1}^{n} y_i)/n$ are the harmonic mean and arithmetic means of the y's, for which we have $A_n \geq H_n$, almost surely. Thus, the MLE cannot, almost surely, be obtained at $\psi = 0$.

3.7 Numerical Optimization of the Log-likelihood

In certain situations, and this includes some well-known standard ones, the likelihood equation (3.2) can be solved to give the ML estimators explicitly. This is preferable when it can be done. However, in general the likelihood equations are not very tractable. Then numerical optimization has to be used. There then seems no advantage in attempting to solve (3.2), as one may as well use a numerical optimizing algorithm that maximizes $L(\boldsymbol{\theta}, \mathbf{y})$ directly. Though algorithm choice might be important in particular problems, our own preference has been the Nelder-Mead search algorithm, proposed by Nelder and Mead (1965). This is frowned on by some workers who point to the lack of theoretical foundation and to elementary examples where the method can be shown to fail. However, our experience is that in practice the method is surprisingly robust, with perhaps its greatest appeal its ease of implementation, involving just function evaluations. The reader preferring to use some alternative numerical optimization choice is of course free to do so.

It should be mentioned that not only Nelder-Mead, but also a conjugate gradient algorithm and the EM algorithm were tried when numerically applying the Bayesian *maximum a posteriori* (MAP) method of parameter estimation, closely related to MLE, in fitting finite mixture models to be discussed in Chapter 17; these perhaps were the most challenging models that were fitted in this book.

A good review of conjugate gradient methods is given in in Burley (1974, Chapter 2). We used the BFGS (Broyden - Fletcher - Goldfarb - Shanno) method of gradient-based optimization introduced by Davidon (1959) to minimize both the negative of the posterior and negative of the logposterior, as the gradient of the posterior distribution can be calculated. For the negative of the posterior, the algorithm did not move far from its starting point, as the gradients calculated at the initial points were very small. For the negative logposterior, the algorithm frequently moved to areas of parameter space associated with a very low posterior probability. The errors causing this originated in the routine updating H, the estimate of the covariance matrix, and seemed due to the surface being a long way from being quadratic.

We also considered the EM algorithm. A good introduction to the algorithm, including its application to mixture models, is given in Bilmes (1997). We used the version introduced in Dempster et al. (1977) to find the mode of the posterior, slightly modified to find the maximum of a posterior distribution. The algorithm did not converge to as good an optimum as the Nelder-Mead, being more sensitive to the starting point, and also was unstable for some initial solutions. Often this occurred when a large number of components was being fitted to a data set for which only a small number of components might be required, and took the form of one of the standard deviation parameters for a component with a very small weighting tending to infinity. The sensitivity of the limiting solution to the initial solution and the convergence to local maxima, or saddle points, are drawbacks that have been discussed elsewhere in the literature, for example, in Diebolt and Ip (1996).

There is surprisingly little theory available for the Nelder-Mead routine. However, this was the most robust of the three methods tried for finite mixture model fitting, and in addition it was certainly the simplest to implement. This was therefore the method used in the examples of this book. Though not taken further here, there is scope for further research in this area. In the finite mixture model case, the EM algorithm, when it did converge, was generally more efficient than the Nelder-Mead algorithm. An adaptation of a more sophisticated version of the EM algorithm, such as that put forward in Arcidiacono and Bailey Jones (2003), or the use of a stochastic EM algorithm, which has previously been applied to mixture models in Diebolt and Robert (1994), might be worth investigation. If a sufficiently good optimum could be obtained without a significant increase in the number of runs required, this method could outperform the Nelder-Mead.

There is one feature of the Nelder-Mead method that is worth stressing. Not only is Nelder-Mead very easy to implement, requiring only function value evaluations, it is also easily modified to handle constraints, this last being a frequent concern in the non-standard problems that we will be considering. In some situations, there can be many parameter constraints, some involving several parameters simultaneously. Such constraints are easily handled using Nelder-Mead, which can be seen if we examine how the method works.

Nelder-Mead operates in an easy-to-understand geometric way, by forming a simplex of $(d+1)$ points $\boldsymbol{\theta}_1, \boldsymbol{\theta}_2, \ldots, \boldsymbol{\theta}_{d+1} \in \boldsymbol{\Theta}$, the d-dimensional parameter space. In ML estimation, it compares the log-likelihood $L(\boldsymbol{\theta}_1), L(\boldsymbol{\theta}_2), \ldots, L(\boldsymbol{\theta}_{d+1})$ obtained at the points of the simplex, and then, typically, updates the simplex by discarding the point $\boldsymbol{\theta}_i$ with the lowest $L(\boldsymbol{\theta}_i)$ value, replacing it by a new point

$$\boldsymbol{\theta}' = \boldsymbol{\theta}_i + \beta(\boldsymbol{\theta}^c - \boldsymbol{\theta}_i), \tag{3.27}$$

where $\beta > 1$, this being the projection of $\boldsymbol{\theta}_i$ in a straight line through the centroid $\boldsymbol{\theta}^c = \Sigma_{i=1}^{d+1} \boldsymbol{\theta}_i/(d+1)$ of the simplex, and likely to give an improved value of the log-likelihood $L(\boldsymbol{\theta}')$ at the new point. This method for selecting $\boldsymbol{\theta}'$ may appear to be ad hoc, but leads to a remarkably robust algorithm. Examples have been given in the literature which show that Nelder-Mead can fail to converge even to a local optimum, see McKinnon (1998), for example; however, such conditions where failure occurs have to be precisely met and so are rather specialized, and we did not find it an issue in practice. The implication of this is that the rule for selecting the new point $\boldsymbol{\theta}'$ is easily modified to handle constraints, whilst still adhering to the general projection idea. All that is needed is to first obtain the point $\boldsymbol{\theta}'$ as in the original algorithm, but then where necessary to adjust the point $\boldsymbol{\theta}'$ so that all constraints remain satisfied. For example, if the new point $\boldsymbol{\theta}' \notin \boldsymbol{\Theta}$ when $\boldsymbol{\Theta}$ is convex and includes its boundary $\partial\boldsymbol{\Theta}$, then we simply adjust its value as given in (3.27) so that β is the (largest) possible value with $\boldsymbol{\theta}'$ still in $\boldsymbol{\Theta}$, i.e. $\boldsymbol{\theta}' \in \partial\boldsymbol{\Theta}$. Simpler variations of such an adjustment can be employed. For example, if a particular component ϕ of $\boldsymbol{\theta}$ has to be positive, and its value with an unconstrained choice of $\boldsymbol{\theta}'$ is $\phi' < 0$, we simply set $\phi' = 0$ instead.

We have used the Nelder-Mead method in most of the examples discussed in the book, handling problems with up to 30 parameters in the process without difficulty.

However, with specific models, we have calculated the MLEs of parameters explicitly if there is a simple way of doing this, not then using Nelder-Mead. The following is such a case.

3.8 Toll Booth Example

We illustrate the discussion in this chapter with an example where ML is used to estimate the parameters of the two-parameter gamma distribution with PDF as given in (3.28), and calculation of CIs using both formulas (3.7) and (3.8) to be discussed in the next chapter. The example does not involve anything non-standard, but simply illustrates the reliance of a conventional analysis on asymptotic theory. The problem arose in a study made of the operation of toll booths of the original Severn River Bridge in the UK, see Griffiths and Williams (1984). Each toll booth is modelled as a single-server queue, and data were collected of the service time of vehicles, that is, the time taken for a vehicle to pay at the toll booth before crossing the bridge. As the example is for illustration only, we use a small sample with sample size $n = 47$. The observations are in seconds and displayed in Table 3.1.

The service times are treated as independent random variates with a gamma distribution $G(a, b)$ with PDF

$$f(y) = \frac{1}{\Gamma(a)b^a} y^{a-1} \exp(-y/b), \quad a, b > 0, \; 0 < y < \infty, \tag{3.28}$$

where the parameters a and b are unknown. We could use Nelder-Mead optimization to obtain the ML estimators from the sample. However, for the two-parameter gamma

Table 3.1 47 Vehicle service times (in seconds) at a toll booth

4.3	4.7	4.7	3.1	5.2	6.7	4.5	3.6	7.2	10.9	6.6	5.8
6.3	4.7	8.2	6.2	4.2	4.1	3.3	4.6	6.3	4.0	3.1	3.5
7.8	5.0	5.7	5.8	6.4	5.2	8.0	4.9	6.1	8.0	7.7	4.3
12.5	7.9	3.9	4.0	4.4	6.7	3.8	6.4	7.2	4.8	10.5	

distribution, accurate values of the MLEs can be obtained using the following approach. The likelihood equations, see Law (2007), can be written as

$$A(a) = [\ln a - \mathrm{Psi}(a)]^{-1} = \left(\ln\left(\frac{\sum y_i}{n}\right) - \frac{\sum \ln y_i}{n}\right)^{-1} = c, \text{ say,} \tag{3.29}$$

and

$$b = \frac{\sum y_i}{n} a^{-1}, \tag{3.30}$$

where $\mathrm{Psi}(\cdot)$ is the digamma function. The function $A(a)$ is close to being a straight line with

$$\lim_{a\to\infty}\left(2a - \frac{1}{3} - A(a)\right) = 0.$$

Equation (3.29) can be solved for a using, say, a six-step modified Newton-Raphson procedure,

$$a_0 = c/2,\ a_i = a_{i-1} - \frac{1}{2}A(a_{i-1}),\ i = 1, 2, \ldots, 6,$$

where the denominator 2 is a simple approximation to the derivative $dA(a)/da$ used in the full Newton-Raphson method, thereby avoiding trigamma function calculations. The suggested six iterations is to cover the worst-case situation, which is when $c < 0.1$. The six iterations more than guarantee agreement of a with the tabulated values $a(c)$ given in Law (2007, Table 6.2.1), see Choi and Wette (1969). In fact, a fairly accurate inversion of (3.29), avoiding any polygamma function calculations, is given by

$$a(c) = A^{-1}(c) \simeq \frac{c}{2} + \frac{c^p + r\sqrt{c}}{6c^p + s\sqrt{c} + q},$$

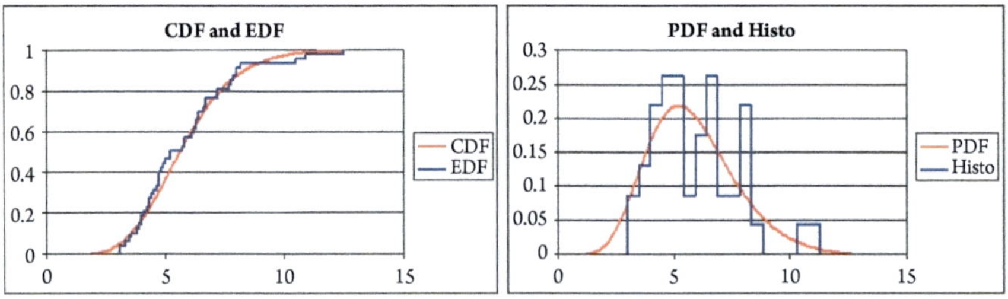

Figure 3.1 Toll booth service time data. Gamma model: fitted CDF, EDF, fitted PDF, and frequency histogram.

where $p = 0.9815$, $q = 3.0619$, $r = 0.04493$, and $s = -0.04471$. This gives $a(c)$ with an absolute error of less than 0.00065 for any $c \geq 0.01$. This error is negligible compared with sampling error for all but the largest samples; those where $n > 10^6$, say. For the Toll Booth example, $c = 18.082$. This gave $\hat{a} = 9.20$ and $\hat{b} = 0.63$, with this latter value obtained from (3.30). The 90% CIs for a and b, calculated from (4.1), were (6.14, 12.27) and (0.41, 0.85), respectively. The first graph in Figure 3.1 compares the fitted CDF with the EDF, and the second compares the fitted PDF with a frequency histogram of the sample.

The main interest in this example is not the parameter estimates themselves but the steady-state expected overall waiting time, $w(x, a, b)$, of customers in the queue, regarded as a function of the average arrival rate x. For instance, if the arrival pattern is Poisson with rate x, then from the well-known Pollaczek-Khinchine formula, see for example Taha (2003), we have

$$w(x, a, b) = \frac{(1 + a)ab^2x}{2(1 - abx)}. \tag{3.31}$$

We would be interested in examining the value of $w(x, a, b)$ over a range, $x_0 \leq x \leq x_1$, of arrival rates. The MLE of w is simply $w(x, \hat{a}, \hat{b})$, $x_0 \leq x \leq x_1$, with a CI for any given x that can be calculated using asymptotic theory by (4.2). We postpone presenting the results on estimating w until after we have considered an alternative approach for calculating CIs using bootstrapping, to be given in the next chapter, so as to directly compare the results of the two approaches.

4 Bootstrap Analysis

This chapter summarizes the use of parametric resampling, commonly known as parametric *bootstrapping* (BS), for calculating confidence intervals (CI) and confidence bands (CB), and for making goodness-of-fit (GoF) and regression lack-of-fit (LoF) tests.

Parametric bootstrapping provides an attractive alternative to asymptotic theory for constructing confidence intervals for unknown parameter values and functions involving such parameter values. Numerically, it is arguably a far preferable approach than that provided by standard asymptotic theory. It is very easily implemented, and is known theoretically to have more rapidly convergent properties than standard asymptotic normality results. In fact, it is probabilistically *exact* for location-scale models.

Parametric bootstrapping can be used also for calculating critical values of EDF statistics used in GoF tests, such as the Anderson-Darling A^2 statistic. This latter is known, see Stephens (1974), to give a GoF test that clearly out-performs better-known tests such as the chi-squared test, but is hampered by having a null distribution that varies with different null hypotheses, including whether parameters are estimated or not. Parametric bootstrapping offers an easy way round the difficulty, so that GoF tests using statistics like A^2 can be routinely applied.

Formal tests of regression lack-of-fit usually require replication of observations. We examine how parametric bootstrapping can overcome this so that a formal test of lack-of-fit can be made without replications.

4.1 Parametric Sampling

The fundamental statistical problem we face is as follows. We view a statistic T as a function of a sample $\mathbf{Y} = (Y_1, Y_2, \ldots, Y_n)$, so that $T = T(\mathbf{Y})$. The Y_i values are assumed to be independent observations from the same distribution with cumulative distribution function (CDF) $F(\cdot, \boldsymbol{\theta})$. To make any statistical inference with T, we will need to know the distribution of T itself. Let $G_n(\cdot, \boldsymbol{\theta})$ be its CDF, the notation making it clear that this will depend on n as well as $\boldsymbol{\theta}$. The objective of parametric sampling is to estimate $G_n(\cdot, \boldsymbol{\theta})$.

Non-Standard Parametric Statistical Inference. Russell Cheng.

We distinguish two forms of parametric sampling: (i) *Monte Carlo estimation* and (ii) *parametric bootstrapping*, depending on whether $\boldsymbol{\theta}$ is known or not.

4.1.1 Monte Carlo Estimation

We call the version of parametric sampling, when $\boldsymbol{\theta}$ is known, *Monte Carlo estimation.* In this version, we generate B independent values of $T : T_1, T_2, \ldots, T_B$, then estimate $G_n(t, \boldsymbol{\theta})$ by the empirical distribution function (EDF) formed from the $T_1, T_2, \ldots, T_B$, namely,

$$\tilde{G}_n(t, \boldsymbol{\theta}) = \frac{\#\text{ of } T_j \leq t}{B}, \quad -\infty < t < \infty.$$

The Glivenko-Cantelli lemma, see Billingsley (1986), for example, states that $\sup_t \left|\tilde{G}_n(t, \boldsymbol{\theta}) - G_n(t, \boldsymbol{\theta})\right| \to 0$ with probability 1 as $B \to \infty$. Thus, in principle, $G_n(t, \boldsymbol{\theta})$, for all practical purposes, can be found numerically to any given accuracy by choosing B sufficiently large. This holds for any given n, no matter how large or small.

In what follows we shall speak of this use of the EDF to estimate the CDF as being *probabilistically exact*, or just simply as being *exact*, in the sense that any desired accuracy is achievable in principle, whatever the n, by taking B sufficiently large. For the purposes of statistical inference, a value of $B = 1000$ is usually sufficient. For exploratory work or for illustrative purposes, where results only need to be indicative rather than sufficiently precise for formal decision taking, a value of B of just a few hundred is often sufficient.

4.1.2 Parametric Bootstrapping

The second case is when $\boldsymbol{\theta}$ is unknown, so that Monte Carlo estimation cannot be applied directly, but we suppose that we have an actual sample $\mathbf{y} = (y_1, y_2, \ldots, y_n)$ from which to estimate $\boldsymbol{\theta}$. In what follows, we will use the *maximum likelihood estimator* (MLE), denoted by $\widehat{\boldsymbol{\theta}}$.

Parametric bootstrapping is simply Monte Carlo estimation of $G_n(t, \widehat{\boldsymbol{\theta}})$ *with the assumption that this will then provide an estimate of* $G_n(t, \boldsymbol{\theta})$. We generate, by computer, B independent random samples $\mathbf{Y}_j^* = (Y_{ij}^*, i = 1, 2, \ldots, n), j = 1, 2, \ldots, B$, with the Y_{ij}^* all generated from the fitted distribution $F(\cdot, \widehat{\boldsymbol{\theta}})$. We then calculate the statistic of interest T from each sample $\mathbf{Y}_j^*$: $T_j^* = T(\mathbf{Y}_j^*), j = 1, 2, \ldots, B$. Here the superscript * indicates a computer-generated 'bootstrap' quantity obtained in a statistically identical way to the corresponding quantity obtained in the original process, except that $\widehat{\boldsymbol{\theta}}$ replaces the unknown true value $\boldsymbol{\theta}$. The EDF $\tilde{G}_n^*(t, \widehat{\boldsymbol{\theta}})$ of the $T_1^*, T_2^*, \ldots, T_B^*$ estimates $G_n(t, \widehat{\boldsymbol{\theta}})$ to whatever accuracy desired, by making B sufficiently large. We have written $\tilde{G}_n^*(t, \widehat{\boldsymbol{\theta}})$ with the added superscript *, instead of simply $\tilde{G}_n(t, \widehat{\boldsymbol{\theta}})$, to emphasize that this EDF has been obtained by bootstrapping. The EDF $\tilde{G}^*(t, \widehat{\boldsymbol{\theta}})$ only estimates $G(t, \widehat{\boldsymbol{\theta}})$, but as $\widehat{\boldsymbol{\theta}}$ estimates the unknown true $\boldsymbol{\theta}$, it is reasonable to consider $\tilde{G}^*(t, \widehat{\boldsymbol{\theta}})$ as an estimator of $G(t, \boldsymbol{\theta})$ and not just of $G(t, \widehat{\boldsymbol{\theta}})$.

Much of bootstrap theory is given to establishing conditions under which $\tilde{G}_n^*(t,\widehat{\boldsymbol{\theta}})$ is a satisfactory estimator of $G_n(t,\boldsymbol{\theta})$. The ideal case is when $\lim_{B\to\infty} \tilde{G}_n^*(t,\widehat{\boldsymbol{\theta}}) = G_n(t,\boldsymbol{\theta})$ for all t, $\boldsymbol{\theta}$, and n. Like Monte Carlo estimation, the parametric bootstrap is then (*probabilistically*) *exact*. We emphasize that only $B \to \infty$ is needed to achieve actual exactness, which will then hold for all n. Thus $n \to \infty$ is *not* required as well.

Under suitable regularity conditions, the MLEs of $\boldsymbol{\theta}$ and smooth functions $g(\cdot,\boldsymbol{\theta})$ will be asymptotically normally distributed, so that $|\widehat{\boldsymbol{\theta}} - \boldsymbol{\theta}| = O_p(n^{-1/2})$ as $n \to \infty$, and, assuming $G_n(t,\boldsymbol{\theta})$ is twice differentiable in $\boldsymbol{\theta}$, also that $|\tilde{G}_n(t,\widehat{\boldsymbol{\theta}}) - G_n(t,\boldsymbol{\theta})| = O_p(n^{-1/2})$, as $n \to \infty$. In this case, $\tilde{G}_n^*(t,\widehat{\boldsymbol{\theta}})$ can be expected to match, at least asymptotically, the behaviour of any theoretically derived estimator of $G_n(t,\boldsymbol{\theta})$. We shall not discuss bootstrap theory in much detail. Hjorth (1994) provides a clear and succinct account from a viewpoint well suited to that of this book. A clear but brief summary is provided by Young and Smith (2005). There is now a vast literature on the subject. Chernick (2008) provides over 140 *pages* of selected references covering research and practice up to 2007.

We discuss the use of parametric sampling for calculating confidence intervals (CIs) and for GoF tests, including an important case where it is exact rather than only approximate.

For simplicity we will omit the subscript n from now on, writing $\tilde{G}_n(t,\widehat{\boldsymbol{\theta}})$, $G_n(t,\boldsymbol{\theta})$, and so on simply as $\tilde{G}(t,\widehat{\boldsymbol{\theta}})$, $G(t,\boldsymbol{\theta})$, when the context makes clear the dependence on n.

4.1.3 Bootstrap Confidence Intervals

In standard asymptotic theory, the simplest two-sided $(1-\alpha)100\%$ confidence interval for φ_i, the ith component of $\boldsymbol{\theta}$ (where we have used φ_i rather than θ_i for symbolic clarity), takes the well-known form

$$\left(\hat{\varphi}_i + z_{\alpha/2}\sqrt{\hat{V}_{ii}},\ \hat{\varphi}_i + z_{(1-\alpha/2)}\sqrt{\hat{V}_{ii}}\right), \tag{4.1}$$

where $\hat{V}_{ii}$ is the ith main diagonal entry of $\hat{V} = [\mathrm{J}(\widehat{\boldsymbol{\theta}})]^{-1}$, the estimated covariance matrix of $\widehat{\boldsymbol{\theta}}$ obtained by inverting the observed information matrix $\mathrm{J}(\widehat{\boldsymbol{\theta}})$, and z_α is the α quantile of the standard normal distribution.

Note that here and in the rest of the book we have used z_α to denote the α quantile. Often z_α is used to denote the *upper* α quantile, that is, $z_{1-\alpha}$ in our notation, to take advantage of the fact that, for a distribution symmetric about zero, the lower quantile can then be written as $-z_\alpha$. We have not used this notation, as we will be dealing with quantiles of some distributions not symmetric about zero.

The formula (4.1) can be extended to a function $\lambda(x,\boldsymbol{\theta})$, $a \le x \le b$, that is dependent on the parameter values. A first-order correct two-sided $(1-\alpha)100\%$ confidence interval for $\lambda(x,\boldsymbol{\theta})$ is

$$(\lambda(x,\widehat{\boldsymbol{\theta}}) + z_{\alpha/2}\delta,\quad \lambda(x,\widehat{\boldsymbol{\theta}}) + z_{(1-\alpha/2)}\delta), \tag{4.2}$$

where

$$\delta = \sqrt{\left.\frac{\partial\lambda(x,\boldsymbol{\theta})}{\partial\boldsymbol{\theta}}\right|_{\widehat{\theta}}^{T} V(\widehat{\boldsymbol{\theta}}) \left.\frac{\partial\lambda(x,\boldsymbol{\theta})}{\partial\boldsymbol{\theta}}\right|_{\widehat{\theta}}}.$$

Under the same standard conditions, we can construct a CI using bootstrapping. Several forms of CIs have been proposed, see Hjorth (1994) for a clear description. We describe just the simplest form.

We assume that the situation is exactly as just described, with the observations in the data sample $\mathbf{y} = (y_1, y_2, \ldots, y_n)$ assumed to have the distribution with CDF $F(y, \boldsymbol{\theta})$, and that estimation of $\boldsymbol{\theta}$ by ML yields the estimator $\widehat{\boldsymbol{\theta}}$. We then obtain B BS samples, $\mathbf{y}_j^*, j = 1, 2, \ldots, B$, each of size n, by sampling from the fitted distribution $F(y, \widehat{\boldsymbol{\theta}})$. Now take T equal to the MLE $\widehat{\boldsymbol{\theta}}$ itself, that is, $T(\mathbf{Y}) = \widehat{\boldsymbol{\theta}}(\mathbf{Y})$, so that from each BS sample $\mathbf{y}_j^*$, $j = 1, 2, \ldots, B$, we calculate $T_j^* = \widehat{\boldsymbol{\theta}}_j^*$, the MLE of $\boldsymbol{\theta}$ based on that BS sample. A CI with confidence level $(1-\alpha)$ for an individual component, denoted by φ for symbolic clarity, can be constructed by arranging the bootstrap ML estimates of $\varphi : \hat{\varphi}_j^*, j = 1, 2, \ldots, B$, in rank order: $\hat{\varphi}_{(1)}^* \le \hat{\varphi}_{(2)}^* \le \ldots \le \hat{\varphi}_{(B)}^*$, then taking the $(1-\alpha)100\%$ confidence inequality $\hat{\varphi}_{(l)} \le \hat{\varphi}^* \le \hat{\varphi}_{(m)}^*$, where $l = \lfloor(\alpha/2)B\rfloor$ and $m = \lfloor(1-\alpha/2)B\rfloor$, and replacing $\hat{\varphi}^*$ by the unknown true φ. The resulting confidence interval for φ takes the form

$$\hat{\varphi}_{(l)}^* \le \varphi \le \hat{\varphi}_{(m)}^*.$$

This is called the *percentile CI.*

The bias of $\hat{\varphi}^*$ can be estimated by

$$\hat{\beta} = B^{-1}\sum_{j=1}^{B}\hat{\varphi}_j^* - \hat{\varphi},$$

though it has to be said that this bias correction gives rather mixed results in practice.

The following flowchart summarizes the steps of the percentile CI calculation just described, including the bias correction. We use the notation $Y \sim F(y, \boldsymbol{\theta})$ to indicate that Y has the distribution whose CDF is $F(y, \boldsymbol{\theta})$, and $\mathbf{y} = (y_i \sim F(y, \boldsymbol{\theta}), i = 1, 2, \ldots, n)$ to denote a random sample of n (independent) observations all drawn from the distribution whose CDF is $F(y, \boldsymbol{\theta})$.

Flowchart for calculating a $(1-\alpha)100\%$ BS CI for a component φ of θ (4.3)

Given $\mathbf{y} = (y_1, y_2, \ldots, y_n)$, a random sample drawn from CDF $F(y, \boldsymbol{\theta})$,

$\downarrow$

$\widehat{\boldsymbol{\theta}} = \widehat{\boldsymbol{\theta}}(\mathbf{y})$ MLE

For $j = 1$ to B,

$$\mathbf{y}_j^* = (y_{ij}^* \sim F(y, \widehat{\boldsymbol{\theta}}), i = 1, 2, \ldots, n) \quad j\text{th BS sample}$$

$$\downarrow$$

$$\widehat{\boldsymbol{\theta}}_j^* = \widehat{\boldsymbol{\theta}}(\mathbf{y}_j^*) \qquad j\text{th BS MLE}$$

$$\downarrow$$

$$\hat{\varphi}_j^* \qquad j\text{th BS MLE of } \varphi\text{, the component of interest}$$

$$\downarrow$$

$$\hat{\beta} = n^{-1} \textstyle\sum_{j-1}^{n} \hat{\varphi}_j^* - \hat{\varphi} \qquad \text{Estimate of bias due to bootstrapping}$$

End j

$$\hat{\varphi}_{(1)}^* \leq \hat{\varphi}_{(2)}^* \leq \ldots \leq \hat{\varphi}_{(B)}^* \quad \text{Ordered BS MLEs of component of interest}$$

$$\downarrow$$

$$l = \lfloor \max(1, (\alpha/2)B) \rfloor, m = \lfloor (1 - \alpha/2)B) \rfloor \quad \text{Rounded integer subscripts}$$

$$\downarrow$$

$$\hat{\varphi}_{(l)}^* - \hat{\beta} \leq \varphi \leq \hat{\varphi}_{(m)}^* - \hat{\beta} \quad \text{Bias-corrected } (1 - \alpha)100 \text{ percentile CI for } \varphi$$

4.1.4 Toll Booth Example

To illustrate the bootstrap discussion so far, we return to the Toll Booth example of Section 3.8. Recall that the MLEs were $\hat{a} = 9.20$ and $\hat{b} = 0.63$, and that the 90% CIs calculated from (4.1) gave $(6.14, 12.27)$ for a and $(0.41, 0.85)$ for b.

For comparison, we calculated the BS percentile CIs for a and b, with $\alpha = 0.1$, $B = 1000$. This gave the BS CI for a as $(7.06, 13.93)$, showing this to be biased to the right compared with the CI using the asymptotic formula. The BS CI for b was $(0.42, 0.85)$, which is almost identical to the asymptotic CI.

4.1.5 Coverage Error and Scatterplots

The quality of CIs is usually discussed in terms of their *coverage error*, that is, the difference between the actual confidence level achieved and the target value of $1 - \alpha$. If its coverage error tends to zero as $n \to \infty$, a CI is said to be *consistent*. The coverage error is usually given in terms of a probabilistic order of magnitude, which can be determined by asymptotic theory that also extends to cover bootstrap CIs, especially for parametric bootstrapping where the MLE $\widehat{\boldsymbol{\theta}}$ is used. In general, the coverage error is typically $O(n^{-1/2})$, see Young and Smith (2005, Section 11.1), a disappointing result given that $\left| \widehat{\boldsymbol{\theta}} - \boldsymbol{\theta}^0 \right| = O_p(n^{-1/2})$, as previously discussed.

However, there is the proviso that for balanced two-sided CIs where the target confidence level is the same (i.e. $\alpha/2$) at both ends of the interval, the coverage error is then usually reduced to $O(n^{-1})$. We do not go into details, but this improved performance of a balanced CI arises because coverage error comes mainly from bias in the estimators. However, the effect of this bias is in opposite directions at the two ends of the CI, so that when the CI is balanced and the distribution of $\hat{\theta}$ is asymptotically normal and so symmetric, they then cancel sufficiently precisely to reduce the coverage error to $O(n^{-1})$.

Prepivoting, as described by Beran (1987), is a method for accomplishing this, but we do not go into details here.

Coverage error can be usefully viewed as the discrepancy between the coverage actually achieved and the target value when the latter is derived under asymptotic normality assumptions. As a general guide, one would therefore expect the coverage error to be lower the closer that the distribution of the estimated parameter of interest, $\widehat{\boldsymbol{\theta}}$, say, is to being normal.

Scatterplots are a useful feature of BS analysis for indicating if the distribution of $\widehat{\boldsymbol{\theta}}$ is normal or not.

As an illustration, consider the Toll Booth example again, but where we consider two alternative parametrizations. Figure 4.1 gives the scatterplots of the $B = 1000$ BS MLEs $(\hat{a}_j^*, \hat{b}_j^*), j = 1, 2, \ldots, B$, and of $(\hat{\mu}_j^*, \hat{\sigma}_j^*), j = 1, 2, \ldots, B$, where $\mu = ab$ and $\sigma = \sqrt{a}b$ are the mean and standard deviation (SD) of the gamma distribution. The points in each plot are divided into two types. The black points correspond to $(1 - \alpha)100\%$ (where $\alpha = 0.1$, in the plots) of the total number of points. These points estimate what is called a $(1 - \alpha)100\%$ *likelihood-based confidence region* as discussed by Hall (1987), the construction of which we will describe in Section 4.3. The green points lie outside the confidence region. For now, we simply look at all the points in each plot as a whole and note that the scatterplot of the original $(\hat{a}_j^*, \hat{b}_j^*)$ has a very noticeable asymmetrically curved 'banana' form. This is typical of such plots, see Hjorth (1994, Fig 6.1) or Davison and Hinkley (1997, Fig 7.11, left lower plot). In contrast, the scatter of $(\hat{\mu}_j^*, \hat{\sigma}_j^*)$ points, though indicating strong correlation between $\hat{\mu}_j^*$ and $\hat{\sigma}_j^*$, has a much more elliptically symmetrical form, indicative of normality.

In Section 4.3, we will consider a practical use of such scatterplots not often discussed.

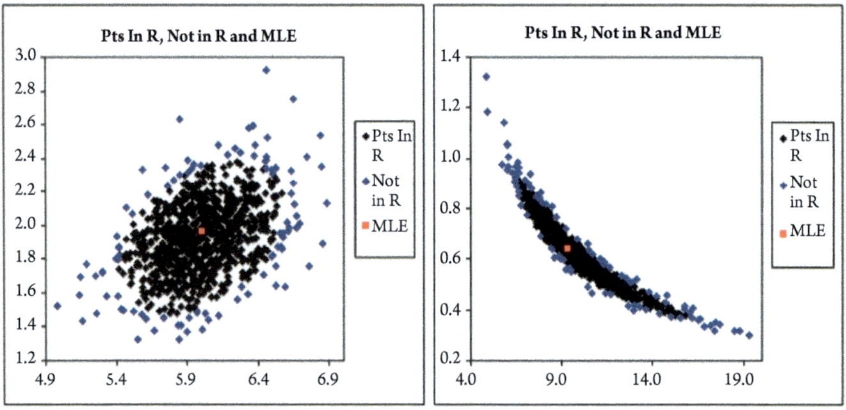

Figure 4.1 Toll Booth example. Left: scatterplots of 1000 BS MLEs $\hat{\mu}$ and $\hat{\sigma}$. Right: 1000 BS MLEs $\hat{a}$ and $\hat{b}$. The MLE from the original sample is the red square. Black points: those in $R_{1-\alpha}$. Green points: those not in $R_{1-\alpha}$. $\alpha = 0.1$.

Table 4.1 Coverage error results: actual coverage of 1000 CIs for parameters a, b, μ, σ of G(1,1) distribution with nominal coverage level $(1 - \alpha) = 0.9$

		a	b	μ	σ
BS not bias corrected	In CI	862	858	887	863
	Not in CI	138	142	113	137
BS bias corrected	In CI	916	871	884	866
	Not in CI	84	129	116	134
Asymptotic	In CI	899	860		
	Not in CI	101	140		

In general, the coverage error of the percentile CI and the simple CI can be noticeable, especially when n is relatively small, but perhaps not so large as to preclude their use. As an example, we consider fitting the gamma distribution $G(a, b)$ with PDF of eqn (3.28) to a sample of size $n = 50$ drawn from the standard exponential distribution, so that the true parameter values are $a = 1$ and $b = 1$. The value of b is unimportant, being a scale parameter, so that the form of the distribution of the MLEs depends only on the value of a. The value $a = 1$ is a fairly extreme choice, making the joint distribution of $(\hat{a}, \hat{b})$ distinctly non-normal when $n = 50$, so provides a difficult coverage error test. Table 4.1 gives the results. It will be seen that the bias-corrected BS CIs behave reasonably well, taking the coverage of both a and b into account.

4.2 Confidence Limits for Functions

We now extend comparison of the asymptotic theory and bootstrapping approaches to estimation of a function $\lambda(x, \boldsymbol{\theta})$, $x_0 \leq x \leq x_1$, taking as our example the waiting time function distribution in the Toll Booth example, $w(x, a, b)$, $x_0 \leq x \leq x_1$, as given in eqn (3.31).

Consider first a two-sided CI for a given x. Using asymptotic theory, the MLE is $w(x, \hat{a}, \hat{b})$, $x_0 \leq x \leq x_1$, and we can apply formula (4.2) directly to construct an asymptotic $(1 - \alpha)$ level CI for $w(x, a, b)$ at any *fixed* x. The left-hand plot in Figure 4.2 shows the MLE of the function $w(x, \hat{a}, \hat{b})$ (red line) calculated at 10 equally spaced values of x in the range $0 \leq x \leq 0.1$. The solid (green) upper line and solid (blue) lower line give the corresponding limits of the CI at level $(1 - \alpha)$, which allows the limits to be read off for any *one* given x. It should be stressed that the two lines do *not* enable CIs to be calculated at *several different* x's with confidence level $(1 - \alpha)$ holding *simultaneously* across all values. A simple but very conservative (simultaneous) Bonferroni confidence level (see, for example, Miller, 1981) can be constructed for M different x's simultaneously by setting the level to be $(1 - M^{-1}\alpha)$ for each x, with overall level $\geq (1 - \alpha)$.

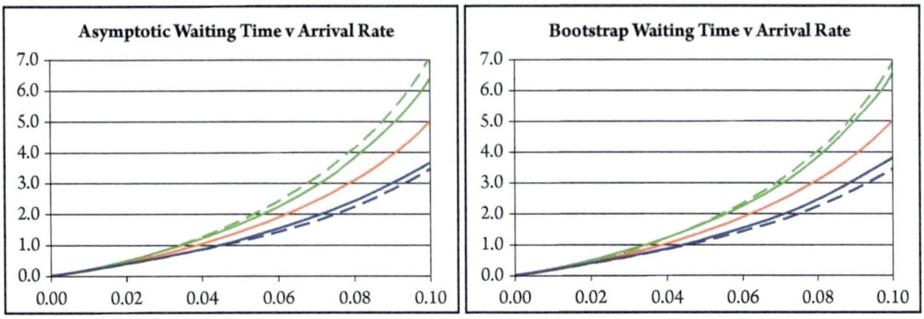

Figure 4.2 Toll Booth example. Plots of expected waiting time versus arrival rate. Upper CB limit - dashed green, upper CI limit - green, MLE - red, lower CI limit - blue, lower CB limit - dashed blue.

The right-hand plot in Figure 4.2 shows the CI limits calculated by bootstrapping. As in the asymptotic formula case, the BS CI limits were calculated over ten equally spaced x– intervals. At each x_i, the $B = 1000$ BS waiting time values $w_j^*(x_i) = w(x_i, \hat{a}_j^*, \hat{b}_j^*)$, $j = 1, 2, \ldots, B$, were calculated, then ordered so that $w_{(1)}^*(x_i) \le w_{(2)}^*(x_i) \le \ldots \le w*_{(B)}$ (x_i), and the confidence interval at that x_i taken as

$$w_{(l)}^*(x_i) \le w(x_i, a, b) \le w_{(m)}^*(x_i),$$

where l and m are as given in the flowchart (4.3).

It will be seen that in the bootstrap case, the CI is slightly asymmetric at each x value.

4.3 Confidence Bands for Functions

For simultaneous CIs, rather than use a Bonferroni CI, it is more efficient to calculate a *confidence band* which, with given confidence level, will *entirely contain* the unknown curve $w(x, \boldsymbol{\theta})$, $x_0 \le x \le x_1$. Simultaneous confidence intervals are discussed by Miller (1981). The construction of confidence bands based on asymptotic theory has been described by Cheng and Iles (1983) for continuous CDFs, and by Cheng (1987) more generally. We again illustrate with the Toll Booth example, so that $\boldsymbol{\theta} = (a, b)$, we have asymptotically as $n \to \infty$, that

$$(\widehat{\boldsymbol{\theta}} - \boldsymbol{\theta})^{\mathrm{T}} \mathrm{J}(\widehat{\boldsymbol{\theta}})(\widehat{\boldsymbol{\theta}} - \boldsymbol{\theta}) \sim \chi_2^2, \tag{4.4}$$

where χ_2^2 is the chi-squared distribution with two degrees of freedom, that is, the exponential distribution. Writing $\chi_2^2(\alpha)$ for the α quantile, inversion of the probability statement $\Pr((\widehat{\boldsymbol{\theta}} - \boldsymbol{\theta})^{\mathrm{T}} \mathrm{J}(\widehat{\boldsymbol{\theta}})(\widehat{\boldsymbol{\theta}} - \boldsymbol{\theta}) \le \chi_2^2(1 - \alpha)) = 1 - \alpha$ in the usual way gives the $(1 - \alpha)$ confidence region

$$R_{1-\alpha} = \{\boldsymbol{\theta} : (\widehat{\boldsymbol{\theta}} - \boldsymbol{\theta})^{\mathrm{T}} \mathrm{J}(\widehat{\boldsymbol{\theta}})(\widehat{\boldsymbol{\theta}} - \boldsymbol{\theta}) \leq \chi_2^2(1 - \alpha)\}. \tag{4.5}$$

Thus, if

$$w_{\min}(x) = \min_{\boldsymbol{\theta} \in R_{1-\alpha}} w(x, \boldsymbol{\theta}),\ w_{\max}(x) = \max_{\boldsymbol{\theta} \in R_{1-\alpha}} w(x, \boldsymbol{\theta}),\ x_0 \leq x \leq x_1 \tag{4.6}$$

then the condition

$$w_{\min}(x) \leq w(x, \boldsymbol{\theta}) \leq w_{\max}(x),\ x_0 \leq x \leq x_1 \tag{4.7}$$

holds *simultaneously* for all $\widehat{\boldsymbol{\theta}} \in R_{1-\alpha}$ with asymptotic confidence level no less than $(1 - \alpha)$, so that (4.7) is an *asymptotic confidence band* containing $w(x, \boldsymbol{\theta})$, for all $x_0 \leq x \leq x_1$, with confidence level no less than $(1 - \alpha)$ as $n \to \infty$.

We can obtain a bootstrap version of the confidence band by using B BS MLEs $\widehat{\boldsymbol{\theta}}_j^*$, $j = 1, 2, \ldots, B$, to construct a bootstrap equivalent of $R_{1-\alpha}$. When $\widehat{\boldsymbol{\theta}}$ is exactly normally distributed, the region $R_{1-\alpha}$ has the property that all points in $R_{1-\alpha}$ have higher likelihood values than all those outside it. Cox and Hinkley (1974, p. 218) call such a region a *likelihood-based region*. Methods for calculating such a region are described by Hall (1987), who recommends the percentile-t bootstrap method of generating a set of $\boldsymbol{\theta}_j^*$ points, and then use of a nonparametric kernel smoothing method to identify the boundary of $R_{1-\alpha}$. This method will usually have the attractive property of making the coverage error in $R_{1-\alpha}$ of order $O(n^{-1})$.

In our example, we use a significantly simpler BS approach to obtain an estimate of $R_{1-\alpha}$, which nevertheless has the same motivation as the approach proposed by Hall. First note that we are free to choose the parametrization $\boldsymbol{\theta}$ in constructing the confidence band. We therefore choose $\boldsymbol{\theta}$ to make the distribution of its MLE $\widehat{\boldsymbol{\theta}}$ as close to being normal as possible when (4.4) will be a good approximation. In the Toll Booth example, a suitable reparametrization to use is the mean and standard deviation (SD) of the gamma distribution, given by $\mu = ab$ and $\sigma = \sqrt{a}b$. If we compare the scatterplot of $(\hat{a}^*, \hat{b}^*)$ with that of $(\hat{\mu}^*, \hat{\sigma}^*)$ in Figure 4.1, the latter is clearly much more ellipsoidal in shape. To construct an estimate of $R_{1-\alpha}$, we note that the quadratic in (4.5) is the leading term in the Taylor expansion of the usual log-likelihood ratio $2(L(\widehat{\boldsymbol{\theta}}) - L(\boldsymbol{\theta}))$, corresponding to distributions that are asymptotically normal. The equation

$$q(\boldsymbol{\theta}) = (\widehat{\boldsymbol{\theta}} - \boldsymbol{\theta})^{\mathrm{T}} \mathrm{J}(\widehat{\boldsymbol{\theta}})(\widehat{\boldsymbol{\theta}} - \boldsymbol{\theta}) = c,$$

with c a constant, thus gives loci where $L(\boldsymbol{\theta})$ is constant, with smaller c corresponding to larger $L(\boldsymbol{\theta})$. Therefore, if we calculate $q_j^* = (\widehat{\boldsymbol{\theta}} - \widehat{\boldsymbol{\theta}}_j^*)^{\mathrm{T}} \mathrm{J}(\widehat{\boldsymbol{\theta}})(\widehat{\boldsymbol{\theta}} - \widehat{\boldsymbol{\theta}}_j^*)$, $j = 1, 2, \ldots, B$, corresponding to the BS points $\widehat{\boldsymbol{\theta}}_j^*$, and order these points so that

$$q_{(1)}^* \leq q_{(2)}^* \leq \ldots \leq q_{(B)}^*,$$

then the points $\widehat{\boldsymbol{\theta}}^*_{(j)}$ corresponding to the first $m = \lfloor (1-\alpha)B \rfloor$ of the $q^*_{(j)}$ values, that is, $q^*_{(1)} \le q^*_{(2)} \le \ldots \le q^*_{(m)}$, will estimate a likelihood-based region of level $(1-\alpha)$. These points can be used directly in (4.6) to represent $R_{1-\alpha}$ in calculating $w_{\min}(x)$ and $w_{\max}(x)$.

The black points in the right-hand plot of Figure 4.1 are for the case $B = 1000, \alpha = 0.1$ in the Toll Booth example. The left-hand plot uses the *same* selection of bootstrap points, but in their original parametrization $\boldsymbol{\theta} = (a, b)$. These are the points used in calculating the CB. Figure 4.2 shows the upper and lower CIs and CB limit curves (solid for CI, dashed for CB) using both the asymptotic and BS approaches. It will be seen that the BS curves compared with the asymptotic curves are skewed slightly higher, and the CB limits are wider than the CI limits for either approach.

4.4 Confidence Intervals Using Pivots

Coverage error can be entirely eliminated if the CI can be constructed using a *pivotal quantity*. This has interesting repercussions not only for CI construction but for parametric bootstrap GoF tests, so we will discuss this rather more fully here.

A *pivotal statistic* is usually defined somewhat cryptically and rather unspecifically as a function of a random sample, $\mathbf{Y}$, that also depends on $\boldsymbol{\theta}$, the parameter of interest, but whose distribution does not depend on the value of $\boldsymbol{\theta}$.

In practice, the form of pivotal statistics tends to follow the two used in constructing CIs for the parameters μ and σ^2 from a sample $\mathbf{Y}$ drawn from the normal distribution $N(\mu, \sigma^2)$, namely, the celebrated studentized statistics T and W defined by

$$T = \frac{\bar{Y} - \mu}{S/\sqrt{\nu n}}, \ W = \frac{S^2}{\sigma^2}, \text{ where } S^2 = \sum_{i=1}^{n}(Y_i - \bar{Y})^2, \text{ with } \nu = n - 1. \tag{4.8}$$

Note that we have used S^2 to denote a sum of squares. This is not to be confused with the lower-case notation s^2 conventionally used to denote the sample variance.

If μ and σ^2 are the true values, then T has the Student's t-distribution and W has the chi-squared distribution, both with ν degrees of freedom. We shall write $t_\nu(\alpha)$ and $\chi^2_\nu(\alpha)$ for the α quantiles of these two distributions. As with z_α, and because the t-distribution is symmetric about zero, some authors use $t_\nu(\alpha)$ and $-t_\nu(\alpha)$ to denote the upper and lower t_ν quantiles of the t-distribution when $\alpha < 0.5$. We have not used this notation in view of our discussion of non-symmetric pivot distributions shortly to follow. An exact, balanced, two-sided $(1-\alpha)100\%$ CI for μ and for σ^2 is obtained by replacing the random variables in the probability statements (assuming $\alpha < 0.5$)

$$\Pr\{t_\nu(\alpha/2) \le \frac{\bar{Y} - \mu}{S/\sqrt{\nu n}} \le t_\nu(1-\alpha/2)\} = 1 - \alpha$$

and

$$\Pr\{\chi^2_\nu(\alpha/2) \le \frac{S^2}{\sigma^2} \le \chi^2_\nu(1-\alpha/2)\} = 1 - \alpha$$

by their sample values, and inverting the inequalities to give the well-known *studentized* confidence intervals for μ and σ^2:

$$(\bar{y} - t_\nu(1-\alpha/2)S/\sqrt{\nu n},\ \bar{y} - t_\nu(\alpha/2)S/\sqrt{\nu n}) \text{ and } \left(\frac{S^2}{\chi^2_\nu(1-\alpha/2)}, \frac{S^2}{\chi^2_\nu(\alpha/2)}\right). \quad (4.9)$$

In this particular case, where the Ys are normal, the distributions of T and W are well known, with Student's t-quantiles and χ^2_ν quantiles well tabulated and available in standard computer routines. There is therefore no need to use Monte Carlo estimation to estimate them.

However, there is a more general situation where Y is not normal, but the same statistics T and W can still be used to construct CIs, if their CDFs are known. If not known, they can easily be estimated by Monte Carlo estimation.

Consider a sample $Y_1, Y_2, \ldots, Y_n$ drawn from the location-scale model $Y = \mu + \sigma X$, where μ and σ are fixed but unknown parameters, and X is a random variable with CDF $F(x)$, whose form $F(\cdot)$ is completely known and not dependent on unknown parameters. The normal model $N(\mu, \sigma^2)$ just considered is an example, with X simply the standard normal distribution $N(0,1)$. Other examples include the exponential, logistic, extreme value, and Weibull distributions (the last under an appropriate transformation so that it becomes the extreme value distribution). Consider

$$T' = \frac{\bar{Y} - \mu}{S/n},\ W = \frac{S^2}{\sigma^2} \text{ where } S^2 = \sum_{i=1}^{n}(Y_i - \bar{Y})^2,$$

these being essentially the T and W statistics defined in (4.8) with a slight simplification in T where the factor ν is replaced by n. Now write $Y_i = \mu + \sigma X_i$, and we get

$$T' = \frac{\bar{X}}{\sqrt{\sum_{i=1}^{n}(X_i - \bar{X})^2}},\ W = \sum_{i=1}^{n}(X_i - \bar{X})^2, \quad (4.10)$$

showing that T' and W are pivotal quantities. The confidence sets for μ and σ^2 take the same form as (4.9) for the normal model, except that appropriate quantiles of $G_{T'}(\cdot)$, the CDF of T', and of $G_W(\cdot)$, the CDF of W, need to be substituted for the Student's t and χ^2-quantiles appearing in (4.9).

If these CDFs are not known, Monte Carlo estimation, as described in subsection 4.1.1, provides a simple way to produce samples $T'_j, W_j, j = 1, 2, \ldots, B$, from (4.10), with the EDF of the $\{T'_j\}$ estimating $G_{T'}(\cdot)$ and of the $\{W_j\}$ estimating $G_W(\cdot)$, from which the required quantiles can be read off.

The EDFs are probabilistically exact. Therefore, for location-scale models, CIs calculated in this way have no coverage error, whatever the sample size n.

4.5 Bootstrap Goodness-of-Fit

Once we have fitted a model, the natural question is: Does the model that we have fitted actually match the data very well? For instance, when we fitted a gamma distribution to the toll booth service time data, does the fitted gamma distribution capture the characteristics of the data properly? We consider the situation for random samples in this section, and the regression case in the next.

For random samples, the classical way to answer this question is to use a GoF test. Let $Y = (Y_1, Y_2, \ldots, Y_n)$, where our *null hypothesis*, denoted by H_0, is that $Y_i \sim F(\cdot, \boldsymbol{\theta})$, with the form of F known. We consider GoF tests based on an EDF test statistic of the form $T = T(Y)$ that measures the difference between the EDF $\tilde{F}(y)$ of the sample values $(Y_1, Y_2, \ldots, Y_n)$ and the fitted $F(y, \boldsymbol{\theta})$. We will consider both the case where $\boldsymbol{\theta}$ is known and where it is not known. In the latter case, we use the MLE $\widehat{\boldsymbol{\theta}}$ to estimate it. The distribution of T under the null hypothesis has to be known to apply the GoF test. We can then simply calculate $T = T(\mathbf{Y})$ for the given sample, and compare the value with a high quantile $T_{1-\alpha}$ of the null distribution, for which $\Pr(T \leq T_{1-\alpha}) = 1 - \alpha$. A small value of α is used, $\alpha = 0.1$ or 0.05 or 0.01 is typical, so that if the null hypothesis is true then it is unlikely that the T value calculated from the actual observations $\mathbf{Y}$ will be greater than $T_{1-\alpha}$. If $T > T_{1-\alpha}$, the GoF test has failed at the $(1 - \alpha)$ level, and H_0 is rejected with the inference that the Y_i have *not* been drawn from $F(\cdot, \boldsymbol{\theta})$.

Two well-known EDF test statistics are the Cramér-von Mises W^2 statistic and the Anderson-Darling A^2 statistic, both of which have the form

$$T = \int_{-\infty}^{\infty} \psi(y, \boldsymbol{\theta})[\tilde{F}(y) - F(y, \boldsymbol{\theta})]^2 dF(y, \boldsymbol{\theta}),$$

where $\psi(y, \boldsymbol{\theta}) = 1$ for the W^2 statistic and $\psi(y, \boldsymbol{\theta}) = [F(y, \boldsymbol{\theta})(1 - F(y, \boldsymbol{\theta}))]^{-1}$ for the A^2 statistic. Equivalent versions that are computationally more convenient are, for the W^2 case,

$$W^2 = \frac{1}{12n} + \sum_{i=1}^{n} \left[Z_i - \frac{2i - 1}{2n} \right]^2, \tag{4.11}$$

and for the A^2 case

$$A^2 = -n - \sum_{i=1}^{n} (2i - 1)[\ln Z_i + \ln(1 - Z_{n+1-i})]/n, \tag{4.12}$$

where, in either case, $Z_i = F(y_{(i)}, \boldsymbol{\theta})$ if $\boldsymbol{\theta}$ is known and $\boldsymbol{\theta}$ is replaced by $\widehat{\boldsymbol{\theta}}$ if not, with $y_{(i)}$ the ith-order statistic.

Another well-known EDF statistic is the Kolmogorov-Smirnov D statistic. Perhaps the best-known test is the chi-squared GoF test. The main reason for its popularity is that it is relatively easy to implement. The test statistic is easy to calculate and, moreover, it has a known chi-squared distribution under the null, which makes critical values easy to obtain. However, the chi-squared test has two obvious weaknesses. It is significantly less powerful than EDF tests, and it has a certain subjective element, because the user has to divide the data into groups of her/his own choosing.

The work done, notably by Stephens, see Stephens (1970, 1974) and D'Agostino, and Stephens (1986), indicates that there are good reasons why A^2 is most likely to yield the most powerful test, and so should be the test statistic of choice. However, from our admittedly more limited experience, the W^2 is also attractive, especially as its behaviour can be more stable in situations where inappropriate models have been fitted to data sets.

The problem with the W^2 and A^2 statistics is that their distributions do not remain the same but change depending on the null. Thus, different critical values are needed. D'Agostino, and Stephens (1986) have tabulated a range of critical values for different GoF statistics and different nulls, including those for the exponential, gamma, Weibull, extreme value, logistic, and Cauchy distributions. Though the situation has improved, statistical packages that offer GoF tests do not always make it clear what critical values are implemented.

The parametric bootstrap offers a very practical way out of the difficulty, whether parameters have to be estimated or not. Stute, Mantega, and Quindimil (1993) show that under regularity conditions, so that the MLE, $\widehat{\boldsymbol{\theta}}$, is a consistent estimator, and provided $F(y, \boldsymbol{\theta})$ is a sufficiently smooth function of y and $\boldsymbol{\theta}$, then the distribution of an EDF GoF test statistic is consistently estimated by its bootstrap version. Stute, Mantega, and Quindimil (1993) demonstrate this specifically for the D and W^2 statistics. Babu and Rao (2004) give a rigorous proof of the weak consistency of the parametric bootstrap, examining a number of examples, including the normal and Cauchy distributions. (See also Babu and Rao, 2003.)

In what follows, we focus on the Anderson-Darling statistic A^2. In view of its similarity, our discussion applies also to W^2, so that the flowchart at (4.13) given for A^2 can be used for W^2, simply by replacing A^2 by W^2. For reasons of space, we do not explicitly discuss W^2 further.

We will write $A^2(\mathbf{y}, \boldsymbol{\theta})$ for the value of A^2 as calculated in (4.12) from the EDF $\tilde{F}$ of a sample $\mathbf{y}$ drawn from $F(\cdot, \boldsymbol{\theta})$. For the case when the MLE $\widehat{\boldsymbol{\theta}}$ is used instead of $\boldsymbol{\theta}$, we write $A^2(\mathbf{y}, \widehat{\boldsymbol{\theta}})$.

In the case where $\boldsymbol{\theta}$ is known, we use Monte Carlo estimation to generate samples $\mathbf{y}_j = (y_{1j}, y_{2j}, \ldots, y_{nj}), j = 1, 2, \ldots, B$, with all $y_{ij} \sim F(\cdot, \boldsymbol{\theta})$ drawn from the known distribution, from which we calculate test statistic values $T_j = A_j^2 = A^2(\mathbf{y}_j, \boldsymbol{\theta}), j = 1, 2, \ldots, B$. The EDF of the A_j^2 is (probabilistically) exact for estimating the null distribution of the test statistic in this case.

If parameters have to be estimated, we still generate B samples $\mathbf{y}_j^*, j = 1, 2, \ldots, B$, each of size n, but now with sampled values $y_{ij}^* \sim F(\cdot, \widehat{\boldsymbol{\theta}})$, where the MLE $\widehat{\boldsymbol{\theta}}$, calculated from the original sample, replaces the unknown $\boldsymbol{\theta}$. For each j, we have to again estimate $\boldsymbol{\theta}$, calculating the MLE $\widehat{\boldsymbol{\theta}}_j^* = \widehat{\boldsymbol{\theta}}(\mathbf{y}_j^*)$ based on the jth sample. We can then calculate $T_j^* = A_j^{*2} = A^2(\mathbf{y}_j^*, \widehat{\boldsymbol{\theta}}_j^*), j = 1, 2, \ldots, B$, using the formula for A^2 with $z_{ij} = F(y_{(i),j}^*, \widehat{\boldsymbol{\theta}}_j^*)$, where

$y^*_{(1),j}, y^*_{(2),j}, \ldots, y^*_{(n),j}$ is the ordered jth BS sample. The EDF of the sample $A_j^{*2} = A^2(\mathbf{y}_j^*, \widehat{\boldsymbol{\theta}}_j^*), j = 1, 2, \ldots, B$, estimates the null distribution of A^2 for the case where the parameters have been estimated by MLE. Writing the GoF critical value at level $(1 - \alpha)$ as $A^2_{(1-\alpha)}$, this is estimated by the $(1 - \alpha)$ quantile of the EDF, so if the A_j^{*2} are placed in order $A^{*2}_{(1)} \leq A^{*2}_{(2)} \leq \ldots, \leq A^{*2}_{(B)}$, the estimate is then $A^{*2}_{(1-\alpha)} = A^{*2}_{(m')}$, where $m' = \lfloor (1 - \alpha)B \rfloor$. These calculations are set out in the GoF flowchart (4.13).

Flowchart for the A^2 GoF test when θ is estimated by the MLE $\widehat{\theta}$ (4.13)

Given $\mathbf{y} = (y_1, y_2, \ldots, y_n)$, a random sample drawn from CDF $F(y, \boldsymbol{\theta}), \boldsymbol{\theta}$ unknown,

$$\downarrow$$

$$\widehat{\boldsymbol{\theta}} = \widehat{\boldsymbol{\theta}}(\mathbf{y}) \qquad \text{MLE}$$

$$\downarrow$$

$$A^2(\mathbf{y}, \widehat{\boldsymbol{\theta}}) \qquad A^2 \text{ GoF test statistic}$$

For $j = 1$ to B,

$$\mathbf{y}_j^* = (y_{ij}^* \sim F(y, \widehat{\boldsymbol{\theta}}), i = 1, 2, \ldots, n) \quad j\text{th BS sample}$$

$$\downarrow$$

$$\widehat{\boldsymbol{\theta}}_j^* = \widehat{\boldsymbol{\theta}}(\mathbf{y}_j^*) \qquad j\text{th BS MLE}$$

$$\downarrow$$

$$A_j^{*2} = A^2(\mathbf{y}_j^*, \widehat{\boldsymbol{\theta}}_j^*) \quad j\text{th BS } A^2 \text{ GoF test statistic}$$

End j

$$A^{*2}_{(1)} \leq A^{*2}_{(2)} \leq \ldots \leq A^{*2}_{(B)} \quad \text{Ordered null-distributed } A_j^{*2} \text{ values}$$

$$\downarrow$$

$$m = (1 - \alpha)B \qquad \text{Rounded integer subscript}$$

$$\downarrow$$

$$A^{*2}_{(1-\alpha)} = A^{*2}_{(m)} \quad \text{BS estimate of GoF critical value at level } (1 - \alpha)$$

$$\downarrow$$

GoF test: If $A^{*2}_{(1-\alpha)} < A^2(\mathbf{y}, \widehat{\boldsymbol{\theta}})$, reject null hypothesis H_0 at level $(1 - \alpha)$.

In general, the bootstrap at each j introduces an approximation error, as $\widehat{\boldsymbol{\theta}}$ has to be used in place of the unknown true $\boldsymbol{\theta}^0$ in forming the sample $\mathbf{y}_j^*$. This introduces an error of order $|\widehat{\boldsymbol{\theta}} - \boldsymbol{\theta}^0| = O_p(n^{-1/2})$. It is relatively easy to incorporate a sensitivity assessment into the procedure to gauge the importance of this error. We will discuss how this is done, but before we do so, we discuss an important family of models where the bootstrapping remains exact even when parameters are estimated.

For location-scale models, the use of MLEs in EDF GoF tests is equivalent to studentization, so that the GoF test is exact. This property has been recognized and commented on in Stephens (1974) and D'Agostino, and Stephens (1986), however, their focus was on the calculation of tables of critical values, and so attention was not drawn to the usefulness of the property for carrying out BS GoF tests without tables. The property is also pointed out by Babu and Rao (2004), but their proof is quite technical. We give a simpler, we hope more transparent, proof that covers the case where the location and scale parameters are both unknown.

Consider the location-scale model where the observations have the form $Y_i = \mu + \sigma X_i$. We write X_i as $X_i = F^{-1}(U_i)$, with F the completely known continuous CDF of X. We can think of the $U_i \sim U(0,1)$, $i = 1, 2, \ldots, n$, as being the fundamental random variables from which the X_i and then the Y_i are formed. Thus, there is a one-to-one correspondence between each sample of U_i's and the observed sample $\{Y_i\}$ that it gives rise to. Our key result is that whatever the underlying sample of uniforms $\{U_i\}$, using MLE to estimate μ and σ, yields an estimated standardized set of Y_i's, $(Y_i - \hat{\mu})/\hat{\sigma}$, $i = 1, 2, \ldots, n$, each value of which, whatever the i, is the same whether calculated for the original sample or under bootstrapping. We show this as follows.

Let $\{u_i^0\}$ be the uniforms that give rise to $\{y_i^0\}$ our observed sample, so that $y_i^0 = \mu^0 + \sigma^0 x_i^0$, where $x_i^0 = F^{-1}(u_i^0)$, and μ^0, σ^0 are the true but unobserved μ and σ values. Let f be the PDF. The log-likelihood in terms of the y_i^0 is $L = -n \ln \sigma + \sum_{i=1}^n \ln f[(y_i^0 - \mu)/\sigma]$, for which the likelihood equations $\partial L/\partial \mu = 0$ and $\partial L/\partial \sigma = 0$ can be written as

$$-\sum_{i=1}^{n} \frac{1}{f[(y_i^0 - \mu)/\sigma]} \left.\frac{\partial f}{\partial y}\right|_{\frac{y_i^0 - \mu}{\sigma}} = 0 \text{ and } -n + \sum_{i=1}^{n} \frac{(y_i^0 - \mu)/\sigma}{f[(y_i^0 - \mu)/\sigma]} \left.\frac{\partial f}{\partial y}\right|_{\frac{y_i^0 - \mu}{\sigma}} = 0, \tag{4.14}$$

showing they depend on the y_i^0, μ, and σ only through the standardized values $(y_i^0 - \mu)/\sigma$ for $i = 1, 2, \ldots, n$.

Let the MLEs obtained from the original sample be $\hat{\mu}^0$ and $\hat{\sigma}^0$ so that they satisfy (4.14). In other words, using $(y_i^0 - \hat{\mu}^0)/\hat{\sigma}^0$ for the standardized forms in (4.14) satisfies both equations.

Now, suppose that, using bootstrap sampling with $\hat{\mu}^0$ and $\hat{\sigma}^0$ in place of the unknown μ^0, σ^0, we had generated the *same* set of uniforms $\{u_i^0\}$ as in the original sample. The BS y_i^* would take the form $y_i^* = \hat{\mu}^0 + \hat{\sigma}^0 x_i^0$. The MLEs $\hat{\mu}^*$ and $\hat{\sigma}^*$ corresponding to these y_i^* must also satisfy (4.14), with y_i^0 replaced by y_i^*. Thus, in this case, use of the standardized forms $(y_i^* - \hat{\mu}^*)/\hat{\sigma}^*$ in (4.14) will again satisfy both equations. However, with the same $\{u_i^0\}$ used, we have $y_i^0 = \mu^0 + \sigma^0 x_i^0$ and $y_i^* = \hat{\mu}^0 + \hat{\sigma}^0 x_i^0$ for all i. We then find that eliminating x_i^0 between these pairs of equations and simply setting

$$\hat{\mu}^* = \hat{\mu}^0 + \frac{\hat{\sigma}^0}{\sigma^0}(\hat{\mu}^0 - \mu^0) \text{ and } \hat{\sigma}^* = \frac{\hat{\sigma}_0^2}{\sigma_0},$$

for every i, we obtain

$$(y_i^0 - \hat{\mu}^0)/\hat{\sigma}^0 = (y_i^* - \hat{\mu}^*)/\hat{\sigma}^* \; i = 1, 2, \ldots, n. \tag{4.15}$$

Therefore, this equality of the standardized versions holds for *any* given uniform sample $\{u_i^0\}$. Allowing $\{u_i^0\}$ to vary over its sample space $[0,1]^n$ clearly generates the entire joint distribution of both $(Y_i^0 - \hat{\mu}^0)/\hat{\sigma}^0$, $i = 1, 2, .., n$, and $(Y_i^* - \hat{\mu}^*)/\hat{\sigma}^*$, $i = 1, 2, \ldots, n$. The equality (4.15) therefore shows that the joint distribution of the BS standardized quantities is identical to that of the original.

We now apply this result to the A^2 statistic. This depends only on the quantities $Z_i = F(y_{(i)}, \widehat{\boldsymbol{\theta}})$. For the location-scale model, we have from (4.15) that $Z_i^* = F((y^*_{(i)} - \hat{\mu}^*)/\hat{\sigma}^*) = F((y_i^0 - \hat{\mu}^0)/\hat{\sigma}^0) = Z_i$ for all i. Thus, for any uniform sample $\{u_i^0\}$, the bootstrap version A^{*2} is identical in value to the A^2 value obtained for the original sample. The bootstrap EDF of a BS sample of A^{*2} values therefore remains an exact estimator of the null distribution of A^2 for any n (> 2) when unknown parameter location and scale parameters are replaced by their MLEs.

The same result holds for W^2 as this also depends only on the Z_i quantities.

For the general case where F may not be a location-scale model, and when $\boldsymbol{\theta}^0$ is unknown, we need to identify the error that arises from using the BS estimated critical value $A^{*2}_{(1-\alpha)}$ in the GoF test instead of the true value $A^2_{(1-\alpha)}$. As $A^{*2}_{(1-\alpha)}$ is a random quantity based on sampling from $F(\cdot, \widehat{\boldsymbol{\theta}})$, we therefore need to estimate its distribution.

The flowchart (4.13) shows how the critical value $A^{*2}_{(1-\alpha)}$ actually used in the GoF test can be calculated as the $(1 - \alpha)$ *quantile* of the EDF formed from a bootstrap random sample, $A^2(\mathbf{y}_j^*, \widehat{\boldsymbol{\theta}}_j^*)$, $j = 1, 2, \ldots, B$, of A^2 values; this EDF being a consistent estimator of the desired null distribution. In this calculation, each BS sample $\mathbf{y}_j^*$ is drawn from $F(\cdot, \widehat{\boldsymbol{\theta}})$ so that $\widehat{\boldsymbol{\theta}}$ is the underlying parameter value determining just this *one* value of $A^{*2}_{(1-\alpha)}$. We can obtain an estimate of the distribution of $A^{*2}_{(1-\alpha)}$ by adding an *additional level of bootstrapping* in the flowchart to form a *random sample* of critical values, $A^{*2}_{(1-\alpha),j}$, $j = 1, 2, \ldots, B'$, where each $A^{*2}_{(1-\alpha),j}$ is calculated in the same way as the original $A^{*2}_{(1-\alpha)}$, but using a BS value $\widehat{\boldsymbol{\theta}}_j^*$ that varies with j as the underlying parameter value instead of $\widehat{\boldsymbol{\theta}}$ to reflect the additional variation in calculating $A^{*2}_{(1-\alpha)}$ from bootstrap samples rather than from the true samples. This calculation can be regarded as an *inner bootstrap* executed at each step j of the now *outer bootstrap* of (4.13). The flowchart for the calculation is as follows.

Flowchart for BS estimation of the distribution of $A^{*2}_{(1-\alpha)}$, the BS estimate of the critical value when θ is estimated by the MLE $\widehat{\theta}$ (4.16)

Given $\mathbf{y} = (y_1, y_2, \ldots, y_n)$, a random sample drawn from CDF $F(y, \boldsymbol{\theta})$, $\boldsymbol{\theta}$ unknown,

$\downarrow$

$\widehat{\boldsymbol{\theta}} = \widehat{\boldsymbol{\theta}}(\mathbf{y})$ MLE

$\downarrow$

$m = (1 - \alpha)B$ Rounded integer subscript

For $j = 1$ to B' Outer bootstrap [typically $B' \leq B$]

$\mathbf{y}_j^* = (y_{ij}^* \sim F(y, \widehat{\boldsymbol{\theta}}), i = 1, 2, \ldots, n)$ jth BS sample

$\downarrow$

$\widehat{\boldsymbol{\theta}}_j^* = \widehat{\boldsymbol{\theta}}(\mathbf{y}_j^*)$ jth BS MLE

For $k = 1$ to B Inner bootstrap

$F(\cdot, \widehat{\boldsymbol{\theta}}_j^*)$

$\downarrow$

$$
\begin{array}{lll}
\mathbf{y}^{**}_{jk} = (y_{1jk}, y_{2jk}, \ldots, y_{njk}) & (j,k)\text{th BS sample with all } y_{ijk} \sim F(\cdot, \widehat{\boldsymbol{\theta}}^{*}_{j}) \\
\downarrow & \\
\widehat{\boldsymbol{\theta}}^{**}_{jk} & \text{MLE of } \boldsymbol{\theta}\text{, calculated from } \mathbf{y}^{**}_{jk} \\
\downarrow & \\
A^{**2}_{jk} = A^{2}(\mathbf{y}^{**}_{jk}, \widehat{\boldsymbol{\theta}}^{**}_{jk}) & (j,k)\text{th BS } A^2 \text{ GoF test statistic} \\
\text{Next } k & \\
A^{**2}_{j,(1)} \le A^{**2}_{j,(2)} \le \ldots \le A^{**2}_{j,(B)} & B \text{ ordered null-distributed BS } A^{**2}_{j} \text{ values} \\
\downarrow & \\
A^{**2}_{(1-\alpha),j} = A^{**2}_{j,(m)} & j\text{th GoF critical value at level } (1-\alpha) \\
\text{Next } j & \\
\text{EDF of } A^{**2}_{(1-\alpha),1}, A^{**2}_{(1-\alpha),2}, \ldots, A^{**2}_{(1-\alpha),B'} & \text{BS estimate of distribution of } A^{2}_{(1-\alpha)}
\end{array}
$$

A double superscript '**' is used to denote quantities calculated in the inner BS loop.

At the end of the outer BS loop, there are B' BS estimated critical values, $A^{**2}_{(1-\alpha),j}$, $j = 1, 2, \ldots, B'$, the EDF of which estimates the distribution of $A^2_{(1-\alpha)}$. If a formal assessment of the accuracy of $A^2_{(1-\alpha)}$ is needed, this can be provided by calculating a percentile CI from this EDF. We have not included this calculation in the flowchart.

Taken together, the outer and inner bootstraps are commonly called a *double bootstrap*.

The double bootstrap is a slight misnomer, as the effort required is not $2B$ but $B'B$, so is computationally expensive if B' is large. In practice, B' only needs to be large for a formal estimate of the BS error. The tabulated critical values given in D'Agostino, and Stephens (1986) for different distributions suggest that this BS error will not be large in general. For instance, in the case of the gamma distribution, when parameters are estimated, the GoF critical values are affected by the shape parameter a, but the effect is not great with, for instance, the critical values when $\alpha = 0.1$, only varying from 0.657 at $a = 1$ to 0.631 at $a = \infty$. Therefore a less formal, more flexible approach to the sensitivity analysis is simply to increase B' by adding replications one at a time, stopping once the magnitude of the error is clear.

We conducted such a double bootstrap experiment for the case $\boldsymbol{\theta}^0 = (a, b) = (1, 1)$, with $B = 1000$ and in fact with $B' = 10$, small but fixed. As we know the value of $\boldsymbol{\theta}^0$, and the example is for illustration only, we generated each initial $\mathbf{y}$ directly from $F(\cdot, \boldsymbol{\theta}^0)$ rather than $F(\cdot, \widehat{\boldsymbol{\theta}})$. The $B' = 10$ BS EDFs of A^{*2} are depicted in Figure 4.3.

The ten critical $A^{*2}_{(1-\alpha)}$ values corresponding to the $(1-\alpha) = 0.9$ critical value calculated from the EDFs are shown in red. These values of $A^2_{(1-\alpha)}$ fell in the range 0.629 – 0.660, straddling the value of 0.657 tabulated in D'Agostino, and Stephens (1986, Table 4.21). As an added indication that there is nothing untoward about the replications, Figure 4.3 also shows, in green, the ten 'test' $A^2(\mathbf{Y}, \widehat{\boldsymbol{\theta}}(\mathbf{Y}))$ values calculated from the initial sample $\mathbf{Y}$ and MLE $\widehat{\boldsymbol{\theta}}(\mathbf{Y})$, this $\widehat{\boldsymbol{\theta}}$ being used to generate the BS samples in the replication.

For the toll booth data, $\hat{a} = 9.20$ and $\hat{b} = 0.63$, for which the BS estimated $(1-\alpha) = 0.9$ critical value, using $B = 1000$, was $A^2 = 0.631$. This is greater than the test value of $A^2 = 0.497$, so that the null hypothesis that the data are gamma distributed is not

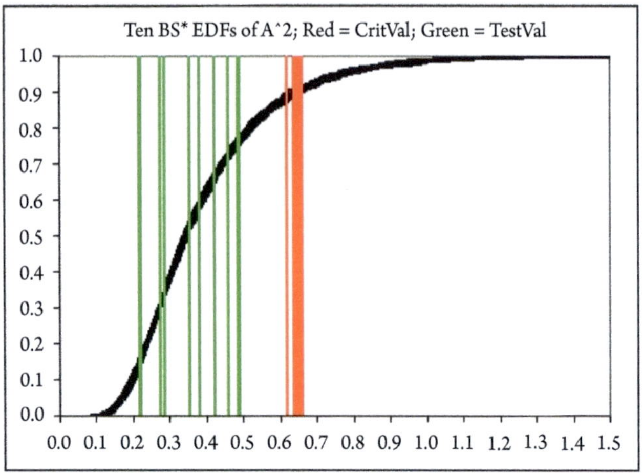

Figure 4.3 Ten green lines: BS null test values. Red lines: BS critical $\alpha = 0.1$ values calculated from EDFs (black graphs) of double BS null test values.

rejected. In comparision, the tabulated critical values in D'Agostino, and Stephens (1986, Table 4.21) are 0.634 at $\hat{a} = 8$ and 0.633 at $\hat{a} = 10$.

In addition, we tested fitting the normal distribution. The MLEs were $\hat{\mu} = 5.80$ and $\hat{\sigma} = 2.04$, with a test value of $A^2 = 1.109$. Bootstrapping using $B = 1000$ gave a critical test value at the $(1 - \alpha) = 0.9$ level of $A^2 = 0.620$. The test value is greater, so we would reject the hypothesis that the data are normally distributed. The D'Agostino, and Stephens (1986, Table 4.7) tabulated critical value at $\alpha = 0.1$ is $A^2 = 0.631$, but to apply the test with this critical value, the test value has to be inflated by a factor of $(1 + 0.75/n + 2.25/n^2) = 1.017$ at $n = 47$. This is equivalent to modifying the critical value to $0.631/1.017 = 0.620$ if the test value is retained at $A^2 = 1.109$, making it the same as the BS version.

4.6 Bootstrap Regression Lack-of-Fit

The EDF tests discussed in the previous section focus on GoF of distributions, and are not directly applicable in regression fitting. In this latter case, we assume the regression takes the univariate form $y = \eta(x, \boldsymbol{\theta}) + \varepsilon$, where x and y are scalar, and that the fitted model is $y = \eta(x, \widehat{\boldsymbol{\theta}})$, where $\widehat{\boldsymbol{\theta}}$ is the ML estimate obtained from a set of n observations (x_j, y_j), $j = 1, 2, \ldots, n$. The standard advice on investigating lack-of-fit (LoF) is clearly set out by Ritz and Streibig (2008), who give some interesting examples. As we refer to a number of their comments, it will be convenient to cite them in the form R&S in this section.

In considering the fit, we concentrate on how well the fitted regression function calculated at the observed $x_j : \eta(x_j, \widehat{\boldsymbol{\theta}})$, $j = 1, 2, \ldots, n$, matches the observations y_j. The obvious quantities to focus on are the residuals

$$r_j = y_j - \hat{y}_j, \text{ where } \hat{y}_j = \eta(x_j, \widehat{\boldsymbol{\theta}}),$$

and it is easiest to inspect these visually, plotting the points, either (x_j, r_j) or $(\hat{y}_j, r_j)$, in either case with the r_j forming the vertical ordinate deviations from the horizontal line $r = 0$. If there are any systematic deviations of the r_j from this line, then this is a clear indication that the fitted model is not fully representing all the systematic variation in the true regression. As discussed in R&S, if there are replicates, m, say, then the visual graphical analysis can be formalized by statistical testing. One can fit a one-way ANOVA model

$$y_{ij} = a_i + \varepsilon_{ij}, \quad i = 1, 2, \ldots, m, \; j = 1, 2, \ldots, n,$$

and then compare the residuals obtained when fitting the regression function η with the residuals obtained from the ANOVA model. Bates and Watts (1988) give a formal F-test method for making this comparison.

An alternative formal test is the likelihood ratio test, see, for example, Huet et al. (2004).

Details of both tests are set out in R&S and are not repeated here.

As discussed in Section 3.3, there exist a number of formal statistical tests for *comparing* model fits, whether nested or non-nested, when we just have a single sample. For single random samples, GoF tests stand out as providing a method for deciding if the sample comes from some given distribution or not. As discussed in Section 4.5, the problem with the best EDF GoF tests having null distributions that are difficult to obtain theoretically is fairly easily overcome by bootstrapping.

Bootstrapping offers a possible means of assessing whether a fitted regression model is an adequate representation of a *single* sample (x_j, y_j), $j = 1, 2, \ldots, n$. We consider only the case where x and y are both scalar. When a fitted regression function fails to adequately represent the functional relationship between x and y, then the residual plot of (x_j, r_j) will display *systematic* deviation from the horizontal line $r = 0$.

Consider the following measure of this deviation.

Suppose the (x_j, r_j) are ordered first by x, and if there are points with a tied x value—i.e. replicated x values—these are then ordered by their r_j values. Now consider what we shall call *drifts*, where each drift is made up of a set of consecutive points (x_j, r_j), where the r_j all have the same sign. The entire sequence of points is thus split into a sequence of drifts d_i, $i = 1, 2, \ldots, M$, with the points (x_j, r_j) in each drift all with r_j of the same sign, but with the sign alternating between consecutive drifts in the sequence. Suppose

$$d_i = \{(x_j, r_j) : j = j_i, j_i + 1, \ldots, j_{i+1} - 1\} \text{ and } \Delta_i = \sum_{j=j_i}^{j_{i+1}-1} \left| r_j \right|, \; i = 1, 2, \ldots, M.$$

Thus Δ_i is the magnitude of drift d_i; for simplicity, we shall also refer to Δ_i as the *drift* when there is no ambiguity. We shall assume that the Δ_i have been ordered so that $\Delta_1 \geq \Delta_2 \geq \ldots \geq \Delta_M$. Let

$$\Delta_0(m) = n^{-1} \sum_{i=1}^{m} \Delta_i, \tag{4.17}$$

where $m \leq M$. Then the drift statistic $\Delta_0(m)$ is a measure of the lack-of-fit (LoF) of $\eta(x, \widehat{\boldsymbol{\theta}})$ to the sample. In practice, we might choose $m = 1, 2$, or 3, say. Flexibility in the choice of m may be appropriate, especially in exploratory work, where the residual plot can be examined to see how the largest drifts occur to ensure that they are included in $\Delta_0(m)$.

As with A^2, parametric bootstrapping provides a ready means of calculating the null distribution of $\Delta_0(m)$ under the hypothesis that $\eta(x, \boldsymbol{\theta})$ is the correct model. Based on this we have the following

Drift Lack-of-Fit (LoF) Test of a Univariate Regression Function (4.18)

Follow precisely the same calculations as given in the GoF flowchart (4.13), except that we replace calculation of A^2 by $\Delta_0(m)$ as given in (4.17). Thus:

(i) For the original sample $(x_j, y_j), j = 1, 2, \ldots, n$, obtain the ML estimate $\widehat{\boldsymbol{\theta}}$, of the parameters $\boldsymbol{\theta}$ of the regression function $\eta(x, \boldsymbol{\theta})$.

(ii) Calculate the residuals $r_j = y_j - \eta(x_j, \widehat{\boldsymbol{\theta}})$, and order the points (x_j, r_j) by x, resolving ties by (secondary) ordering of r.

(iii) Calculate $\Delta_0(m)$ as in (4.17). This is the test value.

(iv) Obtain B BS samples $\{(x_j, y^{*(k)}(x_j)), j = 1, 2, \ldots, n\}$, $k = 1, 2, \ldots, B$, where $y^{*(k)}(x_j) = \eta(x_j, \widehat{\boldsymbol{\theta}}) + \varepsilon_j^{*(k)}$ with $\varepsilon_j^{*(k)} \sim N(0, \hat{\sigma}^2)$, and carry out the calculations (i)-(iii) for each bootstrap sample to obtain B BS overall drift statistics $\Delta_0^{*(k)}(m), k = 1, 2, \ldots, B$.

(v) Estimate $\Delta^*_{0,(1-\alpha)}(m)$, the critical (null distribution) $100(1-\alpha)\%$ value of the overall drift from the EDF of the $\Delta_0^{*(k)}(m)$.

(vi) The LoF is deemed significant if the test value $\Delta_0(m) > \Delta^*_{0,(1-\alpha)}(m)$.

Note that calculation of $\Delta_0(m)$ is very easy, requiring just one pass of the ordered (x_j, r_j), with each Δ_i being built up by summing r_j's until there is a change of sign in the r_j's, which acts as an indication of completion of the current Δ_i.

4.7 Two Numerical Examples

We end this chapter with two numerical examples in which we illustrate the discussion of this chapter by fitting selected nonlinear regression functions to two data sets, both taken from R&S. Though we treat both as 'standard' problems, the fitted models are not actually entirely satisfactory, so our discussion here should be regarded as merely illustrative

Table 4.2 VapCO dataset from R package nls()

x	y	x	y	x	y
51.2	4.890	65.1	8.987	89.7	12.219
56	6.503	67.5	9.498	102.5	13.136
58.2	7.195	71.9	10.191	112.2	13.829
60.4	7.888	76.9	10.884	123.5	14.522
63.2	8.582	81.9	11.526	131.3	14.927

and not a definitive treatment of the data sets. We shall return to both examples, analysing them further in Section 5.7, pointing out non-standard aspects in each.

4.7.1 VapCO Data Set

Table 4.2 gives the VapCO data set discussed by R&S, with x corresponding to temperature, and $y = \ln p$, where p is the pressure tabulated in the **R** nls() package.

We followed R&S and fitted the three-parameter Antoine model

$$y = \ln p = a - \frac{b}{c + x}, \tag{4.19}$$

assuming homoscedastic normally distributed errors. The fitted model is depicted in the upper chart of Figure 4.4, matching Figure 5.2 in R&S, to which it directly corresponds. The error variance is very small in this data set, so that though the fit looks good, as remarked by R&S, a more careful investigation is needed. A clearer picture of what is happening is provided by the standardized residual (versus Temp) plot, which is given in the lower chart of Figure 4.4. This displays a noticeably regular and growing oscillation.

The first four ordered drift values are $\Delta_1 = \sum_{j=11}^{13} |r_j| = 4.231$, $\Delta_2 = \sum_{j=4}^{8} |r_j| = 3.500$, $\Delta_3 = \sum_{j=14}^{15} |r_j| = 0.2.531$, $\Delta_4 = \sum_{j=2}^{3} |r_j| = 1.736$, accounting for 12 out of the 15 observations; these drifts closely match the oscillations of the residuals. The four drifts are shown in different colours in the charts.

We carried out the drift LoF test based on 500 BS samples of $\Delta_0(m)$ for $m = 1, 2, 3, 4$. The results for case $m = 1$ are depicted in Figure 4.5, which shows the EDF of the 500 bootstrap $\Delta_0^*(1)$ values and the test value of $\Delta_0(1) = 4.231$, calculated from the model fitted to the original sample. This indicates a lack-of-fit that is just significant at the 95% level. The plots for the cases $m = 2, 3, 4$ are not shown, as they are very similar, except that the level of significance of the lack-of-fit rises. The test value $\Delta_0(m)$ calculated from the original sample, its p value, and the critical 95% value of $\Delta_0(m)$ for $m = 1, 2, 3, 4$ are given in in Table 4.3, showing this increase in significance of the LoF.

A much more flexible model is needed before $\Delta_0(m)$ becomes insignificant. We will return to this example in Chapter 5, to discuss how components might be added to the model and also to consider how fitting a model to this data set can become non-standard.

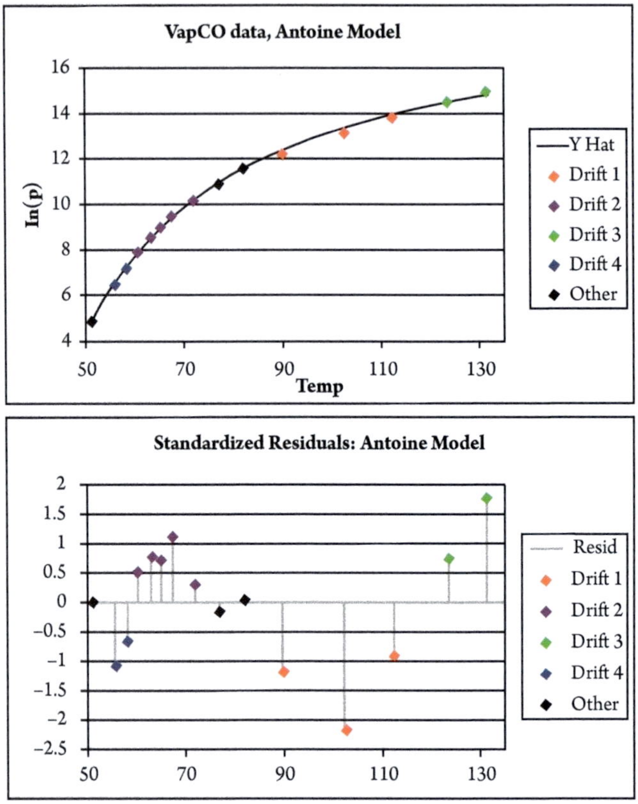

Figure 4.4 Upper chart: Antoine model fit to VapCO data set with four drifts shown in different colours. Lower chart: Standardized residuals, Antoine model.

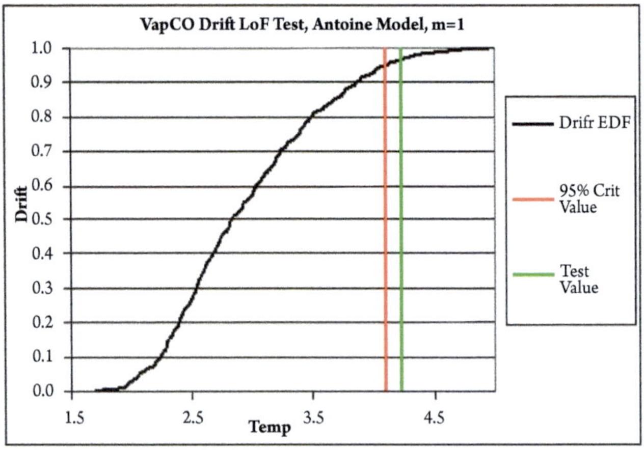

Figure 4.5 Null distribution of the maximum drift $\Delta_0(m)$, $m = 1$, for the Antoine model fitted to the VapCO data, with 95% critical value and test value.

Table 4.3 Drift LoF test results and Antoine model fit to VapCO data

m	Drift Test Value	p-value	95% Crit value
1	4.231	0.038	4.107
2	7.731	0.022	7.346
3	10.262	0.008	9.178
4	11.997	0.002	10.712

Table 4.4 Lettuce data: 14 (x, y) values where x=ln(concentration) in mg/litre, y=biomass (in g)

x	y	x	y	x	y
1.0000	0.833	1.3133	1.336	3.5473	0.488
1.0000	1.126	1.7780	0.754	3.5473	0.560
1.1113	1.096	1.7780	0.985	4.6320	0.344
1.1113	1.106	2.5430	0.683	4.6320	0.375
1.3133	1.163	2.5430	0.716		

4.7.2 Lettuce Data Set

Table 4.4 gives another data set analysed by R&S showing how y, lettuce plant biomass (in grams), is affected by x, the concentration of isobutyl alcohol (in mg/litre), which acts as a herbicide.

As pointed out by R&S, models used for dose-response data are usually monotonic, such as the three-parameter loglogistic model

$$\eta = \frac{c}{(1 + (x/a)^b)}, \tag{4.20}$$

but the lettuce data possibly show what is called hormesis, meaning that the effect of small herbicide doses appears to increase the biomass, before the principal effect with higher doses kicks in to reduce the biomass. The following four-parameter model proposed by Brain and Cousens (1989),

$$\eta = \frac{c + dx}{(1 + (x/a)^b)}, \tag{4.21}$$

is designed to allow for this hormetic effect.

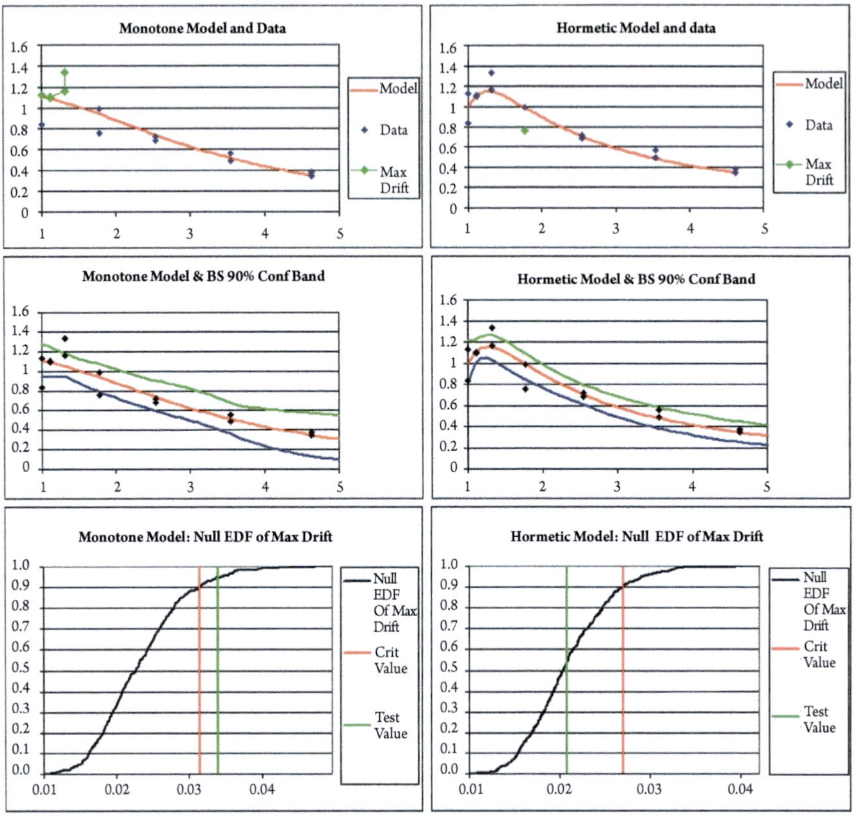

Figure 4.6 Monotone and hormetic model fits to the lettuce data. Top charts: ML fits with maximum drift runs. Middle charts: BS 90% confidence bands calculated from 500 BS samples. Bottom charts: Null EDFs of the drift value $\Delta_0(1)$, calculated from the same 500 BS samples. Also shown: 90% critical and $\Delta_0(1)$ test GoF values.

The fits of the two models are shown in Figure 4.6. The drifts corresponding to the maximum drift values are highlighted in each case. It will be seen that for the monotone dose-response model of (4.20), the drift is located at the points where the modelling of the hormetic region is possibly inadequate.

Figure 4.6 also displays the null distribution EDF plots of $\Delta_0(1)$, the maximum drift value for each model marked with the upper 90% critical value, as obtained in BS lack-of-fit tests based on 500 BS samples. These show that there is a significant lack of fit of the simple dose-response model, whilst the Brain and Cousens hormetic model of (4.21) is satisfactory.

Figure 4.6 also shows the BS 90% confidence bands for the entire regression function for both models. The bands were calculated in the way described in Section 4.3 from the same 500 parametric BS samples used in the lack-of-fit tests.

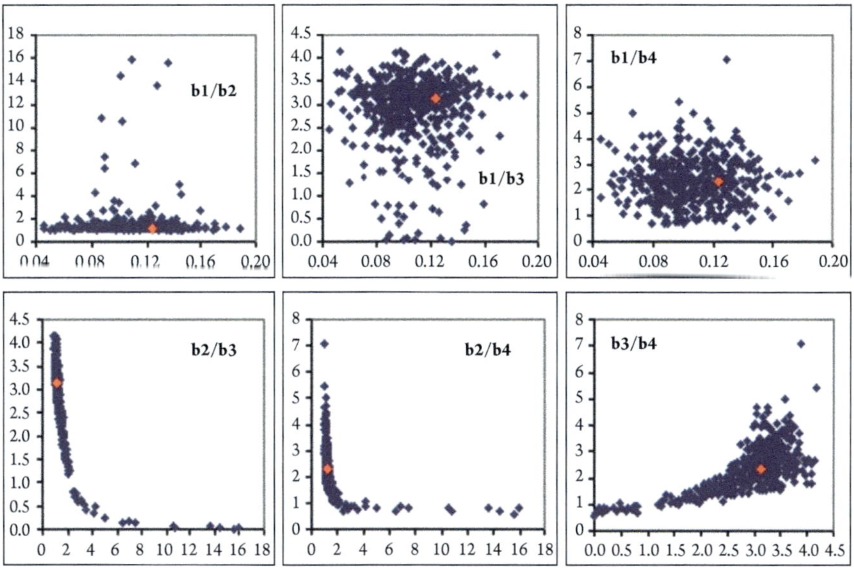

Figure 4.7 Scatterplots of 500 BS parameter ML estimates of monotone dose-response model fitted to the lettuce data. Red dot is the MLE for the original data; $b_1 \equiv \sigma$, $b_2 \equiv a$, $b_3 \equiv b$, $b_4 \equiv c$.

The band for the hormetic model appears more stable and satisfactory than that for the monotone model. Indication of whether confidence interval and confidence band calculations will actually be satisfactory is provided by the behaviour of parameter estimate scatterplots. If the scatterplots do not appear normally distributed, then confidence intervals and bands based on asymptotic theory may be unreliable. BS confidence intervals and bands are probably more robust, but if the scatterplots are uneven, then the calculation of likelihood-based confidence regions in the way described in Section 4.3 may also be unreliable, as the calculation supposes a normal distribution for the parameter estimates.

Figures 4.7 and 4.8 show the scatterplots for the two fitted models in our example. It will be seen that those corresponding to the monotone model are not very nornal. Those corresponding to the hormetic model, though possibly more compact, are not, however, entirely satisfactory either.

The performance of parameter estimates can depend on the parametrization used. There is an interesting aspect of the hormetic model (4.21) that is non-standard and which therefore falls within the remit of this book. The particular parametrization of (4.21) hides a simpler three-parameter model that is actually an equally good fit. However, this model is only obtained if $a \to 0$, and $c, d \to \pm\infty$. We will return to this example in the next chapter, after we have defined and discussed such models, which we shall be calling *embedded models*.

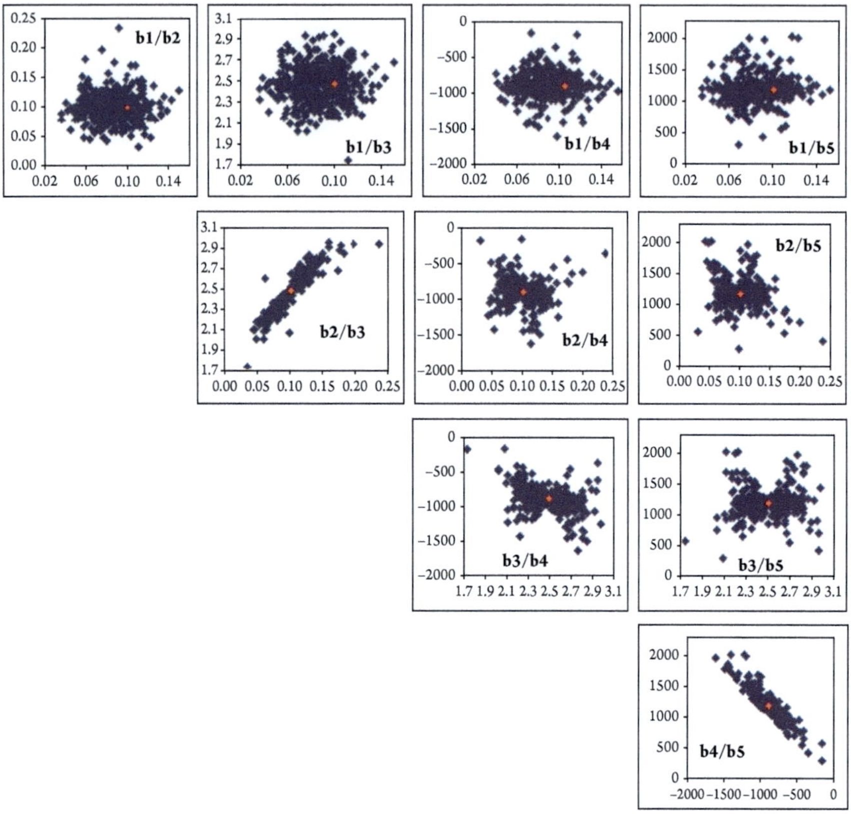

Figure 4.8 Scatterplots of 500 BS parameter ML estimates of hormetic dose-response model fitted to the lettuce data. Red dot is the MLE for the original data; $b_1 = \sigma$, $b_2 \equiv a$, $b_3 \equiv b$, $b_4 \equiv c$, $b_5 \equiv d$.

5 Embedded Model Problem

When fitting multi-parameter models to data, the best fit is sometimes obtained at the boundary of the parameter space $\boldsymbol{\Theta}$.

Boundary models can be straightforward in that though they are at the boundary, their mathematical form is identical to that of models that correspond to interior points. Indeed, they might only be regarded as boundary models because, in the context of the practical application, it is useful to think of them as such.

However, boundary models can have a mathematical form that is significantly different from those of models corresponding to interior points of $\boldsymbol{\Theta}$, moreover, where certain parameters have to be zero or are infinite. This can mean that the mathematical form of the boundary model is not only different from that corresponding to interior points, but their form is not directly expressible in the original parametrization, making a proper identification of such models difficult. In this chapter, we consider the situation where what appears to be a natural parametrization of a family of models results in certain models not appearing to be present at all, because they are hidden in what we call an *embedded form*. This will be formally defined in Section 5.1. Such embedded models are not invariant, but occur because of the form of parametrization. We show how reparametrization of the family can remove an embedding so that such boundary models can be identified and their parameters estimated.

In our general discussion of embedded models in this chapter, we use $f(x, \boldsymbol{\theta})$ to represent either the PDF of a random variable X, or the regression function in the regression

$$y = f(x, \boldsymbol{\theta}) + \varepsilon,$$

where ε is a random 'error' variable. The error has a distribution that may depend on further parameters, but our focus is on $\boldsymbol{\theta}$, so for simplicity we assume that $\varepsilon \sim N(0, \sigma^2)$ with σ^2 unconnected with $\boldsymbol{\theta}$.

Most of the examples of embedded models in this book come from distribution fitting where we wish to fit a distribution with PDF $f(x, \boldsymbol{\theta})$ to a given sample of observations $y_1, y_2, \ldots, y_n$. The three-parameter Weibull distribution (2.1) discussed in Chapter 2 is an example where the extreme-value distribution is an embedded special case.

However, in this chapter, though we discuss embeddedness in general terms, our examples come from the fitting of regression functions.

5.1 Embedded Regression Example

To motivate our discussion of embeddedness, we consider the following example of a regression function model

$$\eta(x, a, b) = b[1 - \exp(-ax)]. \tag{5.1}$$

This example was considered by Simonoff and Tsai (1989) (their Example 1), and has also been discussed by Seber and Wild (2003) as their Example 3.4.

Though outwardly simple, the example already contains several unusual features, showing how non-standardness can arise even in the simplest cases.

If we restrict consideration to the positive parameter quadrant $a \geq 0, b \geq 0$, then the model 5.1 has the submodel

$$\eta = b,$$

obtained by letting $a \to \infty$. The situation is standard in that the submodel, albeit trivial, is obtained by allowing a to tend to a particular value. The particular value happens not to be finite, but if we replace a by $\alpha = a^{-1}$, then the model $\eta = b$ would have been obtained as $\alpha \to 0$. With α tending to the specific value 0, the resulting submodel, $\eta = b$, is what we would expect in terms of the remaining parameter count, i.e. one parameter is fixed, and we have one remaining parameter b to determine.

The situation along the boundaries $a = 0$ and $b = 0$ is different. On these boundaries, we obtain the submodel $\eta = 0$. Though this boundary model is even more trivial than the submodel $\eta = b$, it nevertheless illustrates the situation where setting a parameter to a specific value entirely eliminates another parameter from the model. In this example, not only does setting $a = 0$ eliminate b from the model, but we also find that setting $b = 0$ eliminates a from the model. We call elimination of parameters in this way *indeterminacy*, and the problem is non-trivial, as it can lead to serious inferential anomalies. We shall discuss indeterminacy more fully in Chapter 14.

Embeddedness, though we shall later show is connected with indeterminacy, manifests itself in a different way. If in eqn (5.1) we reparametrize by setting $b = c/a$, then $\eta(x, a, b)$ becomes

$$\eta(x, a, c) = c[1 - \exp(-ax)]/a. \tag{5.2}$$

Letting $a \to 0$, so that $b \to \pm\infty$ depending on the sign of c, gives the model

$$\eta(x, 0, c) = cx. \tag{5.3}$$

Thus, whilst *both* original parameters a and b take special values, the limiting model *still* depends on a parameter of dimension one. This model is not obtainable in the

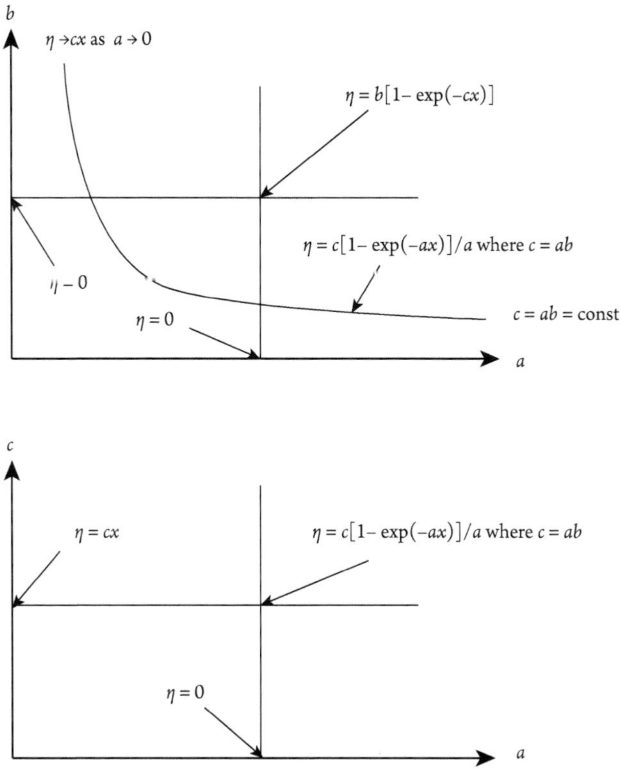

Figure 5.1 Example of a simple regression model with a linear model as an embedded model in its original parametrization, with the embeddedness removed by a reparametrization.

original parametrization, and we call it an *embedded model*. Figure 5.1 gives a geometrical interpretation of what is happening in the (a, b) and (a, c) spaces, showing that the models $\eta = cx$ for all c are obtained at the unbounded limit $a \to 0$ and $b \to \infty$ in the (a, b) parametrization, but at bounded boundary points $a = 0$, $0 \leq c < \infty$ in the (a, c) parametrization.

The example shows that embeddedness is not an invariant property, but depends on the form of the parametrization. The model (5.1) has the linear model (5.3) as an embedded model, but in the reparametrized version (5.2), the linear model is just a simple special case.

The instability where the linear model is only obtained as $b \to \infty$ in the original formulation results in curious, and potentially very unsatisfactory behaviour if one attempts to fit the model to data using a numerical search procedure. If we expand the regression function (5.1) as a series in x, we have

$$\eta = b[1 - \exp(-ax)] = abx - \left(a^2b/2\right)x^2 + O\left(a^3bx^3\right). \tag{5.4}$$

Thus the model, in addition to including a limiting linear model, also encompasses, at least approximately for small a, quadratic models that can be concave or convex depending only on the sign of b. Suppose the best fit (using maximum likelihood, say) is a slightly convex model, i.e. with $\hat{a}^2\hat{b}$ small and $\hat{b} < 0$, but we start a numerical search for the estimates with an initial point in the half-space $b > 0$. Here the fits are concave, and the best fit will be the linear model corresponding to $b \to \infty$. To get to the other half-space $b < 0$, we need to cross the line $b = 0$. But the model on this line is $\eta = 0$, which in general will be a poor fit. Thus any search that happens to start with a point in the 'wrong' half-space, where $b > 0$, could well move away from $a = 0$ with $b \to \infty$, ending with the linear model as the supposed best fit.

As a numerical example, suppose the true parameter values are $a_0 = -0.2, b_0 = -5$, and we have just three essentially exact 'observations':

$$x_1 = 1, y_1 = b_0(1 - \exp(-a_0)) = 1.107\,013\,791,$$

$$x_2 = 2, y_2 = b_0(1 - \exp(-2a_0)) = 2.459\,123\,488,$$

$$x_3 = 3, y_3 = b_0(1 - \exp(-3a_0)) = 4.110\,594\,002.$$

These three points lie nearly exactly on the quadratic

$$\eta = x + 0.1x^2.$$

Assuming normal errors, the ML estimates $\hat{a}, \hat{b}$ will minimize the sum of squares

$$S^2 = \sum_{j=1}^{3} [y_i - \eta(x_i, a, b)]^2.$$

Figure 5.2 is a contour plot of $20 + S^2$. The maximum is at $a = -0.2$, $b = -5$, the top of the boomerang-shaped ridge in the negative quadrant. The contours in the halfplane $b > 0$ indicate another ridge that falls away as $a \to 0$ and that rises as $b \to \infty$, so that the maximum is obtained in this halfplane only as $b \to \infty$, yielding $\eta = x$ as the best fit. This is in accordance with what is indicated in Figure 5.1.

Figure 5.3 is a contour plot of $20 + S^2$ for the reparametrized form of the model as in eqn (5.2):

$$\begin{aligned}\eta(x, a, c) &= (c/a)[1 - \exp(-ax)]\\ &= cx - \frac{1}{2}acx^2 + O\left(ca^2x^3\right).\end{aligned}$$

It will be seen that the maximum is unambiguous in this case, at $a = -0.2$ and $c = 1$.

Finally, we note that the indeterminacy where b becomes undefined if $a = 0$ is effectively removed by the reparametrization, with a meaningful value of c obtained as $a \to 0$. The other indeterminacy, where a becomes indeterminate if $c = 0$, is not removable.

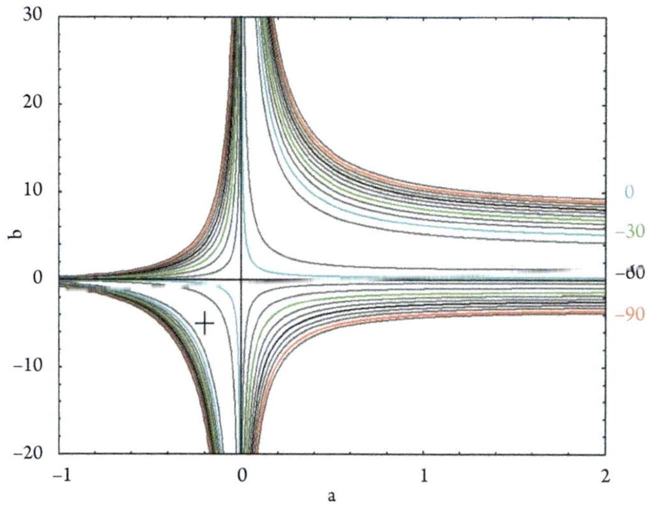

Figure 5.2 Contours of the sum of squares of the simple exponential model in its original parametrization (5.1) for a three-observation data sample displaying bimodal behaviour. The cross indicating the position of the global optimum is at $a = -0.2$, $b = -5$. Contours where $L <= -120$ have been suppressed for clarity.

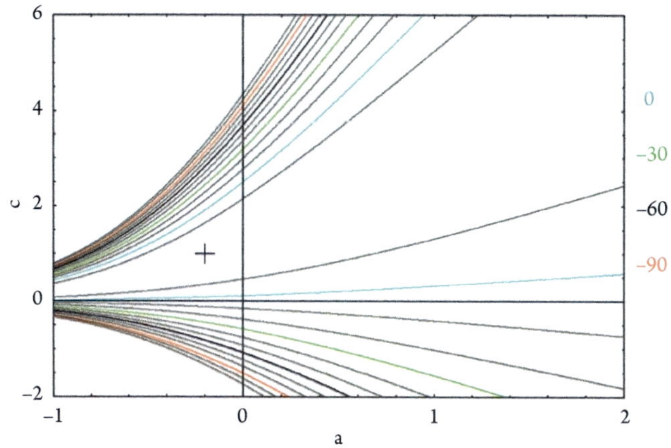

Figure 5.3 Contours of the sum of squares of the same simple exponential model, but reparametrized as in (5.2) to remove the embedded linear model. The unambiguous single maximum is at $a = -0.2$, $c = 1$. Again, for clarity, contours where $L <= -120$ are not displayed.

5.2 Definition of Embeddedness

In general, we discuss embedded models by taking $\boldsymbol{\theta}$ to be a vector of dimension $q = l + r$, partitioned so that $\boldsymbol{\theta} = (\boldsymbol{\theta}^L, \boldsymbol{\theta}^R)$, where $\boldsymbol{\theta}^L = (\theta_1, \theta_2, \ldots, \theta_l)$ and $\boldsymbol{\theta}^R = (\theta_{l+1}, \theta_{l+2}, \ldots, \theta_{l+r})$. We allow $r = 0$, in which case $\boldsymbol{\theta} = \boldsymbol{\theta}^L$, and there is no $\boldsymbol{\theta}^R$. In the example (5.1), $\boldsymbol{\theta}^L = (\theta_1, \theta_2)$ and $\boldsymbol{\theta}^R$ is void.

Now let $\boldsymbol{\theta}^L \to \mathbf{a} = (a_1, a_2, \ldots, a_l)$, a specified constant vector, and let $\boldsymbol{\theta}^R$ be fixed. Without loss of generality, we can assume $\mathbf{a} = \mathbf{0}$, as we can reparametrize, replacing θ_i by $\theta_i - a_i$ if a_i is finite, and replacing θ_i by $1/\theta_i$ if $a_i = \infty$. We therefore consider the limit simply as

$$\boldsymbol{\theta}^L \to \mathbf{0}.$$

We assume that $\boldsymbol{\theta} \in \boldsymbol{\Theta}$, a q-dimensional region containing $(\boldsymbol{\theta}^L, \boldsymbol{\theta}^R) = (\mathbf{0}, \boldsymbol{\theta}^R)$.

We now define an embedded model.

Definition of an Embedded Model. Consider the model $f(x, \boldsymbol{\theta})$, where $\boldsymbol{\theta} = (\boldsymbol{\theta}^L, \boldsymbol{\theta}^R)$ is a partition of the parameters, with $\boldsymbol{\theta}^L = (\theta_1, \theta_2, \ldots, \theta_l)$ and $\boldsymbol{\theta}^R = (\theta_{l+1}, \theta_{l+2}, \ldots, \theta_{l+r})$. Let ϕ_i be functions of $\boldsymbol{\theta}$,

$$\phi_i = g_i(\boldsymbol{\theta}), \quad i = 1, 2, \ldots, k, \tag{5.5}$$

which we can write in vector form as $\boldsymbol{\phi} = \mathbf{g}(\boldsymbol{\theta})$, with $\boldsymbol{\phi} = (\phi_1, \phi_2, \ldots, \phi_k)$. We can regard $\boldsymbol{\phi}$ as a vector of parameters obtained from a reparametrization of $\boldsymbol{\theta}$, with $\boldsymbol{\phi}$ taking values in $\boldsymbol{\Phi}$, some k-dimensional region. Suppose that for fixed $\boldsymbol{\phi} \in \boldsymbol{\Phi}$, it is possible to let $\boldsymbol{\theta}^L \to \mathbf{0}$ with $\boldsymbol{\theta}^R$ held fixed, while keeping $\mathbf{g}(\boldsymbol{\theta}) = \boldsymbol{\phi}$, and that

$$\lim_{\boldsymbol{\theta}^L \to \mathbf{0}} f(x, \boldsymbol{\theta}) = f_0(x, \boldsymbol{\phi}), \tag{5.6}$$

some well-defined function. Then, if $r < k < l + r$, we define $f_0(x, \boldsymbol{\phi})$ to be an *embedded model* of $f(x, \boldsymbol{\theta})$.

In the example (5.1), $l = 2$, $r = 0$, and $k = 1$, with $\boldsymbol{\phi} = \mathbf{g}(\boldsymbol{\theta})$ being just the single condition $\phi = \theta_1\theta_2$, and $f(x, \phi) = \phi x$ the embedded model.

Note that the condition $k > r$ ensures that although l parameters tend to 0 with just r of the original parameters free to vary, we end up with $k > r$ parameters free to vary, and this is typical of an embedding.

5.3 Step-by-Step Identification of an Embedded Model

An elementary way of finding embedded models is to use the definition of embeddedness and simply examine the form of the model $f(x, \boldsymbol{\theta})$ to find functions $g_i(\boldsymbol{\theta})$ that enable us to take as parameters $\phi_i = g_i(\boldsymbol{\theta})$ as in eqn (5.5), so that $f(x, \boldsymbol{\theta}) \to f_0(x, \boldsymbol{\phi})$, a well-defined function, when $\boldsymbol{\theta}^L \to \mathbf{0}$. A direct way of doing this is to consider $\mathbf{g}(\boldsymbol{\theta}) = \boldsymbol{\phi}$ in its inverted form, viewing it instead as a condition that defines $\boldsymbol{\theta}^L$ in terms of the new parameters $\boldsymbol{\phi}$, and to build up this inverted version step-by-step. This is anyway more convenient for identifying the embedded model $f_0(x, \boldsymbol{\phi})$, which is defined in terms of $\boldsymbol{\phi}$.

The simplest situation is where $k = l - 1$. In this case, we introduce an *auxiliary* parameter α on which all the components of $\boldsymbol{\theta}^L$ will depend, identifying the components of $\boldsymbol{\theta}^L$ recursively so that θ_i is a function of α, $\boldsymbol{\theta}^R$, and only the first $(i-1)$ new parameters $\phi_1, \phi_2, \ldots, \phi_i$, that is,

$$\theta_1 = \lambda_1(\alpha, \boldsymbol{\theta}^R),\ \theta_i = \lambda_i(\alpha, \phi_1, \phi_2, \ldots, \phi_{i-1}, \boldsymbol{\theta}^R),\ i = 2, \ldots, l \tag{5.7}$$

The λ_i must satisfy the condition

$$\lambda_1(\alpha, \boldsymbol{\theta}^R) \to 0,\ \lambda_i(\alpha, \phi_1, \phi_2, \ldots, \phi_{i-1}, \boldsymbol{\theta}^R) \to 0,\ i = 2, \ldots, l, \text{ as } \alpha \to 0, \tag{5.8}$$

so that $\boldsymbol{\theta}^L \to \mathbf{0}$ as $\alpha \to 0$.

Frequently, one of the components of $\boldsymbol{\theta}^L$ can be used for α. When we do this, we shall with no loss of generality take this to be θ_1, calling this the *leading parameter*, with the understanding that this is then synonymous with the name *auxiliary*.

The form of (5.7) ensures the original function $f(x, \theta)$ can be written as

$$f(x, \boldsymbol{\theta}) = f[x, \boldsymbol{\theta}^L(\alpha, \boldsymbol{\phi}, \boldsymbol{\theta}^R), \boldsymbol{\theta}^R] = f(x, \alpha, \boldsymbol{\phi}, \boldsymbol{\theta}^R). \tag{5.9}$$

If $f(x, \alpha, \boldsymbol{\phi}, \boldsymbol{\theta}^R)$ and its first two derivatives with respect to α

$$f^{(m)}(x, \alpha, \boldsymbol{\phi}, \boldsymbol{\theta}^R) = \frac{\partial^m}{\partial \alpha^m} f(x, \alpha, \boldsymbol{\phi}, \boldsymbol{\theta}^R),\ m = 1, 2$$

exist, are continuous, and remain finite in some region $\boldsymbol{\phi} \in \boldsymbol{\Phi}$ and $\alpha \in [0, \delta]$, then f can be written as a Maclaurin series in the auxiliary parameter α:

$$f(x, \alpha, \boldsymbol{\phi}, \boldsymbol{\theta}^R) = \sum_{i=0}^{1} \frac{1}{m!} f^{(m)}(x, 0, \boldsymbol{\phi}, \boldsymbol{\theta}^R)\alpha^m + \frac{1}{2} f^{(2)}(x, \alpha^*, \boldsymbol{\phi}, \boldsymbol{\theta}^R)\alpha^2,$$

where $0 < \alpha^* < \alpha$, with the simple limit

$$\lim_{\alpha \to 0} f(x, \alpha, \boldsymbol{\phi}, \boldsymbol{\theta}^R) = f(x, 0, \boldsymbol{\phi}, \boldsymbol{\theta}^R) \tag{5.10}$$

being the embedded model.

As an example of the step-by-step approach just outlined, we consider

$$f(x, \boldsymbol{\theta}) = \left(\frac{x-a}{b}\right)^c. \tag{5.11}$$

We take the logarithm

$$\ln f = c \ln(x-a) - c \ln b,$$

and consider a reparametrization where $\ln h$ involves x only through a term of the form $g_1(\alpha)\ln[1+g_2(\alpha)x]$, with $g_1(\alpha) = O(\alpha^{-1})$ and $g_2(\alpha) = O(\alpha)$ when $\alpha \to 0$. This then allows a series expansion in x. Using the step-by-step approach, our first step is to consider the first term $c\ln(x-a)$. Setting $a = -\sigma/\alpha$, we have

$$\ln f = c\ln[(\sigma/\alpha)(1+\alpha x/\sigma)] - c\ln b \tag{5.12}$$

$$= [c\ln(1+\alpha x/\sigma)] + [c\ln(\sigma/\alpha) - c\ln b] \tag{5.13}$$

$$= g_1 + g_2, \text{ say.} \tag{5.14}$$

An obvious choice is then $c = 1/\alpha$, so that

$$\begin{aligned} g_1 &= c\ln(1+\alpha x/\sigma) \\ &= \alpha^{-1}\ln(1+\alpha x/\sigma) \\ &= (x/\sigma) - (x^2/2)\alpha + O(\alpha^2) \text{ as } \alpha \to 0. \end{aligned}$$

This allows our second step, which is then to take $b = \phi\sigma/\alpha$, so that the remaining terms in $\ln f$ become

$$\begin{aligned} g_2 &= c\ln(\sigma/\alpha) - c\ln b \\ &= -\alpha^{-1}\ln\phi \\ &= -\mu/\sigma, \end{aligned}$$

if we make the obvious choice $\phi = e^{\alpha\mu/\sigma}$.

Summarizing, if $a = -\sigma\alpha^{-1}$, $b = e^{\alpha\mu/\sigma}\sigma\alpha^{-1}$, and $c = \alpha^{-1}$, then

$$\ln f = [(x-\mu)/\sigma] - (x^2/2)\alpha + O(\alpha^2) \text{ as } \alpha \to 0,$$

so that if $\alpha \to 0$, we get the embedded exponential model

$$h = \exp[(x-\mu)/\sigma]. \tag{5.15}$$

5.4 Indeterminate Forms

Using the same definition of embedded models, Shao (2002) proposes a more formal approach to identifying an embedding through indeterminate forms. Though Shao claims some generality for this use of indeterminate forms, the overall way that the approach works is actually very similar to the step-by-step approach described in the previous section. By way of comparison, we summarize the Shao approach here.

In Shao (2002, Corollary 1), the function $f(\cdot)$ is assumed to take the special form

$$f(x,\boldsymbol{\theta}) = f_1[x, h_1(x,\boldsymbol{\theta}^R,\theta_1,\theta_2),\ldots,h_{l-1}(x,\boldsymbol{\theta}^R,\theta_1,\theta_2,\ldots,\theta_l), h_0(x,\boldsymbol{\theta}),\boldsymbol{\theta}^R], \tag{5.16}$$

where $f(\cdot)$ will contain one or more embedded models if $h_i(x, \boldsymbol{\theta})$, $i = 1, 2, \ldots, l-1$, take *indeterminate* forms. Shao (2002) does not define an indeterminate form explicitly, but the context makes it clear that $h_i(x, \boldsymbol{\theta})$ is indeterminate if, in our notation,

$$h_i(x, \boldsymbol{\theta}) \to \frac{0}{0} \text{ as } \boldsymbol{\theta}^L \to \mathbf{0}.$$

Embedded models are simply special cases that are hidden by each indeterminacy. Shao (2002) uses de l'Hôpital's rule to systematically remove each indeterminate form. For example, in the case $f(x) = [(x-a)/b]^c$, taking logarithms

$$\begin{aligned}\ln f(x, \boldsymbol{\theta}) &= \ln\{[(x-a)/b]^c\} \\ &= \ln\{[(1 - x/a)(-a/b)]^c\} \\ &= c\ln[1 - (x/a)] + c\ln(-a/b) \\ &= h_1(x, a, c) + h_2(x, a, b, c),\end{aligned}$$

where $h_1(\cdot)$ and $h_2(\cdot)$ can be indeterminate, De l'Hôpital's rule can then be used to obtain determinate limits for each h_i, thereby identifying embedded models. Done systematically, this is equivalent to the step-by-step approach already described. The only difference is that whereas we used a series expansion approach in removing each indeterminacy, use of de l'Hôpital's rule requires the solution of a set of differential equations instead. Details of this use of de l'Hôpital's rule are given in Shao (2002) and not repeated here.

With either approach, the main problem is identification of all the sets of indeterminate forms that will lead to different embedded models.

Potential reparametrizations of the form (5.7) can sometimes be directly identified using a series expansion in x of the model $f(\cdot)$. Indeed, this was implicitly done in (5.4) in our opening example for the chapter. We discuss this more fully next.

5.5 Series Expansion Approach

In simple cases, Cheng, Evans, and Iles (1992) show how a Taylor series expansion in x of $f(x, \boldsymbol{\theta})$ in x can be used to identify appropriate $\mathbf{g}(\boldsymbol{\theta}) = \boldsymbol{\phi}$ conditions. Suppose that $f(x, \boldsymbol{\theta})$ has a valid Taylor series expansion

$$f(x, \boldsymbol{\theta}) = \sum_{m=0}^{\infty} f^{(m)}(b, \boldsymbol{\theta})(x-b)^m/m!$$

about $x = b$, where

$$f^{(m)}(b, \boldsymbol{\theta}) = \left.\frac{\partial^m f(x, \boldsymbol{\theta})}{\partial x^m}\right|_{x=b}.$$

Then select k of the derivatives $f^{(m_1)}(b,\boldsymbol{\theta}),\ldots,f^{(m_k)}(b,\boldsymbol{\theta})$ and set

$$\phi_i = f^{(m_i)}(b,\boldsymbol{\theta}) = g_i(\boldsymbol{\theta}),\ \ i = 1,2,\ldots,k \tag{5.17}$$

$$\begin{aligned}\boldsymbol{\theta}^I &= (\theta_1,\theta_2,\ldots,\theta_{q-k})\\ \boldsymbol{\theta}^D &= (\theta_{q-k+1},\theta_{q-k+2},\ldots,\theta_l).\end{aligned}$$

We now treat $(\boldsymbol{\theta}^D,\boldsymbol{\theta}^R)$ as being dependent on $(\boldsymbol{\theta}^I,\boldsymbol{\phi})$, with $\boldsymbol{\theta}^I,\boldsymbol{\phi}$ being allowed to vary freely, and try to solve (5.17) for $(\boldsymbol{\theta}^D,\boldsymbol{\theta}^R)$ in terms of $(\boldsymbol{\theta}^I,\boldsymbol{\phi})$,writing such a solution as

$$(\boldsymbol{\theta}^D,\boldsymbol{\theta}^R) = [\boldsymbol{\lambda}^D(\boldsymbol{\theta}^I,\boldsymbol{\phi}),\boldsymbol{\lambda}^R(\boldsymbol{\theta}^I,\boldsymbol{\phi})].$$

If now $\boldsymbol{\lambda}^D(\boldsymbol{\theta}^I,\boldsymbol{\phi}) \to \mathbf{0}$ as $\boldsymbol{\theta}^I \to \mathbf{0}$, then

$$f_0(x,\boldsymbol{\phi}) = \lim_{\boldsymbol{\theta}^I\to 0} f(x,\boldsymbol{\theta}^I,\boldsymbol{\phi})$$

is an embedded model.

An elementary example is the regression function where

$$f(x,\boldsymbol{\theta}) = \theta_1 + \theta_2\exp(\theta_3 x). \tag{5.18}$$

The Taylor series expansion is

$$f(x,\boldsymbol{\theta}) = \theta_1 + \theta_2 + \theta_2\theta_3 x + \theta_2\sum_{j=2}^{\infty}\theta_3^j x^j/j!$$

Then consider the reparametrization

$$\begin{aligned}\phi_1 &= f^{(0)}(x,\boldsymbol{\theta}) = \theta_1 + \theta_2\\ \phi_2 &= f^{(1)}(x,\boldsymbol{\theta}) = \theta_2\theta_3,\end{aligned}$$

where we then let $\theta_3 \to 0$, subject to ϕ_1,ϕ_2 remaining fixed, with $\phi_1 = \theta_1 + \theta_2$ and $\phi_2 = \theta_2\theta_3$. We have

$$\begin{aligned}f(x,\boldsymbol{\theta}) &= \phi_1 + \phi_2 x + \phi_2\sum_{j=2}^{\infty}\theta_3^{j-1}x^j/j!\\ &\to \phi_1 + \phi_2 x \text{ as } \theta_3 \to 0.\end{aligned}$$

So $f_0(x,\phi_1,\phi_2) = \phi_1 + \phi_2 x$ is an embedded model in this case, obtained in the original parametrization when $\theta_1 \to -\infty$, $\theta_2 \to \infty$, and $\theta_3 \to 0$.

5.6 Examples of Embedded Regression

Table 5.1 lists some examples of regression models containing embedded models. The table is based on that of Cheng, Evans, and Iles (1992), but includes examples given by Shao (2002).

Such problems have been discussed by Clarkson and Jennrich (1991). Examples are also given by Seber and Wild (2003, §3.4).

The embedded model is the linear model in several of the examples in Table 5.1. One such is model IV, which was first discussed by Meyer and Roth (1972), who showed that fitting it to a data set giving the log resistance of a thermistor at 16 different temperatures produced very unstable estimates. The example was analyzed by Cheng, Evans, and Iles (1992), who showed that the embedded linear model is a good fit and stable.

Simonoff and Tsai (1989) interpret the problem in terms of multicollinearity and propose a *guided reformulation* that allows for model collinearity, applying this to several models in Table 5.1. We will not discuss guided reformulations, as, though they can be effective, a reformulation is not always directly related to the original model, so that statistical inference in terms of the reformulation may not translate readily to the original model.

The table is limited just to examples where the crucial limiting behaviour of *all* the original parameters $\boldsymbol{\theta}$ is governed by a leading parameter in the transformed parameters set, labelled α in the table, with $\alpha \to 0$. Thus, the condition (5.8) is satisfied, with $k = l - 1$.This condition is actually easily relaxed if we note that more complicated situations can be obtained by combining different examples. For example, the Hershel-Bulkley and shifted power examples are naturally combined to give the model

$$\eta = \theta_1 + \theta_2(x - \theta_4)^{\theta_3}. \tag{5.19}$$

In this model, two distinct embedded models are then possible. We will encounter several such examples when we consider embedded PDFs in the next chapter.

The table layout is slightly different from that of Cheng, Evans, and Iles (1992) in that the transforms in the third column are given as $\boldsymbol{\theta} = \boldsymbol{\theta}(\alpha, \boldsymbol{\phi})$, with the original parameters expressed in terms of the transformed parameters. The instabilities listed in the second column are called 'typical' in that an instability given as $\theta_1 \to \infty$, say, is dependent on the signs of the transformed parameters α and $\boldsymbol{\phi}$. For the examples in the table, the signs of the transformed parameters are in general unrestricted. The only restrictions are that x in the Herschel-Bulkley and $(x - \theta_3)$ in the shifted power models should be positive.

When fitting models, ill-conditioning of the parameter estimates will usually occur if there is an embedded model and it is the best fit. Cheng, Evans, and Iles (1992) discuss how the parameters of a fitted model can behave very unstably in this situation. They show, admittedly in cases where the sample size is small, that parameter estimates can become very sensitive to a change in just a single data value. This occurs in the example considered by Meyer and Roth (1972). Two simple approaches are as follows.

Table 5.1 Examples of embedded regressions

Example	$f(x,\theta)$ Typical instability	Transform: $\theta = \theta(\alpha,\phi)$ Reparametrized model: $f(x,\alpha,\phi)$	Embedded model: $\alpha \to 0$ S&T reform.
I	$\theta_1(1-e^{-\theta_2 x})$ θ_1,θ_2 $\to \infty, 0$	$\theta_1 = \phi_1/\alpha, \theta_2 = \alpha$ $\phi_1(1-e^{-\alpha x})/\alpha$	$\phi_1 x$ $\dfrac{\phi_1 x}{(\phi_2 x + 1)}$
II	$\theta_1 + \theta_2 e^{-\theta_3 x}$ $\theta_1,\theta_2,\theta_3$ $\to -\infty, \infty, 0$	$\theta_1 = \phi_1 - \dfrac{\phi_2}{\alpha}$, $\theta_2 = \dfrac{\phi_2}{\alpha}$, $\theta_3 = \alpha$ $\phi_1 + \phi_2(e^{\alpha x}-1)/\alpha$	$\phi_1 + \phi_2 x$ $\phi_1 + \dfrac{\phi_2}{(1-\phi_3 x)}$
III	$\theta_1 e^{\theta_2/(x+\theta_3)}$ $\theta_1,\theta_2,\theta_3$ $\to 0, -\infty, \infty$	$\theta_1 = \phi_1 e^{\frac{\phi_2}{\phi_1}\alpha^{-1}}$, $\theta_2 = -\frac{\phi_2}{\phi_1}\alpha^{-2}$, $\theta_3 = \alpha^{-1}$ $\phi_1 e^{\frac{\phi_2 x}{\phi_1(1+\alpha x)}}$	$\phi_1 e^{\phi_2 x/\phi_1}$ $\phi_1^{-1}(x+\phi_3)$ $\times e^{\frac{\phi_2}{x+\phi_3}}$
IV	$\theta_1 +$ $\theta_2/(x+\theta_3)$ $\theta_1,\theta_2,\theta_3$ $\to -\infty, \infty, \infty$	$\theta_1 = \phi_1 - \phi_2\alpha^{-1}$, $\theta_2 = \phi_2\alpha^{-2}$, $\theta_3 = \alpha^{-1}$ $\phi_1 - \phi_2 x/(1+\alpha x)$	$\phi_1 - \phi_2 x$ $\phi_1 + \phi_2 x$
Log-logistic	$\theta_1 -$ $\ln(1+\theta_2 e^{-\theta_3 x})$ $\theta_1,\theta_2 \to$ ∞, ∞	$\theta_1 = \phi_1 + \ln(1+\alpha^{-1})$, $\theta_2 = \alpha^{-1}, \theta_3 = \phi_3$ $\phi_1 - \ln(\frac{\alpha + e^{-\phi_3 x}}{\alpha+1})$	$\phi_1 + \phi_3 x$ Not given
Herschel-Bulkley	$\theta_1 + \theta_2 x^{\theta_3}$ $\theta_1,\theta_2,\theta_3$ $\to -\infty, \infty, 0$	$\theta_1 = \phi_1 - \phi_2\alpha^{-1}$, $\theta_2 = \phi_2\alpha^{-1}$, $\theta_3 = \alpha$ $\phi_1 + \phi_2(x^\alpha - 1)/\alpha$	$\phi_1 + \phi_2 \ln x$ Not given
Shifted Power	$\theta_1(x-\theta_2)^{\theta_3}$ $\theta_1,\theta_2,\theta_3$ $\to 0, -\infty, \infty$	$\theta_1 = \phi_1(\frac{\phi_2}{\alpha})^{-\frac{1}{\alpha}}$, $\theta_2 = -\frac{\phi_2}{\alpha}$, $\theta_3 = \alpha^{-1}$ $\phi_1(1+\alpha\frac{x}{\phi_2})^{1/\alpha}$	$\phi_1 \exp(\dfrac{x}{\phi_2})$ Not given

continued

Table 5.1 (Continued)

Brain-Cousens	$\frac{\theta_1 + \theta_2 x}{1 + (x/\theta_3)^{\theta_4}}$ $\theta_1, \theta_2, \theta_3$ $\to \infty, \infty, 0$	$\theta_1 = \phi_1 \alpha^{-1}$, $\theta_2 = \phi_2 \alpha^{-1}$, $\theta_3 = \alpha^{1/\phi_3}$, $\theta_4 = \phi_3$ $\frac{\phi_1+\phi_2 x}{\alpha + x^{\phi_3}}$	$\frac{\phi_1 + \phi_2 x}{x^{\phi_3}}$ Not given

(i) If embedded models have already been identified, one can treat them as being simpler special cases of the full model, and fit them first in their own right, followed by tests for LoF (lack-of-fit) of the fitted embedded models. There is not even a need to fit the full model if an embedded model is acceptable, though of course there is the wider issue of selecting a best fit.

(ii) If there is uncertainty whether an embedded model may be present, one can still attempt to fit the full model directly, but with built-in checks to identify if any parameters appear to be near zero, or are becoming unbounded. If this then occurs, a check can be made to see if this corresponds to a convergence towards an embedded model.

We shall describe more fully the effect of embedded models on systematic building of complex models once we have discussed another problem, indeterminacy, that is related to embeddedness, which has also to be taken into consideration. We shall discuss model building more fully in Chapter 15, after indeterminacy has been discussed in Chapter 14.

5.7 Numerical Examples

We end this chapter by revisiting the VapCO and Lettuce examples introduced in Section 4.7 to consider how embedded models can arise in them.

5.7.1 VapCO Example Again

The VapCO data set is an example where, though the sample size is small, the observations are quite accurate. At first sight, the Antoine model of eqn (4.19) with its three parameters gives an apparently good fit, but which the LoF test showed was not really satisfactory. The oscillatory nature of the residuals in Figure 4.4 suggests that a significantly more complicated model is needed to fully identify the systematic behaviour that appears to be present in the sample. To try to confirm this, we first investigated what improvement might be achieved with a four-parameter distribution by fitting the shifted Herschel-Bulkley (HB) regression of eqn (5.19).

The left-hand charts in Figure 5.4 show the fit obtained with the shifted HB model, with the drifts plotted in different colours. It will be seen from the standardized residuals in the lower left-hand chart that the drifts clearly identify a damped oscillatory pattern to the residuals. For comparison, we also fitted the three-parameter shifted-log model

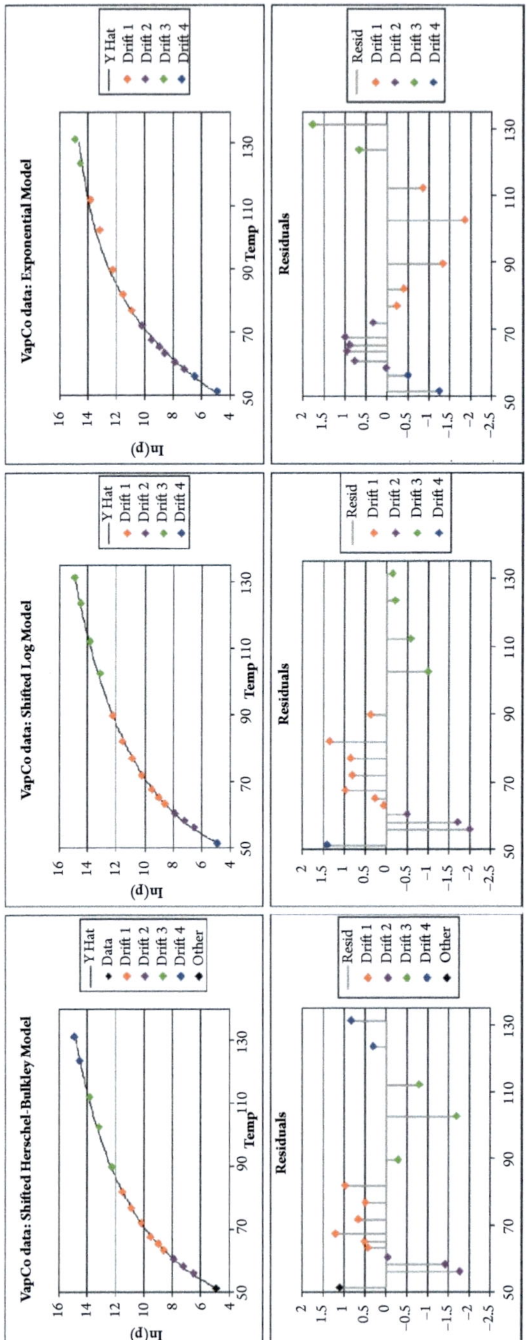

Figure 5.4 Left-hand charts: Plot of the shifted Herschel-Bulkley model fitted to the VapCO data and plot of the residuals. The first four drifts are highlighted in colour. The middle and right-hand charts give the plots of the shifted log and exponential models and the corresponding plots of the residuals.

Table 5.2 BS Lack-of-Fit results for shifted Herschel-Bulkley model and its embedded models

Shifted Herschel-Bulkley				Shifted Log			Exponential		
m	Δ_0	p-val	$\Delta^*_{0,0.95}$	Δ_0	p-val	$\Delta^*_{0,0.95}$	Δ_0	p-val	$\Delta^*_{0,0.95}$
1	4.32	0.052	4.35	4.72	0.010	4.22	4.70	0.004	4.08
2	7.55	0.048	7.46	8.92	0.004	7.41	8.67	0.004	7.31
3	10.33	0.020	9.45	10.86	0.002	9.51	11.13	0.000	9.26
4	11.46	0.024	10.85	12.29	0.002	10.79	12.85	0.000	10.79
$L_{SHB}(\hat{\theta}) = 17.53$				$L_{SL}(\hat{\theta}) = 13.76$			$L_E(\hat{\theta}) = 5.48$		

$$\eta_0(x) = \phi_1 + \phi_2 \ln(x - \phi_3) \tag{5.20}$$

and the three-parameter exponential model

$$\zeta(x) = \phi_1 + \phi_2 \exp(\phi_3 x), \tag{5.21}$$

the two embedded models of the shifted Herschel-Bulkley model (5.19). The middle and right-hand charts in Figure 5.4 show the fits for these models. It will be seen that both residual charts display the same very pronounced oscillatory behaviour highlighted by the drift colouring already seen in the shifted HB residuals. Table 5.2 gives results of LoF drift tests based on 500 BS samples. Here there is a difference between the performance of the shifted HB model and its two embedded models. There is a significant lack-of-fit at the 5% test level of the shifted HB model for $m = 2$, 3, and 4, with the case $m = 1$ marginally not significant. The shifted log and exponential models display definite lack of fit, with, for both models, the p values being very small for all m.

We can also formally compare the shifted HB model fit with the shifted log fit using the likelihood ratio (LR) statistic (3.9). We have

$$LR = 2[L_{SHB}(\hat{\theta}) - L_{SL}(\hat{\phi})] = 2(17.53 - 13.76) = 7.54 > \chi^2_1(0.99) = 6.63,$$

so that the shifted HB is better at the 99% level. Similarly, comparing shifted HB model fit with the exponential model fit, we have

$$LR = 2[L_{SHB}(\hat{\theta}) - L_E(\hat{\phi})] = 2(17.53 - 5.48) = 24.1 > \chi^2_1(0.999) = 10.83.$$

So, the shifted HB is definitely to be preferred to its embedded models. Note that the LR test cannot be used to compare the two embedded models, as neither is nested relative to the other. However, the LoF results indicate that the shifted log model is perhaps to be preferred.

As none of the models give a reassuringly good fit, we consider how the oscillatory behaviour might be modelled, treating this simply as an exercise to illustrate how bootstrapping helps the modelling process, with non-standard behaviour not involved.

Looking at the residuals suggests that the three-parameter model

$$\eta_1(x) = \frac{\phi_4}{x - \phi_5} \cos[\phi_6(x - x_{(1)})^{1/2}],$$

where $x_{(1)}$ is the smallest observed x, might be an appropriate addition to the shifted log model, so that we would be fitting the six-parameter model

$$\eta(x) = \eta_0(x) + \eta_1(x). \tag{5.22}$$

This might be considered a rather arbitrary choice, but in fact required some experimentation, with simpler polynomial functions for $\eta_1(x)$ not proving at all satisfactory. An issue that did arise is that six parameters seems rather a large number given there are only 15 observations. However, it is perhaps an indication that the experiment, being presumably a well-controlled scientific one, has resulted in rather accurate observations, so that the sample is genuinely very informative.

Some care was needed in fitting the model (5.22), due to its complexity. The fitting procedure was sensitive to the initial values used for the parameters in the numerical optimization. Ritz and Streibig (2008) give good advice on the use of 'starter functions', where a grid of initial points is deployed to ensure convergence. We did not investigate multiple global optima closely, but it is clear that the trigonometric form we assumed for $\eta(x) = \eta_0(x) + \eta_1(x)$ is susceptible to this.

Figure 5.5 shows the results. The left-hand charts display the estimated overall regression $\hat{\eta}(x) = \hat{\eta}_0(x) + \hat{\eta}_1(x)$ and the standardized residuals. The most important aspect of the residuals is that there are no extreme values. The upper right-hand chart plots $\hat{\eta}_0(x)$ and the data points, showing that $\hat{\eta}_0(x)$ taken on its own appears to be a reasonable fit. The estimate $\hat{\eta}_1(x)$ can be therefore be regarded as a fit to just the residuals $y_i - \hat{\eta}_0(x_i)$, $i = 1, 2, \ldots, n$, these residuals representing what is left of y_i not explained by $\hat{\eta}_0(x_i)$. The lower right-hand chart plots $\hat{\eta}_1(x)$, comparing it with these residuals $y_i - \hat{\eta}_0(x_i)$, and showing it has captured the right kind of oscillatory behaviour apparent in the residuals.

Table 5.3 shows the drift results obtained from 500 BS samples drawn from the fitted model $y = \hat{\eta}_0(x) + \hat{\eta}_1(x) + \varepsilon$. The p values are all very large, indicating no evidence of lack-of-fit.

To complete the analysis, we compare the $\eta_0 + \eta_1$ model with the shifted HB model using the LR test. We have

$$\begin{aligned} LR &= 2(L_{\eta_0+\eta_1}(\hat{\phi}) - L_{SHB}(\hat{\theta})) \\ &= 2(31.17 - 17.53) = 27.28 > \chi_3^2(.999) = 16.266, \end{aligned}$$

confirming that the $\eta_0 + \eta_1$ model is significantly different from the shifted HB model.

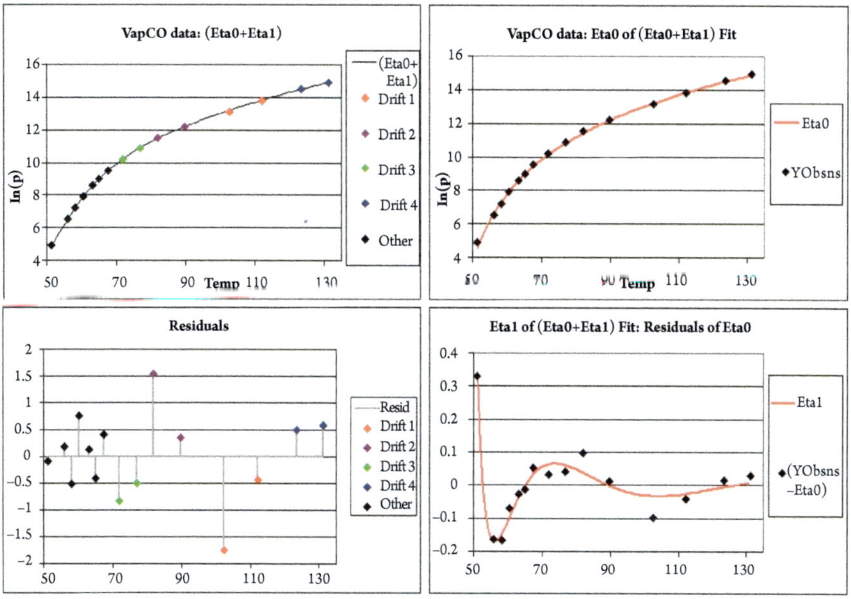

Figure 5.5 VapCO data: Results of fitting the $\eta_0 + \eta_1$ model.

Table 5.3 VapCO data: BS drift results for $\eta_0 + \eta_1$ model

$\eta_0 + \eta_1$ **Model**			
m	Δ_0	**p-val**	$\Delta^*_{0,0.95}$
1	2.16	0.728	4.57
2	4.06	0.726	8.71
3	5.39	0.794	5.39
4	6.49	0.832	11.74
$L_{\eta_0+\eta_1}(\hat{\phi}) = 31.17$			

5.7.2 Lettuce Example Again

In Section 4.7.2, we fitted the Brain-Cousens model (4.21) $\eta = (c + dx))/(1 + (x/a)^b)$ to the lettuce data of Table 4.4 and showed it gave a very satisfactory fit.

As mentioned at the end of Section 4.7.2, if $a \to 0$, and $c, d \to \pm\infty$, we get a simpler three-parameter model, which we now can see from Table 5.1 is the embedded model

$$\eta = \frac{\phi_1 + \phi_2 x}{x^{\phi_3}}, \tag{5.23}$$

where $a \to 0$, and $c, d \to \pm\infty$ in such a way that

$$ca^b \to \phi_1,\ da^b \to \phi_2,\ b = \phi_3,$$

with ϕ_1, ϕ_2, ϕ_3 remaining finite.

We did not give the parameter estimates for the fitted Brain-Cousens model, which were

$$(\hat{a}, \hat{b}, \hat{c}, \hat{d}, \hat{\sigma}) = (0.1092,\ 2.497,\ -766.4,\ 1019.2,\ 0.1007),$$

where σ is the SD of the assumed normal error. These estimates were obtained using the approach (ii) given in Section 5.6 with the numerical calculations terminated when it became clear that $-c$ and d were increasing indefinitely, which we recognized was an indication of the embedded model being the best fit. With these estimates we get

$$\begin{aligned} \tilde{\phi}_1 &= \hat{c}\hat{a}^{\hat{b}} = -3.040 \\ \tilde{\phi}_2 &= \hat{d}\hat{a}^{\hat{b}} = 4.043 \\ \tilde{\phi}_3 &= \hat{b} = 2.497. \end{aligned} \tag{5.24}$$

Fitting the embedded model directly gave

$$(\hat{\phi}_1, \hat{\phi}_2, \hat{\phi}_3, \hat{\sigma}) = (-3.027,\ 4.026,\ 2.492,\ 0.1006), \tag{5.25}$$

which is in good agreement.

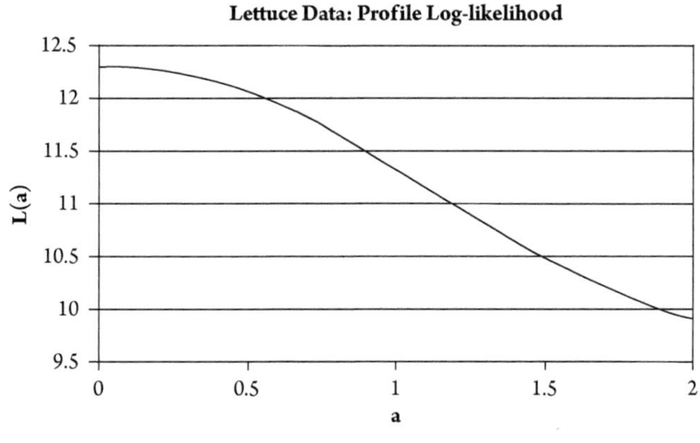

Figure 5.6 Lettuce data: profile log-likelihood plot, with *a* the profiling parameter.

The embedded model turns out to be the best fit in this example, and this is most easily seen in Figure 5.6, showing the profile likelihood

$$L(a) = \max_{b,c,d,\sigma} L(a, b, c, d, \sigma \,|\, \mathbf{y})$$

increasing monotonically to a maximum value as the profiling parameter $a \to 0$.

Incidentally, the profile log-likelihood in this case was easily obtained by fitting the embedded model first, and then calculating the profile log-likelihood by carrying out numerical maximization at each a using the starting values

$$b = \phi_3,\; c = \phi_1 a^{-\phi_3},\; d = \phi_2 a^{-\phi_3}$$

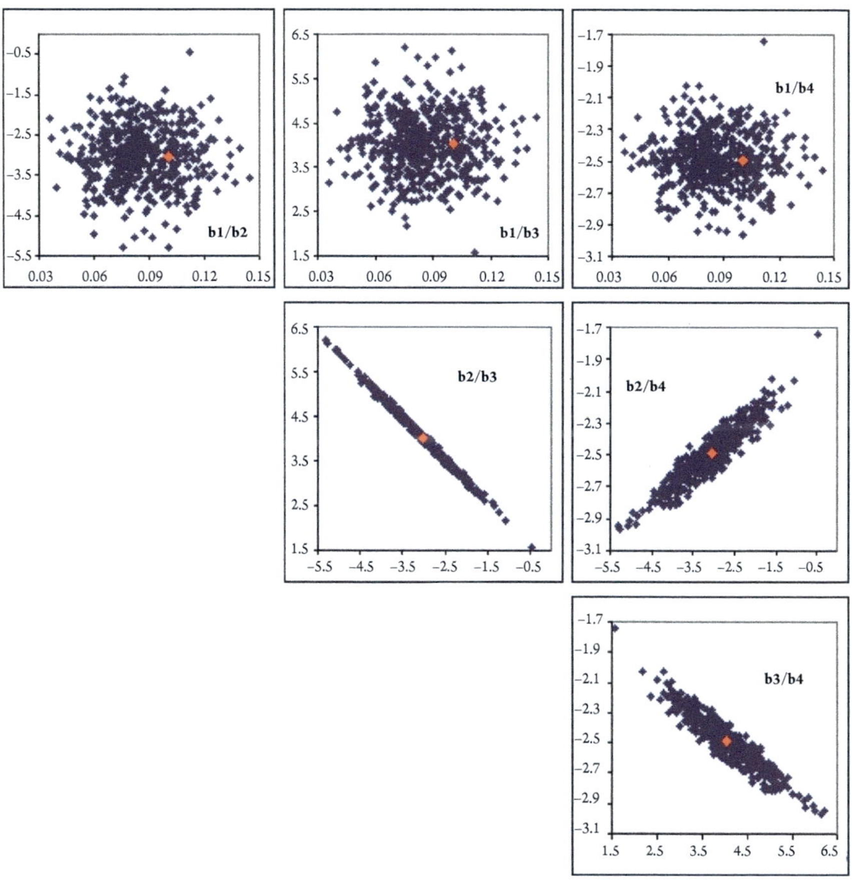

Figure 5.7 Lettuce data: Scatterplot of 500 BS samples of the embedded model in the Brain-Cousens model. Red dot is the MLE for the original data; $b_1 \equiv \sigma$, $b_2 \equiv \phi_1$, $b_3 \equiv \phi_2$, $b_4 \equiv \phi_4$.

obtained by inverting the expressions ϕ_i given in (5.24), but with the embedded model estimates $\hat{\phi}_i$ as given in (5.25).

We also carried out a BS analysis of the embedded model fit using 500 BS samples drawn from the fitted embedded model (5.23). The results are not shown, but the drift test did not reveal any significant lack-of-fit, with large p-values obtained. The BS point estimate scatterplots are given in Figure 5.7, indicating that the estimates are much more normally distributed than the estimates for the full Brain-Cousens model, whose scatterplots were given in Figure 4.8. However, this has not noticeably affected the confidence band results. The BS 90% confidence bands obtained from the fitted embedded model are not displayed separately, being visually identical to that obtained from the full Brain-Cousens model, which is displayed in Figure 4.6.

6 Examples of Embedded Distributions

In this chapter, we give examples of probability distribution models which, in their conventional parametrization, include special cases that are embedded models. A significant issue when fitting such a model is to decide if any of the submodels might be the best fit to the data. Unless there are specific reasons not to do so, we apply Occam's razor, preferring a more parsimonious model with fewer parameters to one with a larger number of parameters.

As with the regression examples discussed in the previous chapter, probability distributions that are special cases which have fewer parameters than the more general model, whether embedded or not, can, under suitable conditions, be regarded as *nested*, so that formal comparisons like the likelihood ratio test as described in Section 3.3 can be used. We shall use hypothesis tests only in the context of comparing an individual submodel with a specific more general model. Thus, we do not attempt multiple comparisons capable of handling more complex relations between models.

An advantage of this simple approach will become evident when we consider the problem of indeterminate parameters, as it is both easy to apply and to understand, compared with other technically much more complex solutions that have been proposed in the literature.

The basic situation that we consider is where the full model takes the form $f(x, \alpha, \boldsymbol{\lambda})$, as in (5.9), where we have written $\boldsymbol{\lambda} = (\boldsymbol{\phi}, \boldsymbol{\theta}^R)$, and α is an auxiliary parameter that highlights the embedded model $f(x, 0, \boldsymbol{\lambda})$, as in (5.10), which is obtained when $\alpha \to 0$. In the case where we are fitting a distribution to a random sample, so that $f(\cdot)$ is a PDF, the relation between the full model and embedded model can usually (but not invariably) be made more evident by expanding the log-likelihood as a Maclaurin power series in α, with the (one-observation) log-likelihood taking the form

$$L(\alpha, \boldsymbol{\lambda}) = L(0, \boldsymbol{\lambda}) + \frac{\partial L(0, \boldsymbol{\lambda})}{\partial \alpha}\alpha + \frac{1}{2}\frac{\partial^2 L(0, \boldsymbol{\lambda})}{\partial \alpha^2}\alpha^2 + O_p(\alpha^3) \tag{6.1}$$

$$= L_0(\boldsymbol{\lambda}) + L_1(\boldsymbol{\lambda})\alpha + L_2(\boldsymbol{\lambda})\alpha^2 + O_p(\alpha^3), \text{ say,} \tag{6.2}$$

Non-Standard Parametric Statistical Inference. Russell Cheng.

where $\boldsymbol{\lambda} = (\boldsymbol{\phi}, \boldsymbol{\theta}^R)$, and the term $L_0(\boldsymbol{\lambda})$ is simply the log-likelihood of the embedded model.

This expansion can be used to develop formal comparisons of the full and embedded models. Let $\hat{\boldsymbol{\lambda}} = \arg\max_{\boldsymbol{\lambda}}(L_0(\boldsymbol{\lambda}))$ be the MLE of the embedded model. Then

$$L(\alpha, \hat{\boldsymbol{\lambda}}) = L_0(\hat{\boldsymbol{\lambda}}) + L_1(\hat{\boldsymbol{\lambda}})\alpha \tag{6.3}$$

is the first-order (linear) approximation, giving the tangent of the profile log-likelihood at $\boldsymbol{\lambda} = \hat{\boldsymbol{\lambda}}$. This gives a simple and immediate graphically motivated check of whether the embedded model might be preferred to the full model. At $\alpha = 0$, $L(\alpha, \hat{\boldsymbol{\lambda}}) = L_0(\hat{\boldsymbol{\lambda}})$ is the value of the log-likelihood when the embedded model is fitted, so that if the sign of $L_1(\hat{\boldsymbol{\lambda}})$ is negative, this indicates that increasing α from zero will decrease the log-likelihood, showing that the embedded model is best locally.

The quantity $L_1(\hat{\boldsymbol{\lambda}})$ converts easily to the score statistic T_S of eqn (3.14), so can be used to provide a more formal test of whether the embedded model is an adequate fit. An alternative is to use T_W, the Wald statistic of eqn (3.11). We discuss this fully in Section 6.2, but note here that in either case, we need the information matrix $\mathrm{I}(\cdot)$ evaluated at the parameter value $(0, \hat{\boldsymbol{\lambda}})$, using the reparametrized form of the model. From the negative Hessian definition for I, it follows that

$$\mathrm{I}(0,\hat{\boldsymbol{\lambda}}) = \begin{bmatrix} \mathrm{I}_{\alpha\alpha}(0,\hat{\boldsymbol{\lambda}}) & \mathrm{I}_{\alpha\lambda}(0,\hat{\boldsymbol{\lambda}}) \\ \mathrm{I}_{\lambda\alpha}(0,\hat{\boldsymbol{\lambda}})) & \mathrm{I}_{\lambda\lambda}(0,\hat{\boldsymbol{\lambda}}) \end{bmatrix} = \begin{bmatrix} -2E(L_2(0,\hat{\boldsymbol{\lambda}})) & -E\left(\frac{\partial L_1(0,\hat{\boldsymbol{\lambda}})}{\partial \boldsymbol{\lambda}}\right) \\ -E(\frac{\partial L_1(0,\hat{\boldsymbol{\lambda}})}{\partial \boldsymbol{\lambda}}) & -E\left(\frac{\partial^2 L_0(0,\hat{\boldsymbol{\lambda}})}{\partial \boldsymbol{\lambda}^2}\right) \end{bmatrix}, \tag{6.4}$$

with the factor 2 appearing in $\mathrm{I}_{\alpha\alpha}(0, \hat{\boldsymbol{\lambda}})$ because of the way $L_2(\boldsymbol{\lambda})$ is defined.

The formula for T_S is just

$$T_S = \mathrm{I}^{\alpha\alpha}(0,\hat{\boldsymbol{\lambda}})[L_1(0,\hat{\boldsymbol{\lambda}})]^2,$$

and this can be made even simpler if the parameters $\boldsymbol{\lambda}$ can be chosen to be locally orthogonal to α in the neighbourhood of $\alpha = 0$, so that $E(\partial L_1(0,\hat{\boldsymbol{\lambda}})/\partial \boldsymbol{\lambda}) = \mathbf{0}$, as then $\mathrm{I}^{\alpha\alpha}(0,\hat{\boldsymbol{\lambda}}) = -\left[2E(L_2(0,\hat{\boldsymbol{\lambda}}))\right]^{-1}$. We shall illustrate the orthogonalization procedure in a number of our examples in Section 6.1.4.

Note that in this chapter, the observations may be written as x, y, or z to allow standardizations like $y = (x - \mu)/\sigma$ and other transformations to be used where convenient.

6.1 Boundary Models

Most boundary models are easily identified. When d, the dimension of $\boldsymbol{\Theta}$, is large, the boundary is typically made up of many regions with different dimensions, with the functional form of model varying between regions. The number of different model forms to be considered can increase exponentially with d. A systematic search of all boundaries is necessary if all the boundary models possible are to be identified.

A simple example is the Weibull model (2.1), already considered in Chapter 2. This was shown to have an embedded model, but otherwise it does not have non-degenerate boundary models. It does include the exponential distribution as an important special case, but this corresponds to interior points, specifically, the region where $c = 1$, of $\mathbf{\Theta}$.

We give an example of a PDF containing both boundary models and embedded models.

6.1.1 A Type IV Generalized Logistic Model

Prentice (1975) considers fitting of the four-parameter probability distribution, where $y = \mu + \sigma z$ and z has the PDF

$$f(z; p, q) = (p/q)^p e^{pz}(1 + pe^z/q)^{-(p+q)}/B(p, q), \tag{6.5}$$

where $\sigma > 0$, $p > 0$, $q > 0$, and $B(\cdot, \cdot)$ is the beta function. So (μ, σ, p, q) are the four parameters, where we have written p, q for Prentice's m_1, m_2. Including the Jacobian, the full four-parameter model has PDF

$$f(y; p, q, \mu, \sigma) = \sigma^{-1}(p/q)^p e^{p(y-\mu)/\sigma}(1 + pe^{(y-\mu)/\sigma}/q)^{-(p+q)}/B(p, q), \tag{6.6}$$

where $-\infty < y < \infty$. This model is one variation of a *Type IV generalized logistic* model which has been fully reviewed by Johnson, Kotz, and Balakrishnan (1995, Chap. 23, §10). As pointed out by Prentice (1975, 1976), this version includes a number of special cases. So $(p, q) = (1, 1)$ is the standard (type I) logistic, and letting $p \to \infty$ gives the special case

$$f(z, q) = \lim_{p\to\infty} f(z, p, q) = \frac{1}{\Gamma(q)} q^q e^{-qz-qe^{-z}},$$

which, for $q = 1$, is the extreme value distribution for maxima, while letting $q \to \infty$ gives the special case

$$f(z, p) = \lim_{q\to\infty} f(z, p, q) = \frac{1}{\Gamma(p)} p^p e^{pz-pe^z},$$

which, for $p = 1$, is the extreme value distribution for minima. In these last two cases, if it is thought possible they might provide the best fit to data, numerical instability is easily avoided by setting $p = r^{-1}$ and $q = s^{-1}$ so that $p \to \infty$ becomes $r \to 0$, and $q \to \infty$ becomes $s \to 0$. All these cases are unexceptional in that either p or q takes a special value, with the model then becoming a non-degenerate three-parameter model.

Further special cases occur if $(p \neq 0, q \to 0)$ or $(p \to 0, q \neq 0)$ or $(p = q \to 0)$, corresponding to the exponential, negative, and double exponential distributions. However, the limits are degenerate forms of the z density, requiring reparametrization of y with $\sigma = q/a, p/a, p/a, (a > 0)$, so that the PDFs of y are, respectively, $a\exp(-a(y - \mu))$, $y > \mu$; $a\exp(-a(\mu - y)), y < \mu$; $(1/2)a\exp(-a\,|y - \mu|)$.

The first two of these special cases, involving the exponential and negative exponential limits, exhibit a form of degeneracy where one of the parameters p or q simply vanishes once the other is set to zero. We call this phenomenon *parameter indeterminacy*. We do not discuss this here, as it is the subject of a separate chapter, Chapter 14.

The last case involving the double exponential model is an embedded model, because though we have four parameters initially, we allow three parameters p, q, and σ to tend to zero, but end up with two parameters still to be determined.

Prentice (1975) discusses one further special case of note, which is also an embedded model, obtained by letting $p, q \to \infty$, but not in any particularly related way. Prentice considers the reparametrization

$$a = (p^{-1} - q^{-1})/(p^{-1} + q^{-1})^{1/2}, b = 2/(p + q),\ \eta = \sigma(p^{-1} + q^{-1})^{1/2}, \tag{6.7}$$

so that the parameters are now (a, b, μ, η). Prentice gives the inversion formulas

$$p = 2/((b^2 + 2a) + b(b^2 + 2a)^{1/2}) \text{ and } q = 2/((b^2 + 2a) - b(b^2 + 2a)^{1/2})$$

for expressing the original parameters p and q in terms of the new parameters a and b; but these are incorrect. From (6.7) we have $b = 2/(p + q)$ so that $q = (2b^{-1} - p)$. Substituting this into the expression for a gives

$$a = (p^{-1} - (2b^{-1} - p)^{-1})/(p^{-1} + (2b^{-1} - p)^{-1})^{1/2}.$$

Solving this quadratic in p and noting the symmetry in the definitions yields the correct formulas for p and q as

$$\begin{aligned} p &= \tfrac{1}{2(2b^2+a^2b)}\left(4b + 2a^2 - 2\sqrt{2a^2b + a^4}\right), \\ q &= \tfrac{1}{2(2b^2+a^2b)}\left(4b + 2a^2 + 2\sqrt{2a^2b + a^4}\right). \end{aligned} \tag{6.8}$$

It is readily verified that these formulas substituted into (6.7) give a and b, whereas the formulas given by Prentice do not.

In terms of the reparametrized parameters, σ is simply

$$\sigma = (2b + a^2)^{-1/2}\eta,$$

with p and q as given in (6.8). We now expand the (one observation) log-likelihood of $f(y; p, q, \mu, \sigma)$, (6.6), as a double power series in the parameters a and b to get

$$L(f(y, p, q, \mu, \eta)) = \sum_{i=0}^{2}\sum_{j=0}^{2} L_{ij}a^i b^j + R,$$

where $L_{00} = \left(-\ln\eta - \frac{1}{2}\ln 2\pi - \frac{1}{2}\frac{(y-\mu)^2}{\eta^2}\right)$, $L_{10} = -\frac{1}{6}\frac{(y-\mu)^3}{\eta^3}$, $L_{01} = \frac{1}{24}\left(\frac{(y-\mu)^4}{\eta^4} - 3\right)$, $L_{02} = -\frac{1}{180}\frac{(y-\mu)^6}{\eta^6}$, $L_{11} = \frac{1}{30}\frac{(y-\mu)^5}{\eta^5}$, $L_{20} = -\frac{1}{12} - \frac{25}{24}\frac{(y-\mu)^4}{\eta^4}$. The leading term L_{00} is the logarithm of a normal log-likelihood. Thus, the normal distribution is the special case obtained when $p, q \to \infty$, so that $a, b \to 0$. But we also have $\sigma \to \infty$ if η is kept fixed, so that we can let three of the original four parameters tend to infinity, but still have two parameters μ and η to determine. The normal model is therefore an embedded model.

6.1.2 Burr XII and Generalized Logistic Distributions

Shao (2002) examines a number of examples of embedded models that involve the function given in equation (5.11), namely,

$$f(x, \boldsymbol{\theta}) = \left(\frac{x-a}{b}\right)^c,$$

which has the embedded exponential model, (5.15), $\exp[(x-\mu)/\sigma]$.

Shao (2002) applies this result to investigate how embedded models can arise in the Burr XII distribution introduced by Burr (1942), which, including an extra location and scale parameter a and b not in the original definition given by Burr, has CDF

$$F(x, a, b, c, d) = 1 - \left[1 + \left(\frac{x-a}{b}\right)^c\right]^{-d}, \quad x > a, \tag{6.9}$$

where $b, c, d > 0$, with a unrestricted. This has a very flexible shape range, making it useful in a wide variety of practical applications. Though not completely comprehensive compared with the Pearson family (to be discussed more fully in Chapter 9), which covers all possible combinations of (β_1, β_2) values, where $\beta_1 = \mu_3^2/\mu_2^3$ and $\beta_2 = \mu_4/\mu_2^2$ are the well-known squared-skewness and kurtosis moment ratios, μ_r being the rth moment about the mean, the Burr XII model nevertheless covers a wide range of values of (β_1, β_2) useful for practical data fitting, starting inside the Pearson Type (PT) I, including part of the PT VI, and ending inside the PT IV regions; see Burr and Cislak (1968) and Rodriguez (1977). In view of its usefulness, we examine how embedding can arise in the model.

Firstly, as $\exp((x-\mu)/\sigma)$ is an embedded model of $((x-a)/b)^c$, we have immediately that $F(x, a, b, c, d)$ has the embedded model whose CDF is

$$F(x, \mu, \sigma, d) = 1 - \left[1 + \exp\left(\frac{x-\mu}{\sigma}\right)\right]^{-d}.$$

This is the Type II logistic distribution, as given by Johnson, Kotz, and Balakrishnan (1995, Chapter 23, §10) and though having one fewer parameter than the Burr XII, is a very useful practical model in its own right.

The Burr XII distribution actually has three three-parameter embedded models. The log-likelihood expansion (6.2) of the one-observation log-likelihood

$$L_{BurrXII} = \ln(d) + \ln(c) + (c-1)\ln(x-a) - (d+1)\ln\left(1 + (x-a)^c b^{-c}\right) - c\ln(b),$$

highlights each embedded model as follows.

Using the substitution

$$a = -\sigma/\alpha, b = (\sigma/\alpha)e^{\mu(\alpha/\sigma)}, c = \alpha^{-1}, \tag{6.10}$$

the coefficients in the log-likelihood expansion, writing $y = (x-\mu)/\sigma$, are

$$\begin{aligned} L_0 &= \ln d - \ln\sigma + y - (1+d)\ln(1+\exp(y)), \\ L_1 &= \frac{x^2}{2\sigma^2}\frac{de^y - 1}{(e^y+1)} - \frac{x}{\sigma}, \\ L_2 &= \frac{1}{2}\frac{x^2}{\sigma^2} + \frac{1}{3}\frac{x^3}{\sigma^3}\left(\frac{1-de^y}{1+e^y}\right) - \frac{(d+1)}{8}\frac{x^4}{\sigma^4}\frac{e^y}{(1+e^y)^2}. \end{aligned}$$

Here L_0 is the log-likelihood of the Type II logistic, the case already identified.

Using the substitution $b = \sigma\alpha^{-1/c}, d = \alpha^{-1}$ and writing $y = (x-a)/\sigma$, we get

$$\begin{aligned} L_0 &= \left(-c\ln\sigma + \ln c + (c-1)\ln(x-a) - y^c\right), \\ L_1 &= \left(\frac{1}{2}y^{2c} - y^c\right), \quad L_2 = \left(-\frac{1}{3}y^{3c} + \frac{1}{2}y^{2c}\right). \end{aligned}$$

Here L_0 is the log-likelihood of the three-parameter Weibull with CDF

$$F(x, a, \sigma, c) = 1 - \exp\left[-\left(\frac{x-a}{\sigma}\right)^c\right], \; x > a, \; \sigma > 0, \; c > 0,$$

establishing this as an embedded model of the original parametrization.

The substitutions $c = \gamma\alpha^{-1}, d = \alpha$ yield the Pareto case. Here the log-likelihood takes the different form

$$L = L_0 - \lambda_1, \tag{6.11}$$

where

$$L_0 = \ln\gamma + \gamma\ln b - (1+\gamma)\ln(x-a), \tag{6.12}$$

$$\lambda_1 = (\alpha+1)\ln\left(1 + \left(\frac{b}{x-a}\right)^{\gamma\alpha^{-1}}\right). \tag{6.13}$$

The leading coefficient L_0 is the log-likelihood of the Pareto distribution with CDF

$$F(x, a, b, \gamma) = 1 - \left(\frac{b}{x-a}\right)^{\gamma}, \quad x > a + b, \; b > 0, \; \gamma > 0. \tag{6.14}$$

Under the constraint $x > a + b$ with $a, b, \gamma > 0$, $(\frac{b}{x-a})^{\gamma\alpha^{-1}} \to 0$ as $\alpha \to 0$; thus $\lambda_1 \to 0$ from above as $\alpha \to 0$ from above. This shows that the Pareto model always corresponds to a local maximum of the log-likelihood; however, this may not be the global optimum.

Table 6.1 collects together the transformations giving the three embedded models.

Note that reparametrizations allowing the log-likelihood of an embedded model to be displayed as the leading term L_0 of a series expansion are not unique. The only requirement is that the reparametrization allows the embedded model to be approached in a stable way. The Burr XII model is a good example. As well as the reparametrization given by eqn (6.10) already considered and listed in Table 6.1, the alternative reparametrization using

$$a = -\alpha^{-1},\ b = (\mu + \alpha^{-1}),\ c = (\mu + \alpha^{-1})/\sigma \tag{6.15}$$

will also yield the Type II generalized logistic case, only now with

$$L_0 = \ln d - \ln \sigma + y - (d+1)\ln(1 + e^y) \text{ and } L_1 = \frac{\sigma}{2}y^2\frac{(de^y - 1)}{(e^y + 1)} - \sigma y,$$

where $y = (x - \mu)/\sigma$. Though obtainable in explicit form, we have omitted L_2 as it is rather messy.

In this latter version, μ has the straightforward interpretation of being a translation applied to all the observations x_i. However, the reparametrization in Table 6.1 allows the definition of the Burr XII distribution to be extended to include α being negative. This will be discussed in Section 6.3.

Table 6.1 Burr XII distribution: first level embedded models

Burr XII	Reparam	Embedded CDF
CDF: $1 - \left[1 + \left(\frac{x-a}{b}\right)^c\right]^{-d}$	Pars: α, a, c, σ $b = \sigma\alpha^{-1/c}, \sigma > 0$ $d = \alpha^{-1}$	Weibull $1 - \exp\left[-\left(\frac{x-a}{\sigma}\right)^c\right]$ $x > a$
PDF: $\frac{cd}{b}\left(1 + \left(\frac{x-a}{b}\right)^c\right)^{-d-1} \times \left(\frac{x-a}{b}\right)^{c-1}$	Pars: α, d, μ, σ $a = -\sigma\alpha^{-1}, \sigma > 0$ $b = (\sigma/\alpha)e^{-\mu(\sigma/\alpha)}$ $c = \alpha^{-1}$	Type II Gen. Logistic $1 - \left[1 + \exp\left(\frac{x-\mu}{\sigma}\right)\right]^{-d}$ x unrestricted
Parameters: $x > a$, a unrestricted $b, c, d > 0$	Pars: α, a, b, γ $c = \gamma\alpha^{-1}, \gamma > 0$ $d = \alpha$	Pareto $1 - \left(\frac{b}{x-a}\right)^\gamma$ $x > a + b$

Table 6.2 Burr XII distribution: second-level embedded models

Distribution	PDF a unrestricted $b, c, d, \gamma > 0$	Reparam. pars, all cases: α, μ, σ	Embedded PDF
Weibull	$\frac{c}{b}\left(\frac{x-a}{b}\right)^{c-1} \times$ $e^{-\left(\frac{x-a}{b}\right)^c}$ $x > a$	$a = \mu - \sigma\alpha^{-1}$ $b = \sigma\alpha^{-1}$ $c = \alpha^{-1}$	Extreme value (EVMin) $\sigma^{-1}e^{(y-e^y)}, y = \frac{x-\mu}{\sigma}$ y unrestricted
Gen. Type II Logistic	$\frac{d}{b}e^{\frac{x-a}{b}} \times$ $\left(1 + e^{\frac{x-a}{b}}\right)^{-d-1}$ x unrestricted	$a = \mu - \sigma \ln \alpha$ $b = \sigma$ $d = \alpha^{-1}$	Extreme value $\sigma^{-1}e^{(y-e^y)}, y = \frac{x-\mu}{\sigma}$ y unrestricted
Gen. Type II Logistic	As in the cell entry immediately above	$a = \mu$ $b = \sigma\alpha$ $d = \alpha$	Two-par. exponential $\sigma^{-1}e^{-(y-\mu)/\sigma}$, $y > \mu$
Pareto	$\frac{\gamma}{x-a}\left(\frac{b}{x-a}\right)^\gamma$ $x - a > b > 0$	$a = -\sigma\alpha^{-1}$ $b = e^{\mu\alpha/\sigma}\sigma\alpha^{-1}$ $\gamma = \alpha^{-1}$	Two-par. exponential $\sigma^{-1}e^{-(y-\mu)/\sigma}$, $y > \mu$

In Section 7.2, we illustrate the implications of the results just obtained in a numerical example involving fitting the Burr XII model, where we discuss the comparison of embedded model fits with the full model fit.

The Burr XII model is also a good example for illustrating the possible hierarchical nature of embedded models. Table 6.2 shows that the Weibull and Pareto embedded models of the Burr XII model each has its own embedded two-parameter model, and that the Type II generalized logistic embedded model has two embedded models. As with the embedded models of Table 6.1, the log-likelihood can be expanded as a series in α, where the leading term L_0 is the log-likelihood of the embedded model.

The following schematic summarizes our discussion of the Burr XII distribution by indicating the hierarchical structure of its embedded models.

$$\begin{array}{ccccc} & & \text{four-parameter Burr XII} & & \\ & \swarrow & \downarrow & \searrow & \\ \text{three-par. Weibull} & & \text{Type II Gen. logistic} & & \text{three-par. Pareto} \\ \searrow & \swarrow & & \searrow & \swarrow \\ \text{extreme value} & & & \text{two-par. exponential} & \end{array} \tag{6.16}$$

6.1.3 Numerical Burr XII Example: Schedule Overruns

Table 6.3 gives the schedule overruns of 276 construction projects reported by Love et al. (2013), who discuss fitting the Burr XII (using a commercial statistical package), amongst other distributions to the data set, arguing that better inference would be achieved than using the less flexible normal distribution. This laudable and sensible approach is, however, hampered by three things.

Table 6.3 276 construction schedule overruns expressed as a percentage

0.0	40.0	0.0	25.7	25.7	29.4	3.7	0.0	13.3
44.4	8.0	13.3	18.5	18.8	58.5	-14.3	34.2	10.3
48.6	-12.3	0.0	0.0	45.5	6.0	4.8	0.0	9.3
12.8	16.7	4.4	-25.0	8.3	14.9	7.1	0.0	0.0
13.3	8.2	14.3	29.7	20.5	9.1	6.3	7.7	5.6
14.3	34.2	0.0	0.0	35.0	20.0	12.5	11.8	4.0
3.3	0.0	3.8	7.0	23.1	7.7	16.7	10.7	5.7
22.6	0.0	5.9	30.0	19.4	0.0	5.6	16.2	7.1
50.0	-19.8	15.0	37.5	4.3	18.5	6.7	25.7	8.0
0.0	3.7	20.0	23.9	7.0	13.5	23.5	15.0	0.0
0.0	20.7	3.8	7.1	11.1	4.3	0.0	6.3	7.4
50.0	15.4	-6.7	7.1	48.7	0.0	21.1	5.0	4.3
5.7	25.7	33.3	23.5	20.0	6.3	11.1	9.4	21.8
16.7	-25.0	3.5	13.0	55.6	29.7	5.5	5.6	11.1
4.5	-10.0	4.0	27.8	-40.5	7.1	3.6	7.1	11.1
4.7	6.3	6.5	7.0	40.0	0.0	0.0	13.2	20.0
17.2	-12.9	13.3	3.7	9.1	4.3	17.2	14.5	23.8
0.0	11.4	5.4	33.3	33.3	0.0	10.3	-10.0	0.0
17.4	8.8	17.5	7.1	20.0	3.3	4.4	17.2	13.3
-18.2	-30.0	33.3	11.5	20.0	0.0	21.7	0.0	6.9
33.3	38.7	22.2	7.1	2.4	0.0	5.6	10.0	7.0
-6.3	7.0	33.3	0.0	26.7	0.0	7.7	25.9	29.4

continued

Table 6.3 (Continued)

0.0	15.4	18.5	0.0	17.6	28.6	3.9	6.0	2.3
24.2	30.7	0.0	11.1	45.5	7.1	17.2	−1.3	2.5
17.6	39.2	0.0	−14.3	40.0	12.5	0.0	9.5	12.7
17.2	0.0	5.6	18.1	23.6	27.8	0.0	4.1	19.4
20.0	0.0	7.1	0.0	2.1	0.0	0.0	4.7	17.3
−5.4	−10.4	0.0	25.7	0.0	13.3	16.0	29.2	23.3
25.0	11.5	0.0	0.0	0.0	22.6	0.0	−12.3	–
−8.7	33.3	25.0	6.7	35.0	10.0	13.3	8.5	–
7.1	0.0	0.0	13.3	11.1	4.2	16.7	2.3	–

(i) An embedded model problem arises in this example. This results in three of the four fitted parameter values being extremely large, see Love et al. (2013).

(ii) Goodness-of-fit tests are carried out, including the Anderson-Darling test. However, the critical test values given by Love et al. (2013, Table 2) are those suitable only for what Stephens (1974) calls 'Case 0' problems when the parameters have known values, so do not have to be estimated. The critical values quoted are far too high for when parameters are estimated. Even so, the test value is itself so high that goodness-of-fit is only acceptable if high critical test values are used corresponding, say, to where $\alpha = 0.02$ or 0.01. Using more accurate critical values, which we can obtain by bootstrapping, it is clear that the lack-of-fit is actually very severe, so that even the increased flexibility of the Burr XII model does not provide an adequate fit for the data set.

(iii) As the Burr XII model does not provide a good representation of the data, the question arises as to what would be a good model. Examination of the sample indicates the data splits naturally into three groups, depending on whether an observation is positive, so that there is an actual overrun, or is zero, meaning that the project is completed exactly on time, or whether an observation is negative, indicating underrun, that is, early completion. Probably the neatest way of analysing the sample is to split it into these three groups, fitting separate continuous distributions to the overrun and underrun groups, and treating the count of projects completed on time as a binomial random variable. A quick approximation to this is to use a finite mixture model with PDF as in eqn (2.6), with the two groups of observations corresponding to projects completed early and late each modelled by one or more continuously distributed components, with a further continuous component, with small variance, to approximate the binomially distributed discrete component counting the zero-value observations of projects completed on time.

We discuss each of these points more fully in the following three subsections.

Burr XII Fit

Love et al. (2013) fit the Burr XII distribution given in Table 6.1 with PDF

$$f_{Burr}(x; a, b, c, d) = \frac{cd}{b}\left(1 + \left(\frac{x-a}{b}\right)^c\right)^{-d-1}\left(\frac{x-a}{b}\right)^{c-1},$$

giving the ML estimates as

$$\hat{a} = -5.6456 \times 10^8, \hat{b} = 5.6456 \times 10^8, \hat{c} = 2.1247 \times 10^8, \hat{d} = 0.19541. \quad (6.17)$$

The values of $\hat{a}$, $\hat{b}$, $\hat{c}$ have no obvious practical interpretation given that the schedule overruns are expressed as a percentage so that all observations must lie between −100 and 100.

If we use the reparametrized form of the Burr XII model, given in Table 6.1, that removes the embedding of the Type II generalized logistic, we find the indicator parameter α has estimated value $\hat{\alpha} = -\hat{a}^{-1} = 1.771 \times 10^{-9}$. This is very near zero, indicating that the Burr XII fit is actually extremely close to being a generalized Type II logistic model with CDF $1 - \left[1 + \exp\left((x - \hat{\mu})/\hat{\sigma}\right)\right]^{-\hat{d}}$ where, using Table 6.1, we have $\hat{\mu} = \hat{a} + \hat{b} = 0$ and $\hat{\sigma} = \hat{b}/\hat{c} = 2.657$, with an unchanged $\hat{d} = 0.19541$.

The two left-hand plots in Figure 6.1 show the fitted CDF and PDF of the Burr XII model using the $(\hat{a}, \hat{b}, \hat{c}, \hat{d})$ values of eqn (6.17). The generalized Type II logistic model using the corresponding $(\hat{\mu}, \hat{\sigma}, \hat{d})$ values gives a visually identical fit so that the figure represents either fit. The log-likelihood value was $L(\hat{a}, \hat{b}, \hat{c}, \hat{d}) = -1132.6$.

As a check, we carried out ML estimation using Nelder-Mead, and obtained a somewhat less extreme set of parameter estimates

$$\tilde{a} = -121.297, \ \tilde{b} = 127.534, \ \tilde{c} = 20.6438, \ \tilde{d} = 0.6745.$$

The fitted Burr XII model is still close to the fit of the embedded generalized Type II logistic model, as the indicator parameter α is still small with a value of $\tilde{\alpha} = -\tilde{a}^{-1} = 0.008\,24$. This fit is displayed in the two right-hand charts of Figure 6.1, showing that the fit takes more account of the schedules completed early, which make up the left tail. The log-likelihood corresponding to this set of estimates is $L(\tilde{a}, \tilde{b}, \tilde{c}, \tilde{d}) = -1112.5$, a somewhat higher value than the log-likelihood value for the fit given by Love et al. (2013).

Goodness-of-Fit

Love et al. (2013) give the value of the Anderson-Darling GoF statistic for the fitted Burr XII model as 3.0801, comparing this with critical values like 1.9286, 2.5018, 3.2892, and 3.9074, corresponding to significance levels $\alpha = 0.1$, 0.05, 0.02, and 0.01, respectively; with the conclusion that goodness-of-fit is not rejected if we reject only at a high test level like $\alpha = 0.02$ or 0.01. However, the critical values used are only appropriate when the parameter values of the supposed true distribution are known. We discussed the problem in Section 4.5, and recommended using bootstrapping to obtain appropriate critical values as given in the flowchart (4.13). This was done as follows in our example.

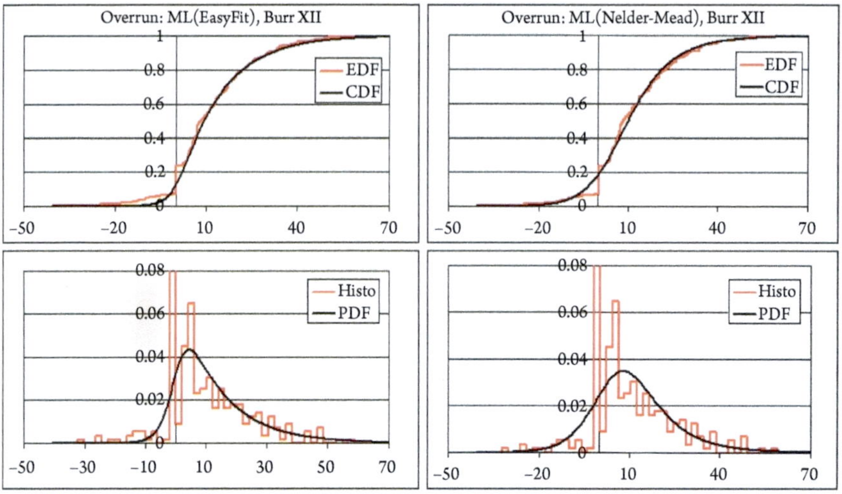

Figure 6.1 ML fits of the Burr XII distribution to the schedule overrun data. Left-hand charts: Burr XII model CDF and PDF fits obtained by Love et al. (2013) using a commercial package. The embedded generalized Type II logistic model is visually indistinguishable. Right-hand charts: same models fitted by Nelder-Mead ML optimization, again visually indistinguishable.

As we know, the fit is effectively a Type II generalized logistic with parameters $\hat{\mu} = 0$, $\hat{\sigma} = 2.657$, and $\hat{d} = 0.1954$; we shall work with this model instead. With these parameter values, the Anderson-Darling statistic is $A^2 = 2.537$. This value should if anything have been greater than the $A^2 = 3.0801$ obtained by Love et al. (2013) for the Burr XII model, as this latter should be a better fit, indicated by a smaller A^2 value. We carried out the bootstrap GoF test given in the flowchart (4.13) with $B = 500$. This gave critical values of 0.767, 0.984, 1.181, and 1.268, at the 0.1, 0.05, 0.02, and 0.01 significance levels, respectively, easily obtained from the EDF of the BS A^{*2} values. Admittedly, with only $B = 500$ replications, the critical values will have limited accuracy, but should be adequate at least to one decimal place. The test value of 2.537 is nevertheless well above all these values, indicating quite clearly that the fit is not good statistically and can be firmly rejected.

Finite Mixtures Fit

In just this section, we digress from giving examples of embedded models, which is the main thread of this chapter, in order to bring the analysis of the schedule overrun data to a more satisfactory conclusion. We briefly outline how a finite mixture model might be used to represent the different components in the schedule overrun data. Fitting finite mixtures involves a non-standard problem which we call *indeterminacy*, to be discussed in Chapter 14, where we show it is related to embeddedness. We also discuss finite mixtures in detail in Chapter 17. Here we simply illustrate its application in modelling the overrun data.

We fitted, by maximum likelihood, a finite mixture distribution of the form given in eqn (2.6), with individual components that are *EVMax* extreme value distributions. This distribution is listed in Table 18.4, and will be discussed later in this chapter. We fitted $k = 1, 2, \ldots, 6$ components. To compare fits, we used the Bayesian information criterion (BIC), given later in eqn (17.4), where it will be more fully discussed, taking it in the form where a larger value indicates a better fit. The calculated BIC_k values were −1149, −1121, −862, −850, **−848**, −855 for $k = 1, 2, \ldots, 6$, respectively, with a maximum at $k = 5$, highlighted in bold. The values are quite similar from $k = 3$ onwards, and in fact all fits from $k = 3$ onwards included components clearly representing all three groups of underrun, on-time, and overrun observations. Figure 6.2 shows the fits obtained for the

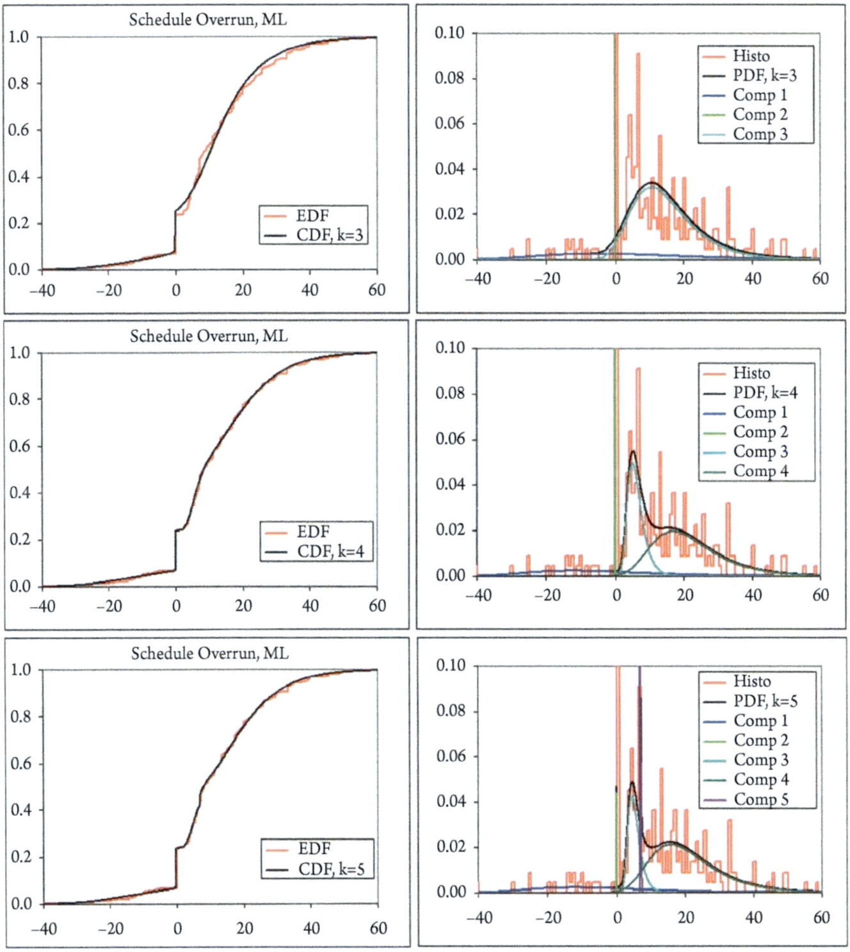

Figure 6.2 ML fits of the EVMax extreme value finite mixture distribution to the schedule overrun data, for $k = 3, 4$, and 5 components.

cases $k = 3$, 4, and 5. The CDF of the $k = 4$ model matches the EDF very closely with the four components, which are depicted in the figure, being meaningfully placed and interpretable. The $k = 5$ fit is also a good fit, but there is evidence of overfitting with one component, the fifth, that seems to have been fitted to a random cluster rather than a true component, so that it is probably superfluous. This is a typical problem when using ML to fit an *overfitted* finite mixture model—a model which has k components when there are only $k_0 < k$ actual components. Overfitting will be discussed in detail in Chapter 17.

6.1.4 Shifted Threshold Distributions

Table 6.4 is similar to that given in Cheng and Iles (1990, Table 1), with the addition of the log-gamma model and with the Weibull model entry omitted, as this has already been given in Table 6.2. All the examples in the table are well-known two-parameter distributions, but with an added threshold (shifted origin) parameter, a.

The other two parameters b, c are those of the conventionally parametrized form of the distribution. The table is slightly different from that given in Cheng and Iles (1990) in that the transform column gives the original parameters (a, b, c) in terms of the parameters (α, μ, σ) used in the reparametrization, rather than the other way round as in the article.

In all cases in Table 6.4, the parameters α and μ are unrestricted in sign, whilst $\sigma > 0$. In the table, α is the leading parameter as in (5.7), so that the embedded model is obtained when $\alpha = 0$, this model not being directly obtainable in the original parametrization. The signs of the original parameters must satisfy the restrictions imposed by the transforms, so that, for example, in the gamma we must have $c > 0$, but b is unrestricted, and in the loglogistic and lognormal cases, b and c must have the same sign.

It should be stressed that in the table, all the PDFs are defined to be *positively skewed* when $\alpha > 0$ in the reparametrized form, with the model that is embedded in the original form obtained as $\alpha \to 0$.

Moreover, the PDFs are written in a generalized form in the original parametrization, so that $\alpha < 0$ gives a valid distribution.

In all cases, the embedded model is symmetric and is approached as $\alpha \to 0$ from above or below, moreover, with the model corresponding to $\alpha = \alpha_0 > 0$ being the mirror reflection about $x = \mu$ of the model corresponding to $\alpha = -\alpha_0 < 0$. Thus, for example, in the gamma distribution, the usual form is where the condition $X > a$ applies, with the other parameters satisfying $b > 0, c > 0$, so that the threshold is a lower bound with the support for X unbounded to the right. However, our definition is valid if $X < a$, $b > 0$, and $c > 0$, so that the threshold is an upper bound for X with its support unbounded to the left. For the same value for c, the PDF has the same shape in either case, except that one is the reflection of the other. Using the reparametrization of column 3 allows both versions to be included with the sign of α determining which is selected. The embedded symmetric model is the $N(\mu, \sigma^2)$ in all the cases listed, except for the loglogistic, where the limit is the logistic distribution.

It will be seen that the reparametrizations are all very similar, with $l = 3$ and $k = 2$, where l and k are as defined in Section 5.2. In all the cases, the log-likelihood can be

Table 6.4 Embedded models in shifted threshold distributions

Distribution	PDF *	Reparam. all cases: $-\infty < \alpha < \infty$ $\mu > 0, \sigma > 0$	Embedded PDF
Gamma	$\frac{1}{\lvert b\rvert\,\Gamma(c)}\left(\frac{x-a}{b}\right)^{c-1} \times e^{-(x-a)/b}$	$a = \mu - \sigma\alpha^{-1}$ $b = \sigma\alpha$ $c = \alpha^{-2}$	$N(\mu, \sigma^2)$
Inverted Gamma	$\frac{1}{\lvert b\rvert\,\Gamma(c)}\left(\frac{x-a}{b}\right)^{-(c+1)} \times \exp\left(-\frac{b}{x-a}\right)$	$a = \mu - \sigma\alpha^{-1}$ $b = \sigma\alpha^{-3}$ $c = \alpha^{-2}$	$N(\mu, \sigma^2)$
Inverse Gaussian	$\left(\frac{bc}{2\pi(x-a)^3}\right)^{1/2} \times e^{-\frac{c(x-a-b)^2}{2b(x-a)}}$	$a = \mu - \sigma\alpha^{-1}$ $b = \sigma\alpha^{-1}$ $c = \alpha^{-2}$	$N(\mu, \sigma^2)$
Loglogistic	$\frac{c}{b}\left(\frac{x-a}{b}\right)^{c-1} \times \left(1 + \left(\frac{x-a}{b}\right)^{c}\right)^{-2}$	$a = \mu - \sigma\alpha^{-1}$ $b = \sigma\alpha^{-1}$ $c = \alpha^{-1}$	Logistic $\sigma^{-1}\frac{e^y}{(1+e^y)^2}$, $y = \frac{x-\mu}{\sigma}$
Lognormal	$\frac{c}{(2\pi)^{1/2}(x-a)} \times e^{-\frac{c^2\left(\ln\left(\frac{x-a}{b}\right)\right)^2}{2}}$	$a = \mu - \sigma\alpha^{-1}$ $b = \sigma\alpha^{-1}$ $c = \alpha^{-1}$	$N(\mu, \sigma^2)$
Gen. Log-gamma	$\frac{1}{\lvert b\rvert\,\Gamma(c)} e^{-c\frac{(x-a)}{b}} \times e^{-e^{-\frac{(x-a)}{b}}}$	$a = \mu + \sigma\alpha^{-1}\psi(\alpha^{-2})$ $b = \sigma\alpha^{-1}$ $c = \alpha^{-2}$ $\psi(\cdot) \equiv$ digamma function	$N(\mu, \sigma^2)$

* In all PDFs, a, b are unrestricted, and $c > 0$. The PDFs in col. 2 are in standard form if $b > 0$. For the log-gamma PDF, $-\infty < x < \infty$. All other PDFs are as given, provided $x > (<)\ a$ according as $b > (<) a$, or in reparametrized form as $\alpha > (<)\ 0$, with $-\infty < x < \infty$ if $b = 0$ or if $\alpha = 0$. The PDFs are zero otherwise.

expanded in a power series in α as in (6.2). We give details of just a few, including the information matrix, and illustrate how the α parameter can be made locally orthogonal to the other parameters. As mentioned in the introduction to this chapter, this simplifies testing of whether the embedded model is an acceptable fit.

It should be noted that though Table 1 in Cheng and Iles (1990) lists a number of examples of shifted threshold models and their embedded models, it is in terms of a series expansion of the form

$$L = L_0 + L_1\alpha^{-1} + O_p(\alpha^{-2}),$$

so that the embedded model is obtained by letting $\alpha \to \infty$. Their α is essentially the threshold value in their models, whilst in our parametrizations here the threshold typically has the form $a \approx -s\alpha^{-1}$, where α is the leading parameter that is small and allowed to tend to zero to reveal the embedded model, with s a positive quantity depending on the other parameters. The L_1 expressions we give here in our examples to follow are therefore for reparametrizations that are not exactly same as those used by Cheng and Iles. However, they will nevertheless be essentially just a positive multiple of the L_1 listed in Table 1 of Cheng and Iles (1990). They therefore have the same use and interpretation in calculations.

We give details of some of the distributions from Table 6.4.

Gamma Model

The one-observation log-likelihood in the gamma model is

$$L = -c\ln b - \ln[\Gamma(c)] + (c-1)\ln(x-a) - \frac{x-a}{b}.$$

For c large, we have from Stirling's formula, see Abramowitz and Stegun (1965, equation 6.1.41),

$$\ln[\Gamma(c)] = \left(c - \frac{1}{2}\right)\ln(c) - c + \frac{\ln(2\pi)}{2} + \frac{1}{12c} - \frac{1}{360c^3} + O(c^{-5}) \text{ as } c \to \infty.$$

This suggests that having $c \to \infty$ as $\alpha \to 0$ would be appropriate. In fact, if we take $\boldsymbol{\theta} = (\theta_1, \theta_2, \theta_3) = (c^{-1}, b, a^{-1})$, then $\theta \to 0$ as $\alpha \to 0$, so that the condition (5.8) is satisfied. Using the series expansion approach, we reparametrize to α, μ, σ with

$$c = \alpha^{-2},\ b = \sigma\alpha,\ a = \mu - \sigma\alpha^{-1}. \tag{6.18}$$

Using Stirling's formula for $\ln[\Gamma(c)]$ and replacing (a, b, c) by (6.18), we can expand L in a Maclaurin series as in (6.2) with coefficients

$$L_0 = -\frac{1}{2}\ln 2\pi - \ln\sigma - \frac{1}{2}\frac{(x-\mu)^2}{\sigma^2},$$

$$L_1 = \left(-\frac{x-\mu}{\sigma} + \frac{1}{3}\frac{(x-\mu)^3}{\sigma^3} \right), \tag{6.19}$$

$$L_2 = \left(-\frac{1}{12} + \frac{1}{2}\frac{(x-\mu)^2}{\sigma^2} - \frac{1}{4}\frac{(x-\mu)^4}{\sigma^4} \right),$$

where $\lambda = (\mu, \sigma)$, and for simplicity we have written $L_i(\boldsymbol{\lambda})$ as L_i.

L_0 is the logarithm of the embedded PDF model, the $N(\mu, \sigma^2)$ distribution in this case.

The one-observation information matrix in the neighbourhood of $\alpha = 0$ can be evaluated as the expected values of the derivatives of $L(\alpha, \boldsymbol{\lambda})$ as given in eqn (6.4). We have in this gamma case

$$\mathrm{i}(0, \mu, \sigma) = \begin{bmatrix} \frac{2}{3} & 0 & 0 \\ 0 & \frac{1}{\sigma^2} & 0 \\ 0 & 0 & \frac{2}{\sigma^2} \end{bmatrix}, \tag{6.20}$$

so that the parameters are already locally orthogonal in the neighbourhood of $\alpha = 0$.

Generalized Log-gamma Model

Prentice (1974) considers the distribution of $Y = \ln X$, where X has the generalized gamma distribution introduced by Stacy (1962). Y then has a *generalized log-gamma* distribution with PDF

$$f(y; a, b, k) = \frac{1}{\Gamma(c)b} \exp\left(c\left(\frac{y-a}{b}\right)\right) \exp(-e^{(\frac{y-a}{b})})dy,$$

where we have followed the notation of Prentice, except that we have replaced his $\ln a$, b, and k by a, b^{-1}, and c, respectively. The log-likelihood is

$$L = \ln b - \ln(\Gamma(c)) + cb(y-a) - e^{b(y-a)}.$$

Prentice points out that the normal model is a limiting special case as $c \to \infty$. This is an embedded model in terms of our definition, so that a reparametrization is needed for the normal model to be a non-degenerate limit. To do this, Prentice uses the following reparametrization α, μ, σ, where α and μ are unrestricted in sign but $\sigma > 0$. The reparametrization can be inverted, allowing the original parameters to be written as

$$a = \mu - \sigma\alpha^{-1}\psi(\alpha^{-2}),\ b = \sigma\alpha^{-1}, c = \alpha^{-2}, \tag{6.21}$$

where $\psi(\cdot)$ is the digamma function, and we have used μ and α for Prentice's α and q. Prentice (1974, eqn 3) gives the PDF in terms of the reparametrized parameters, which we do not reproduce here.

From the viewpoint of embedding, we can regard α as the auxiliary variable, as in (5.9), that is allowed to tend to zero. Letting $\alpha \to 0$, we have, from (6.21), that $c \to \infty$,

$b \to -\infty$, and $a \to -\infty$ as $\alpha \to 0$, so that in the notation of our discussion of embedded models, the number of parameters in the original model tending to special values is $l = 3$. The log-likelihood L can be expanded as a series in α as in (6.2) using the expressions (6.21). After some simplification, involving use of

$$\psi(c) = \ln c - \frac{1}{2}c^{-1} - \frac{1}{12}c^{-2} + \frac{1}{120}c^{-4} + O(c^{-6})$$

and

$$\Gamma(c) = \left(c - \frac{1}{2}\right)\ln c - c + \frac{1}{2}\ln(2\pi) + \frac{1}{12c} - \frac{1}{360c^3} + O(c^{-5})$$

as $c \to \infty$, we find that the log-likelihood expansion $L = L_0 + L_1\alpha + L_2\alpha^2 + O(\alpha^3)$ of (6.2) has coefficients

$$\begin{aligned}
L_0 &= \ln\frac{1}{\sigma} - \frac{1}{2}\ln 2\pi - \frac{1}{2}\frac{(y-\mu)^2}{\sigma^2} \\
L_1 &= \frac{1}{2}\frac{y-\mu}{\sigma} - \frac{1}{6}\frac{(y-\mu)^3}{\sigma^3} \\
L_2 &= -\frac{1}{24}\frac{(y-\mu)^4}{\sigma^4} + \frac{1}{4}\frac{(y-\mu)^2}{\sigma^2} - \frac{5}{24}.
\end{aligned}$$

Letting $\alpha \to 0$, $L \to L_0$, the log-likelihood of the $N(\mu, \sigma^2)$ distribution, which is therefore an embedded model. Note that, though α is unrestricted, taking negative values does not produce anything additional to the range of possible shapes taken by the PDF as the model satisfies a reflection principle, where if Y has PDF $f(y, \alpha, \mu, \sigma)$ then $Z = -Y + 2\mu$ has the same PDF, only reflected about the $y = \mu$ axis, this latter PDF being obtained by reversing the sign of α in $f(y, \alpha, \mu, \sigma)$.

For this model, the information matrix in the neighbourhood of $\alpha = 0$ is

$$\mathrm{i}(0, \mu, \sigma) = \begin{bmatrix} \frac{1}{6} & 0 & 0 \\ 0 & \frac{1}{\sigma^2} & 0 \\ 0 & 0 & \frac{2}{\sigma^2} \end{bmatrix},$$

showing the parameters to be locally orthogonal in the neighbourhood of $\alpha = 0$.

Inverted Gamma Model

This model, listed in Table 6.4, is also known as the Pearson Type V distribution, to be discussed in Chapter 9. The PDF is

$$f_V(y) = \frac{1}{b\Gamma(c)}\left(\frac{b}{y-a}\right)^{c+1}\exp\left(-\frac{b}{y-a}\right), \quad x > a.$$

The transformation $a = \mu - \sigma\alpha^{-1}$, $b = \sigma\alpha^{-3}$, $c = \alpha^{-2}$ given in Table 6.4 is probably the simplest for showing that the normal model is an embedded model in the original parametrization. But as already remarked, and shown with the Burr XII, different reparametrizations can be used. Here we show that an alternative is possible by replacing a, b, c with α, μ, σ using the reparametrization

$$a = \mu - \sigma^{2/3}\alpha^{-1}, b = \alpha^{-3}, \text{and } c = \sigma^{-2/3}\alpha^{-2}.$$

We then get in eqn (6.2), the Maclaurin series of the one-observation log likelihood with coefficients

$$L_0 = \left(-\frac{1}{2\sigma^2}(x-\mu)^2 - \ln\sigma - \frac{1}{2}\ln 2\pi\right),$$

$$L_1 = \frac{1}{3}\sigma^{-\frac{8}{3}}(x-\mu)\left[2(x-\mu)^2 - 3\sigma^2\right],$$

$$L_2 = -\frac{13}{12}\sigma^{\frac{2}{3}} - \frac{3}{4}\sigma^{-\frac{10}{3}}(x-\mu)^4.$$

The local information matrix in the neighbourhood of $\alpha = 0$ calculated from eqn (6.4) is

$$\mathrm{i}(0,\mu,\sigma) = \begin{bmatrix} \frac{20}{3}\sigma^{\frac{2}{3}} & \frac{2}{\sigma^{\frac{2}{3}}} & 0 \\ \frac{2}{\sigma^{\frac{2}{3}}} & \frac{1}{\sigma^2} & 0 \\ 0 & 0 & \frac{2}{\sigma^2} \end{bmatrix}.$$

To orthogonalize, we need to reparametrize μ so that the $\mathrm{i}^0_{\alpha\mu}$ entry becomes zero. This is done by solving the differential equation (3.17), which here takes the simple form

$$\mathrm{i}^0_{\alpha\mu} + \mathrm{i}^0_{\mu\mu}\frac{\partial\mu}{\partial\alpha} = 0, \text{ i.e. } \frac{1}{\sigma^2}\frac{\partial\mu}{\partial\alpha} = -\frac{2}{\sigma^{2/3}},$$

yielding

$$\mu = \rho - 2\sigma^{4/3}\alpha,$$

where ρ, the arbitary constant of integration, is the new parameter to replace μ. In terms of the new parameters (α, ρ, σ), we have

$$a = \rho - 2\sigma^{4/3}\alpha - \sigma^{2/3}\alpha^{-1}, \; c = \sigma^{-2/3}\alpha^{-2}, \; b = \alpha^{-3},$$

and this yields, after some algebra,

$$\mathrm{i}(0,\rho,\sigma) = \begin{bmatrix} \frac{8}{3}\sigma^{\frac{2}{3}} & 0 & 0 \\ 0 & \frac{1}{\sigma^2} & 0 \\ 0 & 0 & \frac{2}{\sigma^2} \end{bmatrix}.$$

Loglogistic Model

This has PDF as given in Table 6.4

$$f = \frac{c}{b}\left(\frac{y-a}{b}\right)^{c-1}\left(1+\left(\frac{y-a}{b}\right)^{c}\right)^{-2}.$$

If we reparametrize to α, μ, σ using

$$a = \mu - \sigma\alpha^{-1}, b = \sigma\alpha^{-1}, c = \alpha^{-1},$$

we find that the Maclaurin series expansion in α has the terms

$$L_0 = (z - \ln\sigma - 2\ln(1+e^z)),\ L_1 = \left(-\frac{1}{2}z^2 - z + e^z\frac{z^2}{(1+e^z)}\right),$$

$$L_2 = 2\left(\frac{1}{3}z^3 + \frac{1}{2}z^2 - \frac{1}{12}\frac{e^z z^3}{1+e^z}(8+3z) + \frac{1}{4}e^{2z}\frac{z^4}{(1+e^z)^2}\right),$$

where $z = (y-\mu)/\sigma$, showing the logistic to be embedded in the original parametrization. The one-observation information matrix is

$$\mathrm{i}(0,\mu,\sigma) = \begin{bmatrix} \frac{7}{180}\pi^4 & \frac{g}{\sigma} & 0 \\ \frac{g}{\sigma} & \frac{h}{\sigma^2} & 0 \\ 0 & 0 & \frac{1}{9}\frac{\pi^2+3}{\sigma^2} \end{bmatrix},$$

where $g = \frac{1}{18}\pi^2 - \frac{1}{3}$, $h = \frac{1}{3}$. To orthogonalize, we solve the orthogonalizing equation (3.17), which reduces in this case to

$$\frac{g}{\sigma} + \frac{h}{\sigma^2}\frac{d\mu}{d\alpha} = 0, \text{ giving } \lambda = \mu + \frac{g\sigma}{h}\alpha,$$

where the constant of integration λ replaces μ, so that $a = \lambda + (\frac{\pi^2}{6} - 1)\alpha\sigma - \alpha^{-1}\sigma$. Using $(\alpha, \lambda, \sigma)$ as the parameters, we have, after some algebra, that the one-observation information matrix is locally orthogonal with the form

$$\mathrm{i}(0,\lambda,\sigma) = \begin{bmatrix} \frac{2}{135}\pi^4 + \frac{1}{18}\pi^2 - \frac{1}{6} & 0 & 0 \\ 0 & \frac{h}{\sigma^2} & 0 \\ 0 & 0 & \frac{1}{9}\frac{\pi^2+3}{\sigma^2} \end{bmatrix}.$$

6.1.5 Extreme Value Distributions

Generalized Extreme Value Distributions

The Weibull entry in Table 6.2 deserves some special mention. In its transformed form, it is just the well-known *generalized extreme value* (GEV) distribution, but as applied to minima of observations, which we denote by *GEVMin*. By convention, the GEV is usually discussed in the form when it is applied to maxima, see Prescott and Walden (1980), Hosking (1984), where it is just called the GEV distribution. For clarity, we will denote this maximum version by *GEVMax*. A tiresome aspect of all EV distributions is the ambiguity in their definition, which we will avoid by appending either *Min* or *Max* to *EV*.

In view of their importance, we summarize some properties of EV distributions, if only to clarify the varied and somewhat confusing terminology associated with them, and to justify the nomenclature GEVMin for the distribution. We will illustrate its use by fitting the GEVMin model to data in a numerical example given in Section 7.1.

The Weibull entry in Table 6.2 shows that the distribution has what is usually called the extreme value (EVMin in this case) distribution as a special case. (EVMax has the same PDF as given for EVMin in Table 6.2, except that y is replaced by $-y$.) From its inception, the Weibull distribution has been recognized as a very useful form of extreme value distribution in its own right. Johnson, Kotz, and Balakrishnan (1995, Chapters 21 and 22, two of the longest chapters) give detailed accounts of the Weibull and of the EV distribution. A more recent reference for the Weibull distribution is Rinne (2008), which gives particularly useful practical advice on fitting the distribution.

The *EVMin* and *EVMax* distributions are also known as the *Gumbel* distribution. There are two other types of EV distributions: the *Fréchet* distribution and the *Weibull* distribution itself. All three types come in max and min form. Their maximum forms were denoted by Fisher and Tippett (1928) as Types I, II, and III, respectively, and this nomenclature is commonly adhered to, but it seems reasonable to extend this to the minimum case as well. The Fréchet and Weibull distributions are simply related, each being the reciprocal of the other. Thus, if X is Weibull, then $Y = 1/X$ is Fréchet. On its own, reciprocation turns a minimum into a maximum, so if X is Weibull appropriate for minima, then $Y = 1/X$ is Fréchet appropriate for maxima.

For $c > 0$, the Weibull entry in Table 6.2 is just the Weibull PDF defined in its most common standard form, which is the *Type III EVMin* version. In this case, the EVMin distribution is embedded, and obtained as $c \to \infty$, this being equivalent to letting $\alpha \to 0$. The entry is still valid for $c < 0$, that is, for $\alpha < 0$. The distribution then becomes the reciprocal Fréchet, but the transformation used in the table also *negates* X by changing the sign of the scale parameter b. This again switches minima and maxima, so that the Fréchet model is the one appropriate for minima also. In its transformed form, the Weibull thus covers all three EVMin distributions, so that it is indeed *GEVMin*. Numerically, it is much easier to use the reparametrized form with parameters α, μ, σ. The CDF is then

$$F(x) = \begin{cases} 1 - \exp(-(1 + \alpha(x - \mu)/\sigma)^{1/\alpha}) \text{ if } \alpha \neq 0 \\ 1 - \exp(-\exp((x - \mu)/\sigma)) \text{ if } \alpha = 0 \end{cases}, \tag{6.22}$$

and the PDF is

$$f(x) = \sigma^{-1} (1 + \alpha(x - \mu)/\alpha)^{\frac{1}{\alpha}-1} \exp\left(-(1 + \alpha(x - \mu)/\alpha)^{\frac{1}{\alpha}}\right), \tag{6.23}$$

encompassing all three Type I, II, and III EVMin distributions according as $\alpha = 0, \alpha < 0$, or $\alpha > 0$, all with $\sigma > 0$. Note that $x > \mu - \sigma\alpha^{-1}$ if $\alpha > 0$ and $x < \mu - \sigma\alpha^{-1}$ if $\alpha < 0$. A numerical example is given in Section 7.1.

The GEVMax version, the one discussed by Prescott and Walden (1980) and by Hosking (1984), has CDF

$$F(y) = \begin{cases} \exp(-(1 - \alpha(y - \mu)/\sigma)^{1/\alpha}) \text{ if } \alpha \neq 0 \\ \exp(-\exp(-(y - \mu)/\sigma)) \text{ if } \alpha = 0 \end{cases}, \tag{6.24}$$

covering all three Type I, II, and III EVMax distributions, according as $\alpha = 0, \alpha < 0$, or $\alpha > 0$, with $y < \mu + \sigma\alpha^{-1}$ if $\alpha > 0$ and $y > \mu + \sigma\alpha^{-1}$ if $\alpha < 0$.

GEVMax and GEVMin are related by $X - \mu = \mu - Y$, showing that they are reflections of each other in the $X = \mu$ axis.

Prescott and Walden (1980) give the information matrix for GEVMax explicitly, in terms involving the gamma and digamma functions.

Weibull-Related Distributions

Johnson, Kotz, and Balakrishnan (1995), in the introduction to their chapter on the extreme value distribution, comment on the extensive work carried out on extreme value distributions, a 'testimony of the vitality and applicability' of such distributions, but also warn that it also reflects on 'the lack of coordination between researches and the inevitable duplication (or even triplication) of results appearing in a wide range of publications'. We concur with this observation, offering, as a simple example, the wide variety of different names used for the Fréchet distribution, including inverse Weibull, reciprocal Weibull, complementary Weibull, even reverse Weibull (see Rinne, 2008), with the discussion often giving the impression that it is a recently defined new distribution.

A more specific concern is raised by Nadarajah and Kotz (2005) over distributions whose CDF have the form

$$F(x) = 1 - \exp[-aG(x)], \tag{6.25}$$

where $G(x)$ is a monotonically increasing function of x, with $G(x) \geq 0$. Nadarajah and Kotz (2005) point out that this functional form was proposed by Gurvich et al. (1997) for the study of the strength of brittle materials, but that several subsequent papers—Chen (2000), Xie et al. (2002) and Lai et al. (2003)—do not cite the Gurvich et al. (1997) paper, the last being perhaps the most notable omission, as Lai et al. (2003) studied exactly the same model

$$F(x) = 1 - \exp[-ax^b \exp(\lambda x)] \tag{6.26}$$

already examined by Gurvich et al. (1997) as a special case of (6.25). The model discussed by Chen (2000) takes the form

$$F(x) = 1 - \exp\{-\lambda[\exp(x^b) - 1]\}. \tag{6.27}$$

If an extra scaling parameter is included, we have

$$F(x) = 1 - \exp\left\{-\lambda a\left[\exp\left(\left(\frac{x}{a}\right)^b\right) - 1\right]\right\}, \tag{6.28}$$

and this is the version proposed by Xie et al. (2002), who point out that this model includes the Weibull as a special case. This latter is actually an embedded model obtained by setting $\lambda = a^{1-b}\sigma^{-b}$ and letting $a \to 0$, yielding the Weibull CDF

$$F(x) = 1 - \exp\left[-\left(\frac{x}{\sigma}\right)^b\right].$$

Nadarajah and Kotz (2005) point out that the model with CDF

$$F(x) = 1 - \exp\{-ax^b[\exp(cx^d) - 1]\} \tag{6.29}$$

includes the models (6.26), (6.27), and (6.28) as special cases. These are all non-embedded. The extra flexibility in (6.29) given by the extra parameter does add a complication, however. If we reparametrize by setting $a = \lambda/c$, then if $c \to 0$, we get

$$F(x) = 1 - \exp\{-\lambda c^{-1}x^b[\exp(cx^d) - 1]\} \to 1 - \exp(-\lambda x^{b+d}),$$

which is the Weibull CDF but with b and d not separately determinate, but only as $b + d$.

This issue of determinacy becomes quite common as the number of parameters increases. Another example is the following case.

Kumaraswamy (1980) introduced the distribution for double-bounded random processes with hydrological applications with PDF

$$f(x) = abx^{a-1}(1 - x^a)^{b-1}$$

and CDF

$$F(x) = 1 - (1 - x^a)^b,$$

which has similar properties to the beta distribution, but has some advantages in tractability. The model is discussed fully by Jones (2009). However, Nadarajah (2008) points out it is capable of being usefully extended, giving various examples. Such an instance is given by Cordeiro and de Castro (2011), who extend its definition to the form

$$F(x) = 1 - [1 - G^a(x)]^b,$$

where $G(x)$ is a CDF so that X can be unbounded above. When $G(x)$ is the Weibull CDF, we obtain the distribution whose CDF is

$$F(x) = 1 - \{1 - [1 - \exp(-\beta^c x^c)]^a\}^b,$$

which we prefer to call the *Ks-Weibull* distribution, as this seems more etymologically appropriate than the name *Kw-Weibull* used by Cordeiro and de Castro (2011). If we set $b = \lambda\beta^{-ac}$, then as $\beta \to 0, b \to \infty$, and we get the (embedded) model

$$F(x) = 1 - \exp(-\lambda x^{ac}),$$

which is the Weibull, but not in a fully determinate form, as a and c are only determinate as ac.

We end this section with a note of caution.

As will be evident, the Weibull model is capable of generalization in many ways. Authors are usually careful, when proposing new generalizations, to give reasons why a particular generalization might a worthwhile addition to the corpus of different models available for practical use. However, only experience will show which models are a worthwhile addition.

As a warning, we give two examples in which published models are clearly *not* usable because they contain indeterminacies which are demonstrably not removable.

A model is proposed by de Gusmão, Ortega, and Cordeiro (2011), which they name the *generalized inverse Weibull distribution* (GIW), with PDF (equation (3) in their paper)

$$f(x) = \gamma\alpha^{\beta}\beta x^{-(\beta+1)} \exp\left(-\gamma\alpha^{\beta}x^{\beta}\right)$$

that is an intended generalization of the *inverse Weibull* (i.e. Fréchet) distribution, where γ is a third parameter not present in the two-parameter IW distribution. However, α and γ are clearly not ever simultaneously estimable, as they only ever appear as $\lambda = \gamma\alpha^{\beta}$ in the PDF. If they were simultaneously estimable, say, by ML estimation, then we would have $\hat{\lambda} = \hat{\gamma}\hat{\alpha}^{\beta}$, but clearly setting $\tilde{\gamma} = 1$ and $\tilde{\alpha} = \hat{\lambda}^{1/\beta}$ would be exactly equivalent.

The GIW is thus no more general than the IW distribution, with the disadvantage that the likelihood equations are not soluble, as two of the likelihood equations are identical up to a multiplicative factor independent of the sample values, namely, $U_{\alpha} = \alpha^{-1}\beta\gamma U_{\gamma}$ in the notation of the paper.

The numerical results of the paper would need revisiting in view of this non-estimability.

The generalized inverse generalized Weibull (GIGW) model proposed by Jain et al. (2014), with PDF (equation (7) in the paper)

$$f(x) = \alpha\beta\gamma\lambda^{\beta}x^{-(\beta+1)}e^{-\gamma\lambda^{\beta}x^{-\beta}}(1 - e^{-\gamma\lambda^{\beta}x^{-\beta}})^{\alpha-1}$$

suffers from the same problem, with γ and λ not simultaneously estimable, as they always appear together as $\mu = \gamma\lambda^{\beta}$ in the PDF. Again, two of the likelihood equations are mathematically identical.

6.2 Comparing Models in the Same Family

We now discuss more formal tests of whether the embedded model already provides an adequate fit.

As already mentioned, the distributions considered in this chapter have a multiparameter PDF $f(y, \boldsymbol{\theta})$ with $\boldsymbol{\theta} = (\boldsymbol{\alpha}, \boldsymbol{\lambda})$ that can be viewed as a family of distributions, where special cases occur if $\boldsymbol{\alpha} = \boldsymbol{\alpha}_0$, a given value. A natural question that arises when such a model is fitted to a data sample is whether a special case may provide a good or even the best fit. Other things being equal, a model with fewer parameters is usually to be preferred, on the grounds that it not only adequately, but will explain the data more simply than a more elaborate model.

In this section, we consider how the adequacy of specific models that are special cases of a more general model can be formally assessed using the parameter hypothesis testing method described in Section 3.3. Three statistics were discussed for doing this: T_{LR} as in (3.9), T_W as in (3.11), and T_S as in (3.14), all of which can be used to test the null hypothesis, H_0: $\boldsymbol{\alpha} = \boldsymbol{\alpha}_0$. The version of the test given in (3.15) is for the standard situation where $\boldsymbol{\alpha}_0$ is an interior point of the space Ψ to which $\boldsymbol{\alpha}$ belongs. It applies to all three T quantities, and allows the dimension d_0 of $\boldsymbol{\alpha}_0$ to be greater than one.

In this section, we will only consider the case $d_0 = 1$.

If $d_0 > 1$, this can be handled in a staged way; we first test the hypothesis that one of the parameters in $\boldsymbol{\alpha}$ can be assigned its particular value, and if the hypothesis is not rejected we move on to test the hypothesis that the next given parameter can be assigned its given value, conditioning on the previously tested parameters being held at their fixed values.

All three T involve squared quantities. However, in some cases to be considered, $\boldsymbol{\alpha}_0$ is a boundary point of Ψ, and the test needs adjustment. This is most easily handled by using linearized versions of T, which we now describe.

For the case $d_0 = 1$, we can assume that $\boldsymbol{\alpha}$ is scalar with $\boldsymbol{\alpha} = \alpha \in A = [0, b)$. If the hypothesized true α_0 is an interior point of A, this simply means that $\alpha_0 > 0$. The linearized version of the size q test, (3.15), is then

$$\text{Reject } H_0 \text{ if } T^{\frac{1}{2}} > z_{1-q/2} \text{ or } T^{\frac{1}{2}} < z_{q/2}, \tag{6.30}$$

with typically $q = 0.1$, 0.05, or 0.01, where z_q is the q quantile of the standard normal distribution. The quantity $T^{\frac{1}{2}}$ can be any of the linearizations corresponding to T_{LR}, T_W, and T_S, specifically:

$$T_{LR}^{\frac{1}{2}} = \text{signum}\,(\hat{\alpha} - \alpha_0)\,\{2[L(\widehat{\boldsymbol{\theta}}) - L(\alpha_0, \widehat{\boldsymbol{\lambda}}_0)]\}^{1/2}, \tag{6.31}$$

$$T_W^{\frac{1}{2}} = [\mathrm{I}^{\alpha\alpha}(\hat{\alpha}, \widehat{\boldsymbol{\lambda}})]^{-1/2}(\hat{\alpha} - \alpha_0) \text{ or } [\mathrm{I}^{\alpha\alpha}(\alpha_0, \widehat{\boldsymbol{\lambda}}_0)]^{-1/2}(\hat{\alpha} - \alpha_0), \tag{6.32}$$

$$T_S^{\frac{1}{2}} = [\mathrm{I}^{\alpha\alpha}(\alpha_0, \widehat{\boldsymbol{\lambda}}_0)]^{1/2}\frac{\partial L(\alpha_0, \widehat{\boldsymbol{\lambda}}_0)}{\partial \alpha}, \tag{6.33}$$

where we continue to write $[\partial L(\alpha, \widehat{\boldsymbol{\lambda}}_0)/\partial\alpha]|_{\alpha=\alpha_0}$ as $\partial L(\alpha_0, \widehat{\boldsymbol{\lambda}}_0)/\partial\alpha$.

Suppose now the situation where the condition $\alpha \geq 0$ holds, and consider the boundary case where $\alpha_0 = 0$. The effect of not allowing α to be negative is described in Moran (1971, case (i)), Chant (1974, case (i)), and Self and Liang (1987, case (ii)). The easiest version to use in our context is that of Chant (1974, eqn (8)), who shows that asymptotically $\hat{\alpha} = 0$ with probability 0.5, and with probability 0.5 it is positive with distribution exactly like that of $\hat{\alpha}$ in the interior point case when $\hat{\alpha} > 0$. Therefore, the possibility that $T^{\frac{1}{2}} < z_{q/2}$ does not arise, so that, to retain the test of H_0 at level q, we simply adjust the test to

$$\text{Reject } H_0 \text{ if } T^{\frac{1}{2}} > z_{1-q}. \tag{6.34}$$

Note that this test is the same whether α_0 is the boundary point $\alpha_0 = 0$, or if it is an interior point (i.e. $\alpha_0 > 0$), but we wish to test H_0 against the alternative that $\alpha > \alpha_0$, with the possibility that $\alpha < \alpha_0$ not of interest whether it can occur or not.

The calculation of $T_{LR}^{\frac{1}{2}}$ requires maximization of the log-likelihood for both the embedded model and the full model, but is otherwise straightforward.

The calculation of $T_W^{\frac{1}{2}}$ is straightforward. $[\mathrm{I}^{\alpha\alpha}(\hat{\alpha}, \widehat{\boldsymbol{\lambda}})]^{-1/2}$ can be calculated by evaluating the information matrix $\mathrm{I}(\hat{\alpha}, \widehat{\boldsymbol{\lambda}})$ as in (3.4), taking its inverse which is the covariance matrix, and then taking the square root of the reciprocal of the variance $\mathrm{I}^{\alpha\alpha}(\hat{\alpha}, \widehat{\boldsymbol{\lambda}})$ in this covariance matrix. A slightly easier alternative is to compute this variance directly as

$$\mathrm{I}^{\alpha\alpha}(\hat{\alpha}, \widehat{\boldsymbol{\lambda}}) = [\mathrm{I}_{\alpha\alpha}(\hat{\alpha}, \widehat{\boldsymbol{\lambda}}) - \mathrm{I}_{\alpha\boldsymbol{\lambda}}(\hat{\alpha}, \widehat{\boldsymbol{\lambda}})\mathrm{I}_{\boldsymbol{\lambda\lambda}}^{-1}(\hat{\alpha}, \widehat{\boldsymbol{\lambda}})\mathrm{I}_{\alpha\boldsymbol{\lambda}}^{\mathrm{T}}(\hat{\alpha}, \widehat{\boldsymbol{\lambda}})],$$

which involves the inversion of $\mathrm{I}_{\boldsymbol{\lambda\lambda}}(\hat{\alpha}, \widehat{\boldsymbol{\lambda}})$ only.

The simplest case is $T_S^{\frac{1}{2}}$. For simplicity, and with no loss of generality, we consider the case where $\alpha_0 = 0$, when only $\mathrm{I}^{\alpha\alpha}(0, \widehat{\boldsymbol{\lambda}}_0)$ is needed.

As already pointed out in the introduction to this chapter, the information matrix $\mathrm{I}(0, \widehat{\boldsymbol{\lambda}}_0)$ is most easily calculated from the first three terms of the series expansion of the log-likelihood of the full model, assuming this exists,

$$L(\alpha, \boldsymbol{\lambda}) = L(0, \boldsymbol{\lambda}) + \frac{\partial L(0, \boldsymbol{\lambda})}{\partial \alpha}\alpha + \frac{1}{2}\frac{\partial^2 L(0, \boldsymbol{\lambda})}{\partial \alpha^2}\alpha^2 + O_p(\alpha^3) \tag{6.35}$$

$$= L_0(\boldsymbol{\lambda}) + L_1(\boldsymbol{\lambda})\alpha + L_2(\boldsymbol{\lambda})\alpha^2 + O_p(\alpha^3), \text{ say,} \tag{6.36}$$

using the formulas already given in (6.4), namely,

$$I_{\alpha\alpha}(0, \widehat{\boldsymbol{\lambda}}_0) = -2E[L_2(\boldsymbol{\lambda}_0)] \tag{6.37}$$

$$I_{\alpha\boldsymbol{\lambda}}(0, \widehat{\boldsymbol{\lambda}}_0) = -E[\partial L_1(\boldsymbol{\lambda}_0)/\partial \boldsymbol{\lambda}] \tag{6.38}$$

$$I_{\boldsymbol{\lambda\lambda}}(0, \widehat{\boldsymbol{\lambda}}_0) = -E[\partial L_0(\boldsymbol{\lambda}_0)/\partial \boldsymbol{\lambda}\partial \boldsymbol{\lambda}]. \tag{6.39}$$

This is usually considerably less laborious to do than evaluating $\mathrm{I}(\alpha, \boldsymbol{\lambda})$ directly from the Hessian formula (3.4), and then setting $\alpha = 0$.

In the score statistic case,

$$T_S = \mathrm{I}^{\alpha\alpha}(0,\hat{\boldsymbol{\lambda}})[L_1(0,\hat{\boldsymbol{\lambda}})]^2,$$

and this can be made even simpler if the parameters $\boldsymbol{\lambda}$ can be chosen to be locally orthogonal to α in the neighbourhood of $\alpha = 0$, so that $E(\partial L_1(0,\hat{\boldsymbol{\lambda}})/\partial\boldsymbol{\lambda}) = \mathbf{0}$, as then $\mathrm{I}^{\alpha\alpha}(0,\hat{\boldsymbol{\lambda}}) = [-2E(L_2(0,\hat{\boldsymbol{\lambda}}))]^{-1}$.

A simple example is the gamma case for which

$$L_1(0,\hat{\boldsymbol{\lambda}}) = \frac{n}{3}g_1,$$

where g_1 is the sample skewness

$$g_1 = (m_3 - 3m_1m_2 + 2m_1^3)/s_2^{3/2}$$

and

$$s_2 = m_2 - m_1^2,$$

with $m_i = \sum_{j=1}^n y_j^i/n$. Thus $g_1 < 0$ would indicate that a better fit is obtained by fitting the reflected gamma distribution.

For the formal test, we have from eqn (6.20) $\mathrm{I}_{\alpha\alpha} = 2n/3$, so that, as the information matrix is locally orthogonal, $\mathrm{I}^{\alpha\alpha} = 3/(2n)$, and we have

$$T_S^{\frac{1}{2}} = (3/(2n))^{1/2}(n/3)g_1 = \sqrt{\frac{n}{6}}g_1.$$

Statistics such as these were obtained using combinatorial means by Fisher (1930), in more accurate form.

For shifted threshold examples where the embedded model is the normal distribution, use of L_1 is probably satisfactory, as its sign is geometrically related to the symmetry of the distribution. However, in general, the author's experience is that use of L_1 is not all that reliable, as it assumes that L will behave linearly for α sufficiently small near 0, and this cannot be relied on. In general, it is worth checking the overall behaviour of the log-likelihood by plotting the profile log-likelihood

$$L(\alpha) = \max_{\phi} L(\alpha,\phi),$$

and, where possible, this should be carried out automatically as part of an overall fitting procedure.

6.3 Extensions of Model Families

A number of the parameters in the models that we have considered are restricted to being positive. In several of the cases, this restriction is due to convention rather than being strictly necessary. An example already encountered is the power parameter $c = 1/\alpha$ in the GEVMin distribution of equation 6.23, which is unrestricted. However, if we impose the restriction $c > 0$, then we obtain the standard Weibull distribution, for which c is conventionally restricted to being positive, with the embedded EVMin distribution arising as a boundary model. If, however, we take $\alpha = c^{-1}$, we can use the null hypothesis $H_0 : \alpha = \alpha_0(= 0)$ to test for the EVMin model. Moreover, $\alpha_0 = 0$ becomes an interior point of the parameter space if the restriction that α be positive is dropped. EVMin is then simply a special case of GEVMin distribution, with the test of H_0 a standard one as set out in (6.30).

Apart from simplifying hypothesis testing of parameter values, extending a parametric model by relaxing restrictions placed on the values that the original parameters can take increases the ability of a model to fit different data sets. Such an extension is often a rather trivial one numerically, but in other cases the extension can be more significant.

Another example is in the Burr XII model, where the reparametrization that makes the Weibull a non-embedded special case is interesting as it allows the restriction that $d > 0$ to be dropped. Using $b = \sigma\alpha^{-1/c}, \sigma > 0$, and $d = \alpha^{-1}$, we can write the CDF as

$$F(x, a, c, \sigma, \alpha) = 1 - \left[1 + \alpha\left(\frac{x-a}{\sigma}\right)^c\right]^{-1/\alpha} \tag{6.40}$$

and the PDF as

$$f((x, a, c, \sigma, \alpha) = \frac{c}{\sigma}\left[1 + \alpha\left(\frac{x-a}{\sigma}\right)^c\right]^{-\alpha^{-1}-1}\left(\frac{x-a}{\sigma}\right)^{c-1}, \tag{6.41}$$

where $c > 0$ and $\sigma > 0$ but α is unrestricted. However, we do require

$$a < x \text{ if } \alpha > 0 \text{ or } a < x < a + (-\alpha)^{-1/c}\sigma \text{ if } \alpha < 0.$$

This transformation is reversible if $\alpha > 0$, as $b = \sigma\alpha^{-1/c}$, but not if $\alpha < 0$. We shall call the distribution with CDF and PDF given by (6.40) and (6.41) the *extended Burr XII* distribution.

The extended Burr XII is a natural generalization of the GEVMin distribution. As in the GEVMin case, use of the extended (four-parameter) Burr XII rather than the standard (four-parameter) Burr XII simplifies the formal test of whether use of the Burr distribution is significantly better than the Weibull model, because, in the extended form, the parameter points where $\alpha = 0$ are internal rather than boundary points. We will discuss more fully how to handle all three embedded models when we consider a numerical example in Section 7.2, but note here similar comments apply when fitting the slightly simpler three-parameter version of (6.40) and (6.41) in which a is fixed at $a = 0$. This simpler case has been examined by Shao et al. (2004), who show that the Weibull and Pareto embedded limits still exist and have to be allowed for.

The shifted threshold distributions listed in Table 6.4 are all examples where an extension of the model definition is possible. In all but one of the distributions in the table, the extension takes the simple form of adding a *reflection* of the original definition, so that X can have the distribution corresponding to $\alpha = -\alpha'$ for some $\alpha' > 0$

where this is equivalent to $R = -X$ having the distribution corresponding to $\alpha = \alpha' > 0$. Thus the extension is the rather elementary one where the original family is in effect being fitted not only to the sample $\mathbf{y}$ but to the sample $-\mathbf{y}$ as well. This form of extension by reflection requires the embedded model corresponding to $\alpha = \alpha_0$ to be symmetric.

In Table 6.4, the embedded models are all symmetric, with all being the normal distribution, except in the loglogistic case, where the embedded model is the logistic distribution.

An exception is the GEVMin distribution (6.22), where the embedded model is the extreme value distribution. In this case, though having $\alpha < 0$ corresponds to valid distributions, the extension is not a reflection.

We have already pointed out that a small change to the definition of the PDF of the Burr XII allows its use in fitting the Burr III distribution. More interestingly, if we examine the three embedded models given in Table 6.1, we find that reparametrization used to identify the generalized Type II logistic embedded model extends directly to allow $\alpha < 0$ as well as $\alpha > 0$. The embedded model is not symmetric, so that, like the GEVMin model of Table 6.4, the reflection principle does not apply; however, the H_0 hypothesis test is standard.

Using the reparametrization $c = \gamma\alpha^{-1}, \gamma > 0, d = \alpha$, the Burr XII PDF is

$$\frac{\gamma}{b^\gamma}(x-a)^{-\gamma-1}\left(1+\left(\frac{b}{x-a}\right)^{\gamma/\alpha}\right)^{-\alpha-1}, \tag{6.42}$$

while the log-likelihood is as given in (6.11), taking the form $L = L_0 - \lambda_1$, with

$$\begin{aligned} L_0 &= \ln\gamma + \gamma\ln b - (1+\gamma)\ln(x-a), \\ \lambda_1 &= (\alpha+1)\ln\left(1+\left(\frac{b}{x-a}\right)^{\gamma\alpha^{-1}}\right), \end{aligned}$$

where L_0 is the log-likelihood of the Pareto distribution. We have already noted that when $b/(x-a) < 1$ (i.e. $x > a+b$) and $b, \gamma > 0$, then $(b/(x-a))^{\gamma\alpha^{-1}} \to 0$ as $\alpha \to 0$ from above, so that λ_1 also tends to zero as $\alpha \to 0$. Hence, L tends to the Pareto log-likelihood in this case.

However, the hypothesis test of $H_0 : \alpha = 0$ is not straightforward, as the log-likelihood (6.11) does not possess a series expansion in α in the neighbourhood of $\alpha > 0$, so that neither of the hypothesis tests (6.30) or (6.34) can be applied.

We can, however, allow $\alpha < 0$ with the PDF still taken as (6.42). We give details for completeness, though the model behaviour is rather strange, so that the value of allowing $\alpha < 0$ is perhaps debatable.

The CDF, instead of being $1 - \left(1+y^{\alpha^{-1}}\right)^{-\alpha}$ where $y = ((x-a)/b)^\gamma$, which is the case when $\alpha > 0$, now takes the form

$$F(x) = \begin{cases} 0 & \text{if } -\infty < x \le a + \eta^{1/\gamma}b \\ 2 - \left(1+y^{\alpha^{-1}}\right)^{-\alpha} & \text{if } \ a + \eta^{1/\gamma}b < x < \infty \end{cases},$$

with $y = ((x-a)/b)^\gamma$ and

$$\eta = \eta(\alpha) = \exp\left(\left(\ln\left(e^{-\frac{\ln 2}{\alpha}} - 1\right)\right)\alpha\right)$$

defined so that the left-hand end of the support of the distribution is at $x = a + \eta^{1/\gamma} b$ with $F(a + \eta^{1/\gamma} b) = 0$. For $\alpha < 0$, $\eta(\alpha)$ is a monotonically decreasing positive function of α as α increases through negative values to 0, with $\eta(-1) = 1$, and with $\eta(\alpha) \to \frac{1}{2}$ as $\alpha \to 0$.

Thus the PDF (6.42) is only positive to the right of $x = a + \eta^{1/\gamma} b$. Moreover, for $-1 < \alpha < 0$, when $1 > \eta > \frac{1}{2}$, the support of the PDF includes the interval $a + \eta^{1/\gamma} b < x < a + b$, so that the range of the support is greater than that of the Pareto limit. In fact, in this interval, the PDF is numerically close in value to the curve

$$f_{Pareto}(x) = \left(\frac{b}{x-a}\right)^\gamma \frac{\gamma}{x-a} \tag{6.43}$$

defining the Pareto limit obtained when $\alpha \to 0$ from above. But, thereafter, the PDF tends to zero much more rapidly than $f_{Pareto}(x)$ as $x \to \infty$.

The limiting PDF as $\alpha \to 0$ from below is precisely $f_{Pareto}(x)$ *but just in the finite interval* $(a + \frac{1}{2}b, a + b)$.

However, the main curiosity is that Burr XII PDF at $\alpha = -1$ is *precisely* the Pareto distribution obtained as $\alpha \to 0$ from above. The log-likelihood (6.11) possesses a series expansion in the neighbourhood of $\alpha = -1$. Writing $\alpha = -1 + s$, we find the one-observation log-likelihood is

$$\begin{aligned} L &= \ln\gamma - \frac{\gamma}{\alpha}\ln b - (1+\alpha)\ln\left(1 + \left(\frac{x-a}{b}\right)^{\frac{\gamma}{\alpha}}\right) + \left(\frac{\gamma}{\alpha} - 1\right)\ln(x-a) \\ &= (\ln\gamma + \gamma\ln b - (\gamma+1)\ln(x-a)) + \\ &\left(\ln\left(\frac{b}{x-a}\right)^\gamma - \ln\left(1 + \left(\frac{b}{x-a}\right)^\gamma\right)\right)s + \frac{\ln\left(\left(\frac{b}{x-a}\right)^\gamma\right)}{1 + \left(\frac{b}{x-a}\right)^\gamma}s^2 + O_p(s^3), \end{aligned}$$

showing that the same limit is approached continuously as $\alpha \to -1$ from above or below.

The previous analysis would indicate that the Pareto limit is probably not a very practical one to consider in applications. If it is to be considered at all, a bootstrap analysis might be the approach to use, but this recommendation is only a tentative one.

Shao (2002) also gives results for the closely related four-parameter Burr III distribution with CDF

$$F(x, a, b, c, d) = \left[1 + \left(\frac{b}{x-a}\right)^c\right]^{-d},$$

showing that its hierarchical embedded model structure mirrors that of the Burr XII. In fact, the two distributions are easily combined as, if $b, d > 0$, the PDF

$$\frac{|c|\,d}{b}\left(1+\left(\frac{x-a}{b}\right)^{c}\right)^{-d-1}\left(\frac{x-a}{b}\right)^{c-1} \tag{6.44}$$

is that of the Burr Type XII when $c > 0$; whilst it is that of the Burr Type III if $c = -c_+ < 0$, where $c_+ > 0$, as then

$$\begin{aligned}&\frac{|c|\,d}{b}\left(1+\left(\frac{x-a}{b}\right)^{c}\right)^{-d-1}\left(\frac{x-a}{b}\right)^{c-1}\\&=\frac{|c_+|\,d}{b}\left(1+\left(\frac{b}{x-a}\right)^{c_+}\right)^{-d-1}\left(\frac{b}{x-a}\right)^{c_+ +1},\end{aligned}$$

which is the Burr Type III with $b, c_+, d > 0$. Thus, taking the PDF in the form (6.44) allows both versions to be modelled, with the sign of c determining which version is selected. Note, however, that the transition where c moves through zero from a small positive quantity to a small negative one is not a smooth one.

6.4 Stable Distributions

For simplicity, we shall from now on refer to *stable law distributions* simply as *stable distributions.*

The use of stable distributions to model data has grown remarkably. One field where stable distributions are now regularly used is in the modelling of extreme financial events, where the long tail behaviour of stable distributions seems especially appropriate. Nolan (2015) provides a good introduction. Nolan recommends Chapter 9 of Breiman (1968, 1992) for a clear exposition of univariate stable laws. Authoritative references are Ibragimov and Linnik (1971), Zolotarev (1986), and Samorodnitsky and Taqqu (1994).

Much of the early development of stable theory discussed aspects like their infinite divisibility and characterization through characteristic functions, this latter being particularly important as only three special cases of stable distributions are known which have PDFs that can be written down in terms of elementary functions, namely, the normal, the Cauchy with PDF

$$f_{Cauchy}(y) = \frac{1}{\pi}\frac{b}{b^2+(y-a)^2}, \quad -\infty < y < \infty, \tag{6.45}$$

and the Lévy distribution with PDF as in (3.26). This makes challenging the practical use of stable distributions in statistical modelling. As pointed out by Knight (1997), even a standard reference like Samorodnitsky and Taqqu (1994) does not fully address two aspects of stable distribution theory of special interest to practising statisticians: (i) the weak convergence of sums of independently and identically distributed (IID) random variables to a stable limiting distribution and (ii) the modelling of data drawn from distributions with infinite variance.

The parametrizations used in the early literature on stable distributions used a form in which embedding occurs. Though not viewed in the way that we have defined embedding in Chapter 5, the problem was recognized. Notable early work by Zolotarev, summarized by the author in Zolotarev (1986), covers several parametrizations, one of which, version (M), recognizes and handles the problem; we will discuss this version in what follows.

In practical situations, we are usually interested in identifying the stable distribution to which a given sum of IID random variables is converging—this being colourfully described in the stable law literature as the *domain of attraction* to which the given sum belongs. We discuss how to derive the characteristic function which uniquely defines the stable distribution. The PDF and CDF of the distribution are then needed to calculate quantities such as ML estimates and corresponding confidence intervals. As the explicit form of the PDF and CDF is only known for the normal, Cauchy, and Lévy distributions, numerical methods are needed in general for practical calculations, and we discuss this aspect as well. An application of the results collected together in the next section is given in Section 8.5, where we discuss how stable distributions may be useful for constructing confidence intervals in certain situations where the MLE is not always consistent.

6.5 Standard Characterization of Stable Distributions

The form taken by the characteristic function of all stable distributions is well known. Several forms are extant, and discussed by Zolotarev (1986). One form, designated as (A) by Zolotarev, has the logged, sometimes called Cartesian, form

$$\ln \phi_{\alpha,\beta}(t) = \begin{cases} i\delta t - \gamma^{\alpha} |t|^{\alpha} + i\gamma^{\alpha}\beta |t|^{\alpha-1} t \tan\left(\frac{\pi\alpha}{2}\right) & \text{if} \alpha \neq 1 \\ i\delta t - \gamma |t| - i\beta \frac{2}{\pi} t \ln(|t|) & \text{if} \alpha = 1 \end{cases}, \tag{6.46}$$

where δ and γ are not the same as the parameters λ and γ used by Zolotarev, but are simply related to them. Here $0 < \alpha \leq 2$ is called the *index*, whilst $-1 \leq \beta \leq 1$ is a symmetry parameter with the distribution symmetric when $\beta = 0$, positively skew when $\beta > 0$, and negatively skew when $\beta < 0$. The parameter $0 < \gamma$ is treated as a scaling parameter, while $-\infty < \delta < \infty$ is a shift parameter. This arguably is the parametrization that has been most commonly discussed.

It is unfortunate that this form has been been so popular, as it is unnecessarily tangled by a number of issues over the nature and interpretation of the parameters, which can be avoided. Hall (1981) discusses the issues, referring to the situation as 'a comedy of errors'. Many authors, including McCulloch (1996), give careful consideration of how to choose the parameters. Overall, this may give the impression to the casual reader that the problem is perhaps more complicated than it really is.

An immediate watch point is that in much of the early literature, the sign of the term containing β when $\alpha \neq 1$ is negative rather than positive, as it is in (6.46). This is the case, for example, in Lukacs (1969), and in the Holt and Crow (1973) extensive numerical tabulation of the PDFs of the stable distribution; the direction of skewness in the tabulation is therefore reversed when $\alpha \neq 1$ compared with the $\alpha = 1$ case. This particular awkwardness is avoided using (6.46).

However, we will not be using (6.46), as it has two unpleasant features, both of which can be corrected.

The first problem concerns the *standard form* of stable distribution that is obtained when we set $\delta = 0$ and $\gamma = 1$ in (6.46). Denoting the standardized stable random variable by S, we have

$$\begin{aligned} \ln \phi_S(t) &= -|t|^\alpha + i\beta t|t|^{\alpha-1} \tan\left(\frac{\pi\alpha}{2}\right) \quad \alpha \neq 1 \\ &= -|t| - i\frac{2}{\pi}\beta t \ln|t| \qquad \alpha = 1. \end{aligned}$$

Now the most general form of stable random variable possible is a linear transform of S:

$$Y = \gamma S + \delta, \; \gamma > 0,$$

where γ is a simple rescaling, followed by a shift δ. The log characteristic function for $Y = \gamma S + \delta$ is therefore

$$\begin{aligned} \ln \phi_Y(t) &= \ln E(e^{itY}) = it\delta + \ln \phi_S(\gamma t) \\ &= \begin{cases} it\delta - \gamma^\alpha |t|^\alpha + i\gamma^\alpha \beta t|t|^{\alpha-1} \tan(\frac{\pi\alpha}{2}) & \text{if } \alpha \neq 1 \\ it\delta - \gamma\,|t| - i\gamma\frac{2}{\pi}\beta t \ln|\gamma t| & \text{if } \alpha = 1 \end{cases}. \end{aligned} \tag{6.47}$$

However, the representation (6.46) only matches this for the case $\alpha \neq 1$. To hold for $\alpha = 1$ as well, δ in (6.46) for the case $\alpha = 1$ needs to be changed to

$$\delta' = \delta - \gamma\frac{2}{\pi}\beta \ln \gamma.$$

The reason why this is not often pointed out is that, in numerical calculations, once γ and δ are known, the PDF of Y is easily obtained by evaluating the PDF of S, and then transforming this PDF in the usual way to give that of Y.

The problem is most simply handled by including the missing γ in the expression for $\ln \phi_Y(t)$ when $\alpha = 1$, so that we have $\ln \phi_Y(t)$ as defined in (6.47). Then γ and δ really are just scale and translation parameters in $Y = \gamma S + \delta$ for all $0 < \alpha \leq 2$. This is the parametrization suggested by DuMouchel (1975). Hall (1981) suggests an alternative. McCulloch (1996) points out the complications of switching from one parametrization to another.

There is yet another well-recognized problem; actually, one involving embedding. Zolotarev (1986) views the problem as one where the parameters do not behave continuously as $\alpha \to 1$; if $\alpha \to 1, \beta \to \beta_0, \gamma \to \gamma_0, \delta \to \delta_0$ where $\beta_0, \gamma_0, \delta_0$ are fixed values, (6.47) tends to that of a degenerate discrete distribution. Cheng and Liu (1997) show that in the *polar representation*, called version (B) by Zolotarev (1986), if $\beta_0 \neq 0$, then convergence to a non-degenerate stable distribution with parameters $\alpha = 1, \beta_0, \gamma_0, \delta_0$ is (only) possible by letting $\beta \to \pm 1$, whilst simultaneously allowing γ and δ to become unbounded.

We can treat this problem as one of embedding as given in eqn (5.6). Specifically, we have a $p = l + r$ parameter problem, in which $r = 0$ parameters are kept fixed, so that $l = 4$ encompasses all $p = 4$ parameters which we let tend to infinity or specific limits, but where the result is convergence to a model where $k = 3$ parameters are still free to be chosen. The values of $r = 0$, $l = 4$, and $k = 3$ satisfy the condition $r < k < l + r$ under which the model of eqn (5.6) is embedded.

Remarkably, the embedding turns out to be removable. The (M) representation of Zolotarev (1986) is just the Cartesian representation (6.47) of Dumouchel (1975), to which the further shift term $-i\gamma\beta\tan(\frac{\pi\alpha}{2})t$ is included for the case $\alpha \neq 1$. This *shifted Cartesian* representation is

$$\begin{aligned} \ln\phi(t) &= i\delta t - \gamma^{\alpha}|t|^{\alpha} + i\beta\gamma t(|\gamma t|^{\alpha-1} - 1)\tan\left(\frac{\pi\alpha}{2}\right) \text{ if } \alpha \neq 1 \\ &= i\delta t - \gamma\,|t| - i\gamma\left(\frac{2}{\pi}\right)\beta t\ln|\gamma t| \text{ if } \alpha = 1. \end{aligned} \tag{6.48}$$

It is readily verified that $\lim_{\alpha\to 1}(|\gamma t|^{\alpha-1} - 1)\tan(\pi\alpha/2) = -(2/\pi)\ln|\gamma t|$, so that if we let $\alpha \to 1$ in the case $\alpha \neq 1$, we have convergence to the case $\alpha = 1$. Therefore, all four parameters α, β, γ, δ have the same meaning whether α is equal unity or not.

The representation (6.48) is easily the most satisfactory one to use for statistical modelling with stable distributions. In the next section, we consider calculation of the PDF and CDF using this representation.

6.5.1 Numerical Evaluation of Stable Distributions

Tabulations of the PDFs and CDFs of the stable distributions are available but in somewhat limited form. Holt and Crow (1973) give extended tables, but only for the PDF, arguably not as useful as the CDF in practical statistical work. Tables for the CDF are given by Panton (1993) for the symmetric case and by McCulloch and Panton (1997) for the maximally skewed case.

Zolotarev (1964), however, gives integral formulas, which can be used to represent both the PDF and CDF whatever the parameter values. Nolan (1997) considers the Zolotarev approach in detail, obtaining integrals over a finite range that give both the PDF and CDF when parametrized as in the shifted Cartesian representation of the logged characteristic function given in (6.48). Other approaches have been suggested, see Belov (2005), but the formulas given by Nolan (1997) are comprehensive and described in detail. We summarize Nolan's results.

We give the formulas for the PDF and CDF corresponding to the random variable S for which $\ln\phi(t)$ is as given in formula (6.48), but in the standarized form where $\delta = 0$ and $\gamma = 1$. The full four-parameter form for the PDF and CDF of $Y = \gamma S + \delta$ is then

$$f_Y(y) = \gamma^{-1} f_S[(y-\delta)/\gamma], \; F_Y(y) = F_S[(y-\delta)/\gamma].$$

Let

$$\zeta = \zeta(\alpha, \beta) = \begin{cases} -\beta \tan \frac{\pi\alpha}{2} & \text{if } \alpha \neq 1 \\ 0 & \text{if } \alpha = 1, \end{cases}$$

$$\theta_0 = \theta_0(\alpha, \beta) = \begin{cases} \frac{1}{\alpha} \arctan(\beta \tan(\frac{\pi\alpha}{2}) & \text{if } \alpha \neq 1 \\ \frac{\pi}{2} & \text{if } \alpha = 1, \end{cases}$$

$$V(\theta; \alpha, \beta) = \begin{cases} (\cos(\alpha\theta_0))^{\frac{1}{\alpha-1}} \left(\frac{\cos\theta}{\sin(\alpha(\theta_0+\theta))}\right)^{\frac{\alpha}{\alpha-1}} \frac{\cos(\alpha\theta_0+(\alpha-1)\theta)}{\cos(\theta)} & \text{if } \alpha \neq 1 \\ \frac{2}{\pi}\left(\frac{\frac{\pi}{2}+\beta\theta}{\cos\theta}\right) \exp\left(\frac{1}{\beta}\left(\frac{\pi}{2}+\beta\theta\right)\tan\theta\right) & \text{if } \alpha = 1, \end{cases}$$

$$c_1 = c_1(\alpha, \beta) = \begin{cases} \frac{1}{\pi}(\frac{\pi}{2} - \theta_0) & \text{if } \alpha < 1 \\ 0 & \text{if } \alpha = 1 \\ 1 & \text{if } \alpha > 1, \end{cases}$$

$$c_2 = c_2(x, \alpha, \beta) = \begin{cases} \frac{\alpha}{\pi|\alpha-1|(x-\zeta)} & \text{if } \alpha \neq 1 \\ 1/(2|\beta|) & \text{if } \alpha = 1, \end{cases}$$

$$c_3 = c_3(\alpha) = \begin{cases} \frac{\operatorname{sgn}(1-\alpha)}{\pi} & \text{if } \alpha \neq 1 \\ \frac{1}{\pi} & \text{if } \alpha = 1, \end{cases}$$

$$g(\theta) = g(\theta; x, \alpha, \beta) = \begin{cases} (x-\zeta)^{\frac{\alpha}{\alpha-1}} V(\theta; \alpha, \beta) & \text{if } \alpha \neq 1 \\ e^{-\frac{\pi x}{2\beta}} V(\theta; 1, \beta) & \text{if } \alpha = 1. \end{cases}$$

When $\alpha \neq 1$:

$$f(x; \alpha, \beta) = \begin{cases} c_2 \int_{-\theta_0}^{\frac{\pi}{2}} g(\theta) \exp[-g(\theta)] d\theta & \text{if } x > \zeta \\ \frac{\Gamma(1+\frac{1}{\alpha}) \cos\theta_0}{\pi(1+\zeta^2)^{1/(2\alpha)}} & \text{if } x = \zeta \\ f(-x; \alpha, -\beta) & \text{if } x < \zeta, \end{cases} \tag{6.49}$$

$$F(x; \alpha, \beta) = \begin{cases} c_1 + c_3 \int_{-\theta_0}^{\frac{\pi}{2}} \exp[-g(\theta)] d\theta & \text{if } x > \zeta \\ \frac{1}{\pi}(\frac{\pi}{2} - \theta_0) & \text{if } x = \zeta \\ 1 - F(-x; \alpha, -\beta) & \text{if } x < \zeta. \end{cases} \tag{6.50}$$

When $\alpha = 1$:

$$f(x, 1, \beta) = \begin{cases} c_2 \int_{-\frac{\pi}{2}}^{\frac{\pi}{2}} g(\theta) \exp[-g(\theta)] d\theta & \text{if } \beta \neq 0 \\ \frac{1}{\pi(1+x^2)} & \text{if } \beta = 0, \end{cases} \tag{6.51}$$

$$F(x; 1, \beta) = \begin{cases} c_3 \int_{-\frac{\pi}{2}}^{\frac{\pi}{2}} \exp[-g(\theta)] d\theta & \text{if } \beta > 0 \\ \frac{1}{2} + \frac{1}{\pi} \arctan x & \text{if } \beta = 0 \\ 1 - F(-x; 1, -\beta) & \text{if } \beta < 0. \end{cases} \tag{6.52}$$

To keep neat the right-hand sides of formulas, we have suppressed the full list of arguments on which the quantities ζ, θ_0, c_1, c_2, c_3, and $g(\theta)$ depend. In calculating $f(x; \alpha, \beta)$ and $F(x; \alpha, \beta)$, all these quantities must be evaluated at the values x, α, and β as given on the left-hand side of the formula. Particular care is needed with terse formulas like 'If $\alpha \neq 1$ then $f(x; \alpha, \beta) = f(-x; \alpha, -\beta)$ if $x < \zeta$', which we have retained from Nolan (1997), but where the self-referential form is potentially confusing. In this example, the ζ in the condition 'if $x < \zeta$' is $\zeta = \zeta(\alpha, \beta)$. However, ζ *also* appears in the formula $f(-x; \alpha, -\beta)$, as do c_2 and θ_0. These must be calculated as $\zeta = \zeta(\alpha, -\beta)$, $c_2 = c_2(-x, \alpha, -\beta)$, and $\theta_0 = \theta_0(\alpha, -\beta)$ using the formula '$c_2 \int_{-\theta_0}^{\frac{\pi}{2}} g(\theta) \exp[-g(\theta)] d\theta$', which is applicable 'if $x > \zeta$', but which still holds in the form $-x > -\zeta$, as the sign of ζ is changed, because it is calculated as $\zeta = \zeta(\alpha, -\beta)$.

Even with these formulas, it is difficult to maintain precision if α and β are to be allowed to vary over their full ranges. The most problematic regions are when (i) α is close to unity but different from unity, (ii) α near zero, (iii) β near zero. Even away from these regions, the term $g(\theta)$ can contain some factors that are very large and some very small. To avoid underflow when evaluating $\exp[-g(\theta)]$, which appears in all the integrands, it is best to calculate the logarithm of $g(\theta)$ at the different values of θ used in the quadrature. Then underflow is avoided by only exponentiating $\ln(g(\theta))$ if this is less than 700, say, before exponentiating to get $\exp[-g(\theta)]$, with integrand terms treated as negligible when $\ln(g(\theta)) > 0$. Discussion of such issues is given by McCulloch and Panton (1997) and by Nolan (1997).

To illustrate the use of the stable distribution, we give one example in Section 8.5, discussing how stable distributions can be used in a non-standard problem involving estimation of threshold parameters where the estimators are not asymptoticallly normally distributed.

We also consider the implementation of ML estimation to fit the four-parameter stable distribution using the formulas (6.49), (6.50), (6.51), and (6.52) for the PDF and CDF, giving a numerical example using a simple VBA macro where the formulas are implemented using standard iterative (rather than recursive) adaptive Gauss-Legendre quadrature. As one might expect, the estimation can be slow for large data sets, though the fits obtained are satisfactory. This example is described in Section 9.7.3.

7 Embedded Distributions: Two Numerical Examples

In this chapter, we consider in more detail the actual fitting of some of the probability distributions whose properties were examined and discussed in the previous chapter. We fit different models to two data sets and then compare the fits obtained by the formal method of hypothesis testing, using not only asymptotic theory but also bootstrapping to make quantitative comparisons. One of our main aims is to illustrate the usefulness of bootstrapping and graphical displays in conveying information on how robust the results actually are.

7.1 Kevlar149 Fibre Strength Example

The first example involves fitting the GEVMin and EVMin models to a sample, reproduced in Table 7.1, giving breaking strengths of 107 Kevlar149 fibres (source not known).

The GEVMin model with PDF as given in (6.23) is a meaningful distribution to fit to such data, as one might posit breaking strength to be due to a 'weakest link' mechanism, so that a distribution applicable where each observation is the minimum of a set of random variables is appropriate. We compare the fits using the full GEVMin model with that obtained by fitting its embedded model, the EVMin distribution. Our analysis covers the following aspects.

(i) We fit the full model using ML estimation with the parameters defined as in (6.23).

(ii) Separately, we also use ML estimation to fit the embedded model, in this case the EVMin distribution, which is listed as the embedded entry (column 3) corresponding to the Weibull model entry (row 1) of Table 6.2.

Non-Standard Parametric Statistical Inference. Russell Cheng.

Table 7.1 Sample of 107 Kevlar149 fibre strengths

0.600	0.727	0.830	0.900	0.998	1.055	1.113	1.136	1.142	1.199
1.199	1.205	1.222	1.246	1.251	1.251	1.274	1.274	1.298	1.298
1.320	1.326	1.332	1.332	1.332	1.338	1.344	1.344	1.355	1.355
1.355	1.355	1.361	1.361	1.372	1.372	1.390	1.396	1.401	1.401
1.407	1.407	1.407	1.413	1.430	1.430	1.436	1.436	1.436	1.436
1.447	1.459	1.459	1.465	1.476	1.482	1.488	1.488	1.488	1.488
1.488	1.494	1.499	1.505	1.517	1.517	1.522	1.522	1.540	1.545
1.545	1.545	1.545	1.551	1.551	1.557	1.560	1,562	1.574	1.574
1.580	1.580	1.580	1.586	1.586	1.592	1.603	1.603	1.615	1.615
1.620	1.626	1.661	1.661	1.666	1.672	1.684	1.684	1.690	1.701
1.724	1.724	1.736	1.770	1.782	1.793	1.834			

(iii) We can then calculate, using the MLEs and maximized log-likelihood values obtained from the fitted full and embedded models, the $T^{\frac{1}{2}}$ statistics of equations (6.31), (6.32), and (6.33) to evaluate whether there is a significant difference between the fits of the full and embedded models.

(iv) We also illustrate the alternative approach of using GoF tests to evaluate the fits of the full and embedded models. We illustrate the bootstrap method described in Section 4.5 for doing this, where bootstrap samples of the fitted full and embedded models are obtained, and these are then used to calculate appropriate critical values of GoF test statistics taking as our examples the Anderson-Darling statistic and the Cramér von-Mises statistic.

As previously intimated, we use Nelder-Mead optimization of the parameters α, μ and σ in the GEVMin case and μ and σ in the EVMin case with the constraint $\sigma > 0$ in both cases.

The MLEs obtained for the parameters in the GEVMin case were $\hat{\alpha} = 0.026$, $\hat{\mu} = 1.53$, and $\hat{\sigma} = 0.169$. The small value of $\hat{\alpha}$ indicates that the best fit is close to being the EVMin distribution. The ML estimates of the parameters of the EVmin model, $\hat{\mu} = 1.53$, and $\hat{\sigma} = 0.168$, are indeed almost identical to the estimates of the corresponding parameters in the full GEVMin model. Figure 7.1 shows the fitted CDF and PDF of the GEVMin model. The CDF and PDF of the fitted EVMin distribution is visually identical to those of the GEVMin. Figure 7.2 depicts the profile log-likelihood, with the maximum at $\alpha = 0.026$, corroborating the ML estimate of α.

Calculation of the values of the three test statistics of the null hypothesis $H_0 : \alpha = 0$ was discussed in Section 6.2. $T_W^{\frac{1}{2}}$ and $T_S^{\frac{1}{2}}$ requires the information matrix $I(\alpha, \mu, \sigma)$. The GEVMax version is given by Prescott and Walden (1980). The GEVMin version is the

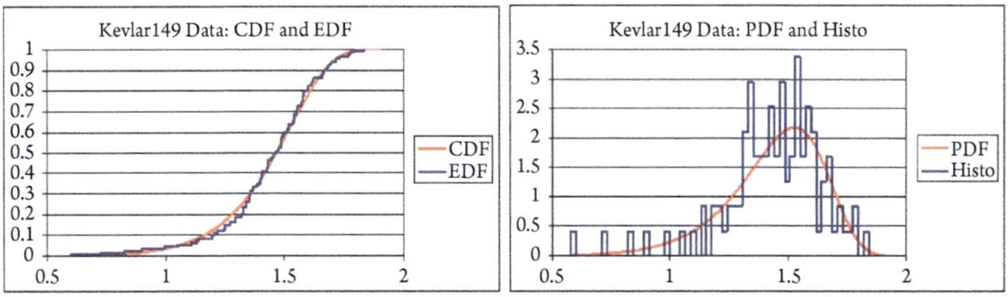

Figure 7.1 CDF with EDF, and PDF with frequency histogram, of the GEVMin and EVMin models fitted by ML to the Kevlar149 data. The two model fits are indistinguishable.

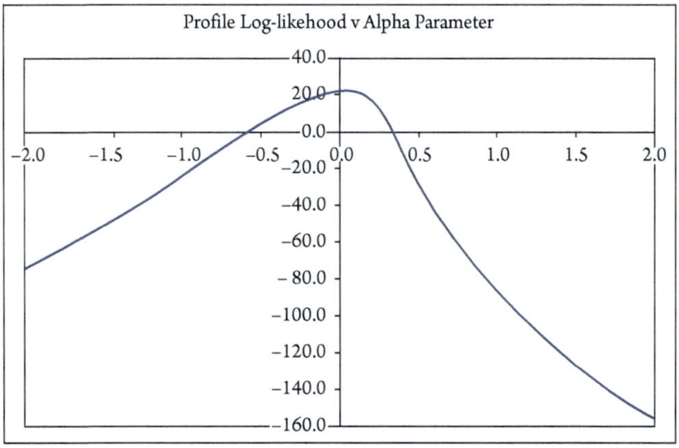

Figure 7.2 GEVMin profile log-likelihood plot for the Kevlar149 data. The profiling parameter is α. The maximum of the profile log-likelihood is at $\alpha = 0.026$.

same except for a change of sign in $i_{\alpha\mu}$ and $i_{\mu\sigma}$. For convenience we give the GEVMin version here in its one-observation form:

$$
\begin{aligned}
i_{\alpha\alpha} &= \alpha^{-2}\left[\frac{\pi^2}{6} + (1-\gamma-\alpha^{-1})^2 + 2\alpha q + \alpha^{-2}p\right], i_{\alpha\mu} = \alpha^{-1}\sigma^{-1}(q + p\alpha^{-1}),\\
i_{\alpha\sigma} &= \alpha^{-2}\sigma^{-1}\{1-\gamma-\alpha^{1}[1-\Gamma(2-\alpha)] - q - \alpha^{-1}p\}, i_{\mu\mu} = \sigma^{-2}p,\\
i_{\mu\sigma} &= -\alpha^{-1}\sigma^{-2}[p-\Gamma(2-\alpha)], i_{\sigma\sigma} = \alpha^{-2}\sigma^{-2}[1-2\Gamma(2-\alpha)+p],
\end{aligned}
$$

where

$$
\begin{aligned}
p &= (1-\alpha)^2\Gamma(1-2\alpha),\\
q &= \Gamma(2-\alpha)[\psi(1-\alpha) - \alpha^{-1}(1-\alpha)],\\
\gamma &\simeq 0.5772157 \text{ is Euler's constant.}
\end{aligned}
$$

Though not immediately obvious from the formulas, the information matrix has a nonsingular limit as $\alpha \to 0$, whose exact mathematical form is fairly tractable, but which we do not display here as it is easily obtained using an algebraic package like *Maple*®.

We give however a numerically based version suitable for practical work, like calculation of confidence intervals. The asymptotic covariance matrix is

$$I^{-1} = n^{-1}\begin{pmatrix} 0.476\,651\,189 & -0.258\,368\,234\,5\sigma & 0.146\,798\,399\,0\sigma \\ -0.258\,368\,234\,5\sigma & 1.248\,713\,088\sigma^2 & -0.336\,593\,939\,3\sigma^2 \\ 0.146\,798\,399\,0\sigma & -0.336\,593\,939\,3\sigma^2 & 0.653\,137\,877\,5\sigma^2 \end{pmatrix}.$$

The asymptotic variance of $\hat{\alpha}$ under the null hypothesis $H_0 : \alpha = 0$ is in fact

$$I^{\alpha\alpha}(0, \mu, \sigma) = n^{-1}360\pi^2[11\pi^6 - 2160\zeta^2(3)]^{-1} = 0.476\,651\,188\,9n^{-1}$$

where $\zeta(\cdot)$ is the Riemann zeta function. This variance is independent of μ and σ. This is the result given in Cheng and Iles (1990); that the asymptotic variance of a sample of size n is $(2.09797n)^{-1}$. As remarked by Cheng and Iles (1990), Hosking (1984) contains a typographical error where the variance is given as $2.098n^{-1}$.

Using this value for $I^{\alpha\alpha}(0, \mu, \sigma)$ gave test statistic values of $T_{LR}^{\frac{1}{2}} = 0.425$, $T_{W}^{\frac{1}{2}} = 0.383$, and $T_{S}^{\frac{1}{2}} = 0.459$ for the fitted EVMin model compared with the GEVMin model. All these values are well below the critical value of $z_{0.9} = 1.282$, the upper 10% point of the standard normal distribution, so that the formal test suggests that the EVMin model is quite satisfactory for this sample.

An alternative is provided by the bootstrap GoF tests using the method given in the flow chart starting at equation line (4.16). Figure 7.3 shows the results using $B = 1000$ BS samples, for the A^2 Anderson Darling and the W^2 Cramér–von Mises statistics, for both the GEVMin and EVMin fits to the Kevlar149 data set.

It will be seen that the BS null distributions of the test statistics behave similarly in all cases, with the GoF test statistic well below the 90% critical values obtained from the BS null distributions. Thus, for example, $A^2 = 0.350$ (green in the chart) for the fitted GEVMin model with an upper 10% critical value of $A^2 = 0.519$ (red). The test values were $A^2 = 0.394$ (green) for the fitted EVMin model with an upper 10% critical value of $A^2 = 0.655$ (red). These results confirm the visual fits displayed in Figure 7.1, showing that fitted EVMin model is as satisfactory as the fitted GEVMin model.

When calculating confidence intervals for threshold models like the GEVMin distribution, attention needs to be paid to regularity conditions. This will be discussed more fully in Chapter 8. For the GEVMin distribution, the key requirement for standard normal asymptotic theory to be applicable turns out to be that $\hat{\alpha}$ needs to satisfy $-0.5 < \hat{\alpha} < 0.5$, which it does in our example. The standard formula given in (4.1) is therefore applicable. Table 7.2 gives the 90% confidence intervals for the GEVMin parameters and also the EVMin parameters.

Confidence intervals can also be obtained for the parameters by BS sampling (with $B = 1000$). The percentile confidence intervals can be calculated using the method given in the flow chart beginning at equation line (4.3). The CIs obtained in this way are also reported in Table 7.2, showing fair agreement between asymptotic and BS approaches, with the BS intervals being fractionally wider in the main. As remarked in Section 4.1.5,

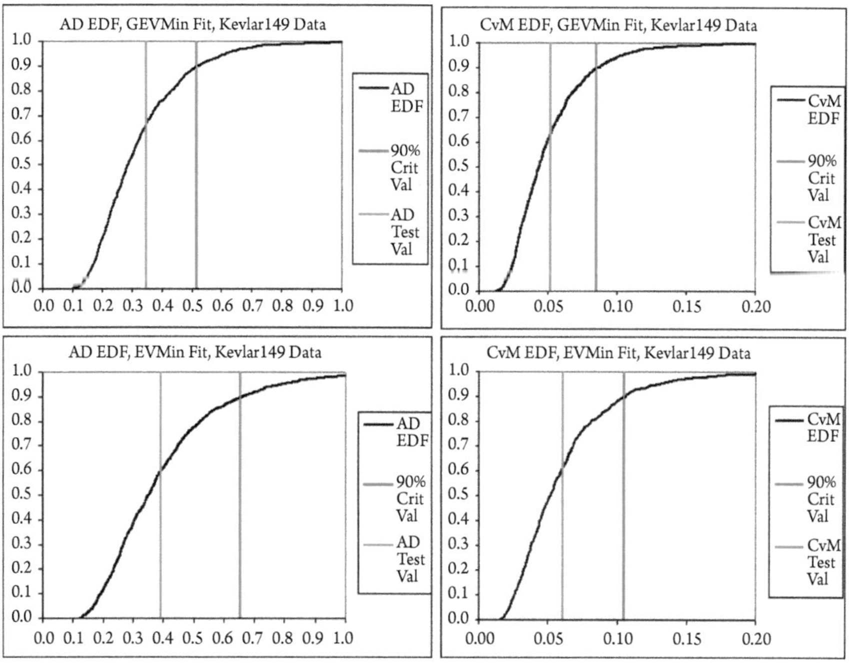

Figure 7.3 Bootstrap null EDFs of the A^2 and W^2 GoF test statistics corresponding to the GEVMin (upper charts) and EVMin (lower charts) fits to the Kevlar149 data. Red: 10% critical values; Green: test values; BS sample size: $B = 1000$.

Table 7.2 90% confidence intervals for the parameters of the GEVMin and EVMin distributions fitted to the Kevlar149 data

Model	Parameter	MLE	Asymp. CI Limits		BS CI Limits	
			Lower	Upper	Lower	Upper
GEVMin	α	0.026	−0.071	0.122	−0.096	0.149
	μ	1.53	1.50	1.56	1.500	1.56
	σ	0.169	0.148	0.190	0.145	0.189
EVMin	μ	1.53	1.51	1.56	1.50	1.56
	σ	0.168	0.147	0.189	0.145	0.189

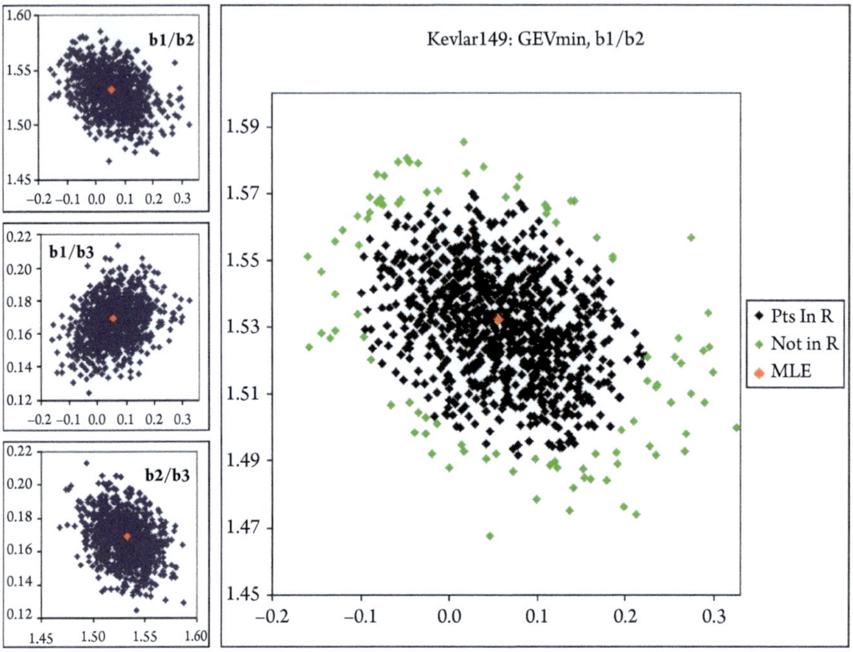

Figure 7.4 Kevlar149 Data: GEVMin BS ML parameter scatterplots; $b_1 \equiv \alpha$, $b_2 \equiv \mu$, $b_3 \equiv \sigma$.

the scatterplots of the BS parameter MLEs provide a useful check of whether the asymptotic results will be satisfactory or not. Figures 7.4 and 7.5 give the scatterplots of the BS parameter MLEs for the GEVMin and EVMin models. It will be seen that the plots display the typical symmetrical oval form of scatter appropriate to a multivariate normal distribution.

Finally, as discussed in Section 4.3, we can use BS sampling to calculate confidence bands or CIs. Figure 7.6 depicts the 90% BS confidence bands calculated for the entire CDFs of the GEVMin and EVMin models. These look satisfactory.

7.2 Carbon Fibre Failure Data

In the second example, we consider the four-parameter Burr XII distribution, fitting it to a sample comprising 66 observations of failure stresses of single carbon fibres of length 50 mm given by Crowder et al. (1991, Table 4.1d). Our main purpose is to illustrate the kind of calculations that can be carried out when fitting this particular model, but we have selected a data set where, as in the first example, there is a contextual basis for doing this—the Burr XII containing the Weibull distribution (for minima) as a special case, making it a possibly suitable distribution to use with such data. The data are reproduced in Table 7.3.

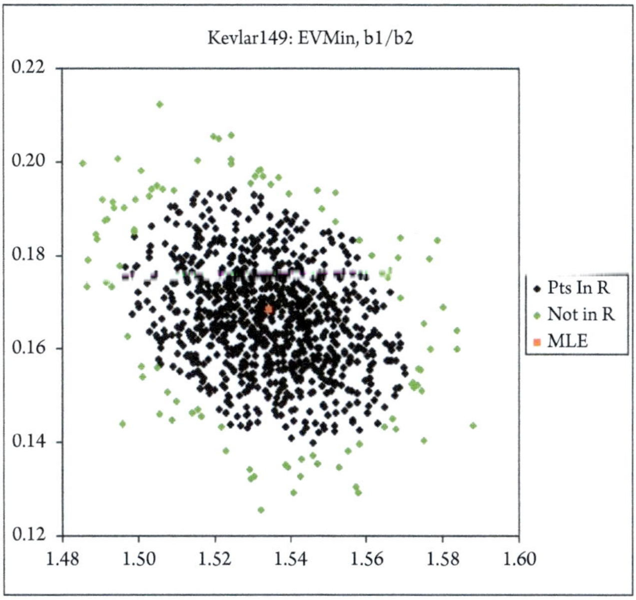

Figure 7.5 Kevlar149 Data: EVMin BS ML parameter scatterplots; $b_1 \equiv \mu$, $b_2 \equiv \sigma$.

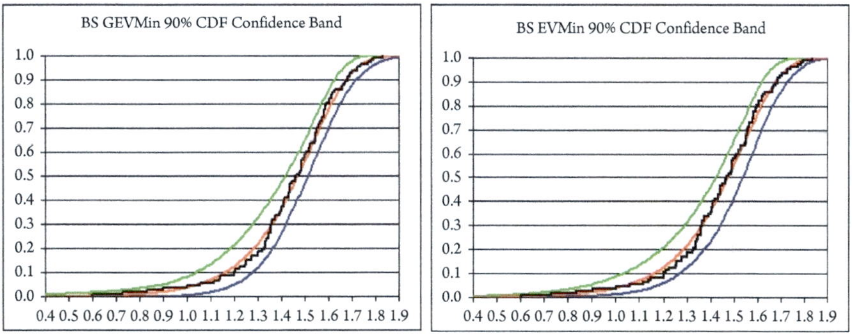

Figure 7.6 90% confidence bands for the GEVMin and EVMin models fitted to the Kevlar149 data.

The Burr XII distribution contains a hierarchy of embedded models, as depicted in the schematic figure at eqn (6.16). The easiest way of systematically fitting the Burr XII model is to fit the different models of the figure from the simplest upwards, testing the goodness-of-fit of each. One could in principle stop immediately at a model judged not unsatisfactory by the goodness-of-fit test. However, it is more robust to run through the whole hierarchy and making appropriate comparisons, before making a final selection of the 'best' model, if this is needed. As shown in Figure 6.16, the four-parameter Burr XII model is at the top level with its three embedded three-parameter models, the GEVMin,

Table 7.3 Failure stresses of single carbon fibre, length 50 mm

1.339	1.434	1.549	1.574	1.589	1.613	1.746	1.753	1.764	1.807
1.812	1.840	1.852	1.852	1.862	1.864	1.931	1.952	1.974	2.019
2.051	2.055	2.058	2.088	2.125	2.162	2.171	2.172	2.180	2.194
2.211	2.270	2.272	2.280	2.299	2.308	2.335	2.349	2.356	2.386
2.390	2.410	2.430	2.431	2.458	2.471	2.497	2.514	2.558	2.577
2.593	2.601	2.604	2.620	2.633	2.670	2.682	2.699	2.705	2.735
2.785	2.785	3.020	3.042	3.116	3.174				

Type II generalized logistic and Pareto next in the hierarchy, with the two simplest models, the EVMin and two-parameter exponential distributions, at the lowest level.

We fitted each of the six models separately, but, in presenting the results, the figures and tables highlight different aspects of interest in the fits, so that the models all appear together under each of the aspects discussed. The parametrizations used for each distribution are those listed in the tables unless stated otherwise. The figures and tables are grouped as follows:

(i) Figures 7.7 and 7.8 show the CDFs and PDFs of the fits.
There is an immediate point of interest here concerning the Pareto model. ML estimation using the three-parameter form given in Table 6.2 always yields

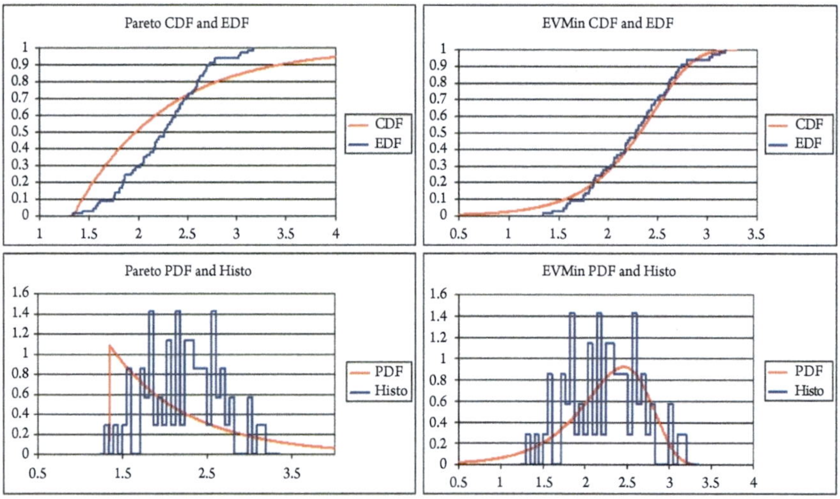

Figure 7.7 CDFs and PDFs of the Pareto/exponential (left charts) and EVMin (right charts) models fitted to the carbon fibre sample.

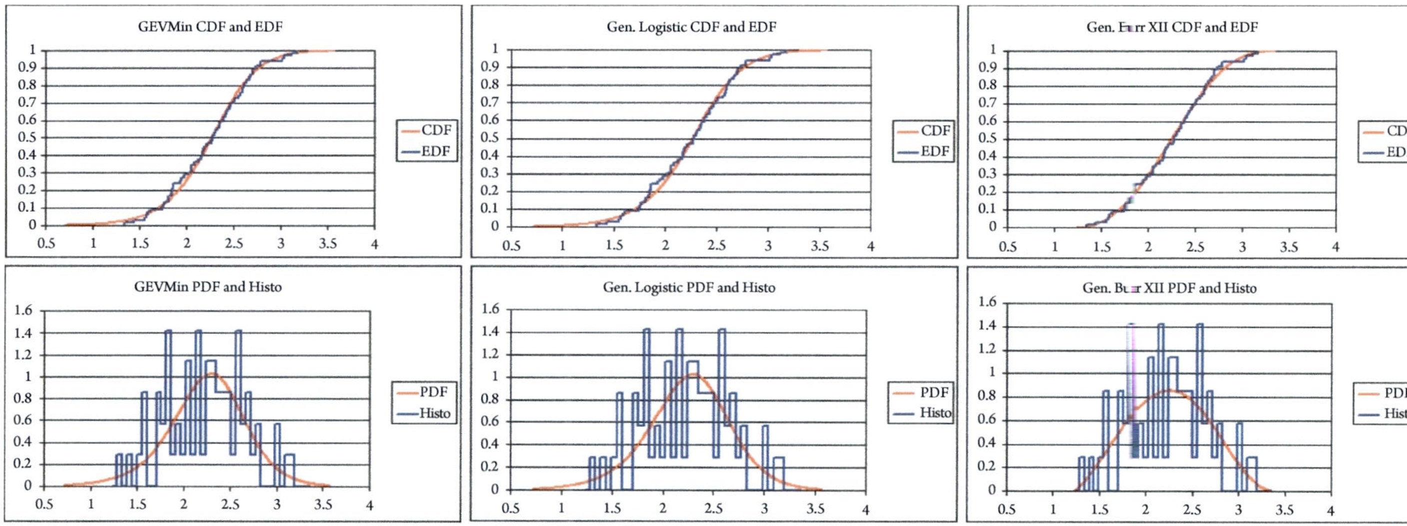

Figure 7.8 CDFs and PDFs of the GEVMin (left charts), Type II generalized logistic (middle charts), and Burr XII (right charts) models fitted to the carbon fibre sample.

the exponential model as being the best fit. We have therefore not fitted the exponential and Pareto models separately. For illustration, we have fitted the Pareto model in the numerical optimization, but, in anticipation of the result, using the transformed parametrization α, μ, σ of Table 6.2. The ML estimates are $\hat{\alpha} = 0$, $\hat{\mu} = 1.339$, and $\hat{\sigma} = 0.913$. In corroboration of this, fitting using the original parametrization, we find $\hat{a} \to -\infty$, $\hat{b} \to \infty$, and $\hat{\gamma} \simeq \hat{b}/\hat{\sigma}$. Concerning the fits, a cursory glance of the EDF and frequency histogram of the sample makes it plain that the Pareto/exponential distribution, with its convex decreasing PDF over its support, is not very appropriate. The EVMin is rather better but still not really a convincing fit. The GEVMin and Type II generalized logistic fits are very similar and look satisfactory, though their PDFs are slightly different from that of the full Burr XII model, which looks to be the best fit.

Table 7.4 90% confidence intervals for the parameters of selected distributions fitted to the carbon fibre data

Model	Parameter	MLE	Asymp. CI Limits		BS CI Limits	
MaxLogLik			Lower	Upper	Lower	Upper
EVMin	μ	2.46	2.37	2.54	2.36	2.55
-39.60	σ	0.399	0.336	0.462	0.331	0.458
Pareto	$\alpha = (1/\gamma)$	0	0	0	0	0
-60.01	μ	1.3389	1.33886	1.33894	1.3395	1.381
	σ	0.913	0.730	1.097	0.722	1.092
GEVMin	α	0.307	0.166	0.448	0.182	0.485
-34.99	μ	2.39	2.30	2.49	2.29	2.49
	σ	0.416	0.349	0.483	0.342	0.478
Type II Gen.	$d = 1/\alpha$	1.42	-0.0231	2.86	0.665	9.21
Logistic	μ	2.39	1.97	2.81	2.11	3.21
-36.81	σ	0.268	0.182	0.355	0.186	0.352
Gen. Burr	α	-0.260	-0.679	0.159	-1.01	-0.003
Type XII	a	1.22	0.930	1.51	1.04	1.54
-34.62	σ	1.26	0.995	1.52	1.00	1.72
	c	2.33	0.979	3.67	0.963	3.36

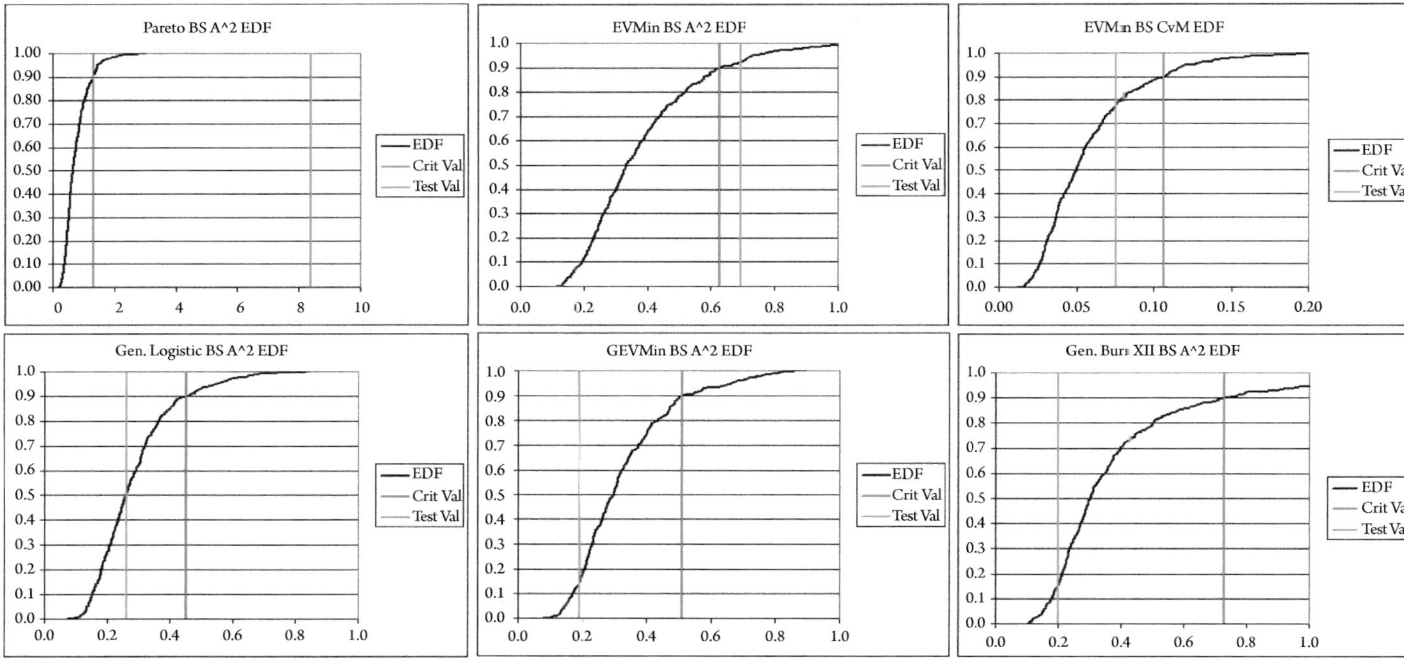

Figure 7.9 Carbon fibre data: BS EDFs of the null distribution of the A^2 GoF statistic for the extended Burr XII (lower right) model, and for its embedded models: GEVMin (lower middle), generalized Type II logistic (lower left), Pareto (upper left) and EVMin (upper middle). Also: The BS EDF of the null distribution of the W^2 GoF statistic for the EVMin model (upper right). Red line - 10% critical value; Green line - test value.

Confidence intervals, calculated both from asymptotic theory and by bootstrapping, for the parameter estimates for all the models are given in Table 7.4. Throughout this example, the bootstrap sample size was just $B = 500$. The confidence intervals were broadly similar, except for the parameter $d = 1/\alpha$ in the Type II generalized logistic fit and for the parameter α in the Burr XII fit. The much more variable CIs for these parameters perhaps reflected the greater uncertainty about the true value of α in these more flexible models.

(ii) Figure 7.9 shows the bootstrap EDF estimating the null distribution of the A^2 GoF statistic for each of the fitted models. The critical 90% value of the null distribution is depicted in red for each null distribution, with the test value based on the model fitted to the original sample shown in green. The Pareto test value indicates strong rejection of the model. The A^2 test in the EVMin

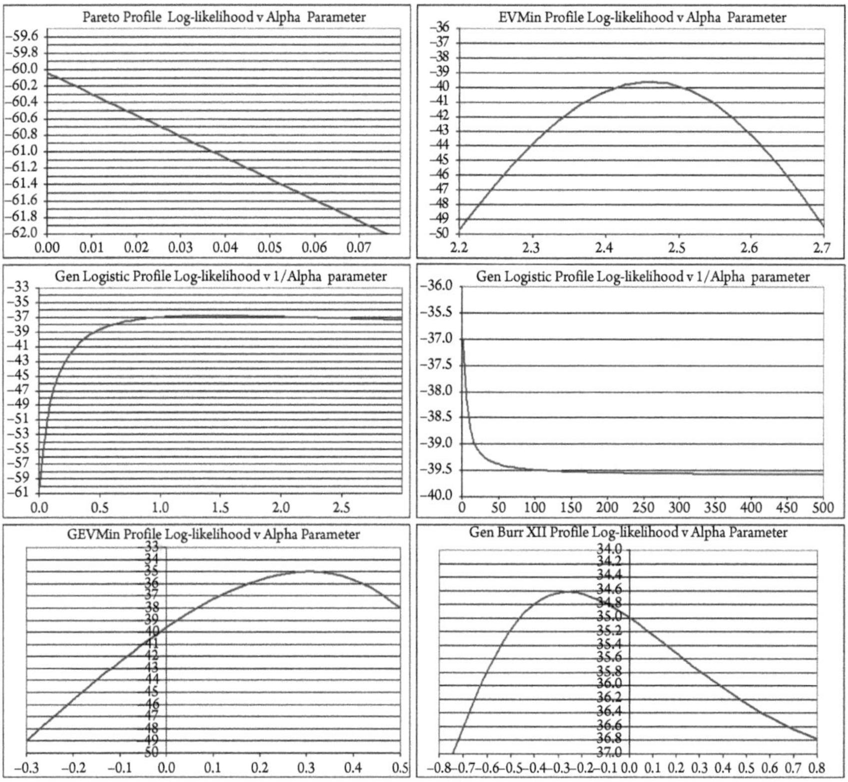

Figure 7.10 Profile log-likelihood plots for selected models fitted to the carbon fibre data. Top two charts: Pareto and EVmin models with α as profiling parameter. Middle two charts: Both are for the generalized Type II logistic model with α^{-1} as profiling parameter but on different scales. Bottom two charts: GEVMin and extended Burr XII models with α as profiling parameter.

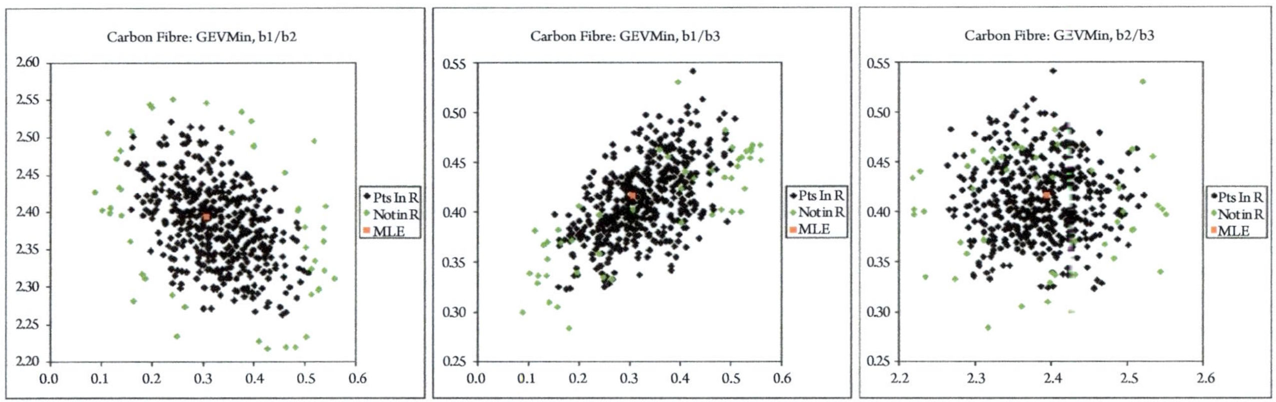

Figure 7.11 Scatterplots of BS parameter estimates for the GEVMin model fitted to the carbon fibre data; b1 $\equiv \alpha$, b2 $\equiv \mu$, b3 $\equiv \sigma$.

case also indicates a significant lack-of-fit. For all the other models, the A^2 value does not indicate a significant lack-of-fit. Though not shown here, similar results were obtained with the W^2 statistic, except in the EVMin case. The W^2 null EDF is shown in the figure, indicating that the W^2 statistic does not quite give a significant lack-of-fit at the 90% level in this case, so that it has produced a different result from the A^2 statistic in this case. This is an example of the finding of Stephens (1974) that A^2 is the more sensitive GoF statistic.

(iii) Figure 7.10 gives selected profile log-likelihood plots. In the Pareto case, we have already shown that the exponential distribution is always the best fit and this is corroborated with the $\max L = -60.01$ obtained at $\alpha = 0$. The 'alpha' parameter in EVMin profile log-likelihood (top right chart) is actually the location parameter μ, showing that the log-likelihood is maximized with $\max L = -39.60$ at $\hat{\mu} = 2.46$. The Type II generalized logistic profile log-likelihood is shown in the middle two charts in the figure with $d = \alpha^{-1}$ as the abscissa; the left-hand one for values of d near zero, with $d = 0$ corresponding to the exponential limit in the Type II generalized logistic case where $L = -60.01$. The log-likelihood is very flat near the maximum at $\hat{d} = 1.416$. The horizontal asymptote of $L = -39.60$ as $d \to \infty$ corresponds to the EVMin fit. The profiling parameter in the GEVMin chart (bottom left chart) is α, the shape parameter in the GEVMin PDF of eqn (6.23). The graph at $\alpha = 0$ corresponds to the EVMin fit, and has the value $\max L = -39.60$ corresponding to the maximum in the EVMin chart. Its maximum is greater with a value of $\max L = -34.99$ at $\alpha = 0.307$. The final plot (bottom right chart) shows the profile log-likelihood for the extended Burr XII,

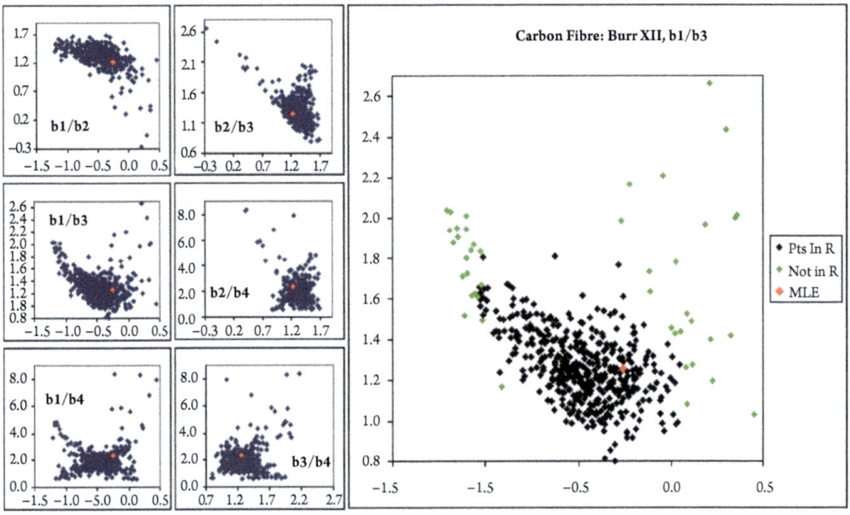

Figure 7.12 Scatterplots of BS parameter estimates for the extended Burr XII model fitted to the carbon fibre data; b1 $\equiv \alpha$, b2 $\equiv a$, b3 $\equiv \sigma$, b4 $\equiv c$.

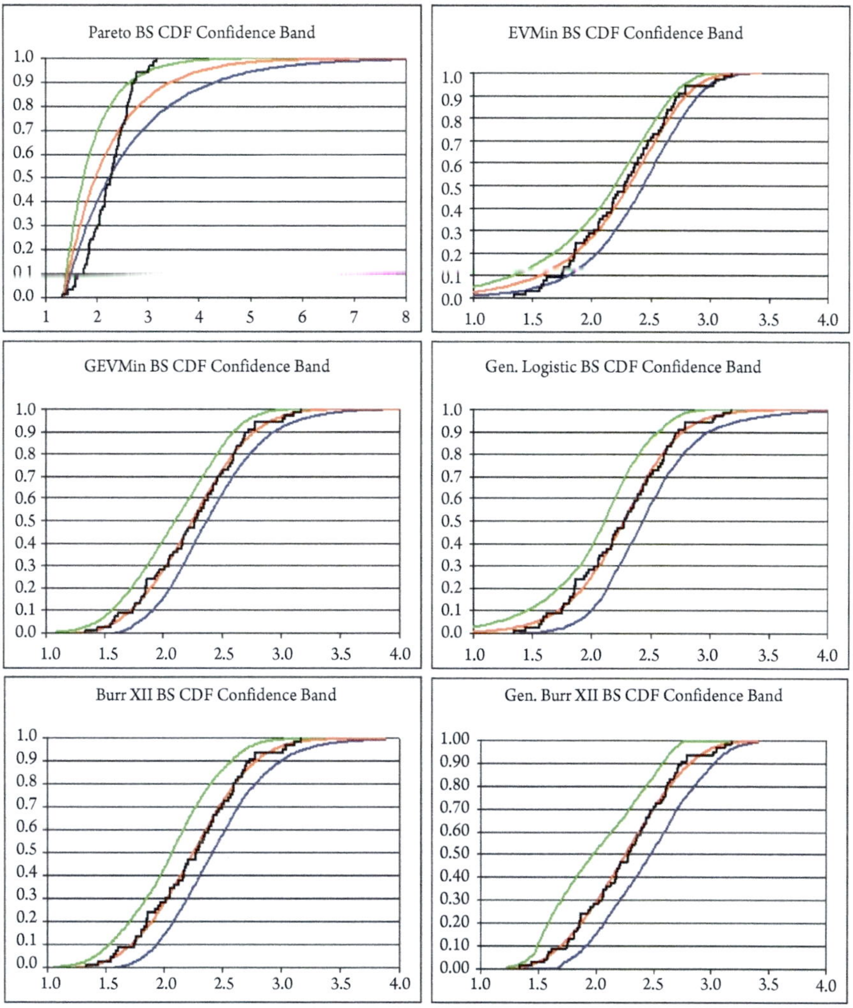

Figure 7.13 BS 90% confidence bands for the entire CDFs of the Pareto (top left), EVMin (top right), GEVMin (middle left), generalized Type II logistic (middle right), Burr XII (bottom left), and generalized Burr XII (bottom right) models fitted to the carbon fibre data.

where α, the shape parameter in the PDF as given in (6.41), is the profiling parameter. In this distribution, α is allowed to be negative. The extension in this case does lead to a fractionally higher maximum of $L = -34.62$ at $\alpha = -0.260$.

(iv) Figure 7.11 and Figure 7.12 give selected scatterplots which provide a simple means of examining whether asymptotic theory is applicable. The two figures are fairly typical in that they show the range of variation in the behaviour of point estimates. In the GEVMin case, the parameter estimates appear reasonably normally distributed, whereas the Burr parameter estimates are clearly non-normal.

(v) Figure 7.13 displays confidence bands within which each model should lie with 90% level of confidence. Assessment of the reliability of such bands should take into account goodness-of-fit results. Calculating a confidence band for the Pareto model is clearly pointless given rejection of fit by the GoF test. Though less dramatic, the fit of the EVMin is not very satisfactory, so that the confidence band is also not reliable. The GEVMin and Type II generalized logistic appear the most satisfactory. The bands for the two versions of the Burr XII model seem conservative, but their calculation is more dependent on normality of the estimates, and, as the scatterplots for the extended Burr model show, the distribution of the estimates is not normal. The effect on the confidence band is particularly noticeable in this model.

8 Infinite Likelihood

8.1 Threshold Models

The Weibull model listed in Table 6.2 is an example of a two-parameter distribution to which has been added a *threshold* or *shifted origin* parameter, a, that is, in the positively skewed form of the model, the left limit of the support of the PDF. When all the parameters, including a, are unknown, a problem can occur with maximum likelihood estimation, because in many (but not all) such models, there are always paths in the parameter space along which the likelihood $L \to +\infty$. Along such a path $a \to y_{(1)}$, the smallest observation, but the other two parameters b and c tend to values that are statistically inconsistent. This problem with the Weibull model was discussed in Section 2.2 of Chapter 2.

Other examples of shifted origin distributions are listed in Table 6.4. The same unbounded likelihood problem occurs in the gamma, lognormal, and loglogistic models, but not in the inverted gamma, inverse Gaussian, or log-gamma models.

The problem also occurs with the Burr XII distribution of eqn (6.9) discussed in Section 6.1.2, and also the Pareto distribution as listed in Table 6.2.

We indicate why the problem arises shortly and then discuss how to estimate the parameters in this situation, but we first illustrate how the problem becomes manifest, using two samples obtained by Steen and Stickler (1976) in a survey of beach pollution. Each sample gives the pollution level measured in number of coliform per 100 ml, observed on 20 different days. The samples are given in Table 8.1 and Table 8.2. We refer to the samples as 'SteenStickler0' and 'SteenStickler1'. The latter was actually considered previously by Cheng and Amin (1983), who showed that fitting the three-parameter Weibull fails using ML estimation. In our discussion that follows, by way of variation, we consider fitting the three-parameter loglogistic model for which SteenStickler1 is the less difficult sample in that the three-parameter loglogistic *can* be successfully fitted using ML, whereas SteenStickler0 is more difficult, as ML estimation fails for this model.

For the loglogistic distribution, we have the CDF

$$F(y) = w/(1+w), \text{ where } w = [(y-a)/b]^c, \tag{8.1}$$

Non-Standard Parametric Statistical Inference. Russell Cheng.

Table 8.1 SteenStickler0: first sample of the pollution level on a South Wales beach observed by Steen and Stickler (1976) on 20 different days

109	111	154	200	282
327	336	482	718	900
918	1045	1082	1345	1454
1918	2120	5900	6091	53600

Table 8.2 SteenStickler1: second sample of the pollution level on a South Wales beach observed by Steen and Stickler (1976) on 20 different days

1364	2154	2236	2518	2527
2600	3009	3045	4109	5500
5800	7200	8400	8400	8900
11500	12700	15300	18300	20400

and PDF as given in Table 6.4, namely:

$$f(y) = \frac{c}{b}\left(\frac{y-a}{b}\right)^{c-1}\left[1+\left(\frac{y-a}{b}\right)^{c}\right]^{-2}, y > a; b,\ c > 0. \tag{8.2}$$

The problem with using ML estimation is most easily illustrated by plotting the profile log-likelihood

$$L^*(a) = \max_{b,c} L(a, b, c|\mathbf{y}),$$

where the threshold a is the profiling parameter. The value of $a = \hat{a}$ at which $L^*(a)$ is maximum is the ML estimator. The two upper charts in Figure 8.1 show the profile log-likelihood plots of the loglogistic distribution for the two samples. The horizontal scale is $x = -\ln(y_{(1)} - a)$, so that $x \to \infty$ as $a \uparrow y_{(1)}$.

Comparing the two loglogistic profile log-likelihoods, it will be seen that in the case of SteenStickler0, the profile log-likelihood $L^*(a)$ is monotonically increasing in x, so that the only possible maximum is that approached as $x \to \infty$, that is, as $a \uparrow y_{(1)}$. In the case of SteenStickler1, there is a local maximum of $L^*(a)$ obtained at $\hat{a} < y_{(1)}$. However, this local maximum is ultimately surpassed with $L^*(a) \to \infty$ as $a \uparrow y_{(1)}$.

As in the Weibull model discussed in Section 2.2, the global maximum of the log-likelihood is obtained at $a = y_{(1)}$ for both SteenStickler samples. The estimator $\hat{a} = y_{(1)}$ is obviously consistent for a. However, the estimators of the two other parameters are not consistent.

We will return to this example to discuss it more fully in Section 8.5.1.

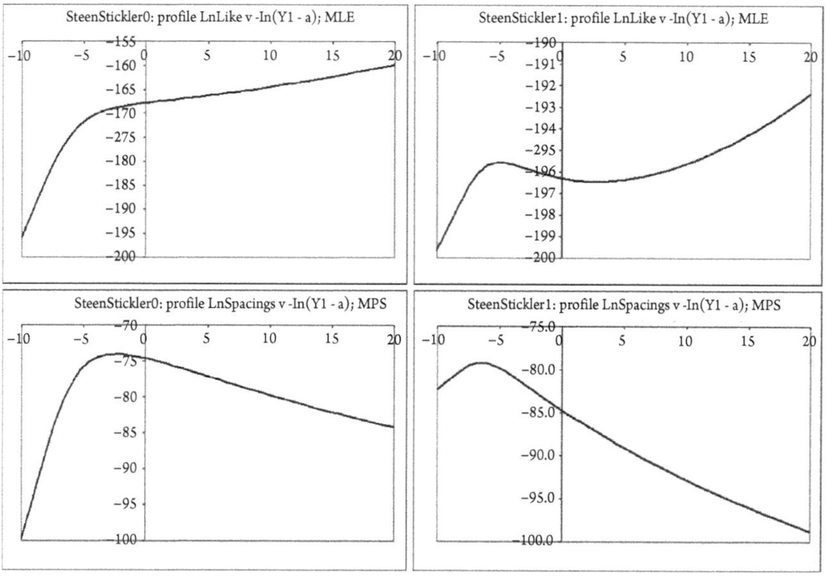

Figure 8.1 Profile log-likelihood (upper charts) and logspacings (lower charts) plots for the loglogistic distribution fitted to the SteenStickler0 (left-hand charts) and SteenStickler1 (right-hand charts) data samples.

8.2 ML in Threshold Models

Harter and Moore (1965, 1966) suggested taking the parameter point giving the local, rather than the global, maximum to be the ML estimator. This method can be used in the SteenStickler1 sample. However, this strategy obviously cannot be used with the SteenStickler0 sample in the loglogistic case where there is no local maximum, so ML obviouosly fails in this case. Methods based on searches such as those given by Wingo (1975, 1976) cannot always avoid the problem in these instances.

Smith (1985) gives the most thorough theoretical account of the behaviour of the ML estimator, showing how this depends on the degree of contact of the PDF with the abscissa at the threshold origin. Specifically, Smith considers PDFs of the form

$$f(y; a, \phi) = (y - a)^{c-1} g(y - a; \phi),\ a < y < \infty, \tag{8.3}$$

where a is the threshold and ϕ is a p-dimensional vector, $p \geq 1$, which can include c but does not include a. We assume that as $y \to a$,

$$g(y - a; \phi) \to g_0(\phi) = cd(\phi), \tag{8.4}$$

where $d(\phi)$ is a positive quantity depending only on ϕ. This formulation includes the Weibull, gamma, and loglogistic models, with ϕ comprising a scale parameter $b > 0$

and c the power parameter, with $g_0(\phi) = c/b^c$ for the Weibull and loglogistic cases, and $g_0(\phi) = 1/[\Gamma(c)b^c]$ in the gamma case.

Let

$$\begin{aligned} M_0 &= (m_{i,j}(\phi)),\ 0 \le i,j \le p \\ &= -E_\phi\left[\left(\frac{\partial^2}{\partial\phi^i\partial\phi^j}\right)\ln f(Y;0,\phi)\right],\ \text{with}\ \frac{\partial}{\partial\phi^0} \equiv -\frac{\partial}{\partial y}. \end{aligned}$$

So M_0 is essentially the $(p+1) \times (p+1)$ matrix of expected values of the negative Hessian of second derivatives with respect to the full set of parameters appearing in the formula (3.4) for the asymptotic variance of the MLE under standard regularity conditions. Define also

$$M_1 = (m_{i,j}(\phi)),\ 1 \le i,j \le p,$$

which is simply M_0 with the zeroth row and column omitted. Smith (1985) gives smoothness assumptions 1, 6, and 8 on $f(y; a, \phi)$ and $g(y-a; \phi)$, including $m_{i,j}(\phi)$ being finite, with

$$\begin{aligned} E_\phi\left[\left(\frac{\partial}{\partial\phi^i}\right)\ln f(Y;0,\phi)\right] &= 0,\ i = 1,2,\ldots,p, \\ E_\phi\left[\left(\frac{\partial}{\partial\phi^0}\right)\ln f(Y;0,\phi)\right] &= 0,\ \text{if}\ c > 1 \end{aligned}$$

and

$$m_{00} > 0,\ \text{if}\ c > 2,$$

under which the following theorems hold.

Theorem 1 (Smith 1985)

If $c > 1$, and

(i) M_0 is positive definite when $c > 2$,
(ii) M_1 is positive definite when $1 < c \le 2$,

then there exists a sequence of solutions $(\hat{a}_n, \hat{\phi}_n)$ as $n \to \infty$ of the likelihood equations such that

$$\hat{a}_n - a_0 < O_p[(n\xi_{n,c})^{-1/2}],\ \hat{\phi}_n - \phi_0 < O_p(n^{-1/2}),$$

where $\xi_{n,c} = 1$ if $c > 1$, $\xi_{n,c} = \ln n$ if $c = 1$, and $\xi_{n,c} = n^{1/c-1}$ if $0 < c < 1$.

Moreover, if $c = 2$,

$$\hat{a}_n - \bar{a}_n < O_p[n^{-1/2}(\ln n)^{-1}],\ \hat{\phi}_n - \bar{\phi}_n < O_p[(n\ln n)^{-1/2}],$$

and if $1 < c < 2$,

$$\hat{a}_n - \bar{a}_n < [O_p(n^{-\frac{2}{c}+\frac{1}{2}})],\ \hat{\phi}_n - \bar{\phi}_n < O_p(n^{-1/c}),$$

where $\bar{\phi}_n$ is the maximum likelihood estimator of ϕ when a_0 is known, and $\bar{a}_n$ is the maximum likelihood estimator of a when ϕ_0 is known.

This theorem gives the order of convergence of estimators but not their asymptotic form. This latter is given by

Theorem 3 (Smith 1985)

Let $(\hat{a}_n, \hat{\phi}_n)$ be a sequence of estimators satisfying Theorem 1, with d as defined in eqn (8.4). Then.

(i) If $c > 2$, $n^{1/2}(\hat{a}_n - a_0, \hat{\phi}_n - \phi_0)$ converges in distribution to a normal random vector with mean $\mathbf{0}$ and covariance matrix M_0^{-1}.

(ii) If $c = 2$, then $[(nd \ln n)^{1/2}(\hat{a}_n - a_0), n^{1/2}(\hat{\phi}_n - \phi_0)]$ converges to a random normal vector with with mean $\mathbf{0}$ and covariance of the form

$$\begin{bmatrix} 1 & 0 \\ 0 & M_1^{-1} \end{bmatrix}.$$

(iii) If $1 < c < 2$, then $[(nd)^{1/c}(\hat{a}_n - a_0), n^{1/2}(\hat{\phi}_n - \phi_0)] \to (W, Z)$ in distribution where W and Z are independent, Z is normal with mean $\mathbf{0}$ and covariance matrix M_1^{-1}, and W is a complicated distribution given by Woodroofe (1974, Theorem 2.4).

In a corollary to Theorem 3, Smith shows that $-(\ln n)^{-1}\partial^2 L_n(\hat{a}_n, \hat{\phi}_n)/\partial a^2 \to d$ (c in Smith's notation), so that when $c = 2$,

$$(\hat{a}_n - a_0)[-n\partial^2 L_n(\hat{a}_n, \hat{\phi}_n)]^{1/2} \to Z \text{ in distribution,}$$

where Z is standard normal. This shows that the variance can be estimated by the *observed* information matrix, even though the *expected* information does not exist.

The two theorems of Smith just presented give a precise insight into the behaviour of MLEs in threshold models of the form (8.3), but are somewhat complicated to apply when $1 < c \leq 2$, and do not cover the case when $0 < c < 1$. Smith (1985) suggests, for the case $0 < c < 2$, what we shall call the *modified-ML* method of taking $\breve{a} = y_{(1)}$, the smallest order statistic, and estimating the remaining parameters from the reduced sample $y_2, \ldots, y_n$, by ML with a set to $\breve{a} = y_{(1)}$. The asymptotic distribution of $\tilde{\phi}_n$, the MLE of ϕ obtained from the reduced sample $y_2, \ldots, y_n$, with a set to $\breve{a} = y_{(1)}$ is regular. We have

Smith (1985, Corollary to Theorem 4)

If $c < 2$, then $n^{1/2}(\tilde{\phi}_n - \phi_0)$ converges to the normal distribution with mean 0 and covariance matrix $M_1^{-1}(\phi_0)$.

This result allows confidence intervals to be calculated for the components of ϕ_0, treating $\breve{a} = y_{(1)}$ as if it were the true value a_0. Moreover, the asymptotic distribution of $n^{1/c}(\breve{a} - a_0)$ is then Weibull, with $W = n^{1/c}[g_0(\phi)]^{1/c}(Y_{(1)} - a)/b$ the standard Weibull random variable with CDF $F_W(w) = 1 - \exp(-w^c)$. Thus, a confidence interval (a_-, a_+) for a can be calculated from

$$n^{1/c}[g_0(\tilde{\phi})]^{1/\tilde{c}}(Y_{(1)} - a_\pm)/\tilde{b} = w_\pm, \tag{8.5}$$

where $w_\pm$ are appropriate quantiles of the standard Weibull distribution with $c = \tilde{c}$.

However, if $c > 2$, it would be unsatisfactory to take $\breve{a} = y_{(1)}$, as ML estimation is then standard for all the parameters including a, with the asymptotic distribution of $(\hat{a}_n, \hat{\phi}_n)$ as given in Theorem 3(i). An overall method would therefore need to first identify a local maximum, even when this is not very pronounced, applying Theorem 3(i) in this case, and only switch to $\breve{a} = y_{(1)}$ if a local maximum cannot be found.

In the next section, we discuss an estimation method using *spacings*, which applies for all $c > 0$.

8.3 Definition of Likelihood

Barnard (1967) and Giesbrecht and Kempthorne (1976) argue that the unbounded likelihood problem arises because the observations are treated as exact values drawn from a continuous model when this is only an approximation. They argue that if rounding error is taken into account, with the observations grouped into cells so that one can be reasonably sure that all observations are in their correct cells, then the model is always discrete with observations drawn from a multinomial distribution whose cell probabilities can be calculated in terms of the parameters of the continuous distribution. ML estimation using the multinomial likelihood then successfully estimates the parameter values.

This does resolve the unbounded likelihood problem, although some loss of efficiency of the threshold parameter estimate may result as the precise position of $y_{(1)}$ is not taken into account. For instance, this approach is not an efficient method of estimating the threshold parameter in J-shaped distributions. In addition, choice of the grouping interval size is often subjective, with the possibility of the results being sensitive to the choice. We shall not discuss this approach further.

Treating the original sample as being discrete and multinomially distributed is a neat practical way of circumventing the problem, but does not provide a genuine explanation of why ML fails. The argument does not deal with the fact that the unbounded likelihood problem will occur with *any* sample $\mathbf{y}$. Even if it is supposed that all observations y_i are subject to error, we can nevertheless assume that each has a true value y_i^0. If, therefore, we replace all the y_i by their true values y_i^0, then we would have a sample $\mathbf{y}^0 = (y_1^0, y_2^0, \ldots, y_n^0)$ that definitely came from the original continuous model without any rounding error. The Barnard or Giesbrecht and Kempthorne argument that the sample has rounding errors cannot therefore be applied to $\mathbf{y}^0$. But the unbounded likelihood problem will *still* occur with the sample $\mathbf{y}^0$, as there are *always* paths in the parameter space where $a \to y_{(1)}$ for which $L \to +\infty$, including $y_{(1)} = y_{(1)}^0$, whatever its value.

Cheng and Iles (1987) argue that the difficulty arises because the conventional definition of the likelihood function is not appropriate in shifted origin models.

For continuous parametric densities, ML estimation is defined as the maximization (with respect to the parameters) of the likelihood function as given by (1.3) or the log-likelihood (1.4). Cheng and Iles (1987) point out that a more fundamental definition of ML estimation is to maximize an actual probability, specifically the *probability differential*

$$dL(\boldsymbol{\theta}, \mathbf{y}) = \lim_{\delta \to 0} \ln \left[\prod_{i=1}^{n} \delta F(y_i, \boldsymbol{\theta}) \right]. \tag{8.6}$$

In regular cases, we can write

$$\delta F(y_i, \boldsymbol{\theta}) = \int_{y_i}^{y_i+\delta y} f(y, \boldsymbol{\theta}) dy,$$

when the log-likelihood can then be approximated by a rectangular area $\delta F(y_i, \boldsymbol{\theta}) = f(y_i, \boldsymbol{\theta})\delta y$, giving a log-likelihood of

$$\sum_{i=1}^{n} \ln(f(y_i, \boldsymbol{\theta})\delta y) = \sum_{i=1}^{n} \ln f(y_i, \boldsymbol{\theta}) + n \ln \delta y.$$

Maximization of (8.6) is then equivalent to maximization of $\sum \ln f(y_i, \boldsymbol{\theta})$, since $n \ln \delta y$ can be disregarded, as it is independent of $\boldsymbol{\theta}$. However, the steps followed in this argument are that

$$\begin{aligned} \arg\max_{\boldsymbol{\theta}} dL(\boldsymbol{\theta}, \mathbf{y}) &= \arg\max_{\boldsymbol{\theta}} \lim_{\delta \to 0} \ln \left[\prod_{i=1}^{n} \delta F(y_i, \boldsymbol{\theta}) \right] \\ &= \arg \lim_{\delta \to 0} \max_{\boldsymbol{\theta}} \ln \left[\prod_{i=1}^{n} \delta F(y_i, \boldsymbol{\theta}) \right] \\ &= \arg\max_{\boldsymbol{\theta}} \left(\sum_{i=1}^{n} \ln f(y_i, \boldsymbol{\theta}) \right) \end{aligned}$$

which assumes that the operations of letting $\delta \to 0$ and of maximization with respect to $\boldsymbol{\theta}$ can be interchanged.

In the case of shifted origin distributions this interchange is not valid.

The situation is equivalent to calculation of the derivative of an integral where the operations of integration and differentiation can be interchanged if the integral is sufficiently wellbehaved. Thus, the interchange is invalid if the limits of the integration depend on the parameter being differentiated with respect to.

The unboundedness problem is avoided by replacing the likelihood function by a closely related function called the *spacings function,* in which the PDF values in the likelihood are replaced by probability increments. Estimators can be obtained by maximizing the spacings function. This way of estimating parameters is called the *maximum product of spacings* (MPS) estimation method, and was suggested by Cheng and Amin (1983) and independently by Ranneby (1984). A good introductory review of the properties of MPS estimation is given by Ekström (2008), who points out the practical usefulness of MPS, the aspect emphasized by Cheng and Amin (1983), but links this in with the more theoretical Kullback-Leibler information approach adopted by Ranneby (1984). We describe the main characteristics of the method in the next section, emphasizing the application aspects.

8.4 Maximum Product of Spacings

The MPS method, suggested by Cheng and Amin (1983) and by Ranneby (1984), avoids the problem of an unbounded likelihood by maximizing the *spacings function*

$$H(\boldsymbol{\theta}) = \frac{1}{n+1}\sum_{i=1}^{n+1} \ln D_i(\boldsymbol{\theta}), \tag{8.7}$$

where $D_i(\boldsymbol{\theta}) = F(y_i; \boldsymbol{\theta}) - F(y_{i-1}; \boldsymbol{\theta})$, $i = 1, \ldots, n+1$, are the spacings, with $y_0 = -\infty$, $y_{n+1} = +\infty$. If the D_i could be freely chosen, subject only to $0 \leq D_i$ all i, and $\sum_i D_i = 1$, with $H = H(D_1, D_2, \ldots, D_{n+1})$ treated simply as a function of the D_i, we have the upper bound

$$H(D_1, D_2, \ldots, D_{n+1}) \leq \max_{D_i} \frac{1}{n+1}\sum_{i=1}^{n+1} \ln D_i = -\ln(n+1), \tag{8.8}$$

with the bound obtained when all the D_i are equal, with $D_i = (n+1)^{-1}$ for all i. The properties of spacings and of H are described by Darling (1953). Now the $D_i(\boldsymbol{\theta})$, as defined in $H(\boldsymbol{\theta})$, depend explicitly on $\boldsymbol{\theta}$, and so will only vary as $\boldsymbol{\theta}$ varies. Moreover, the $D_i(\boldsymbol{\theta})$ also depend on $\mathbf{y}$, making them random variables, with the property that if $\boldsymbol{\theta}^0$ is the true value of $\boldsymbol{\theta}$, then setting $\boldsymbol{\theta} = \boldsymbol{\theta}^0$ will make the $D_i(\boldsymbol{\theta})$ identically distributed. We can view this as the statistical analogue of setting all the D_i equal in the deterministic situation where the D_i can be freely chosen. A simple way of trying to get as close to this deterministic case as possible is to choose $\boldsymbol{\theta}$ so that $H(\boldsymbol{\theta})$ is as large as possible, and this is the basis of our proposed estimator.

Cheng and Amin (1983) therefore define the *maximum product of spacings* (MPS) estimator as

$$\tilde{\boldsymbol{\theta}} = \arg\max_{\boldsymbol{\theta}} H(\boldsymbol{\theta}).$$

Ranneby (1984) proposed the same estimator from a different viewpoint, regarding the estimator as an approximation to Kullback-Leibler information.

In view of the upper bound (8.8), the unboundedness problem clearly cannot occur with MPS estimation. As an illustration of this, consider the SteenStickler samples. The two lower charts of Figure 8.1 depict the profile logspacings likelihood

$$H^*(a) = \sup_{b,c} H(a,b,c),$$

and it will be seen that there is a clear and unambiguous maximum obtained with $\hat{a} < y_{(1)}$. We consider the MPS estimators in these samples in more detail at the end of this chapter.

8.4.1 MPS Compared with ML

We now contrast the different behaviour of MPS and ML estimators in threshold models of the type given in eqn (8.3). This is exemplified by the Weibull and gamma distributions, the MPS estimators of which were discussed explicitly by Cheng and Amin (1983). We use the parametrizations as given in Tables 6.2 and 6.4 for the Weibull and gamma distributions, respectively. Though the results of Cheng and Amin (1983) are not as detailed as those given by Smith (1985) in the ML case, they are sufficient to show the clear advantage of using MPS estimation, identifying the properties of the MPS estimators sufficiently precisely to provide a simple working estimation methodology.

Cheng and Amin (1983, Theorem 1), show that, as opposed to ML, there are only two distinct forms of behaviour of MPS estimators (MPSE), depending on whether the true power parameter is $c < 2$ or $c > 2$. Explicitly, we have

Cheng and Amin 1983, Theorem 1

(i) When $c > 2$, the MPS estimator possesses the same asymptotic properties as ML when $c > 2$, so that there is in probability a solution $\tilde{\boldsymbol{\theta}} = (\tilde{a}, \tilde{b}, \tilde{c})$ of

$$\partial H / \partial \boldsymbol{\theta} = \mathbf{0},$$

for which

$$\sqrt{n}(\tilde{\boldsymbol{\theta}} - \boldsymbol{\theta}) \xrightarrow{d} N\{0, -nE(\partial^2 L / \partial \boldsymbol{\theta}^2)^{-1}\} \text{ as } n \to \infty.$$

(ii) If $0 < c < 2$, then there is in probability a solution $\tilde{\boldsymbol{\theta}}$ where

$$\tilde{a} - a = O_p(n^{-1/c}),$$

and where for $\tilde{\boldsymbol{\phi}} = (\tilde{b}, \tilde{c})$:

$$\sqrt{n}(\tilde{\boldsymbol{\phi}} - \boldsymbol{\phi}) \xrightarrow{d} N\{0, -nE(\partial^2 L / \partial \boldsymbol{\phi}^2)^{-1}\} \text{ as } n \to \infty.$$

Thus, when $c < 2$, there exists in probability a solution where the MPSE $\tilde{a}$ is *hyper-efficient* with $\tilde{a} - a = O_p(n^{-1/c})$, so that its asymptotic variance is of smaller order than n^{-1}, with the property that $\tilde{\boldsymbol{\phi}}$ is asymptotically normally distributed as if a were known.

The proof of the theorem is given in Cheng and Amin (1982). See also Cheng and Amin (1979).

Compared with the MLE, it will be seen that for $c > 2$ and $1 < c < 2$, the asymptotic properties of the MPSE are the same as the MLE. However, the MPSE is still valid when $0 < c < 1$, whilst the MLE $\hat{\boldsymbol{\phi}}$ becomes inconsistent. Thus MPS estimators are asymptotically at least as efficient as ML estimators when these exist. Moreover, even when the distribution is J-shaped and ML fails, asymptotically efficient MPS estimators are still obtainable.

The theorem does not cover the cases $c = 1$ and $c = 2$, whereas the results of Smith (1985) do. Also, in the case $0 < c < 1$, we do not have the theoretical asymptotic distribution of the MPSE $\tilde{a}$.

However, in practical terms, our results for MPS are sufficient to allow a practical implementation covering essentially all $c > 0$. When they are required, the distributions of all MPSEs are easily estimated using bootstrap resampling. Thus, confidence interval calculations can be handled in a straightforward manner.

8.4.2 Tests for Embeddedness when Using MPS

In Section 6.2, we gave tests for the existence of an embedded model for three-parameter distributions with a shifted threshold when ML is used. Thornton (1989) uses a similar approach to obtain test statistics for three-parameter distributions when using MPS.

Nakamura (1991) gives a condition, eqn (3.3) for interval-censored data, to derive a more general spacings statistic to test if the embedded model is the best fit for interval-censored data from three-parameter distributions of power-location-scale type, that is, where the distribution has a CDF of the form $F\{[(y-a)/b]^c\}$.

If $Y_1,\ldots, Y_n$ are n independent observations from such a three-parameter distribution, with each Y_i lying in an interval C_i with non-empty interior, then $\mathbf{C} = \{C_1, \ldots, C_n\}$ are the interval-censored data. The finite endpoints of $\mathbf{C}$ are arranged in order of magnitude and denoted by $y_1, \ldots, y_m$, with $y_0 = -\infty$ and $y_{m+1} = \infty$. Let the cell counts $n_{i,j}$, $0 \le i < j \le m+1$ be defined to be the number of C_k, $1 \le k \le n$, such that the lower and higher endpoints of C_k are y_i and y_j, respectively. Then the log-likelihood for $\mathbf{C}$ is

$$L(\boldsymbol{\theta}) = \text{const} + \sum_{j=1}^{m+1}\sum_{i=0}^{j-1} n_{i,j} \ln\left(F(y_j;\boldsymbol{\theta}) - F(y_i;\boldsymbol{\theta})\right), \tag{8.9}$$

where $F(-\infty) = 0$ and $F(\infty) = 1$. Condition (3.3) of Nakamura (1991) is derived from examination of the behaviour of this log-likelihood near a type of boundary of the parameter space, termed the *probability contents inner boundary* by Nakamura. We do not write out condition (3.3) in full here, but note that if all the terms of the condition are put on the left-hand side, it is then a simple inequality, which we shall refer to as

$$N_1 < 0. \tag{8.10}$$

When satisfied, this means that the best fit is obtained with all parameters finite.

Following the grouped likelihood approach to MPS estimation suggested by Titterington (1985), Nakamura points out that condition (3.3) can be applied to MPS estimators for ordinary random samples by interpreting the method as being that of ML applied to interval-censored data. That is, if $[y_i, y_{i+1}), i = 0, \ldots, n$, are regarded as the cells of an interval-censored sample with one observation in each cell, then the logspacings function is the log-likelihood of this sample. Here, the cell counts are $n_{j-1,j} = 1 \quad (j = 1, \ldots, n+1)$ and zero otherwise, and (8.9) becomes the logspacings function taken in the form

$$N = \sum_{i=1}^{n+1} \ln D_i(\alpha, \mu, \sigma),$$

where

$$D_i = \int_{y_{i-1}}^{y_i} f(y; \alpha, \mu, \sigma)\, dy.$$

These cell counts satisfy the conditions (3.1), (3.2) of Nakamura (1991). We examine the form N_1 in three-parameter models $f(y, \boldsymbol{\theta})$, where $\boldsymbol{\theta} = (a, b, c)$ of the form (8.3) and not simply of power-scale-location form. We suppose that (α, μ, σ) is a reparametrization for which the log-likelihood has the series expansion

$$L = L_0(\mu, \sigma) + L_1(\mu, \sigma)\alpha + R,$$

with $L_0(\mu, \sigma)$ the log-likelihood of the embedded model in the original parametrization, $L_1(\mu, \sigma)$ the basis of the score statistic, and R is $O_p(\alpha^2)$. Expressed in terms of the density function,

$$\begin{aligned} L &= \sum_{i=1}^{n} \ln f(y_i, \alpha, \mu, \sigma) \\ &= \sum_{i=1}^{n} \ln \left(f_0(y_i, \mu, \sigma) + f_1(y_i, \mu, \sigma)\alpha\right) \\ &= \sum_{i=1}^{n} \left(\ln f_0(y_i, \mu, \sigma) + \frac{f_1(y_i, \mu, \sigma)}{f_0(y_i, \mu, \sigma)}\alpha + O(\alpha^2) \right). \end{aligned} \tag{8.11}$$

Thus, for an individual term $L_1(y_i, \mu, \sigma) = f_1(y_i, \mu, \sigma)/f_0(y_i, \mu, \sigma)$, and this relationship can be used when considering the spacings function in the following way.

Analogous to (8.11), the logspacings function can be written as

$$N = N_0(\mu, \sigma) + N_1(\mu, \sigma)\alpha + O_p(\alpha^2).$$

The quantity $N_1(\mu, \sigma)$ is obtained, like $L_1(\mu, \sigma)$, as $\partial N/\partial(\alpha)\ |_{\alpha\to 0}$. Hence, writing

$$f(y; \alpha, \mu, \sigma) = f_0(y; \mu, \sigma) + f_1(y; \mu, \sigma)\alpha + O(\alpha^2)$$
$$N_1(\mu, \sigma) = \sum_{i=1}^{n+1} \frac{\frac{\partial}{\partial\alpha} \int_{y_{i-1}}^{y_i} (f_0(y; \mu, \sigma) + f_1(y; \mu, \sigma)\alpha)\, dy}{D_i(\alpha, \mu, \sigma)}$$
$$= \sum_{i=1}^{n+1} \frac{\int_{y_{i-1}}^{y_i} f_1(y; \mu, \sigma)dy}{F(y_i; \mu, \sigma) - F(y_{i-1}; \mu, \sigma)}.$$

We now use the relationship $f_1(\mu, \sigma) = f_0(\mu, \sigma)L_1(\mu, \sigma)$ to obtain

$$N_1(\mu, \sigma) = \sum_{i=1}^{n+1} \frac{\int_{y_{i-1}}^{y_i} L_1(y, \mu, \sigma) f_0(y, \mu, \sigma)\, dy}{F(y_i; \mu, \sigma) - F(y_{i-1}; \mu, \sigma)}. \tag{8.12}$$

We see therefore that N_1 in the Nakamura (3.3) condition is just $N_1 = N_1(\tilde{\mu}, \tilde{\sigma})$, showing that the MPS estimate for a threshold is finite if $N_1(\tilde{\mu}, \tilde{\sigma}) < 0$.

Asymptotically, we would expect $N_1(\mu, \sigma) \simeq L_1(\mu, \sigma)$, at least in the neighbourhood of $\alpha = 0$, which we can show informally as follows. Let $y_i = y_{i-1} + h_i$; then the ith term in (8.12) is

$$\frac{\int_{y_{i-1}}^{y_i} L_1(y, \mu, \sigma) f_0(y, \mu, \sigma)\, dy}{F(y_i; \mu, \sigma) - F(y_{i-1}; \mu, \sigma)} \simeq \frac{L_1(y_i, \mu, \sigma) f_0(y_i, \mu, \sigma) h_i}{(f_0(y_i, \mu, \sigma) + f_1(y_i, \mu, \sigma)\alpha) h_i} \simeq L_1(y_i, \mu, \sigma)$$

for small α, so that then

$$N_1(\mu, \sigma) \simeq \sum_{i=1}^{n} L_1(y_i, \mu, \sigma) = L_1(\mu, \sigma).$$

Overall, we see that the MPS method can be used satisfactorily both in cases where ML fails due to unboundedness in the likelihood, and in situations where the existence and selection of an embedded model is required.

8.5 Threshold CIs Using Stable Law

The results in Section 8.2 and Section 8.4 show that calculation of a CI is only difficult for the threshold parameter a, and then only when $0 < c < 2$. We consider a possible approach, based on a suggestion made by Smith (1985), on how to carry out hypothesis testing for a in this case using the score statistic. The approach (i) leads to an interesting application of the theory of stable law distributions, for which quite explicit distributional results are obtainable, and (ii) also provides an interesting alternative to the asymptotic theory discussed in the previous section.

Smith (1985) points out that hypothesis testing concerning the threshold parameter a, when $0 < c < 2$, can be carried out using the score statistic

$$\frac{\partial L}{\partial a} = n^{-1} \sum_{j=1}^{n} \frac{\partial}{\partial a} \ln f(y_i, \boldsymbol{\theta}),$$

which is asymptotically equivalent to using

$$S_n(\boldsymbol{\theta}) = \alpha_n^{-1} \sum_{j=1}^{n} W_j^{-1} - \beta_n, \tag{8.13}$$

where

$$W_j = (Y_j - a)/b,$$

with α_n and β_n depending on c. Smith (1985) points out that α_n and β_n can be chosen so that $S_n(\boldsymbol{\theta})$ has a stable distribution, but does not give actual values for α_n and β_n or the stable distribution. However, Smith does show that, when $0 < c < 2$, the asymptotic properties of the interval are unchanged if unknown values of b and c are replaced by $\sqrt{n}$ consistent estimators $\tilde{b}$ and $\tilde{c}$.

Geluk and de Haan (2000) show that α_n and β_n should be chosen to satisfy

$$\lim_{n\to\infty} ns_c\{1 - F_W(\alpha_n) + F_W(-\alpha_n)\} = 1, \tag{8.14}$$

$$\beta_n = \frac{n}{\alpha_n} \int_0^{\alpha_n} \{1 - F_W(s) - F_W(-s)\} ds + \frac{2p-1}{s_\alpha} c_\alpha, \tag{8.15}$$

where

$$s_\alpha = \Gamma(1-\alpha) \cos\frac{\alpha\pi}{2} \text{ and } c_\alpha = \Gamma(1-\alpha) \sin\frac{\alpha\pi}{2} - \frac{1}{1-\alpha},$$

and

$$p = \lim_{t\to\infty} \frac{1 - F_Y(t)}{1 - F_Y(t) + F_Y(-t)},$$

and α is the stable law index of eqn (6.48) or Geluk (2000, eqn (7)), not to be confused with α_n.

For threshold models satisfying (8.3), $\alpha = c$, the power parameter. Also, we have immediately $\lim_{t\to\infty} F_Y(-t) = 0$, so that $p = 1$.

As regards a confidence interval for a, if we have the distribution of S_n, we can, for given $100(1-q)\%$ confidence level, obtain appropriate quantile points s_- and s_+ (not to be confused with s_α), so that

$$\Pr(s_- < S_n < s_+) = 1 - q.$$

We can then solve $S_n = s_-$ and $S_n = s_+$ for a, that is, solve

$$\alpha_n^{-1}\left(\sum_{j=1}^{n}[(Y_j - a_\pm)/b]^{-1}\right) - \beta_n = s_\pm,$$

to give limits a_- and a_+ for a CI for a with confidence level $100(1-q)\%$, with $\sqrt{n}$ consistent estimators $\tilde{b}$ and $\tilde{c}$ used in this formula and with α_n and β_n depending on $\tilde{c}$. This method has been discussed by Cheng and Liu (1995, 1997) for the gamma and Weibull distribution cases. Cheng and Liu (1997) do give values for (α_n, β_n) depending on whether $0 < c < 1$, $c = 1$, or $1 < c < 2$, but point out that if the standard form of stable distribution is used, with characteristic function as given in eqn (6.46), this will degenerate into a delta atom as $c \to 1$. Cheng and Liu (1997, eqn 20) give a transformed version of S_n, where the characteristic function takes the form (6.48), which removes the degeneracy as c passes through $c = 1$, but with incomplete details.

The method just outlined can be applied in principle to any threshold model with PDF as given in eqn (8.3), however, calculation of α_n, and particularly β_n, from eqns (8.14) and (8.15) is in general not straightforward. Geluk and de Haan (2000) give limiting values

$$\beta_n \to (2p-1)\tan(\alpha\pi/2) \text{ for } 0 < \alpha < 1$$

and

$$\beta_n - nE(W)/\alpha_n \to (2p-1)\tan(\alpha\pi/2) \text{ for } 1 < \alpha < 2$$

as $n \to \infty$, but it is not clear how accurate these are even for moderately large n, especially in the case where $1 < \alpha < 2$.

Here we describe fitting the three-parameter loglogistic and Weibull models, as in these cases derivation of α_n and β_n is straightforward.

Consider first the loglogistic CDF as given in eqn (8.1). We can modify S_n of (8.13) by including c in the definition of W, so that

$$W_1(a,b,c) = \left(\frac{Y-a}{b}\right)^c. \tag{8.16}$$

Then W_1 has the specific standardized loglogistic CDF

$$F_{W_1}(w) = \begin{cases} 0 \text{ if } w \le 0 \\ w/(1+w) \text{ if } w > 0 \end{cases} \tag{8.17}$$

and the limiting distribution of

$$S_{1n}(\boldsymbol{\theta}) = \alpha_n^{-1}\sum_{j=1}^{n}[W_{1j}(a,b,c)]^{-1} - \beta_n, \tag{8.18}$$

as $n \to \infty$ no long depends on $\boldsymbol{\theta}$.

It is easy to see that $F_W(w)$ is a *regularly varying* function as given in definition 2 of Geluk and de Haan (2000), with $\gamma = 1$ in that definition. Then, following Geluk and de Haan (2000), we find that $S_{1n}(\boldsymbol{\theta})$ has the stable limit distribution with index $\alpha = 1$, whose characteristic function is

$$\ln \phi(t) = -|t| - i\left(\frac{2}{\pi}\right) t \ln |t|. \tag{8.19}$$

This is the special case of eqn (6.48), with $\alpha = 1$, $\beta = 1$, $\delta = 0$, and $\gamma = 1$ in that formula. This follows from Geluk and de Haan (2000, eqn (7)), which gives the same stable distribution, with $\delta = 0$ and $\gamma = 1$ already assumed. This stable distribution is obtained on noting that inclusion of c in the definition of W_{1j} in our loglogistic case is equivalent to setting $\alpha = 1$ in both (6.48) and in Geluk and de Haan's formula. We already have $\beta = p = 1$.

We also have $\lim_{\alpha\to 1} s_\alpha = \pi/2$ and $\lim_{\alpha\to 1} c_\alpha = -\gamma$, where $\gamma \simeq 0.577\,2156$ is Euler's constant. We find quite easily that

$$\alpha_{1n} = \lim_{\alpha\to 1} \alpha_n = \frac{1}{2}n\pi - 1 \tag{8.20}$$

and

$$\begin{aligned}\beta_{1n} &= \lim_{\alpha\to 1} \beta_n = \frac{2n}{n\pi - 2}\ln\left(\frac{n\pi}{2}\right) - 2\frac{\gamma}{\pi} \\ &= \frac{2}{\pi}\left(\ln\frac{n\pi}{2} - \gamma\right) + \frac{4}{\pi^2}\ln\left(\frac{n\pi}{2}\right)n^{-1} + \frac{8}{\pi^3}\ln\left(\frac{n\pi}{2}\right)n^{-2} + O(n^{-3}\ln n).\end{aligned} \tag{8.21}$$

The series expansion for β_{1n} is simply to show that it only increases logarithmically with n; in practical work, one may as well use the exact formula for β_{1n}.

Summarizing: To obtain a CI with confidence coefficient $1 - q$, we

(i) Find $s_{1,-}$ and $s_{1,+}$ satisfying

$$\Pr(S_1 < s_{1,-}) = q/2 \text{ and } \Pr(S_1 < s_{1,+}) = 1 - q/2, \tag{8.22}$$

where S_1 is a random variable whose distribution has characteristic function (8.19). Values for selected q are given in Table 8.3.

(ii) Then solve $S_{1n}(a_-, \tilde{b}, \tilde{c}) = s_{1,-}$ for a_- and $S_{1n}(a_+, \tilde{b}, \tilde{c}) = s_{1,+}$ for a_+, i.e.

$$\alpha_{1n}^{-1}\left(\sum_{j=1}^{n}[(Y_j - a_\pm)/\tilde{b}]^{-\tilde{c}}\right) - \beta_{1n} = s_{1,\pm}, \tag{8.23}$$

taking

$$(a_-,\ a_+)$$

as the $100(1 - q)\%$ confidence interval for a.

For the Weibull case, the preceding argument can still be used, with the only change being that W_1 of eqn (8.16) now has the specific CDF

$$F_{W_1}(w) = \begin{cases} 0 \text{ if } w \leq 0 \\ 1 - \exp(-w) \text{ if } w > 0, \end{cases} \tag{8.24}$$

which is simply that of the exponential distribution with parameter unity. The limiting distribution of $S_{1n}(\boldsymbol{\theta})$ is again independent of $\boldsymbol{\theta}$. In this case, we find that

$$\alpha_{1n} = [\ln(\pi n/(\pi n - 2))]^{-1} \tag{8.25}$$

and

$$\begin{aligned} \beta_{1n} &= \frac{n}{\alpha_{1n}}\left(\alpha_{1n} - \alpha_{1n}e^{-\alpha_{1n}^{-1}} + E_1\left(\alpha_{1n}^{-1}\right)\right) - 2\frac{\gamma}{\pi} \\ &= \frac{2}{\pi}\left(1 + \ln\left(\frac{n\pi}{2}\right) - 2\gamma\right) + \frac{2}{\pi^2}\left(1 + \ln\left(\frac{n\pi}{2}\right) - \gamma\right)n^{-1} + \\ &\quad \left(\frac{8}{3\pi^3}\left(1 + \ln\left(\frac{n\pi}{2}\right) - \gamma\right) - \frac{1}{3\pi^3}\right)n^{-2} + O(n^{-3}\ln n), \end{aligned} \tag{8.26}$$

where

$$E_1(z) = \int_z^\infty \frac{e^{-t}}{t}dt$$

is the exponential integral. The CI (a_-, a_+) in this case is obtained in exactly the same way as for the loglogistic model using (8.23), only with α_{1n} and β_{1n} obtained from (8.25) and (8.26). For practical purposes, the leading term in the expansion (8.26) for β_{1n} will usually be sufficiently accurate, though in our numerical examples, we used the first two terms.

The quantiles $s_{1,\pm}$ used for the Weibull case are unchanged from the loglogistic case, as in both, the same stable law is used, corresponding to the characteristic function (8.19). Selected quantiles $s(q)$ of S_1, where

$$\Pr\{S_1 \leq s(q)\} = q,$$

are given in Table 8.3, these being taken from McCulloch (1996) and Cheng and Liu (1997).

In using (8.23), we are relying on the distribution of $S_{1n} = \left(\sum_{j=1}^n W_{1j}^{-1}\right)/\alpha_{1n} - \beta_{1n}$, where the $\{W_{1j}, j = 1, 2, \ldots, n\}$ are an IID sample either from the standardized loglogistic (8.17) or Weibull (8.24) distribution, being close to that of S_1, the stable law random variable with characteristic function (8.19).

As a quick check of this, we simulated $B = 50{,}000$ replications of S_{1n}, in each replication calculating S_{1n} from an independent sample $\{W_{1j}, j = 1, 2, \ldots, n\}$, assuming the W_{1j}

Table 8.3 Stable law percentage points, $\alpha = 1$

q	0.005	0.01	0.025	0.05	0.10
$s(q)$	−1.746	−1.628	−1.433	−1.241	−0.983
q	0.90	0.95	0.975	0.99	0.995
$s(q)$	7.129	14.00	27.22	66.02	130.1

are distributed exactly as in (8.17) for the loglogistic case or as in (8.24) for the Weibull case, so that parameter values are actually known in the simulation. The formulas for generating the W_{1j} are particularly simple:

$$W_{1j} = \begin{cases} U_j/(1 - U_j) & \text{in the loglogistic model} \\ -\ln(U_j) & \text{in the Weibull model,} \end{cases}$$

where the U_j are uniform $U(0, 1)$ variates. (Note, incidentally, that in the case of the loglogistic, either W_{1j} or W_{1j}^{-1} can be used in the formula for S_{1n}, as they have the same distribution.) We then calculated sampled percentile CIs as in the bootstrap CI given in (4.3). Thus, if $S^*_{1n,(1)} \leq S^*_{1n,(2)} \leq \ldots \leq S^*_{1n,(B)}$ are the ordered S_{1n} values, the sampled symmetric CI with confidence coefficient $(1 - q)$ is

$$(S^*_{1n,(l)}, \; S^*_{1n,(m)}), \; l = \lfloor \max(1, (q/2)B) \rfloor, \; m = \lfloor (1 - q/2)B) \rfloor.$$

This was done for $n = 10, 100, 1000$, and for $q = 0.2, 0.1, 0.05, 0.02, 0.01$, for both the loglogistic and Weibull models, with the results depicted in Figure 8.2. The figure also depicts the confidence intervals calculated from the $s(q)$ values obtained theoretically as given in Table 8.3, and comparison of the CIs obtained by the simulation sampling with these shows quite good agreement, even when n is as low as $n = 10$.

In the next section, we give a numerical example by fitting just the loglogistic model (8.2) to the SteenStickler data sets, including calculation of CIs. All the ML and MPS estimates turn out to give estimates for c with values $c < 2$, so that for these data sets, calculation of a confidence interval for the threshold a cannot be obtained by standard asymptotic theory. We can, therefore, in the case of the threshold parameter a, compare the CIs obtained by the stable law approach as well as those obtained using the ML, modified ML where $\hat{a} = y_{(1)}$, and MPS methods.

8.5.1 Example of Fitting the Loglogistic Distribution

In this section, we give two examples that numerically compare ML ($\hat{\boldsymbol{\theta}}$), the modified-ML method ($\breve{\boldsymbol{\theta}}$) suggested by Smith (1985) where $\breve{b}$ and $\breve{c}$ are obtained by ML with $\breve{a} = y_{(1)}$, and MPS ($\tilde{\boldsymbol{\theta}}$), by considering their performance in the fitting of the loglogistic model to the two data sets, SteenStickler0 and SteenStickler1, discussed in Section 8.1. As shown by the profile log-likelihood plots of Figure 8.1,

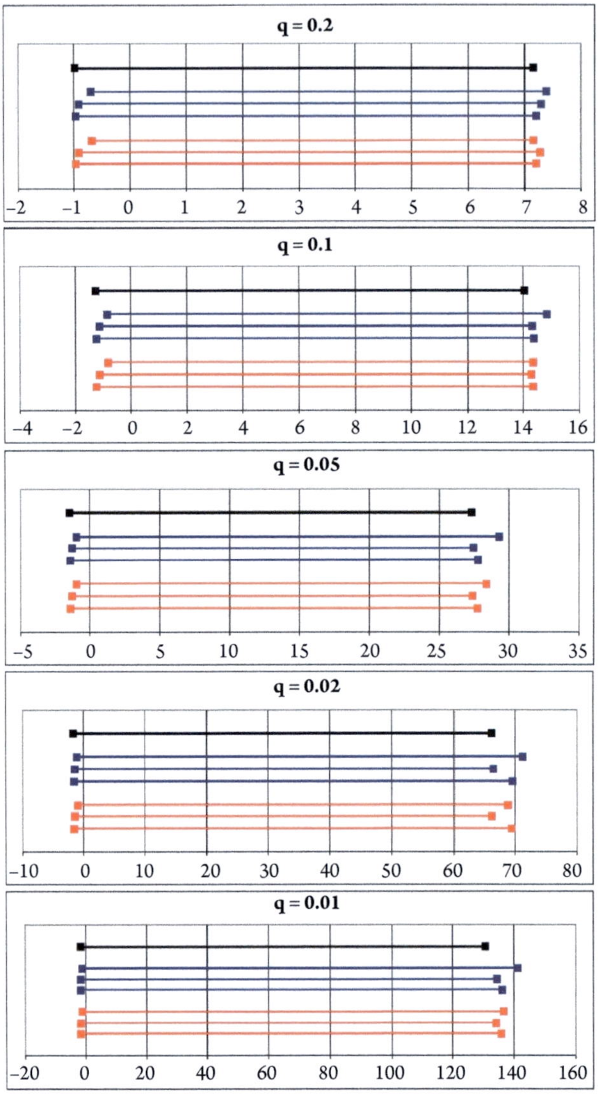

Figure 8.2 Five charts comparing the theoretical and sampled CIs with confidence coefficient $(1-q)$, where $q = 0.2, 0.1, 0.05, 0.02, 0.01$ for the threshold parameter calculated by the stable law method. In each chart, the accurate CI, as calculated from Table 8.3, is shown in black. The three blue CIs are sampled versions, corresponding to the loglogistic case, with each calculated from a sample of 50,000 replications of S_{1n}, where each replication uses a sum of n reciprocal standardized loglogistic variates with, from the top downwards, $n = 10$, $n = 100$, $n = 1000$. The red CIs are the corresponding CIs for the Weibull (exponential) case, with each replicate using a sum of reciprocal standard exponential variates.

Table 8.4 Estimated parameters of the loglogistic model fitted to the SteenStickler samples

	StnStk0	StnStk0	StnStk0	StnStk1	StnStk1	StnStk1
	ML	mod-ML	MPS	ML	mod-ML	MPS
a	109	109	97.17	1220	1364	710
b	497	681	598	3760	4002	4472
c	0.454	0.928	0.795	1.44	1.57	1.56

the infinite likelihood problem is a definite issue with SteenStickler0 and a potential issue with SteenStickler1.

Table 8.4 gives the parameter estimates for all three methods and both samples. It will be seen that $\hat{c}, \breve{c}, \tilde{c} < 1$ for SteenStickler0 and $1 < \hat{c}, \breve{c}, \tilde{c} < 2$ for SteenStickler1, so that the non-standard behaviour just discussed applies.

We now consider the fits in more detail.

Figure 8.3 shows the loglogistic model fitted to the SteenStickler0 sample. The two left-hand charts depict the ML CDF and PDF fits, showing these to be not at all satisfactory, as anticipated by the profile log-likelihood plot in Figure 8.1. In fact, to get any sort of estimate at all, a limit of $a \leq y_{(1)} - 10^{-10}$ was placed on the threshold parameter a to stop convergence to $y_{(1)}$. The first column in Table 8.5 gives the value of the Anderson-Darling GoF $\hat{A}^2$ value of the ML fit and $\hat{A}^2_{0.1}$, the upper 10% critical A^2 value in the ML case, obtained by bootstrap, using a bootstrap sample of size 500. The test shows that the ML fit is not satisfactory at the 10% level, with the $\hat{A}^2 = 1.703$ value that exceeds the 10% critical value of $\hat{A}^2_{0.1} = 1.140$.

In contrast, the MPS CDF and PDF fits for the SteenStickler0 sample depicted in the two right-hand charts of Figure 8.3 look satisfactory, and this is corroborated by the GoF test given in the third column of Table 8.5, showing that the AD value of $\tilde{A}^2 = 0.312 < \tilde{A}^2_{0.1} = 0.500$, so that the lack-of-fit is not significant at the 10% level. That the $\hat{L}_{\max} = -158.2$ value of the ML fit is much higher than the $\tilde{L}_{\max} = -169.14$ value of the MPS fit is explained by the fact the ML fit corresponds to a fit where $\hat{a} = y_{(1)} - 10^{-10}$, when the other parameter estimates are most likely inconsistent and unsatisfactory.

The middle-column charts in Figure 8.3 show the CDF and PDF fits obtained using the modified-ML method. These look reasonable. The GoF values for this case are in the second column of Table 8.5, showing that the AD value of $\breve{A}^2 = 0.268 < \breve{A}^2_{0.1} = 0.529$, so that the lack-of-fit is not significant at the 10% level. The $\hat{L}_{\max} = -162.01$ value of the modified-ML fit is somewhat higher than the $\tilde{L}_{\max} = -169.14$ value of the MPS fit, but lower than the $\hat{L}_{\max} = -158.2$, which is in line with what one would expect, as this latter corresponds to a genuine maximum.

Consider now the SteenStickler1 sample. Figure 8.4 shows the ML, modified-ML, and MPS fits, and all look satisfactory. The results of the GoF tests are given in the right-hand three columns of Table 8.5, showing that the ML and MPS fits are satisfactory, but with the modified-ML fit marginal.

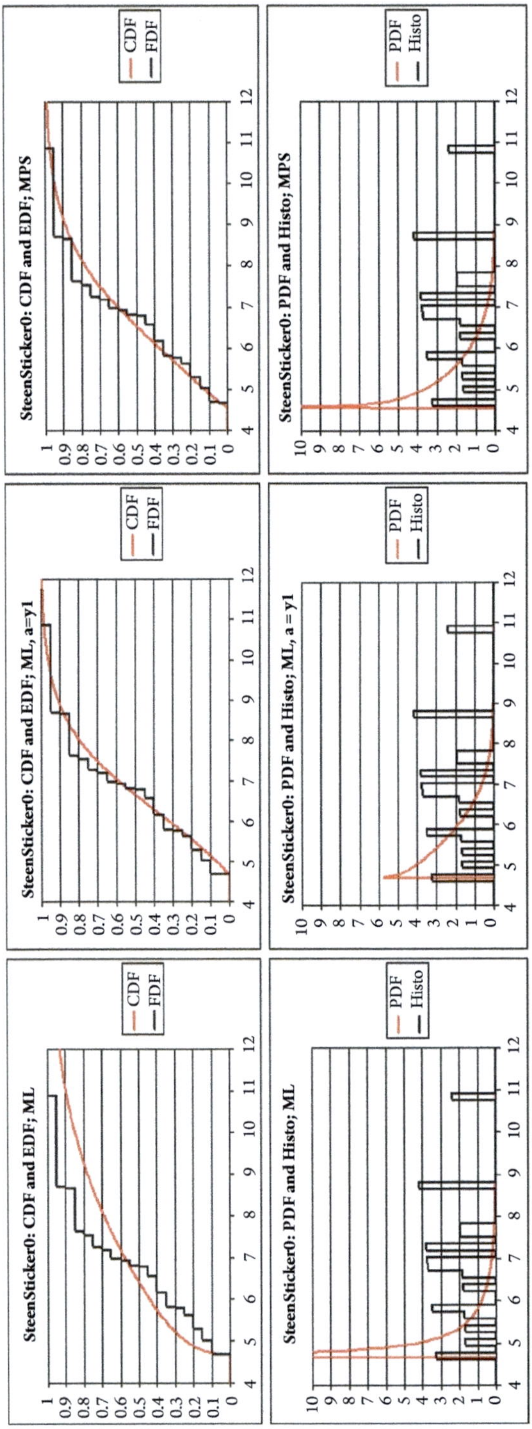

Figure 8.3 Loglogistic model fitted by ML, modified-ML, and MPS to SteenStickler0 sample. Horizontal axis is $x = \ln(\text{pollution level})$.

Table 8.5 GoF results fitting the loglogistic model to the SteenStickler samples

	StnStk0	**StnStk0**	**StnStk0**	**StnStk1**	**StnStk1**	**StnStk1**
	ML	mod-ML	MPS	ML	mod-ML	MPS
A^2	1.703	0.268	0.312	0.418	0.536	0.375
$A^2_{0.1}$	1.140	0.529	0.500	2.266	0.540	0.459
$p - val$	0.006	0.588	0.386	0.500	0.102	0.210
$L_{\max}$	−158.20	−162.01	−169.14	−195.57	−185.74	−196.14

The CIs for the parameters fitted by all three methods, calculated by bootstrapping and by asymptotic normal theory, are given in Figure 8.5 for both SteenStickler data sets.

The left-hand charts in Figure 8.5 show the CIs of the ML, modified-ML, and MPS fitted parameters for the SteenStickler0 sample calculated by bootstrapping.

We have included the CIs given by asymptotic normal theory for comparison, but stress that those for the parameter a are not actually meaningful in view of the theoretical results given in the previous section.

The top-left chart in Figure 8.5 illustrates the difficulties that arise using ML or the modified-ML methods when estimating the threshold parameter a in the SteenStickler0 case. Using ML, one ends up essentially with $\hat{a} = y_{(1)}$, and whether using asymptotic formula or bootstrapping, the CI does not provide much indication of the estimator variance. Using the modified-ML method, we know that $\breve{a} = y_{(1)}$ is actually an upper bound on the true a value if the model is correct. We used the one-sided version of the asymptotic Weibull CI formula of eqn (8.5) in calculating the CI shown in the chart. The bootstrap version behaves quite differently, as the bootstrap samples were all generated with the threshold fixed at $\breve{a} = y_{(1)}$. Thus, each bootstrap sample $\mathbf{y}^{*(j)}$ is bound to have smallest observed value $y^{*(j)}_{(1)} > y_{(1)}$, so that the modified-ML estimate of a from the BS sample will be $\breve{a}^{(j)} = y^{*(j)}_{(1)} > y_{(1)}$. Bootstrapping is, therefore, not very satisfactory in this case, with all BS estimates of a biased high, as occurs in the CI shown in the top-left chart.

The same behaviour occurs with the SteenStickler1 sample.

The method using the stable law to form a CI for a seems to be quite satisfactory with a small width and, when asymmetrically placed relative to $\hat{a}$ or $\tilde{a}$, being positioned so that the upper limit of the CI is at or close to $y_{(1)}$.

8.6 A 'Corrected' Log-likelihood

Calculation of the spacings involves evaluating the CDF, which is often computationally more demanding than evaluating the PDF. In this section and the next, we outline two

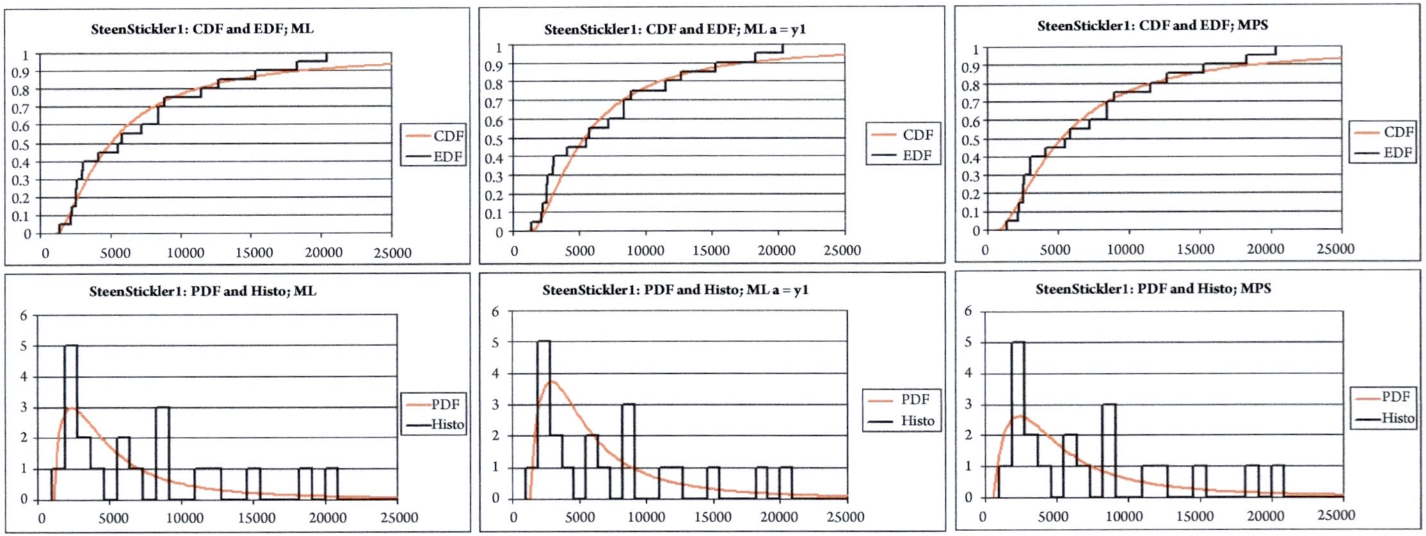

Figure 8.4 Loglogistic model fitted by ML, modified-ML, and MPS to SteenStickler1 sample. Horizontal axis is pollution level plotted on logged-scale.

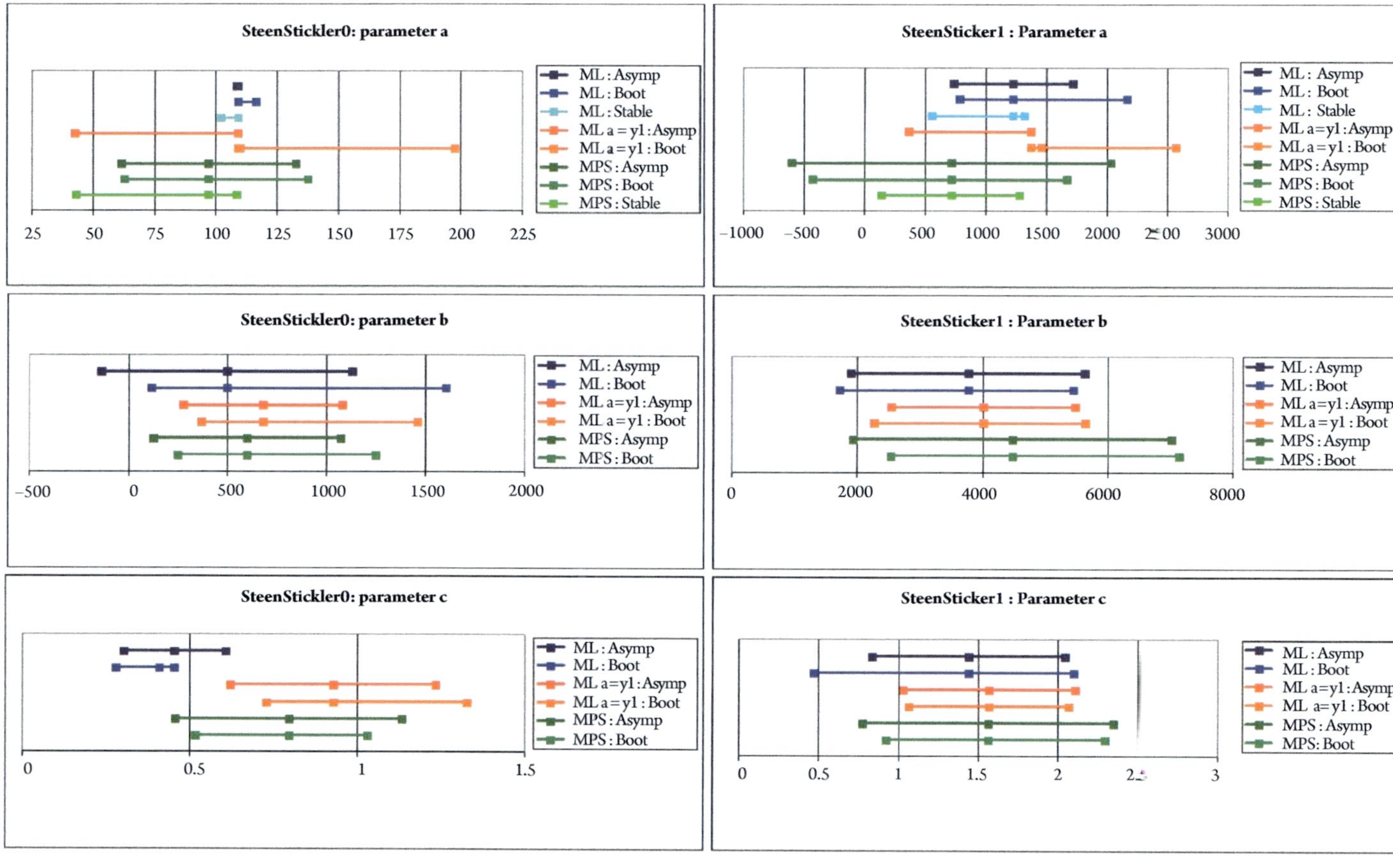

Figure 8.5 CIs of the loglogistic parameters fitted by ML, modified-ML, and MPS to the SteenStickler0 and SteenStickler1 samples.

variants of MPS giving estimators with the same asymptotic properties, but involving less reliance on evaluation of the CDF. In this section we assume the y_j are in rank order.

Cheng and Iles (1987) point out that if $y_1 < y_2$, then it is only the first term $\ln f(y_1, \boldsymbol{\theta})$ that can become unbounded, so that the 'corrected' log-likelihood

$$L_h(\boldsymbol{\theta}) = \ln \int_{y_1}^{y_1+h} f(y, \boldsymbol{\theta})dy + \sum_{i=2}^{n} \ln f(y_i, \boldsymbol{\theta}), \tag{8.27}$$

which differs from the usual log-likelihood only in the first term, then behaves in a regular way for any given $h > 0$. Cheng and Iles (1987) give the following result.

Cheng and Iles (1987, Result 2)

Let $\tilde{\boldsymbol{\theta}}(h)$ maximize $L_h(\boldsymbol{\theta})$ as given in (8.27), subject to $a \le y$. Then $\tilde{\boldsymbol{\theta}}(h)$ can be found so that

(i) If $c > 2$ or $1 < c < 2$, then $\tilde{\boldsymbol{\theta}}(h)$ has the same asymptotic distribution as $\hat{\boldsymbol{\theta}}$. Moreover,

$$\lim_{h \to 0} \tilde{\boldsymbol{\theta}}(h) = \hat{\boldsymbol{\theta}}, \text{ in probabiity.}$$

(ii) If $0 < c < 1$, then $\tilde{a}(h) = y_1$, and is superefficient in the sense that

$$|\tilde{a}(h) - a| = O_p(n^{-1/c}).$$

Moreover, $\tilde{\boldsymbol{\phi}}(h)$ has the same asymptotic distribution as $\hat{\boldsymbol{\phi}}$ when a is assumed known, provided

$$h \ge O_p(n^{-1/c}).$$

Finally,

$$\lim_{h \to \infty} \tilde{\boldsymbol{\theta}}(h) = \hat{\boldsymbol{\theta}}_1 \equiv (y_1, \hat{\phi}_1),$$

where $\hat{\boldsymbol{\theta}}_1$ is the MLE of (b, c) obtained by setting $a = y_1$.

The proof of result 2 is outlined in Cheng and Iles (1987, Appendix). The proof is broadly similar to conventional proofs of existence and asymptotic normality of MLE, moreover, using a very similar argument to that given in Cheng and Amin (1982) in proving Theorem 1 in Cheng and Amin (1983).

8.7 A Hybrid Method

We will usually implement ML and MPS by maximizing the log-likelihood or the log-spacings function. In ML estimation, local maxima that lie in an open neighbourhood of the parameter space Θ can be obtained instead by solving the likelihood equations $\partial L/\partial \boldsymbol{\theta} = 0$, where $\boldsymbol{\theta} = (a, b, c)$. Similarly, the MPS method can be implemented by solving the spacings equations $\partial H/\partial \boldsymbol{\theta} = 0$.

The following *hybrid method* that combines solution of $\partial L/\partial b$=0, $\partial L/\partial c$=0 with $\partial H/\partial a$ =0 was considered by Traylor (1994). The unboundedness problem does not occur here, because ML breaks down simply because of irregularities in the solution of $\partial L/\partial a = 0$, and this hybrid method avoids its use, replacing it with $\partial H/\partial a = 0$.

A convenient approach when applying any of the three methods is to use the profile log-likelihood or spacings function to obtain an estimate of a, instead of solving the log-likelihood or spacings equation for a directly. For the hybrid method, one way of doing this is to fix a and obtain $\hat{b}(a)$, $\hat{c}(a)$ from the likelihood equations. Instead of then examining the profile log-likelihood $L^*(a)$, the profile spacings function

$$K^*(a) = H(a, \hat{b}(a), \hat{c}(a))$$

is searched to find its overall maximum. The maximum point of this profile 'hybrid' spacings function,

$$\tilde{a} = \arg\left(\max_a K^*(a)\right).$$

is a solution of $\partial H/\partial a$ =0, and yields the estimates

$$\check{\boldsymbol{\theta}} = (\tilde{a}, \hat{b}(\tilde{a}), \hat{c}(\tilde{a})).$$

8.7.1 Comparison with Maximum Product of Spacings

We now show that the hybrid estimators possess the same asymptotic properties as those of MPS estimators discussed above. The result is comparable with the one given by Cheng and Iles (1987), which uses the equation

$$\boldsymbol{\theta} - \boldsymbol{\theta}^0 + \varepsilon(\boldsymbol{\theta}; \boldsymbol{\theta}^0, n) = \beta(n), \tag{8.28}$$

where $\beta(n) = O_p(n^{-\frac{1}{2}})$, and where $\boldsymbol{\theta} - \boldsymbol{\theta}^0 = O(n^{-\frac{1}{2}})$ implies $\varepsilon(\boldsymbol{\theta}; \boldsymbol{\theta}^0, n) = o_p(n^{-\frac{1}{2}})$.

When $b^0 > 2$, solution of the likelihood function satisfies an equation of the above form for each of the three parameters. Then the equations are

$$\boldsymbol{\theta}_i - \boldsymbol{\theta}_i^0 + \varepsilon_i(\boldsymbol{\theta}; \boldsymbol{\theta}^0, n) = \beta_i(n) \qquad i = 1, 2, 3.$$

Cheng and Iles (1987) show that, for the Weibull and gamma distributions, $\partial H/\partial \boldsymbol{\theta}$ can be directly written in the same way, and so the previous three equations are shown to have a solution $\tilde{\boldsymbol{\theta}}$, where the asymptotic distribution of $\tilde{\theta}_i$ is that of $\beta_i(n)$. Here, the $\varepsilon_i(\boldsymbol{\theta};\boldsymbol{\theta}^0, n)$ are different from those that appear in the likelihood equation version, but still satisfy the same properties of (8.28). Therefore, the corresponding solution $\tilde{\boldsymbol{\theta}}$ has the same property as $\hat{\boldsymbol{\theta}}$.

When $1 < c^0 < 2$, $\partial L/\partial a = 0$ has to be treated as a special case, since $\hat{a} - a^0$ has a different rate of convergence. Therefore, the previous system of equations cannot be used for all three parameters. However, as $\hat{a}$ is hyper-efficient, the behaviour of the solution of $\partial L/\partial b = 0$ and $\partial L/\partial c = 0$ is asymptotically the same as if a^0 were known. Therefore, these two parameter estimates are found from two equations of the form of (8.28), substituting $\hat{a}$ for a^0.

Comparison of the spacings threshold equation with the likelihood threshold equation shows $\tilde{a}$ is also hyper-efficient, so use of $\tilde{a}$ instead of $\hat{a}$ in two equations of the form of (8.28) again allows the remaining parameters to be estimated.

It is seen from the preceding discussion that the spacings method requires use of two equations of the form (8.28) for estimation of the two non-threshold parameters, regardless of the value of c^0. The ML method also makes use of two such equations. Cheng and Iles (1987) show that, apart from its order of convergence, the precise form of ε in (8.28) is immaterial. Thus our argument shows that either pair of equations can be used. The hybrid method simply takes the solution $\tilde{a}$ obtained from the spacings function and uses it instead of a^0 in the two non-threshold likelihood equations. This results in parameter estimates which, as with MPS estimators, have the required asymptotic properties, namely, that when $c^0 > 2$, there is a hybrid solution $\check{\boldsymbol{\theta}}$ which has the standard ML asymptotic properties. When $c^0 < 2$, there exists, in probability, a solution $\tilde{a}$ which is hyper-efficient, with an improved convergence of $\tilde{a} - a^0 = O_p(n^{-1/c^0})$, with the standard result of asymptotic normality still holding for $\hat{b}(\tilde{a})$ and $\hat{c}(\tilde{a})$.

8.8 MPSE: Additional Aspects

8.8.1 Consistency of MPSE

In an early investigation, Cheng and Amin (1979), considered the family of PDFs $f(y,\boldsymbol{\theta})$, $\boldsymbol{\theta} \in \boldsymbol{\Theta}$, where $\boldsymbol{\Theta}$ is a finite dimensional Cartesian subspace, satisfying $f(y,\boldsymbol{\theta}) > 0$ for $y \in (a,b)$ with $f(y,\boldsymbol{\theta}) = 0$ otherwise, where the end points $a = a(\boldsymbol{\theta})$, $b = b(\boldsymbol{\theta})$ are continuous functions of $\boldsymbol{\theta}$, or, if independent of $\boldsymbol{\theta}$, are possibly infinite. Cheng and Amin (1979) gave general assumptions under which the MPSE $\tilde{\boldsymbol{\theta}}$ is (weakly) consistent for $\boldsymbol{\theta}^0$ as $n \to \infty$, showing that the assumptions included the three-parameter lognormal distribution of Table 6.4, which includes a shifted origin or threshold; a case of particular concern for Hill (1995).

More generally, the consistency of MPSEs has been investigated by Ekström (1996), Shao and Hahn (1999), and Shao (2001). The usual approach of exploring the extent of applicability of the MPS method is to try to establish *weak* regularity conditions under which ML estimation can fail, but where MPSEs are still consistent. Thus Shao and

Hahn (1999) establish very general conditions under which MPSEs are consistent, but which allow situations where ML fails. However, such conditions, as pointed out by Shao (2001), are not *strictly* weaker than the classical regularity conditions, as outlined in Chapter 3 and introduced by Wald (1949), under which he established the consistency of MLEs.

We have omitted giving full details of the conditions discussed by Shao and Hahn (1999), as they are fairly technical. We simply note that the conditions are sufficiently weak to include most situations where ML is known to fail. Thus, they cover the case of unknown thresholds discussed in detail in this chapter, and similar other examples discussed by Stuart and Ord (1987), Cheng and Amin (1983), and Ekström (1996), not discussed here.

Alternative ML consistency theorems are known, but, as pointed out by Le Cam (1986), are variants of the Wald (1949) conditions. It is known that a local dominance-type condition is necessary to guarantee that MLEs are consistent, see Perlman (1972). Shao and Hahn (1999) essentially replace such a condition by a similar condition under which MPS is still consistent, but where ML may not be. In a further development, Shao (2001) gives a different form of condition under which assumptions 2 and 6 of Wald (1949) are not needed.

Shao (2001) also discusses MPS in a nonparametric context, giving in particular the interesting example of how a consistent unimodal MPS estimator can be constructed for a unimodal distribution with unknown mode.

It should be pointed out that additional theoretical aspects of both the spacings and Kullback-Leibler viewpoint have been the subject of additional study, see, for example, Lind (1994), who gives a principle of least information, showing that it gives a reference distribution identical with MPS, and Anatolyev and Kosenok (2005), who have given an alternative derivation of the asymptotics of MPS, using a comparison between its objective function with that in ML estimation. Related methods of estimation using spacings have been studied, see, for example, Ranneby and Ekström (1997) and Ghosh and Jammalamadaka (2001). Roeder (1992) considers use of spacings in finite mixture models for which ML estimation is non-standard, including non-finiteness of the likelihood. We shall consider these models in detail in Chapters 17 and 18. An extension of the spacings approach to multivariate estimation has been considered by Ranneby et al. (2005), using Dirichlet cells.

8.8.2 Goodness-of-Fit

Moran (1951) suggested a statistic denoted by $M(\boldsymbol{\theta})$ for goodness-of-fit tests. Cheng and Stephens (1989) took this statistic in the form

$$M(\boldsymbol{\theta}) = -(n+1)H(\boldsymbol{\theta}), \tag{8.29}$$

so that it is simply a multiple of $H(\boldsymbol{\theta})$. The distribution of $M(\boldsymbol{\theta})$ when $\boldsymbol{\theta}$ is known is asymptotically normal with mean and variance given by

$$\gamma_m = m(\ln m + \gamma) - \frac{1}{2} - \frac{1}{12m} + \ldots$$
$$\sigma_m^2 = m\left(\frac{\pi^2}{6} - 1\right) - \frac{1}{2} - \frac{1}{6m} + \ldots,$$

where $\gamma = 0.57722$ is Euler's. Cheng and Stephens (1989) show that for finite n, the distribution of $M(\boldsymbol{\theta})$ is well approximated by that of

$$A = C_1 + C_2\chi_n^2,$$

where χ_n^2 is a chi-squared variable with n degrees of freedom and

$$\begin{aligned} C_1 &= \gamma_{n+1} - (0.5n)^{1/2}\sigma_{n+1}, \\ C_2 &= (2n)^{-1/2}\sigma_{n+1}. \end{aligned} \tag{8.30}$$

Thus if $\boldsymbol{\theta} = \boldsymbol{\theta}_0$ is known (called Case 0), the GoF test is

(a) calculate $S = [M(\boldsymbol{\theta}_0) - C_1]/C_2$
(b) reject H_0 at significance level α if $S > \chi_n^2(\alpha)$, where $\chi_n^2(\alpha)$ is the upper α level point of χ_n^2.

The main result given by Cheng and Stephens (1989) covers the case where a distribution has CDF $F(y, \boldsymbol{\theta})$ and $f(y, \boldsymbol{\theta})$ *where* $\boldsymbol{\theta}$ *is not known*, but with support (a, b) where a and b are known, with infinite a and b allowed. Suppose the following three assumptions hold:

A(i) F and f are regular in the sense of Cramér (1946), with $\partial f/dy$, $\partial^2 f/\partial\boldsymbol{\theta}\partial y$, $\partial^3 f/\partial\boldsymbol{\theta}^2\partial y$ absolutely continuous on (a, b). Moreover, a $\delta > 0$ exists such that for all $\boldsymbol{\theta}$ for which $|\boldsymbol{\theta} - \boldsymbol{\theta}_0| < \delta$, $F(y, \boldsymbol{\theta})$ satisfies the following:

A(ii) If y_1, y_n are such that

$$\int_a^{y_1} f dy = O(n^{-1}), \int_{y_n}^b f dy = O(n^{-1}),$$

then

$$\frac{\partial}{\partial\boldsymbol{\theta}} \int_a^{y_1} f dy = o(n^{-1/2}) and \frac{\partial}{\partial\boldsymbol{\theta}} \int_{y_n}^b f dy = o(n^{-1/2}),$$

A(iii) There exist $c > a$ and $d < b$ such that each component of $\partial \ln f/\partial\boldsymbol{\theta}$ and $\partial^2 \ln f/\partial\boldsymbol{\theta}^2$ is monotone in y in (a, c) and in (d, b), we have

Cheng and Stephens (1989, Theorem 1)

(i) Assume that $\boldsymbol{\theta}$ is k - dimensional. If $\tilde{\boldsymbol{\theta}}$ and $\hat{\boldsymbol{\theta}}$ are the MPS and ML estimators, then

$$\tilde{\boldsymbol{\theta}} - \hat{\boldsymbol{\theta}} = o_p(n^{-1/2}),$$

so that

(ii) $\tilde{\boldsymbol{\theta}}$ has the same standard asymptotic normal distribution as $\hat{\boldsymbol{\theta}}$, viz.

$$n^{1/2}(\tilde{\boldsymbol{\theta}} - \boldsymbol{\theta}_0) \sim N\left\{0, -nE\left[\left(\frac{\partial^2 L}{\partial \boldsymbol{\theta}^2}\right)^{-1}\right]\right\}.$$

(iii) Moreover,

$$M(\tilde{\boldsymbol{\theta}}) = M(\boldsymbol{\theta}_0) - \frac{1}{2}Q + o_p(1),$$

where Q is distributed as a chi-squared variable with k degrees of freedom. Thus, $n^{-1/2}M(\tilde{\boldsymbol{\theta}})$ and $n^{-1/2}M(\boldsymbol{\theta}_0)$ have the same asymptotic distribution, as $\frac{1}{2}Q$ is of smaller order than $M(\boldsymbol{\theta}_0)$.

The theorem enables the Case 0 GoF test to be modified to give a GoF test for the case where $\boldsymbol{\theta}$ is not known.

(a) Let $\breve{\boldsymbol{\theta}}$ be either the MLE $\hat{\boldsymbol{\theta}}$ or the MPSE $\tilde{\boldsymbol{\theta}}$.

(b) Calculate

$$T(\breve{\boldsymbol{\theta}}) = \frac{M(\breve{\boldsymbol{\theta}}) + \frac{1}{2}k - C_1}{C_2}, \tag{8.31}$$

where C_1 and C_2 are as in eqn (8.30).

(c) Reject H_0 at significance level α if $T(\breve{\boldsymbol{\theta}}) > \chi_n^2(\alpha)$.

8.8.3 Censored Observations

The MPS method can easily handle the inclusion of randomly censored observations by making use of a method described by D'Agostino, and Stephens (1986). Full details of this technique are deferred until Chapter 11, where we examine the fit of a change-point model involving censored observations. The censored observations may be positioned so that the asymptotic distribution of $H(\boldsymbol{\theta}^0)$ is unchanged. Therefore, as in the uncensored case, a confidence region can still be found in the previously discussed way.

8.8.4 Tied Observations

To overcome the difficulty of tied observations in the spacings function, we adopt the interpretation proposed by Cheng and Stephens (1989) that this situation occurs because of rounding, when different observations are recorded as the same value.

The authors suppose that a rounding error δ can be specified so that the true values, say, y_i to y_{i+r-1}, all lie in the range $y \pm \delta$. Cheng and Amin (1983) show that the MPS estimator can still be found in this situation. If $y_i = y_{i+1} = \ldots = y_{i+r-1}$ ($= y$, say) are tied observations, we define an interval $y \pm \delta$ within which all the true values lie, and then adjust their values so that they are positioned equally spaced within this interval. Therefore, the procedure is to reset the tied observations as

$$y_{i+t} = (y - \delta) + 2\delta \frac{(t+1)}{(r+1)}$$

for $t = 0, \ldots, r-1$. For our calculations, we set $\delta = 0.5 \times 10^{-p}$, where p is the number of decimal places taken for tied observation y. Note that the spacings function is very sensitive to small spacings, and so in some instances it might not be unreasonable to adjust a tied observation by one unit of the recorded accuracy.

9 The Pearson and Johnson Systems

9.1 Introduction

In this chapter, we revisit two of the best-known systems of parametric distributions: the Pearson and the Johnson systems, see Pearson (1895) and Johnson (1949). Most readers may well be acquainted, even familiar with the two systems, and might therefore think this chapter unnecessary. However, both systems comprise a number of subfamilies of distributions, which we will refer to as *types* or *models*, each of which has its own separate and distinct functional form, and which are therefore usually discussed separately when being considered as fitted models. Doing this, one loses the greatest benefit of using either system, which is that each offers a model fit that is comprehensive in a well-defined sense that is set out below, if the fitting process allows *any* member of the constituent subfamilies to be the best fit.

The main subfamilies all contain embedded models, so a fitting process that considers all possible subfamilies in its search for a best fit will need to be systematic and to take due account of the possibility that an embedded model is the best fit. This is what is discussed in this chapter.

If we write $\beta_1 = \gamma_1^2$ for the squared skewness and β_2 for the kurtosis, each system offers a range of shapes that is comprehensive in the sense that for any (β_1, β_2) corresponding to a non-degenerate distribution, there is a distribution in each system with the same (β_1, β_2) value. The region in the plane of all (β_1, β_2) points corresponding to all non-degenerate distributions can be divided into three main two-dimensional regions, each covered by one of three types of the Pearson family, namely: types I, IV, and VI, the first and last being the beta distributions of the first and second kind, respectively. Similarly, the (β_1, β_2) plane can be divided into two two-dimensional regions covered by two of the distributions of the Johnson system: the S_B and S_U models. For both systems, models are also defined on the boundaries of the main two-dimensional regions that correspond to further (sometimes called *transitional*) models with a functional form that is different from those defined in the main two-dimensional regions.

Non-Standard Parametric Statistical Inference. Russell Cheng.

There is an immediate point of note here. The basic form of each type or model corresponding to each main two-dimensional region are just two-parameter models, whereas the boundary models are just one-parameter. In all cases, these parameters are shape parameters whose values affect β_1 and β_2. However, in practical applications, the observed Y variable is assumed to be subject to an *additional* location-scale transform which introduces two additional parameters. This provides extra fitting flexibility, but does not alter a model's β_1 and β_2 values. The form all the models that we consider includes these location-scale parameters. Thus the Pearson types I, IV, and VI and Johnson S_B and S_U models that we consider are all four-parameter, with boundary models, to be defined, that are three-parameter. Special models corresponding to a specific individual (β_1, β_2) point, like the normal, for which $(\beta_1, \beta_2) = (0, 3)$, are two-parameter, these being just the additional location and scale parameters.

We will show that with the parametrization usually adopted for the main four-parameter models, the boundary transition models are embedded models in the sense of Chapter 5, so that boundaries cannot be approached without at least one parameter having to become infinite.

For both systems, a parametric model can be fitted to a data sample by the *method of moments*, where the first two, three, or four moments of the model, depending on the subfamily, are set equal to the corresponding sample moments of the data set. However, as is well known, the method of moments is not always efficient, even asymptotically. A detailed examination of this is given by Pearson (1963).

Use of maximum likelihood seems preferable. It has been suggested (for the Pearson case, see Becker and Klöner, 2013, for the Johnson case, see Hill et al., 1976) that use of the parameter estimates obtained from the method of moments might at least serve to provide an initial estimate for a numerical procedure for obtaining the ML estimators. However, a procedure capable of finding the ML estimator without restriction to type, but starting with the moment estimators, will have to allow for the possibility the initial model type will be different from the model type corresponding to the best overall ML fit. Thus, the procedure will have to be able to approach and possibly cross the boundary of the region in a stable way. This chapter focuses on how to handle this problem.

The Pearson distributions involve a shifted origin, so that the infinite likelihood problem can occur. Any numerical algorithm used will need to be able to cope with this.

The Pearson type IV has the reputation of being hard to fit, and as Johnson, Kotz, and Balakrishnan (1994, Chapter 12, §4.1) have remarked, 'It is rarely attempted'. This is perhaps slightly undeserved, though it is not as straightforward to handle as other subfamilies in the family. Heinrich (2004) discusses the problem, and an implementation is given by Becker and Klöner (2013). We give a method for fitting this distribution that is reasonably robust, subject to certain restrictions on the parameter values. In contrast, the Johnson family is rather easier to fit, and with its strong association with the normal distribution, is probably the easier family to use in practice, though expressions for the moments are too complicated to be useful in the S_U case.

There is one additional watchpoint. Setting $\beta_1 = \gamma_1^2$ means that each point (β_1, β_2) corresponds to two distributions, one the mirror image of the other, so that the sign of the skewness switches. Thus any particular aspect of a fitting procedure applied to a positively skew distribution will have a corresponding mirror version applicable to the mirror image. In both Pearson and Johnson systems, the parametrization used in each model can be defined so as to allow the skewness to be either positive or negative, but for simplicity, and to avoid repetition, we focus just on fitting to data samples that are positively skew. We will discuss separately the situation where there may be doubt as to whether the parametric model should be positively skew or not, because the sample is close to being symmetric, in Section 9.5.

9.2 Pearson System

9.2.1 Pearson Distribution Types

Figure 9.1 gives three depictions of the (β_1, β_2) plane (in the way it is conventionally shown, see Pearson (1916), with the β_2 axis pointing downwards) dividing the plane into separate regions corresponding to the main Pearson types.

The middle graph gives relatively restricted ranges for β_1 and β_2 conventionally used in the literature, see Stuart and Ord (1987, §6.2), for example, though Pearson (1916) does use the ranges $0 \leq \beta_1 \leq 10$, $0 \leq \beta_2 \leq 24$. These clearly show the nature of the boundaries of the different regions near the β_2 axis, an area where (β_1, β_2) will commonly be when the sample is not highly asymmetric. It will be seen that there are three main two-dimensional Pearson regions:

The type I region lying between the blue and red lines. Type I is the beta distribution of the first kind, with PDF

$$f_{\mathrm{I}}(y) = \frac{\Gamma(p+q)}{\Gamma(p)\Gamma(q)}(b-a)^{1-p-q}(y-a)^{p-1}(b-y)^{q-1},\ a < y < b, \tag{9.1}$$

where the four parameters satisfy $p, q > 0$, and $-\infty < a < b < \infty$.

The type VI region lying between the red and green lines. Type VI is the beta distribution of the second kind, with PDF

$$f_{\mathrm{VI}}(y) = f(x) = \frac{\Gamma\,(p+q)}{\Gamma\,(p)\,\Gamma\,(q)}\left(\frac{y-a}{b}\right)^{p-1}\frac{b^{-1}}{\left(1+\frac{y-a}{b}\right)^{p+q}},\ a < y < \infty, \tag{9.2}$$

where $p, q > 0$, $-\infty < a < \infty$, and $0 < b < \infty$.

The type IV region lying below the green line. Type IV has PDF

$$f_{\mathrm{IV}}(y) = \frac{|\Gamma(b+bci)|^2\,\Gamma(b)}{[\Gamma(b)]^2\Gamma(b-\frac{1}{2})\Gamma(\frac{1}{2})}\frac{\exp\{2bc\arctan[(y-a)/\tau]\}}{\tau\left[1+(\frac{y-a}{\tau})^2\right]^b},\ -\infty < y < \infty, \tag{9.3}$$

where $-\infty < a < \infty$, $0.5 < b < \infty$. We can take $c, \sigma > 0$ without loss of generality, except for the omission of negatively skewed forms that are merely reflected versions of positively skewed forms.

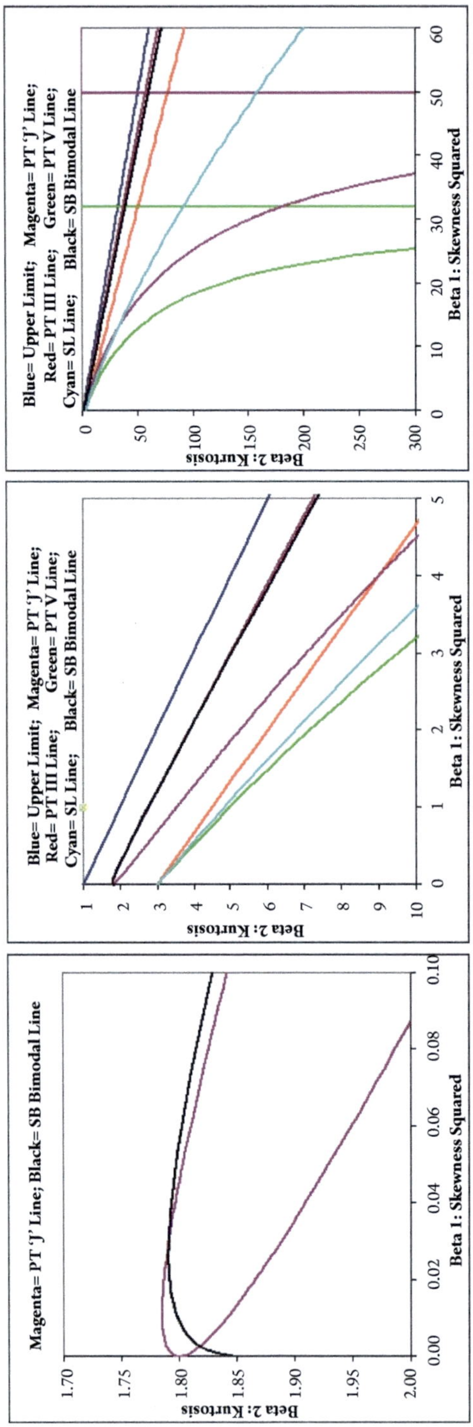

Figure 9.1 Regions in the (β_1, β_2) plane corresponding to different types of Pearson and Johnson distributions.

The red line between the types I and VI is

$$\beta_2 = 3 + 1.5\beta_1.$$

This boundary corresponds to type III, the gamma distribution with PDF which we take in the form

$$f_{III}(y) = \frac{1}{\Gamma(p)\lambda^p}(y-a)^{p-1}\exp\left(-\frac{y-a}{\lambda}\right), \quad a < y < \infty. \tag{9.4}$$

The green line between types VI and IV corresponds to type V, the inverse gamma distribution, with PDF which we take in the form

$$f_V(y) = \frac{1}{\Gamma(p)}\lambda^p(y-a)^{-p-1}\exp\left(-\frac{\lambda}{y-a}\right), \quad a < y < \infty. \tag{9.5}$$

The line is usually given as

$$\beta_1(\beta_2+3)^2 = 4(4\beta_2 - 3\beta_1)(2\beta_2 - 3\beta_1 - 6),$$

see Stuart and Ord (1987). Solving this for β_2 gives

$$\beta_2 = 3(16 + 13\beta_1 + 2\sqrt{(\beta_1+4)^3})/(32-\beta_1), \quad \text{for } \beta_1 < 32, \tag{9.6}$$

showing the line has a vertical asymptote at $\beta_1 = 32$, so that there are no distributions of type V or VI for which $\beta_1 > 32$.

Also shown in Figure 9.1 is the magenta looped curve

$$\beta_1(\beta_2+3)^2(8\beta_2 - 9\beta_1 - 12) = (4\beta_2 - 3\beta_1)(10\beta_2 - 12\beta_1 - 18)^2,$$

termed the I(J) loop in Stuart and Ord (1987). This curve delineates whether a Pearson distribution is bimodal, J-shaped, or unimodal, depending on whether (β_1, β_2) lies respectively 'above', 'within', or 'below' the loop (as depicted in Figure 9.1). In particular, for $\beta_1 \leq 4$, it bounds the type I region where the density is J-shaped, whilst for $\beta_1 > 4$, the lower branch of the I(J) curve bounds where the type IV density is J-shaped. Pearson (1916) gives the parametric representation of this curve as

$$\beta_1 = \frac{4(2\gamma-1)^2(\gamma+1)}{3\gamma-1}, \quad \beta_2 = \frac{3(\gamma+1+\beta_1)}{3-\gamma} = \frac{3(\gamma+1)(16\gamma^2-13\gamma+3)}{(3\gamma-1)(3-\gamma)},$$

which is convenient for plotting purposes. Not immediately obvious from this representation is that the lower branch of the loop has a vertical asymptote at $\beta_1 = 50$. If β_2 is expressed in terms for β_1, we find that the lower and upper branches are given by

$$\beta_{2,L} = \frac{3}{8}\,\frac{-r(\beta_1)\cos[\theta(\beta_1)] - \sqrt{3}r(\beta_1)\sin[\theta(\beta_1)] - 144\beta_1 - 160 + \beta_1^2}{\beta_1 - 50} \tag{9.7}$$

and

$$\beta_{2,U} = \begin{cases} \frac{3}{8}\frac{-r(\beta_1)\cos(\theta(\beta_1))+\sqrt{3}r(\beta_1)\sin[\theta(\beta_1)]-144\beta_1-160+\beta_1^2}{\beta_1-50} & \text{if } \beta_1 < 50, \\ \frac{3}{4(\beta_1-50)}r(\beta_1)\cos[\theta(\beta_1)] + \frac{3}{8}\frac{-144\beta_1-160+\beta_1^2}{\beta_1-50} & \text{if } \beta_1 > 50 \end{cases}$$

where $\beta_{2,U}$ at $\beta_1 = 50$ is defined by continuity, and

$$r(\beta_1) = (\beta_1 + 100)^{1/2}(\beta_1 + 4)^{3/2}, \theta(\beta_1) = \frac{1}{3}\arccos\left(\frac{(\beta_1^2 - 360\beta_1 + 2000)}{(\beta_1 + 100)^{3/2}(\beta_1 + 4)^{1/2}}\right).$$

This version of the I(J) curve is easier to use for checking where the fitted $\hat{\beta}_1, \hat{\beta}_2$ value lies relative to the I(J) loop.

The third graph of Figure 9.1 extends the range to $0 \leq \beta_1 \leq 60$ to show the (green) PT V line and its asymptote at $\beta_1 = 32$, and the (magenta) $\beta_{2,L}$ line with its asymptote at $\beta_1 = 50$.

9.2.2 Pearson Embedded Models

In fitting the Pearson family, both the PT III gamma and the PT V inverted gamma distributions can arise as an embedded model, each in two ways. We discuss the four cases in this section.

We make repeated use of the Stirling approximation

$$\ln\Gamma(z) \simeq H(z) = \left(z - \frac{1}{2}\right)\ln z - z + \frac{\ln(2\pi)}{2} + \frac{1}{12z} - \frac{1}{360z^3} + O(z^{-5}) \text{ as } |z| \to \infty \tag{9.8}$$

given, for example, by Abramowitz and Stegun (1965, eqn 6.1.41).

PT I → PT III

We first show the PT III gamma distribution of equation (9.4) is an embedded model of the PT I beta distribution of equation (9.1). The (one-observation) logarithm of this latter distribution is

$$\ln(f_I) = \ln(\Gamma(p+q)) - \ln(\Gamma(p)) - \ln(\Gamma(q)) +$$
$$(1-p-q)\ln(b-a) + (p-1)\ln(y-a) + (q-1)\ln(b-y).$$

Writing $b = \alpha^{-1}\lambda\ q = \alpha^{-1}$, and using (9.8), we have

$$\begin{aligned}\ln(f_I) = &\left(-\ln(\Gamma(p)) - p\ln\lambda + (p-1)\ln(y-a) - \frac{(y-a)}{\lambda}\right) + \\ &\left(\frac{1}{2}(p-1)p + \frac{y-a}{\lambda} - \frac{1}{2}\frac{(y-a)^2}{\lambda^2}\right)\alpha + \\ &\left(-\frac{1}{12}p(2p-1)(p-1) + \frac{1}{2}\frac{(y-a)^2}{\lambda^2} - \frac{1}{3}\frac{(y-a)^3}{\lambda^3}\right)\alpha^2 + O(\alpha^3) \\ = &\ L_0^I + L_1^I\alpha + L_2^I\alpha^2 + O(\alpha^3), \text{ say.}\end{aligned}$$

We have for simplicity written the coefficients as L_i^I, but it will be useful in what follows to think of these as

$$L_i^I = L_i^I(\boldsymbol{\theta}^{III}) \text{ where } \boldsymbol{\theta}^{III} = (a, \lambda, p)$$

to emphasize that the coefficients are derived from the PT I (full) model, but are dependent only on the parameters of the PT III embedded model. We have immediately that L_0^I is the logarithm of the PT III gamma density (9.4), showing that this is an embedded model of the PT I distribution under its original parametrization. Under the null assumption that PT III is the correct model, we have, when the parameters are at their true values, $E_{III}(L_1^I) = 0$ and $E_{III}(L_2^I) = -\frac{1}{4}p^2 - \frac{1}{2}p^3 - \frac{1}{4}p < 0$, reflecting the standard behaviour of the log-likelihood in the neighbourhood of the true parameter value.

A simple indication of whether the PT I model is to be preferred to the PT III is provided by $L_1^I(\hat{\boldsymbol{\theta}}^{III})$, where $\hat{\boldsymbol{\theta}}^{III}$ is the MLE of the parameters of the PT III model, this being the non-standardized score. If $L_1^I(\hat{\boldsymbol{\theta}}^{III})$ is positive, this shows the likelihood would increase as α increases from zero, indicating that fitting PT I would be preferable to fitting PT III. A more formal version would be to carry out the score test using (6.38).

PT VI → PT III

Consider now the PT VI beta distribution of equation (9.2). The one-observation log-likelihood is

$$\begin{aligned}\ln(f_{VI}) = &\ln(\Gamma(p+q)) - \ln(\Gamma(p)) - \ln(\Gamma(q)) + q\ln(b) \\ &+(p-1)\ln(y-a) - (p+q)\ln(b+y-a).\end{aligned}$$

Writing $b = \lambda\alpha^{-1}$, $q = \alpha^{-1}$, and using (9.8), we have

$$\begin{aligned}\ln(f_{VI}) = &\left(-\ln(\Gamma(p)) - p\ln\lambda + (p-1)\ln(y-a) - \frac{y-a}{\lambda}\right) + \\ &\left(\frac{1}{2}(p-1)p - p\frac{y-a}{\lambda} + \frac{1}{2}\frac{(y-a)^2}{\lambda^2}\right)\alpha +\end{aligned}$$

$$\left(-\frac{1}{12}p\,(2p-1)\,(p-1)+\frac{1}{2}p\frac{(y-a)^2}{\lambda^2}-\frac{1}{3}\frac{(y-a)^3}{\lambda^3}\right)\alpha^2+O\left(\alpha^3\right)$$
$$=L_0^{VI}+L_1^{VI}\alpha+L_2^{VI}\alpha^2+O(\alpha^3).$$

We again have that L_0^{VI} is the log-likelihood of the gamma distribution, showing that this is an embedded model in the original parametrization. Under the assumption that PT III is the correct model, $E_{III}(L_1^{VI})=0$ and $E_{III}(L_2^{VI})=-\frac{1}{4}p^2-\frac{3}{4}p<0$. Again, $L_1^{VI}(\hat{\boldsymbol{\theta}}^{III})>0$ provides an indication that the PT VI would be preferred to the PT III.

PT VI → PT V

Similarly writing $b=\lambda\alpha, p=\alpha^{-1}$ and using (9.8), we get

$$\begin{aligned}\ln(f_{VI}) &= \left(-\ln\left(\Gamma\left(q\right)\right)+q\ln\lambda-(q+1)\ln\left(y-a\right)-\frac{\lambda}{y-a}\right)+\\ &\left(\frac{1}{2}\left(q-1\right)q-q\frac{\lambda}{y-a}+\frac{1}{2}\frac{\lambda^2}{\left(y-a\right)^2}\right)\alpha+\\ &\left(-\frac{1}{12}q\left(2q-1\right)\left(q-1\right)+\frac{1}{2}q\frac{\lambda^2}{\left(y-a\right)^2}-\frac{1}{3}\frac{\lambda^3}{\left(y-a\right)^3}\right)\alpha^2+O\left(\alpha^3\right)\\ &=L_0^{VI}+L_1^{VI}\alpha+L_2^{VI}\alpha^2+O(\alpha^3),\end{aligned} \tag{9.9}$$

with L_0^{VI} the log-likelihood of the PT V distribution given in equation (9.5), showing that this is an embedded model in the original parametrization. Under the assumption that PT V is the correct model, $E_V(L_1^{VI})=0$ and $E_V(L_2^{VI})=-\frac{1}{4}q\,(4q^2+5q+3)<0$. Here, $L_1^{VI}(\hat{\boldsymbol{\theta}}^{V})>0$ provides an indication that the PT VI would be preferred to the PT V.

PT IV → PT V

We now show that the PT V distribution is an embedded model of PT IV. Let $b=(p+1)/2, c=\lambda(p+1)^{-1}\alpha^{-1/2}$ and $\tau=\alpha^{1/2}$ in $\ln(f_{IV})$. We can then write

$$\ln(|\Gamma(b+bci)|^2)=\ln(\Gamma(z))+\ln(\Gamma(w)),$$

where $z=(\frac{1}{2}(p+1+i\lambda\alpha^{-1/2}))$ and $w=(\frac{1}{2}(p+1-i\lambda\alpha^{-1/2}))$. Using this in $\ln(f_{IV})$, and expanding as a series in α, we find, after lengthy algebra, where we use (9.8) with complex argument and $\arctan(z)=\frac{\pi}{2}-\frac{1}{z}+\frac{1}{3z^3}-\frac{1}{5z^5}+O(z^7), |z|>1$ (see Abramowitz and Stegun, 1965, 4.4.42), that

$$\begin{aligned}\ln(f_{IV}) &= \left(-\ln\Gamma(p)-\ln(\lambda)+(p+1)\ln\left(\frac{\lambda}{y-a}\right)-\frac{\lambda}{(y-a)}\right)\\ &+\left(\frac{1}{6}\left(p-1\right)p\frac{p+1}{\lambda^2}-\frac{1}{2}\frac{(p+1)}{(y-a)^2}+\frac{1}{3}\frac{\lambda}{(y-a)^3}\right)\alpha\end{aligned} \tag{9.10}$$

$$+\left(-\frac{\lambda^{-4}}{60}(p-1)p(p+1)(3p^2-7)+\frac{1}{4}\frac{(p+1)}{(x-a)^4}-\frac{1}{5}\frac{\lambda}{(x-a)^5}\right)\alpha^2$$
$$+O(\alpha^3)$$
$$=L_0^{IV}+L_1^{IV}\alpha+L_2^{IV}\alpha^2+O(\alpha^3). \tag{9.11}$$

The leading term L_0^{IV} is the log-likelihood of the PT V distribution given in equation (9.5), confirming this to be an embedded model of the PT IV distribution under its original parametrization. Under the assumption that PT V is the correct model, and when the parameters are at their true values, we have $E_V(L_1^{IV})=0$ and $E_V(L_2^{IV})=-\frac{1}{24}(2b-1)\frac{6b^3+6b^2+5b+1}{b^3}$, which is negative as we must have $b>\frac{1}{2}$. As with the other embedded models given in this section, $L_1^{IV}(\hat{\boldsymbol{\theta}}^V)>0$ indicates that a better fit would obtained using the full model, PT IV in this case, than with the embedded model, PT V in this case.

9.2.3 Fitting the Pearson System

We now consider how to fit the Pearson system to a given sample using ML estimation with numerical optimization carried out by Nelder-Mead. Though we use ML, it is worth first plotting the $(\tilde{\beta}_1,\tilde{\beta}_2)$ calculated from *sample moments* as a point in the $(\beta_1\ \beta_2)$ plane to give a preliminary indication of which Pearson type might be appropriate.

A simple approach begins by fitting whichever is the main four-parameter model, PT I, PT VI, or PT IV, corresponding to the region in which $(\tilde{\beta}_1,\tilde{\beta}_2)$ lies, using an initial parameter point $\boldsymbol{\theta}$ corresponding to a (β_1,β_2) point lying in the same region as $(\tilde{\beta}_1,\tilde{\beta}_2)$. There is no special need to ensure that $(\beta_1,\beta_2)=(\tilde{\beta}_1,\tilde{\beta}_2)$ exactly. In Section 9.4, we suggest an initial point $\boldsymbol{\theta}$ based on the sample that is appropriate for each of the models PT I, PT VI, or PT IV. It is then easy at each step of the Nelder-Mead optimization to check that the parameter $\boldsymbol{\theta}$ satisfies the constraints required for the distribution type being fitted. Morover, the original parametrization can be used without directly worrying about the possibility of embeddedness, provided a check is made to see if the search is converging towards the boundary corresponding to an embedded model. This can be done with any of the four cases discussed in Section 9.2.2. For example, in Section 9.2.2, we showed that when fitting the PT I model with the model parameters $\boldsymbol{\theta}=(a,b,p,q)$ as given in equation (9.1), if the parameters $b,q\to\infty$, then this means that the PT III boundary is being approached. A check can be easily included to see if this is happening, so that if it occurs, a switch can be made to fitting the PT III model directly.

With the initial points suggested in Section 9.4, one could simply separately fit each of the three four-parameter Pearson types, comparing the results to see which is best. However, it is perhaps more insightful to fit the three-parameter PT III and PT V models *first*. In the PT III case, if the MLE of the parameters is $\hat{\boldsymbol{\theta}}^{III}$, then $L_1^{I}(\hat{\boldsymbol{\theta}}^{III})>0$ indicates fitting PT I would give an improvement, whilst $L_1^{VI}(\hat{\boldsymbol{\theta}}^{III})>0$ indicates PT VI would give an improvement. Similarly, if $\hat{\boldsymbol{\theta}}^V$ is the MLE of the PT V model, then $L_1^{VI}(\hat{\boldsymbol{\theta}}^V)>0$ indicates fitting PT VI would give an improvement, whilst $L_1^{IV}(\hat{\boldsymbol{\theta}}^V)>0$ indicates PT IV would give an improvement. We illustrate this kind of fitting process in the numerical examples at the end of this chapter.

9.3 Johnson System

9.3.1 Johnson Distribution Types

Johnson (1949) proposed a system comprising three models, calling the first two S_B and S_U, these being four-parameter. The third model is called S_L, this being the three-parameter lognormal model already listed in Table 6.4, with a slight change of notation, where b and c in that table is replaced by δ^{-1} and μ in the version given later in eqn (9.14). The system is comprehensive, like the Pearson system, in that for every combination of squared-skewness and kurtosis values, (β_1, β_2), for which a non-degenerate continuous distribution is possible, there is one member of the Johnson system with that given (β_1, β_2) value. The regions of the (β_1, β_2) plane covered by each of the three models is illustrated in Figure 9.1. In the middle chart of the figure, the region between the blue and cyan lines is covered by the S_B model. This has PDF

$$f_{SB}(y) == \frac{1}{\sqrt{2\pi}} \frac{(b-a)\delta}{(y-a)(b-y)} \exp\left\{-\frac{1}{2}\left[\gamma + \delta \ln\left(\frac{y-a}{b-y}\right)\right]^2\right\}, a < y < b. \quad (9.12)$$

In addition to $a < b$, it is assumed that $\delta > 0$. The subscript in S_B is a reminder that y has bounded support. The PDF f_{SB} can be unimodal or bimodal. The black line in all three charts is the boundary, above which the PDF is bimodal and below which it is unimodal. The most notable aspect of this boundary is how close it is to the upper I(J) boundary, above which the PT I distribution is bimodal. Johnson (1949) shows that points above the black line satisfy

$$\delta < 1/\sqrt{2}, \ |\gamma| < \delta^{-1}\sqrt{(1-2\delta^2)} - 2\delta \operatorname{arctanh}\sqrt{(1-2\delta^2)}. \quad (9.13)$$

Unfortunately, as is made clear in Johnson (1949, Appendix), moments of the S_B distribution are not easy to compute, with no simple formulas giving β_1 or β_2 in terms of (γ, δ). The bimodal line is therefore not all that easy to plot in the (β_1, β_2) plane.

Draper (1952, eqn 19) used a series expansion. This requires setting a certain constant h that has to be made small. Draper does not give the value(s) of h used in calculating the bimodal line in his Figure 2, the position of which would suggest that the version given in Johnson (1949, Figure 2) is not particularly accurate near $\beta_1 = 0$. Our depiction in the charts of Figure 9.1 is based on calculation of 50 (β_1, β_2) points using the Draper eqn (19), with $h = 0.0008$ and a summation where n runs from –4000 through to 4000.

As a check of the Draper method, we also carried out the calculation using an adaptive version of the well-known Gaussian-Legendre quadrature method with up to $n = 12$

nodes, to evaluate the integral representation of rth moments about zero as given in Johnson (1949, eqn 54), the values of β_1, β_2 obtained with the two methods agreeing to at least four decimal places.

The full bimodal line is not required if only a check is needed of whether a particular fitted S_B model is bimodal or not. Then, all that is required is to check if the parameter condition (9.13) is satisfied or not when evaluated at the ML estimates $\hat{\gamma}$ and $\hat{\delta}$.

The cyan line in Figure 9.1 corresponds to the lognormal model, which we take in the form

$$f_{SL}(y) = \frac{\delta}{\sqrt{2\pi}(y-a)} \exp\left\{-\frac{1}{2}\delta^2\left[\ln(y-a)-\mu\right]^2\right\},\ a < y, \tag{9.14}$$

where $\delta > 0$. The lognormal (cyan) line can be obtained from known expressions for β_1 and β_2 for the lognormal model, given, for example, by Johnson, Kotz, and Balakrishnan (1994, eqns 14.9a and 14.9b), which show that

$$\beta_2 = \omega^4 + 2\omega^3 + 3\omega^2 - 3, \text{ where } \omega = c + c^{-1} - 1$$

with

$$c = \frac{1}{2}\left[8 + 4\beta_1 + 4(4\beta_1 + \beta_1^2)^{1/2}\right]^{1/3}.$$

The region in the (β_1, β_2) plane below the lognormal line corresponds to the S_U distribution whose PDF we take in the form

$$f_{SU}(x) = \frac{\delta \exp\left\{-\frac{1}{2}\left[\gamma + \delta \operatorname{arcsinh}(\frac{x-a}{b})\right]^2\right\}}{(2\pi)^{1/2}\left[(x-a)^2 + b^2\right]^{1/2}},\ -\infty < y < \infty. \tag{9.15}$$

The subscript in the name S_U is a reminder that the distribution has unbounded support.

9.3.2 Johnson Embedded Models

We show that the lognormal model S_L is an embedded model of both the S_B and S_U distributions.

$S_B \to S_L$

The parameters are a, b, γ, δ in the S_B PDF f_{SB} in eqn (9.12). We use the reparametrization where a and δ are unchanged, but b and γ are replaced by α and μ, with $b = \alpha^{-1}$ and $\gamma = \delta(\ln(\alpha^{-1} - a) - \mu)$. Expanding the reparametrized log-likelihood $\ln(f_{SB})$ as a power series in α, we find

$$\begin{aligned}\ln(f_{SB}) &= \left(-\frac{1}{2}\ln(2\pi) + \ln\delta - \ln(y-a) - \frac{1}{2}\delta^2(\ln(y-a)-\mu)^2\right) + \\ &\quad (y-a)\{1+\delta^2[\mu - \ln(y-a)]\}\alpha + \\ &\quad \frac{1}{2}(y-a)\left\{\left[1-\delta^2\left(\ln(y-a)-\mu\right)\right](y+a) - \delta^2(y-a)\right\}\alpha^2 + O(\alpha^3) \\ &= L_0^{SB} + L_1^{SB}\alpha + L_2^{SB}\alpha^2 + O(\alpha^3).\end{aligned}$$

The constant term L_0^{SB} is the log-likelihood of the lognormal PDF f_{SL} given in (9.14). Letting $\alpha \to 0$, we have that $f_{SB} \to f_{SL}$; and, as then $b \to 0$ and $\gamma \to \infty$, this shows that f_{SL} is an embedded model in the original parametrization. Let $L_1^{SB}(\hat{\boldsymbol{\theta}}^{SL})$ be the value of the coefficient of α at $\hat{\boldsymbol{\theta}}^{SL}$, the MLE of the parameters in the S_L model. If then $L_1^{SB}(\hat{\boldsymbol{\theta}}^{SL}) > 0$, this would indicate that fitting the S_B model would improve the fit over the S_L model.

$S_U \to S_L$

As in the S_B model, the parameters in the S_U model are a, b, γ, δ, only now in the PDF of f_{SU} in eqn (9.15). In this case, we reparametrize, replacing b and γ by α and μ, where $b = \alpha^{1/2}$, $\gamma = \delta(\frac{1}{2}\ln\alpha - \ln 2 - \mu)$. Using the reparametrization and writing $\text{arcsinh}(z) = \ln[z + (1+z^2)^{1/2}]$ in (9.15), we can expand the log-likelihood of f_{SU} as a series in α, which gives

$$\begin{aligned}\ln(f_{SU}) &= \\ &\left(-\frac{1}{2}\ln(2\pi) + \ln\delta - \ln(y-a) - \frac{1}{2}\delta^2(\ln(y-a)-\mu)^2\right) + \\ &\left(\delta^2\frac{\mu - \ln(y-a)}{4(y-a)^2} - \frac{1}{2(y-a)^2}\right)\alpha + \\ &\left(\frac{3}{32}\delta^2\frac{\ln(y-a)-\mu}{(y-a)^4} - \frac{\delta^2}{32(y-a)^4} + \frac{1}{4(y-a)^4}\right)\alpha^2 + O(\alpha^3) \\ &= L_0^{SU} + L_1^{SU}\alpha + L_2^{SU}\alpha^2 + O(\alpha^3),\end{aligned}$$

where $L_0^{SU} = \ln(f_{SL})$ is the log-likelihood of the lognormal distribution given in (9.14). Letting $\alpha \to 0$ shows that $f_{SU} \to f_{SL}$; and, as this requires $b \to 0$, $\gamma \to -\infty$, this shows that f_{SL} is an embedded model in the original parametrization. Similar to the S_B case, if $\hat{\boldsymbol{\theta}}^{SL}$ is the MLE of the parameters in the S_L model, then $L_1^{SU}(\hat{\boldsymbol{\theta}}^{SL}) > 0$ would indicate that an improvement would be made in fitting the S_U model.

9.3.3 Fitting the Johnson System

The same methods discussed in fitting the Pearson system can be used for the Johnson system. In this case, as there are only two four-parameter models involved, S_B and S_U, one could simply fit each separately using Nelder-Mead optimization, with a starting parameter point corresponding to a point in the appopriate (β_1, β_2) region. Selection of such a point is discussed in the next section.

However, it is probably simplest just to fit the S_L model first, this giving the MLE, $\hat{\boldsymbol{\theta}}^{SL}$, of the parameters for this model, whose corresponding (β_1, β_2) value lies on the lognormal line in the (β_1, β_2) plane. One can then extend the search to the S_B region if $L_1^{SB}(\hat{\boldsymbol{\theta}}^{SL}) > 0$, and to the S_U region if $L_1^{SU}(\hat{\boldsymbol{\theta}}^{SL}) > 0$. It is conceivable that both additional searches are needed, but this seems only likely to occur in the near symmetric case, when the best fit is the normal model, this being an embedded model of all three models S_L, S_B, and S_U.

9.4 Initial Parameter Search Point

We discuss the choice of an initial parameter point, $\tilde{\boldsymbol{\theta}}$, with which to start the numerical optimization in the specific situation where we want to fit a particular Pearson or Johnson type to the sample, which we write in ordered form: $y_{(1)} < y_{(2)} < \ldots < y_{(n)}$. Each of the parameters of the Pearson and Johnson systems has a specific role in defining the location, scale, and shape characteristics, such as the skewness, of a particular distribution. Our approach will be to select, where possible, a $\tilde{\boldsymbol{\theta}}$ that reflects the sample versions of these characteristics. Even if the parameter values only loosely reflect these characteristics, the Nelder-Mead algorithm seems quite tolerant, with algorithm convergence achieved without requiring very great accuracy in the selected initial values. The following approach is easy to implement, and provides a satisfactory starting parameter point for all the models of both systems.

In specific models, better initial points can be obtained, but at the expense of more elaborate calculations. Though we will be mentioning one or two of these cases, the added effort seems not especially worthwhile in general.

9.4.1 Starting Search Point for Pearson Distributions

We consider first the general approach to the Pearson types PT I, III, IV, V, and VI.

PT I is the only distribution where the support is finite, with $a < y < b$. For PT I with PDF as in eqn (9.1), all the observations are subjected to an initial location-scale transform, with y only appearing as $w = (y - a)/(b - a)$. We therefore set a and b as

$$\tilde{a} = y_{(1)} - r/n,\ \tilde{b} = y_{(n)} + r/n,\ \text{where } r = y_{(n)} - y_{(1)}.$$

We then select the two remaining parameters so that the first two moments of the Pearson distribution match the first two sample moments of the sample $w_i = (y_i - \tilde{a})/(\tilde{b} - \tilde{a})$, $i = 1, 2, \ldots, n$. This gives

$$\tilde{p} = \frac{m^2(1-m)}{s^2} - m,\ \tilde{q} = \frac{\tilde{p}}{m} - 1,$$

with

$$m = \bar{w} = n^{-1}\sum_{i=1}^{n} w_i \text{ and } s^2 = n^{-1}\sum_{i=1}^{n}(w_i - \bar{w})^2, \tag{9.16}$$

which is essentially as given in Johnson, Kotz, and Balakrishnan (1995, eqns 25.28 and 25.29).

For PT VI, with PDF as in eqn (9.2), the support is left-bounded with $a < y$, and with y appearing in the PDF equation only as $w = (y - a)/b$. We set

$$\tilde{a} = y_{(1)} - r/n \text{ and } \tilde{b} = \tilde{\lambda} = (1 + 2/n)(y_{(n)} - y_{(1)})$$

so that $\tilde{a}$ is as in the PT I case, but treating b as a scale parameter which we set to be slightly larger than the sample range r of the y_i. The distribution of the $w_i = (y_i - \tilde{a})/\tilde{b}$ where $\tilde{a}$ and $\tilde{b}$ are treated as known is the standardized version of PT VI, whose first two moments are $E(w) = p/(q-1)$ and $\text{Var}(w) = p(p+q-1)(q-2)^{-1}(q-1)^{-2}$. Equating these to the w-sample versions, and solving for p and q gives

$$\tilde{p} = m(m^2 + m + s^2)s^{-2} \text{ and } \tilde{q} = 1 + \tilde{p}/m,$$

with m and s^2 as in (9.16), only with $w_i = (y_i - \tilde{a})/\tilde{b}$.

In the PT III and PT V cases, with respective PDFs (9.4) and (9.5), the support is again left-bounded with $a < y$. As there are only three parameters, we simply take $w = y - \tilde{a}$, again with $\tilde{a} = y_{(1)} - r/n$. The probability distributions of w are, respectively, the standard two-parameter gamma and inverted gamma distribution. In each case, we can again equate the parametric form of the first and second moments of the distribution to the sample moments of the sample w_i. This gives

$$\tilde{p} = m^2 s^{-2},\ \tilde{\lambda} = s^2/m \text{ for PT III},$$
$$\tilde{p} = 2 + m^2 s^{-2},\ \tilde{\lambda} = m(1 + m^2 s^{-2}) \text{ for PT V}.$$

The PT IV case has unbounded support $-\infty < y < \infty$, but again has y appearing in the PDF of eqn (9.3) only as $(y - a)/\dot{\sigma}$. In this case, we can set $w = (y - \tilde{a})/\tilde{\sigma}$, still with $\tilde{a} = y_{(1)} - r/n$, but now with $\tilde{\sigma} = (1 + 2/n)(y_{(n)} - y_{(1)})$. We then obtain the two shape parameters b and c by equating the distribution first and second moments with the corresponding sample moments of the $w_i = (y_i - \tilde{a})/\tilde{\sigma}$ to get

$$\tilde{b} = \frac{3}{2} + \frac{1 + m^2}{2s^2} \text{ and } c = \frac{(\tilde{b} - 1)}{\tilde{b}}m.$$

Given the difficult reputation of fitting the PT IV, a more refined starting value might give additional reassurance of optimization reliability. Cheng (2011) gives a version which matches the first three moments of the parametric distribution as given in the PDF (9.3) with the moments of the observed sample, together with a fourth condition that ensures the (β_1, β_2) value of the starting point lies on an explicit locus lying just within the PT IV region of the (β_1, β_2) plane.

9.4.2 Starting Search Point for Johnson Distributions

We adopt a similar approach for selecting a parameter value $\tilde{\boldsymbol{\theta}}$ that is a suitable starting value when fitting a Johnson of specific type to a sample $\mathbf{y}$. In the Johnson case, we take advantage of the fact that each model can be defined in terms of a specific monotonic translation of the y observation to a standard $N(0, 1)$ variable.

From equation (9.14) giving the PDF of the S_L model, we have that $z = \delta(\ln(y - a) - \mu)$ is $N(0, 1)$ distributed. In this case, as $a < y$, we set $\tilde{a} = y_{(1)} - r/n$, where $r = (y_{(n)} - y_{(1)})$ as in the Pearson case, and take $w_i = \ln(y_i - \tilde{a})$. We then determine the other two parameters δ and μ by equating the first two sample moments calculated from $z_i = \delta(\ln(w_i) - \mu)$ with the corresponding standard normal moments, which are trivially $E(z) = 0$ and $\text{Var}(z) = 1$. From $\bar{z} = E(z)$, we get $\delta(\bar{w} - \mu) = E(z) = 0$, so that $\mu = \bar{w}$, and from $n^{-1}\Sigma_{i=1}^{n} z_i^2 = \text{Var}(z) = 1$, we get $\delta^2 s^2 = 1$, where $s^2 = n^{-1}\sum(w_i - \bar{w})^2$, so that $\delta = s^{-1}$.

From equation (9.12) giving the PDF of the S_B model, we have that $z = \gamma + \delta \ln\left[(y - a)/(b - y)\right]$ is $N(0, 1)$ distributed. As $a < y < b$ in this case, we set $\tilde{a} = y_{(1)} - r/n$ and $\tilde{b} = y_{(n)} + r/n$, and take $w_i = \ln[y_i - \tilde{a})/(\tilde{b} - y_i)]$ with $z_i = \gamma + \delta w_i$. We determine the remaining parameters γ and δ by equating the first and second sample moments of the z_i with the standard normal moments $E(z) = 0$ and $\text{Var}(z) = 1$. This gives $\tilde{\delta} = s^{-1}$ and $\tilde{\gamma} = -\bar{w}s^{-1}$, where s^2 is the sample variance of the $w_i = \ln[y_i - \tilde{a})/(\tilde{b} - y_i)]$, that is, $s^2 = n^{-1}\sum(w_i - \bar{w})^2$.

From the equation (9.15) giving the PDF of the S_U model, we have that $z = \gamma + \delta \ln(\frac{y-a}{b} + \sqrt{1 + (\frac{y-a}{b})^2})$ is $N(0, 1)$ distributed. Though y is unrestricted in this case, we still set $\tilde{a} = \bar{y} - r/n$, but take $\tilde{b} = (n + 2)r/n$. We then take $z_i = \gamma + \delta w_i$, with $w_i = \ln\left\{(y_i - \tilde{a})/\tilde{b} + \sqrt{1 + [(y_i - \tilde{a})/\tilde{b}]^2}\right\}$. As the linear expression for z_i is the same as in the S_B case, the formulas are the same apart from the change in w_i. We therefore again get $\tilde{\delta} = s^{-1}$ and $\tilde{\gamma} = -\bar{w}s^{-1}$, with $s^2 = n^{-1}\sum(w_i - \bar{w})^2$.

9.5 Symmetric Pearson and Johnson Models

With both Pearson and Johnson systems, we have glossed over the boundary in the (β_1,β_2), where $\beta_1 = 0$, which corresponds to symmetric distributions. The reason is that, except for the normal model, which is an embedded model of all the Pearson and Johnson models, the symmetric models of either system can all be derived as simple boundary models, so that parameter values do not become unbounded or indeterminate. Thus, in the Pearson system, the PT I model approaches the β_1 boundary simply when its p and

q parameters converge to a common value. If we require only that $p > 0$ and $q > 0$, then PT I can be negatively skewed as well as positively skewed.

For the PT IV case, this is symmetric if $c \to 0$, when we get the Pearson type VII PDF

$$f_{VII}(y) = \frac{\Gamma(b)}{\sqrt{\pi}\tau\Gamma(b-\frac{1}{2})}\left[1+\left(\frac{y-a}{\tau}\right)^2\right]^{-b}, \quad -\infty < y < \infty,$$

which includes Student's t-distribution as a special case.

In the Johnson system, the S_B and S_U models both become symmetric as $\gamma \to 0$. The mirror versions of S_B and S_U, which cover the case where these distributions are negatively skew, are obtained when $\gamma < 0$. We can therefore handle both negative and positive cases simply by allowing γ to be unrestricted in sign.

The remaining cases, PT III, PT V, and S_L, all have support $a < y$ in the standard parametrizations given in eqns (9.4), (9.5), and (9.14), but with the normal as an embedded model in all cases. As already suggested, when fitting these cases, one can simply check if convergence of the Nelder-Mead iterations is towards the normal. Alternatively, one can use the reparametrizations given for all these models in Table 6.4, with the additional benefit that, in that table, all the models are defined to cover the mirror versions, where the skewness is negative.

9.6 Headway Times Example

In this section, we give an example where we fit models of the Pearson and Johnson systems to a sample of headway times reported and discussed by Cowan (1975). Headway times are the times between crossings at a fixed road point by the front axles of consecutive vehicles. The sample discussed by Cowan has $n = 1324$ observations, but as our example is for expository purposes only, we have randomly drawn a subsample of just 200 observations, and these are given in Table 9.1.

9.6.1 Pearson System Fitted to Headway Data

Consider first the estimation of the best Pearson model. This gave the following results, which we present in the sequence obtained:

1. (i) The black cross in Figure 9.2 gives the position in the (β_1, β_2) plane of the sample point $(\tilde{\beta}_1, \tilde{\beta}_2) = (4.73, 8.18)$ obtained from the sample moments of the sample given in Table 9.1, showing that the point falls in the 'J' part of the PT I region.

   (ii) We therefore try to fit the PT I model. This was unsuccessful, with the parameters b and q in the PT I PDF of eqn (9.1) both becoming indefinitely large, a clear indication of convergence towards the embedded PT III model.

Table 9.1 A sample of 200 observations from the Cowan 1975 headway data

94	155	53	49	3	8	53	22	24	8	99	5	24	6	49
5	3	3	6	15	24	11	8	56	93	55	5	19	57	6
9	4	54	77	3	68	19	29	34	14	27	27	10	78	16
33	10	16	14	6	22	34	5	9	13	10	74	39	100	38
101	27	21	12	38	46	75	21	40	32	160	48	21	13	23
120	29	34	8	21	29	22	14	7	12	9	31	38	12	111
17	10	14	21	61	12	25	38	4	9	14	24	10	8	9
39	18	7	12	9	9	14	4	6	4	47	16	9	21	8
44	76	10	9	16	6	13	46	23	13	17	21	8	28	10
20	9	38	17	24	6	7	14	24	24	13	17	40	78	13
40	17	10	48	6	7	40	34	8	6	22	16	4	100	148
17	10	14	20	16	42	4	74	10	31	18	17	54	66	9
34	6	7	13	5	11	9	10	9	12	133	56	54	24	29
12	39	17	11	15										

2. (i) Fitting the PT III model gave $\hat{\boldsymbol{\theta}}^{III} = (a, p, \lambda) = (2.999, 0.805, 31.6)$ with $L_1^I(\hat{\boldsymbol{\theta}}^{III}) = -0.0275$ and $L_1^{VI}(\hat{\boldsymbol{\theta}}^{III}) = 0.0275$. The negative value of $L_1^I(\hat{\boldsymbol{\theta}}^{III})$ is an added indication that the best fit is not a PT I model. Positivity of $L_1^{VI}(\hat{\boldsymbol{\theta}}^{III})$ shows that fitting PT VI would improve the fit.
   (ii) A meaningful $(\hat{\beta}_1, \hat{\beta}_2)$ is always possible with PT III, in this case $(\hat{\beta}_1^{III}, \hat{\beta}_2^{III}) = (4.97, 10.5)$. The position of this point is shown by the red cross in Figure 9.2.
   (iii) The fitted model is shown in the top left chart in Figure 9.3. The left-hand spike in the PDF seems a poor fit, confirmed by the significant lack of fit of the BS A^2 GoF statistic test value with p-value= 0.0.
3. (i) Fitting PT VI gave $\hat{\boldsymbol{\theta}}^{VI} = (\hat{a}, \hat{p}, \hat{q}, \hat{b}) = (2.999, 0.84, 15.1, 428)$. The fit is depicted in the middle left chart of Figure 9.3, showing it to be similar to the PT III fit. The BS A^2 GoF test yields a p-value not measurably different from $p = 0.0$, indicating a very strong lack of fit.
   (ii) The corresponding point $(\hat{\beta}_1^{VI}, \hat{\beta}_2^{VI}) = (7.09, 15.1)$ is marked by the green cross in Figure 9.2.

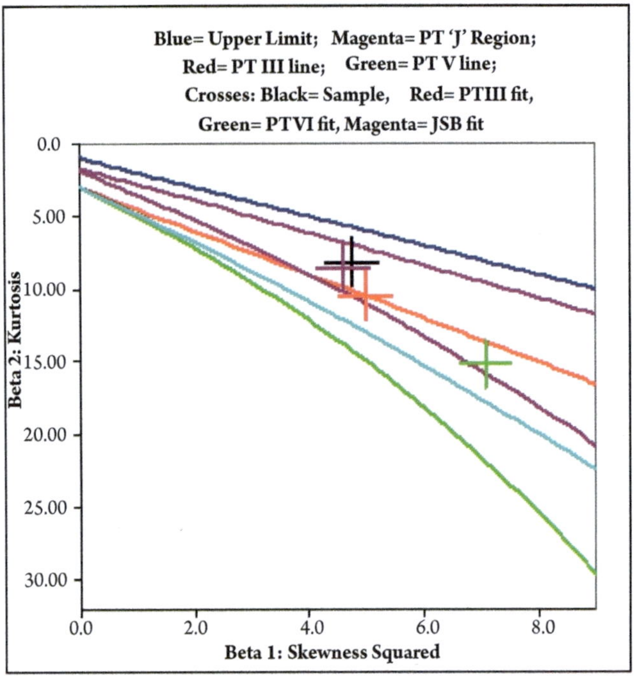

Figure 9.2 Plot of (β_1, β_2) points corresponding to the Pearson PT III (red cross), PT VI (green cross), and Johnson S_B (magenta cross) fits to the Cowan headway data.

4. (i) Fitting the PT V model gave an estimated parameter value of $\hat{\boldsymbol{\theta}}^V = (\hat{a}, \hat{p}, \hat{\lambda}) = (-1.56, 1.82, 28.2)$. The PT V PDF has infinitely flat contact with the abscissa axis. The fit does not display the spike at the left-hand end. The parameter values do not allow a (β_1, β_2) to be calculated.
   (ii) The BS A^2 GoF test gave $p = 0.066$.
   (iii) The coefficients $L_1^{IV}(\hat{\boldsymbol{\theta}}^V) = -0.00033$ and $L_1^{VI}(\hat{\boldsymbol{\theta}}^V) = -7.4E(-6)$ indicate that locally in the neighbourhood of the PT V fit, fitting either PT IV or PT VI is not beneficial.
5. An attempt to fit PT IV results in $c \to \infty$ and $\tau \to 0$, indicating convergence towards the embedded PT V model.

The pronounced left-hand spike in the PDFs of the fitted PT III and PT VI models suggest that the infinite likelihood problem discussed in Chapter 8, though not explictly present, is beginning to exert an effect when fitting these distributions. We therefore calculated the MPS fits for these two models and also for the PT V model. The fitted PDFs are displayed in the right-hand charts of Figure 9.3. These show that MPS estimation makes a difference, reducing the level of the spike in the PT III model and

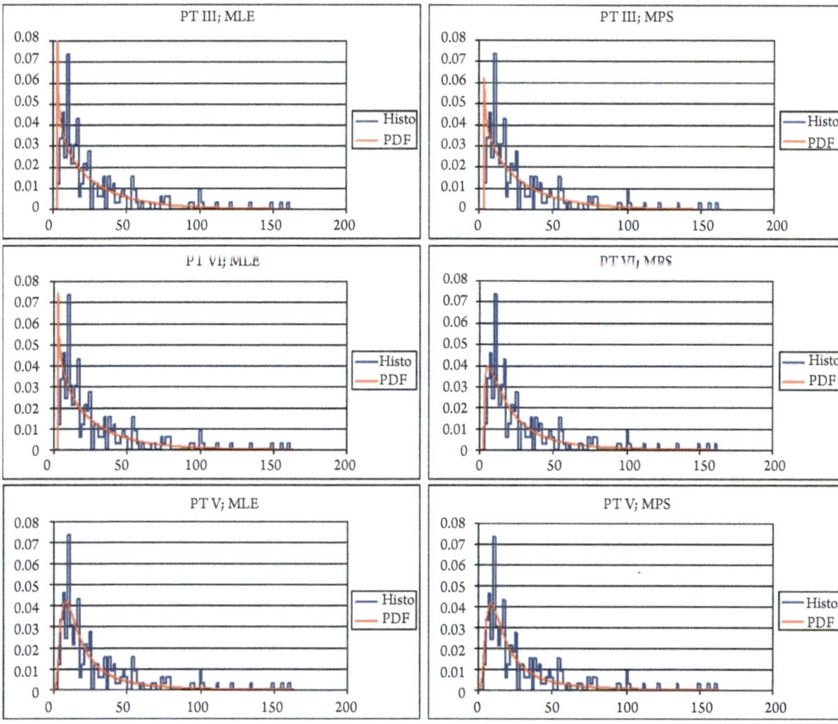

Figure 9.3 Plots of PDFs of the PT III (upper charts), VI (middle charts), and V (lower charts) models fitted to the Cowan headway data, using ML (left-hand charts) and MPS (right-hand charts) estimation.

removing it altogether in the PT VI case. The effect on the PT V fit is not so visually evident. However, the BS p-values of the GoF test for these latter two models of 0.128 and 0.14 now mean that the lack-of-fit is no longer significant at the 10% level for these two models.

The BS p-values and the maximized log-likelihood values of the Pearson fits are summarized in Table 9.2. For this example, MPS is preferable to MLE, with little to choose between the PT VI and PT V models.

9.6.2 Johnson System Fitted to Headway Data

Fitting the Johnson system to the Headway data is rather easier.

1. Figure 9.2 shows that the sample point $(\tilde{\beta}_1, \tilde{\beta}_2) = (4.73, 8.18)$ (black cross) falls in the unimodal part of the S_B region. However, we illustrate the procedure where we fit the S_L model first.

Table 9.2 Maximized log-likelihood values of Pearson PT III, VI, V, Johnson S_L and S_B distribution ML fits to the Cowan headway data, and BS A^2 GoF p-values; acceptable fits in bold. Maximized logspacings values and GoF p-values for the Pearson MPS fits

Distn	PT III	PT VI	PT V	S_L	S_B
MLE p-value	0.0	0.0	0.066	**0.428**	**0.302**
MLE $L_{\max}$	–843.7	–843.5	–850.0	–846.5	–844.5
MPS p-value	0.004	**0.128**	**0.140**		
MPS L_{MPS}	–822.6	–819.9	–822.4		

2. (i) Fitting the S_L model gave $\hat{\boldsymbol{\theta}}^L = (\hat{a} = 1.87, \hat{\theta} = 2.76, \hat{\sigma} = 1.055)$. The gradient coefficients are $L_1^B(\hat{\boldsymbol{\theta}}^L) = 3.088$ and $L_1^U(\hat{\boldsymbol{\theta}}^L) = -0.002$. The positive value of $L_1^B(\hat{\boldsymbol{\theta}}^L)$ is an added indication that the S_B model will give the best fit. However, negativity of $L_1^U(\hat{\boldsymbol{\theta}}^L)$ does not rule out S_U at this stage.

   (ii) The value of (β_1, β_2) corresponding to $\hat{\boldsymbol{\theta}}^L$ is $(\hat{\beta}_1^L, \hat{\beta}_2^L) = (52.09, 167.4)$. The position of this point lies on the cyan-coloured line, but cannot be shown in Figure 9.2 as it is way off the scale used, emphasizing how the MLE $(\hat{\beta}_1, \hat{\beta}_2)$ can be very different from $(\tilde{\beta}_1, \tilde{\beta}_2)$. The maximized log-likelihood is $L_{\max} = -846.5$.

   (iii) The fitted model is shown in the top left chart in Figure 9.4. The BS A^2 GoF test gives $p = 0.428$, indicating a satisfactory fit.

3. An attempt to fit S_U leads to $b \to 0$ and $\gamma \to -\infty$, this taking place rather slowly but still indicating convergence towards S_L. We stopped the Nelder-Mead iterations once $b < 0.05$, at which point $\hat{\boldsymbol{\theta}}^U = (\hat{a} = 1.87, \hat{b} = 0.046, \hat{\gamma} = -6.19, \hat{\delta} = 0.948)$ and $(\hat{\beta}_1^U, \hat{\beta}_2^U) = (52.07, 167.3)$.

4. (i) Fitting S_B gave $\hat{\boldsymbol{\theta}}^B = (\hat{a} = 2.37, \hat{b} = 233.5, \hat{\gamma} = 2.06, \hat{\delta} = 0.788)$.

   (ii) The corresponding point $(\hat{\beta}_1^B, \hat{\beta}_2^B) = (4.601, 8.548)$ is depicted in Figure 9.2 (magenta cross), showing it is quite close to the sample point. The maximum log-likelihood is $L_{\max} = -844.5$.

   (iii) The fitted model is shown in the bottom row of charts in Figure 9.4. The BS A^2 GoF test gives $p = 0.302$, indicating a satisfactory fit.

Summarizing, there seems little to choose between the S_L and S_B fits, the S_L fit doing somewhat better on GoF but slightly worse in its $L_{\max}$ value, which is slightly lower. A notable difference is the $(\hat{\beta}_1, \hat{\beta}_2)$ values, illustrating how variable these can be.

For comparison, the BS p-values and the maximized log-likelihood values of the Johnson fits are included in Table 9.2 together with the Pearson results.

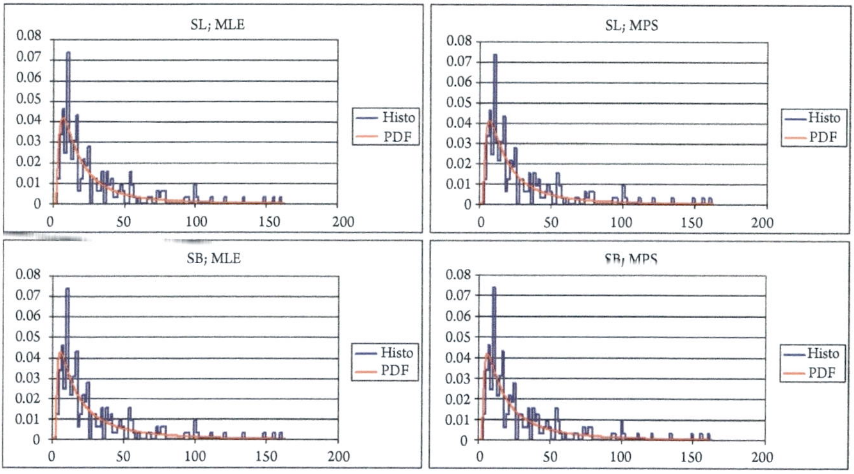

Figure 9.4 Plots of PDFs of the SL (upper charts) and SB (lower charts) models fitted to the Cowan headway data, using ML (left-hand charts) and MPS (right-hand charts) estimation.

9.6.3 Summary

The data are very positively skew, but probably with an underlying PDF that has a mode which is close to, but not, zero. The Pearson system has submodels which can be J-shaped. Fitting such distributions by ML can result in fits that over-emphasize the mode, leading to a measurable lack of fit. Using MPS estimation avoids this, with the PT V and PT VI models providing adequate fits for the given data set.

The models of the Johnson system all have PDFs with a high (actually, infinite degree of) contact with the abscissa axis at the two ends of its support. ML estimation does not overemphasize the mode for this particular data set. Both S_L and S_B models provide adequate fits.

9.7 FTSE Shares Example

Our second example illustrates the situation where the data are quite symmetric, but rather heavy-tailed compared with the normal distribution. The example was discussed by Cheng (2011) in the context of fitting the PT IV distribution, with a comparison with the PT V fit. Here we give a fuller discussion based on fitting either the Pearson or Johnson system as a whole.

Stable distributions are often used for modelling data with long tails. Nolan (2005) gives a clear review and discussion of estimation and goodness-of-fit tests for the stable distribution, from which it is clear that ML estimation of the parameters is the method of choice. So, for extra comparison, and by way of illustration, we have also included fitting the four-parameter stable distribution of the random variable $Y = \gamma S + \delta$, where $S_{\alpha,\beta}$ has the standardized two-parameter distribution with PDF given in eqns (6.49) and (6.51), using ML estimation. For the stable distribution, ML estimation is much more computer intensive than with the Pearson and Johnson families, but our example is simply to demonstrate that the approach is nevertheless viable.

Formal tests of goodness-of-fit of the stable distribution fit using bootstrapping is still prohibitively expensive for the general practitioner, though use of general purpose graphical parallel processing, see Cheng (2014), for example, should overcome this. Nolan (2005) recommends more informal graphical comparisons for goodness-of-fit, however, we have not considered this further, as it is outside the intended scope of this book.

The data come from a financial application arising in a study of the movement of the stock market (with a view to generating similar data for use in a simulation). We consider a data set in the form

$$y_i = \ln(p_i/p_{i-1}), \quad i = 1, 2, \ldots, n, \tag{9.17}$$

where p_i is the closing Financial Times Stock Exchange (FTSE) 100 index on day i. The data set discussed by Cheng (2011) comprises $n = 250$ observations, with the last day observed being 17 March 2011. There is the possibility of correlation between succeeding observations, but the lag-one autocorrelation is fairly small at 0.016. As the example is for illustration only, we have therefore treated it as a random sample.

To simplify presentation of the data, not all of the original sample considered in Cheng (2011) has been included, but a only subsample of size 100, randomly selected without replacement from the original sample. This subsample of y values is displayed in Table 9.3, which shows the observations multiplied by 10^6. In the actual model fitting, we used the original values y_i as given in eqn (9.17).

9.7.1 Pearson System Fitted to FTSE Index Data

We consider fitting a model from the Pearson system to the FTSE index data.

1. For this data set, $(\tilde{\beta}_1, \tilde{\beta}_2) = (0.526, 7.09)$, placing the point well in the PT IV region, as depicted in Figure 9.6.
   - (i) Fitting the PT IV distribution, with PDF as given in eqn (9.3), to this data set, the ML estimates are $\hat{\boldsymbol{\theta}}^{IV} = (\hat{a}, \hat{t}, \hat{b}, \hat{c}) = (0.00054, 0.0129, 2.13, -0.018)$, with an ML value of $L(\hat{\boldsymbol{\theta}}^{IV}) = 319.63$. The lower two charts of Figure 9.5 show the fitted CDF and PDF.
   - (ii) The GoF statistic value was $A^2 = 0.319$ with a p-value of 0.19, so that the lack-of-fit was not significant at the 10% level, the critical value for which is $A^2_{0.1} = 0.402$.
   - (iii) For this sample, $\hat{\beta}_1^{IV} = 0.081$, but the value of $\hat{b} = 2.13$, being less than 2.5, means that $\hat{\beta}_2^{IV}$ does not exist, so that $(\hat{\beta}_1^{IV}, \hat{\beta}_2^{IV})$ cannot be plotted. In fact, for the full sample, $\hat{b} = 3.03$, giving $(\hat{\beta}_1^{IV}, \hat{\beta}_2^{IV}) = (0.004, 8.67)$. This point is

Table 9.3 100 observations of $10^6 ln(p_i/p_{i-1})$, where p_i is the closing FTSE100 index on day i

-111	6621	-11979	7362	4373	6832	8477	1764
10238	-76	-1906	2043	-14003	5142	8059	21929
-8936	-25940	341	4828	9641	5569	-187	16921
-8610	-4164	10106	-2736	8189	-2122	-11162	-723
6061	-3152	-2413	8715	-795	-24083	-2057	-3075
-590	-17424	-3625	1143	-2804	-20982	13490	-4418
5280	-8610	-10203	-2844	-4439	2132	-17670	-3795
11809	9980	-865	-3150	549	-2327	-4493	-1711
5049	11430	50323	-15236	-687	-6244	857	-6747
-16485	-4849	-3724	19138	-3022	-2013	-1310	9282
9121	-2997	26664	-10594	-1583	-12871	1881	4342
5634	11742	1346	3413	-25755	14261	2613	758
4527	11561	5174	2354				

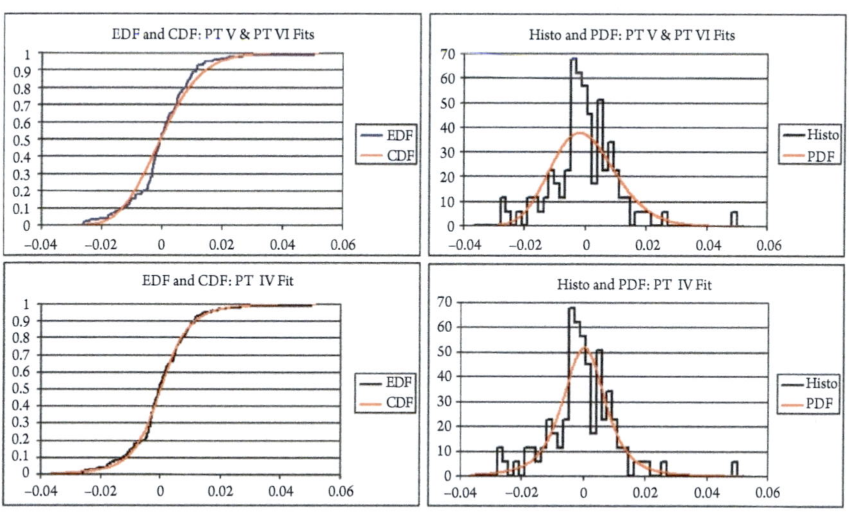

Figure 9.5 FTSE stock market data. Upper charts: CDFs and PDFs of the Pearson PT V and PT VI fits (visually indistinguishable). Lower charts: CDF and PDF of the Pearson IV fit.

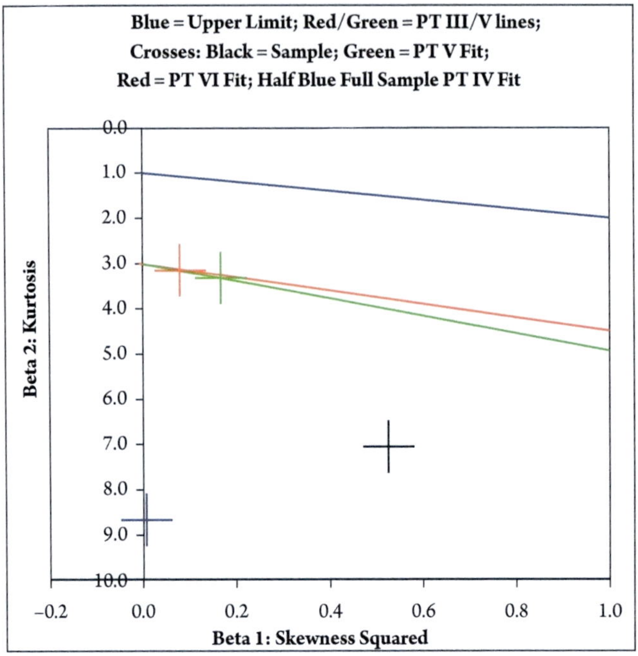

Figure 9.6 Plot of (β_1, β_2) points corresponding to Pearson PT V (green cross), PT VI (red cross), and full sample PT IV (half blue) fits to the FTSE sample.

marked by the faint blue cross in Figure 9.6. The near zero value of $\hat{\beta}_1^{IV}$ shows that the fitted distribution is very close to being symmetrical.

2. As a comparison, we fitted the PT V model with PDF as in eqn (9.5), with parameters (a, λ, p).
   (i) In attempting ML estimation of the parameters, it became clear that the parameter p was increasing without bound. So Nelder-Mead iterations were stopped at $p = 100$, at which point $\hat{\boldsymbol{\theta}}^V = (\hat{a}, \hat{\lambda}, \hat{p}) = (-0.107, 10.60, 100)$, with an ML value of $L(\hat{\boldsymbol{\theta}}^V) = 312.98$.
   (ii) The GoF statistic value $A^2 = 1.58$ has a p-value of 0.0, showing a highly significant lack-of-fit. The BS estimate of the critical value of A^2 at the 10% level is $A^2_{0.1} = 0.548$.

   Both the maximized log-likelihood value and the goodness-of-fit test indicate that the PT V fit is inferior to the PT IV fit.
   (iii) In terms of (β_1, β_2) values, we have $(\hat{\beta}_1^V, \hat{\beta}_2^V) = (0.167, 3.32)$ for the fitted PT V model. This point is marked by the green cross in Figure 9.6. For the normal model fit, we would have $(\hat{\beta}_1, \hat{\beta}_2) = (0, 3)$ precisely, the point identifying all normals. In fact, the fitted PT V model is actually quite close to the normal fit. This is already suggested by the

non-convergence of p, indicating that the model is tending to the embedded normal distribution in this case. The fitted PT V CDF and PDF (and indeed the maximized log-likelihood value) are depicted in the upper charts of Figure 9.5 and are visually indstinguishable from the normal fit.

(iv) The positive gradient coefficient $L^{IV}(\hat{\boldsymbol{\theta}}^V) = 1.14$ corroborates the fact that the PT IV fit is superior. However, we also have $L^{VI}(\hat{\boldsymbol{\theta}}^V) = 0.0017$, indicating a possible improvement if PT VI is fitted. The reason for this is not too difficult to see.

3. Fitting PT VI with PDF as in eqn (9.2) results in a fit that is also essentially normal.

   (i) The parameter MLEs are $\hat{\boldsymbol{\theta}}^{VI} = (\hat{a}, \hat{b}, \hat{p}, \hat{q}) = (-0.108, 0.167, 172, 266)$ with $L(\hat{\boldsymbol{\theta}}^{VI}) = 313.18$, showing that the maximized log-likelihood is indeed very slightly improved compared with the PT V fit. However, the very large values of $\hat{p}$ and $\hat{q}$ indicate that the model is essentially the embedded normal model.

   (ii) That the fitted PT VI model is close to normal is corroborated by the value $(\hat{\beta}_1^{VI}, \hat{\beta}_2^{VI}) = (0.077, 3.14)$ which, like that of the PT V fit, is also close to the value $(\beta_1, \beta_2) = (0, 3)$ for the normal model. This is the red cross in Figure 9.6.

   (iii) The GoF statistic value $A^2 = 1.40$ has p-value of 0.0, showing a highly significant lack-of-fit. The BS estimate of the critical value of A^2 at the 10% level is $A^2_{0.1} = 0.553$, which is similar to the PT V case.

9.7.2 Johnson System Fitted to FTSE Index Data

Fitting the Johnson system to the FTSE index data follows in a very similar way to fitting the Pearson system.

1. The position of the sample value $(\tilde{\beta}_1, \tilde{\beta}_2) = (0.526, 7.09)$ is depicted by the black cross in Figure 9.8 showing it to be well in the S_U region (the region below the cyan-coloured lognormal boundary). The relatively large value of $\tilde{\beta}_1$ might suggest that the sample comes from a distribution that is more skew than it really is. In fact, the full sample of size 250 from which our sample is drawn was actually not very skew, with a sample β_1 of value $\tilde{\beta}_1 = 0.023$.

   (i) Fitting the S_U model with PDF as given in eqn (9.15) yields the MLE $\hat{\boldsymbol{\theta}}^U = (\hat{a}, \hat{b}, \hat{\gamma}, \hat{\delta}) = (0.0002, 0.0090, 0.00295, 1.20)$ with maximized log-likelihood $L(\hat{\boldsymbol{\theta}}^U) = 319.82$. The lower two charts of Figure 9.7 depict the fitted CDF and PDF.

   (ii) The BS GoF statistic value was $A^2 = 0.309$ with a p-value of 0.172, so that the lack-of-fit was not significant at the 10% level, the critical value for which is $A^2_{0.1} = 0.348$.

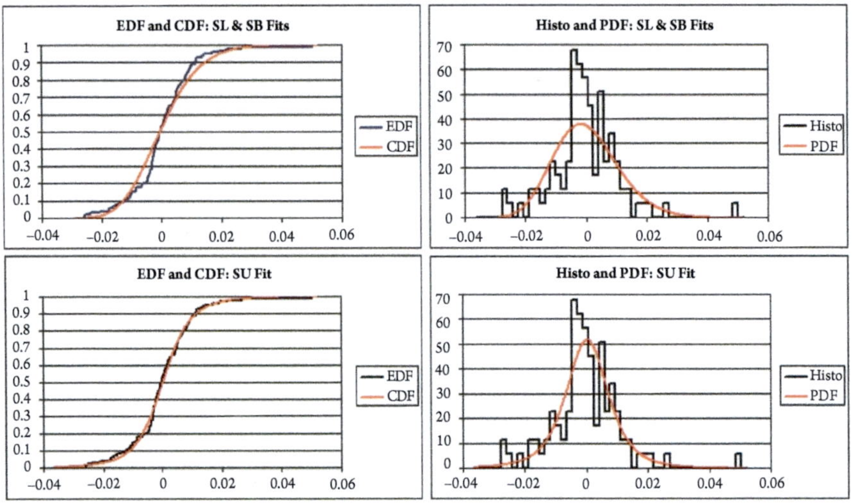

Figure 9.7 FTSE stock market data. Upper charts: CDFs and PDFs of the Johnson S_L and S_B fits (visually indistinguishable). Lower charts: CDF and PDF of the S_U fit.

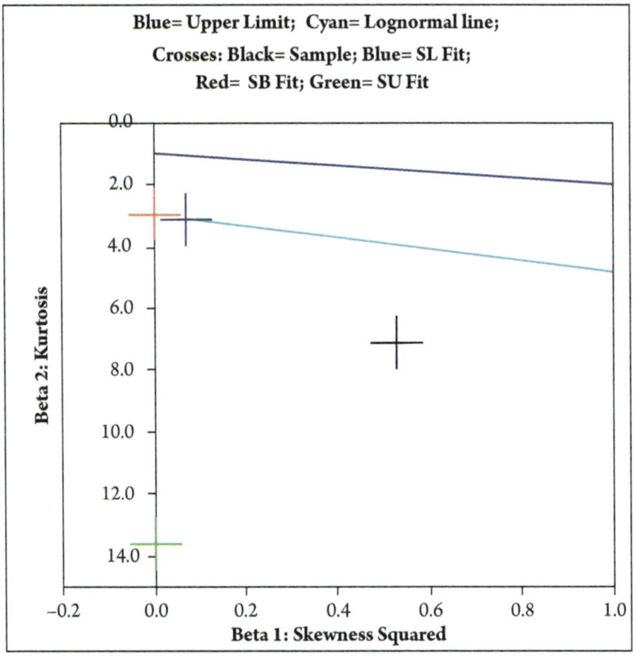

Figure 9.8 FTSE stock market data. Plot of (β_1, β_2) for the Johnson S_L (blue cross), S_B (red cross), and S_U (green cross) fits.

(iii) Finally, we have $(\hat{\beta}_1^U, \hat{\beta}_2^U) = (0.000165, 13.56)$. This point is marked by the green cross in Figure 9.8. The near zero value of $\hat{\beta}_1^U$ shows that the fitted distribution is very close to being symmetrical.

2. As a comparison, we consider the S_L model with PDF as given in eqn (9.14).

(i) The MLE for the parameters of this model is $\hat{\boldsymbol{\theta}}^L = (\hat{a}, \hat{\mu}, \hat{\delta}) = (-0.117, -2.14, 0.090)$ with maximized log-likelihood $L(\hat{\boldsymbol{\theta}}^L) = 313.15$.

(ii) The GoF statistic value was $A^2 = 1.398$ with a p-value of 0, so that the lack-of-fit was highly significant. At the 10% level, the critical value is $A^2_{0.1} = 0.534$.

(iii) Finally, we have $(\hat{\beta}_1^L, \hat{\beta}_2^L) = (0.074, 3.13)$. This point is marked by the blue cross in Figure 9.8, and corresponds almost precisely to the normal model. This is confirmed by the upper two charts of Figure 9.7, showing the fitted CDF and PDF, which visually are indistinguishable from the ML fitted normal model $N(0.00027, 0.0107^2)$.

(iv) The positive gradient coefficient $L^U(\hat{\boldsymbol{\theta}}^L) = 0.211$ corroborates the fact that the S_U fit is superior. We also have $L^B(\hat{\boldsymbol{\theta}}^L) = -0.00077$, suggesting that fitting the S_B model will not give a better fit.

3. Fitting the S_B model of eqn (9.12) gave the following results.

(i) The parameter ML estimates were $(\hat{a}, \hat{b}, \hat{\gamma}, \hat{\delta}) = (-30.1, 20.1, -452, 1125)$, with γ large and negative and δ large and positive, indicating that the embedded normal distribution is being approached. The maximized log-likelihood was $L(\hat{\theta}^B) = 311.60$.

(ii) This is corroborated by $(\hat{\beta}_1^B, \hat{\beta}_2^B) = (0.001, 3.002)$, which is close to the normal point $(\beta_1, \beta_2) = (0, 3)$. The $(\hat{\beta}_1^B, \hat{\beta}_2^B)$ point is indicated in Figure 9.8 by the red cross close to the blue cross corresponding to the S_L fit.

(iii) The GoF statistic value was $A^2 = 1.368$ with a p-value of 0.002, so that the lack-of-fit was highly significant. At the 10% level, the critical value is $A^2_{0.1} = 0.641$.

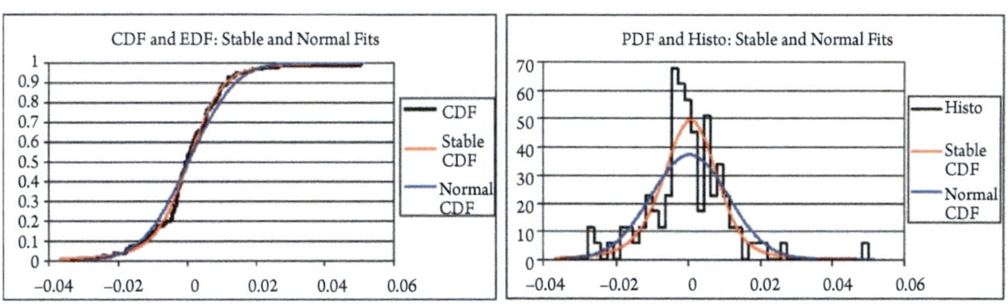

Figure 9.9 Stable and normal distributions fitted by ML to the FTSE shares data.

Table 9.4 Anderson-Darling statistic values of Pearson, Johnson, and stable distributions fitted to FTSE sample by ML

Distribution	PT IV	PT V	PT VI	S_U	S_L	S_B	Stable
A^2	**0.319**	1.58	1.40	**0.309**	1.40	1.37	**0.342**
$A^2_{0.1}$	**0.402**	0.548	0.553	**0.348**	0.534	0.641	
p-value	**0.19**	0.0	0.0	**0.172**	0.0	0.002	
L_{max}	**319.63**	312.98	313.81	**319.86**	313.15	311.60	**318.92**

9.7.3 Stable Distribution Fit

For comparison, we also fitted the four-parameter stable distribution by MLE. Figure 9.9 depicts the CDF and PDF of the fitted stable distribution, together with the normal fit for comparison. Numerical details of the fit were as follows.

(i) The parameter ML estimates were $(\hat{\alpha}, \hat{\beta}, \hat{\gamma}, \hat{\delta}) = (1.61, -0.151, 0.00577, 0.000414)$, with $L_{max} = 318.92$. The fit is therefore actually slightly negatively skewed.

(ii) The GoF statistic value was $A^2 = 0.342$. A bootstrap estimate of the level of significance or p-value was not attempted, as it would have been very computer intensive.

However, the A^2 and L_{max} values do allow a comparison with those of the fitted Pearson and Johnson distributions. Table 9.4 shows the A^2, p-values, 10% critical $A^2_{0.1}$ test values, and L_{max} obtained.

The PT IV and S_U fits are highlighted as providing satisfactory fits. We have also highlighted the stable fit as its A^2 and L_{max} values, though not quite as good, are nevertheless quite similar. It will be seen that the other fits, not highlighted, are all unsatisfactory, and all these have similar A^2, $A^2_{0.1}$, p, and L_{max} values. Though we have not attempted to calculate the $A^2_{0.1}$ and p-values for the stable fit, its A^2 and L_{max} already show the actual fit obtained is on a par with the PT IV and S_U fits, which we know are satisfactory.

9.7.4 Summary

The data are fairly symmetrical, but with a kurtosis greater than that for the normal distribution. Fitting the Pearson or Johnson system to the FTSE index data gives very similar results, with only the PT IV model and the S_U models capable of representing this adequately, as indicated by the results of the A^2 goodness-of-fit test. The fitted stable distribution is very similar, so should also give an adequate fit.

10 Box-Cox Transformations

In this chapter, we revisit the well-known model introduced by Box and Cox (1964) to investigate certain non-standard aspects that occur when fitting it to data, and to discuss how they can be handled. We begin with a brief examination of the rationale of the model, as a main possibility considered in this chapter is the use of an alternative model, which effectively eliminates the non-standardness, but that adheres to the underlying motivation behind the Box-Cox model.

Consider first the linear model

$$Y = \mathbf{X}\boldsymbol{\beta} + \epsilon,$$

where $\mathbf{X}$ is a design matrix of known constants, $\boldsymbol{\beta}$ a vector of parameters, and Y the response. This is a popular model under the usual attendant assumptions:

1. $E(Y)$ has a simple, additive structure
2. the error variance is homogeneous
3. the additive errors are normally distributed
4. the observations are independent.

These assumptions are important, as they allow exact inferences to be obtained irrespective of sample size. However, actual data do not always satisfy all these assumptions. A simple way to include more flexibility in the linear model is to apply a nonlinear transformation of the original observations $\mathbf{y}$, under the assumption that it is the transformed observations which satisfy the previously listed assumptions. Early extensions considered include, for example, Bartlett (1947), who considered transformations to achieve constant error variance, and Tukey (1949), who considered the problem of removing the interaction present in the model. The formulation considered by Box and Cox (1964) is a simple power transform of Y under the assumption that the transformed variable can still be analysed using the linear model. A review of the method focusing on practical aspects is given by Sakia (1992).

Non-Standard Parametric Statistical Inference. Russell Cheng.

The Box-Cox transform, which we introduced briefly in Section 2.9 and which we give again later in eqn (10.4), is similar to the power quantity

$$\left(\frac{y-a}{b}\right)^c \tag{10.1}$$

of equation (5.11), whose non-standard behaviour has already been discussed in Section 5.3 including the shifted power regression model in Table 5.1, and Chapter 6 where examples were given.

Fitting the power model (10.1) in general, and more specifically the Box-Cox model, is non-standard. Embeddedness where $c \to 0$ is already directly allowed for in the Box-Cox transform, however, the infinite likelihood problem discussed in Chapter 8 can and does occur. Moreover, though the aim of the Box-Cox model is to transform a non-normal variable into a normal variable, it does not do this precisely, with the transformed variable being distributed as a truncated normal rather than a full normal.

Such is the popularity of the Box-Cox model, it seems worthwhile reconsidering how these difficulties can be handled when fitting the model.

In the following sections, details of the Box-Cox transformation are given, together with methods proposed for fitting the model. We highlight how normality of the errors may not be satisfied. We will also discuss an alternative model, where the shifted power transformation is used, but simply to obtain a distribution that describes the data well. In particular, a transformed response will be considered, having a PDF that is not normal. Analysis can nevertheless still be carried out fairly easily using GLIM-style methodology, though the transformed data are still non-normally distributed.

10.1 Box-Cox Shifted Power Transformation

Tukey (1957) introduced a family of power transformations for $y > 0$ as

$$\begin{aligned} y(\lambda) &= y^\lambda \qquad && \text{if } \lambda \neq 0 \\ &= \ln y && \text{if } \lambda = 0 \end{aligned}$$

and gave its structural features for $|\lambda| \leq 1$. However, this family has a discontinuity at $\lambda = 0$, which prompted Box and Cox (1964) to alter the family with a linear transformation to achieve continuity over a full range of λ. They proposed the parametric family of power transformations

$$\begin{aligned} y(\lambda) &= (y^\lambda - 1)/\lambda \qquad && \text{if } \lambda \neq 0 \\ &= \ln y && \text{if } \lambda = 0, \end{aligned} \tag{10.2}$$

where

$$Y(\lambda) = \mathbf{X}\boldsymbol{\beta} + \epsilon,$$

so that

$$E(Y(\lambda)) = \mathbf{X}\boldsymbol{\beta}, \tag{10.3}$$

where $\boldsymbol{\beta}$ is a vector of unknown parameters associated. The transformed observations are thus assumed to be independently and approximately normally distributed with constant variance, represented by a model with an additive linear structure.

The preceding power transformation is suitable for responses which are restricted to being non-negative, but are without an upper bound. Alternative parametric families of transformations have been proposed for responses which are restricted in other ways. For instance, Atkinson (1985) discusses analogues to the power transformation for data which are proportions, where the values lie between 0 and 1. Other examples include the folded power transformation due to Mosteller and Tukey (1977), and the Guerrero and Johnson (1982) transformation, where (10.2) is applied to odds. In this chapter, we will be considering the case where the power transformation (10.2) is only appropriate after a shift has been included to all the observed values of the response. For example, in a study of survival times after the administration of a toxic substance, there may be a latent period before the dose begins to act. Thus a transformation, if appropriate at all, will only be applicable after the latent period has been subtracted from the response.

The Box-Cox shifted power transformation is defined as

$$\begin{aligned} y(\mu, \lambda) &= ((y + \mu)^{\lambda} - 1)/\lambda \quad && \text{if } \lambda \neq 0 \\ &= \ln(y + \mu) && \text{if } \lambda = 0 \end{aligned} \tag{10.4}$$

for $y > -\mu$. Royston (1993) considers other transformations to normality derived from this shifted transformation. The following section summarizes the steps required to estimate the parameters λ and μ and hence fit the model.

10.1.1 Estimation Procedure

At first sight, fitting the shifted power transform looks straightforward. Let

$$\mathbf{y}(\mu, \lambda) = (y_1(\mu, \lambda), \ldots, y_n(\mu, \lambda))^T$$

be the shifted power transform (10.4) of a sample of observations $\mathbf{y}$. For a given λ and μ, we simply have the linear model

$$\mathbf{y}(\mu, \lambda) = \mathbf{X}\boldsymbol{\beta} + \boldsymbol{\epsilon}.$$

The log-likelihood for the full model is just the log-likelihood corresponding to this linear model with an added term $\ln J$, where J is the Jacobian of the transformation of the y_i to $y_i(\mu, \lambda)$, namely,

$$J(\mu, \lambda) = \prod_{i=1}^{n} \left| \frac{\partial y_i(\mu, \lambda)}{\partial y_i} \right| = \prod_{i=1}^{n} (y_i + \mu)^{\lambda - 1},$$

which is independent of $\boldsymbol{\beta}$ and σ. Thus, for fixed μ and λ, we have the usual estimators for $\boldsymbol{\beta}$ and σ:

$$\begin{aligned}\hat{\boldsymbol{\beta}}(\lambda,\mu) &= (\mathbf{X}^{\mathrm{T}}\mathbf{X})^{-1}\mathbf{X}^{\mathrm{T}}\mathbf{y}(\mu,\lambda),\\ \hat{\sigma}^2(\lambda,\mu) &= n^{-1}\mathbf{y}^T(\mu,\lambda)\mathbf{R}(\mu,\lambda) \quad = r(\mu,\lambda)/n,\end{aligned} \tag{10.5}$$

where

$$\mathbf{R}(\mu,\lambda) = \left[\mathbf{I} - \mathbf{X}(\mathbf{X}^{\mathrm{T}}\mathbf{X})^{-1}\mathbf{X}^{\mathrm{T}}\right]\mathbf{y}(\mu,\lambda),$$

and we have written $r(\mu,\lambda)$ for the residual sum of squares of $\mathbf{y}(\mu,\lambda)$, and the parameters of the linear model as $\hat{\boldsymbol{\beta}}(\mu,\lambda)$ and $\hat{\sigma}(\mu,\lambda)$ to indicate that they are conditional on the values of μ and λ.

The maximization with respect to λ and μ can therefore be carried out by maximizing the partially maximized log-likelihood

$$L^*(\mu,\lambda) = L[\lambda,\mu,\hat{\boldsymbol{\beta}}(\boldsymbol{\lambda},\mu),\hat{\sigma}(\lambda,\mu)] + (\lambda-1)\sum_{i=1}^{n}\ln(y_i+\mu) \tag{10.6}$$

$$= -n(1+\ln(2\pi\hat{\sigma}^2))/2 + (\lambda-1)\sum_{i=1}^{n}\ln(y_i+\mu) \tag{10.7}$$

with respect to λ and μ. A search method can be employed to find the values $\hat{\lambda}$ and $\hat{\mu}$ which maximize (10.7), and then the MLEs can be substituted into (10.4) to obtain the final transformation.

There has been considerable interest in testing the adequacy of the fitted transformation. Box and Cox examine the Neyman-Pearson L_1 criterion for testing for constancy of variance given the achievement of normality, and the standard F-ratio criterion for testing the absence of interaction, given normality and constancy of variance. Hernandez and Johnson (1980) investigate large-sample behaviour. They consider use of the Kullback-Leibler information to measure the discrepancy between the density of the transformed variable and that of a normal distribution, thus indicating the maximum amount of improvement achievable through the transformation.

However, there is a serious difficulty in estimating the parameters μ and λ. Atkinson (1985) evaluated (10.7) in several examples, and noted that two general types of behaviour are possible. The log-likelihood always tends to infinity as $\mu \to -y_1$, however, either a local maximum is also present or the log-likelihood tends monotonically to infinity. In this latter case, λ cannot be consistently estimated by ML. The problem is of the unboundedness type discussed in Chapter 8. In the next section, we consider this unboundedness more closely and methods which have been proposed to overcome it.

10.1.2 Infinite Likelihood

The one-parameter Box-Cox power transformation (10.2) assumes that a suitable λ can be found for which

$$w = \frac{y^\lambda - 1}{\lambda}, \text{ where } y > 0 \text{ and } \lambda \neq 0,$$

is normally distributed. But w must satisfy the restriction that $w > -\lambda^{-1}$ if $\lambda > 0$ or $w < -\lambda^{-1}$ if $\lambda < 0$. When $\lambda > 0$, the normality assumption is therefore only satisfactory if $\Pr(w < -\lambda^{-1})$ is sufficiently small to be negligible, and similarly when $\lambda < 0$. This truncation issue tends often to be overlooked.

The problem is more serious when a shift is included. In using the shifted power transformation of (10.4), the assumption is that the sample

$$(y_i + \mu)^\lambda = E_i, \; (i = 1, \ldots, n) \tag{10.8}$$

will be approximately normally distributed. However, since we have the restriction $y + \mu > 0$, the range of E_i is restricted to being positive. Therefore, if the PDF of the E is non-zero at $E = 0$, so that $f_E(y) \to k > 0$ as $y \to 0$, then the PDF of Y will be

$$f_Y(y) \simeq \lambda k(y + \mu)^{\lambda - 1} \tag{10.9}$$

in the neighbourhood of $y = -\mu$, with the result that if $0 < \lambda < 1$ then

$$f_Y(y) \to \infty \text{ as } y \to -\mu.$$

In ML estimation, when the roles of y and μ are essentially reversed, with y a fixed observation and μ allowed to vary as an unknown parameter to be estimated, this results in an unbounded likelihood as $\mu \to -y_{(1)}$. For some of the examples given by Atkinson (1985), the profile log-likelihood

$$L^*(\mu) = \max_{\lambda, \boldsymbol{\beta}, \sigma^2} L(\boldsymbol{\theta}) \tag{10.10}$$

displays exactly this unbounded behaviour as $\mu \to -y_{(1)}$, with not even a local maximum present, so that ML estimation is bound to fail.

In the next section, we describe a grouped likelihood approach that has been examined in the literature for handling this unbounded likelihood problem.

Note, however, that the form of (10.9) is essentially the same as that of the threshold models in eqn (8.3) discussed in Chapter 8, with the likelihood becoming unbounded in the same way, as the value of the threshold parameter tends to the smallest observation. We shall therefore show how the spacings methods described in Chapter 8 can also be applied in the shifted Box-Cox power transform case. This saves having to appeal to the assumption of supposed inaccuracy in the observations, whether actually present or not, on which the grouped likelihood approach that is outlined in the next section is based.

10.2 Alternative Methods of Estimation

10.2.1 Grouped Likelihood Approach

Atkinson, Pericchi, and Smith (1991) show that the expression for the residual sum of squares contains a term which will always tend to infinity as $\mu \to -y_1$, although for fixed μ there will exist a λ which gives a local minimum. They propose a grouped likelihood approach to avoid this problem, using the argument given by Barnard (1967) and Kempthorne (1966) that the observations $\mathbf{y}$ can be regarded as being measured only to an accuracy, δ, so that their true value is uncertain in the interval $y_i \pm \delta$. The probability that y lies in an interval is then positive and can be explicitly calculated in the estimation process. Atkinson et al. (1991) apply the method to two examples considered in Atkinson (1985), showing how the unboundedness is prevented, with a unique stationary maximum obtained in both cases.

However, there is a complication with the approach. The grouped likelihood for all n observations includes an approximate correction for the size of the grouping interval chosen. The overall method is dependent on this interval size, with choice of a reasonable group interval size important. If δ is too small, the likelihood may have characteristics that are still similar to those of the original ungrouped likelihood, whilst if it is chosen too large, information about the data in the group with the smallest observations may be lost, affecting the accuracy of the parameter estimates. There is a certain subjectiveness involved in the selection of δ, and Atkinson et al. (1991) suggest considering profile likelihoods for a range of δ values, selecting the δ which makes the function smooth near $\mu = -y_1$. Thus, some effort may be involved in the successful application of this method.

To avoid the unbounded likelihood problem, we suggest application of the spacings-based methods discussed in Chapter 8 to the shifted Box-Cox transformation, specific details of which are now provided.

10.2.2 Modified Likelihood

We consider use of spacings-type methods in fitting the Box-Cox and related models.

Concomitant variables can be awkward to handle using spacings-based methods. If an iterative procedure is needed to estimate $\boldsymbol{\theta}$, then re-ordering of the CDF elements may be necessary during the iterations because of their dependence on $\boldsymbol{\theta}$. To avoid this, we do not use the straight MPS method, but use instead the method employing the modified likelihood $L_h(\boldsymbol{\theta})$ given in eqn (8.27), suggested by Cheng and Iles (1987).

For computational simplicity, we use the modified likelihood in conjunction with the hybrid technique of Section 8.7, which allows concomitant variables to be handled more easily. The MLEs for $\boldsymbol{\beta}$ and σ^2 are still used, but now in $L_h(\boldsymbol{\theta})$. Cheng and Traylor (1991) show that use of these MLEs in the calculations does not affect the asymptotic properties of the estimators. We therefore obtain the partially maximized modified log-likelihood

$$L_h^*(\mu, \lambda) = L_h[\mu, \lambda, \hat{\boldsymbol{\beta}}(\boldsymbol{\lambda}, \mu), \hat{\sigma}(\lambda, \mu)].$$

This can be written as

$$L_h^*(\mu, \lambda) = \sum_{i=1}^{n} l_i(\mu, \lambda), \tag{10.11}$$

where, if $y_i > y_1$,

$$\begin{aligned} l_i(\mu, \lambda) &= \ln\left((2\pi\hat{\sigma}^2)^{-1/2} \exp\{-[y_i(\mu, \lambda) - \mathbf{X}\hat{\boldsymbol{\beta}}]^2/2\hat{\sigma}^2\} \, |J(\mu, \lambda)|\right) \\ &= -\ln(2\pi\hat{\sigma}^2)/2 - R_i^2(\mu, \lambda)/2\hat{\sigma}^2 + (\lambda - 1)\ln(y_i + \mu), \end{aligned}$$

and if $y_i = y_1$,

$$\begin{aligned} l_1(\mu, \lambda) &= \ln\left(\Phi\{[y_1(\mu + h, \lambda) - \mathbf{X}\hat{\boldsymbol{\beta}}]/\hat{\sigma}\} - \Phi\{[y_1(\mu, \lambda) - \mathbf{X}\hat{\boldsymbol{\beta}}]/\hat{\sigma}\}\right) \\ &= \ln\left\{\Phi[R_1(\mu + h, \lambda)/\hat{\sigma}] - [\Phi R_1(\mu, \lambda)/\hat{\sigma}]\right\}, \end{aligned}$$

writing $\hat{\boldsymbol{\beta}} = \hat{\boldsymbol{\beta}}(\boldsymbol{\lambda}, \mu)$ and $\hat{\sigma} = \hat{\sigma}(\lambda, \mu)$ for short. Using the same procedure as for $L^*(\mu)$, we then maximize $L_h^*(\mu, \lambda)$ with respect to λ to give $L_h^*(\mu)$.

10.3 Unbounded Likelihood Example

Atkinson (1985) illustrates the unboundedness problem with an example where a transformation is postulated for chimpanzee learning times considered by Brown and Hollander (1977). The data, given in Table 10.1 below, record the time, in minutes, taken for four chimpanzees to learn each of ten signs. We show how the modified likelihood approach can be used to successfully obtain parameter estimates here, where ML is shown to fail.

We begin by using the Brown and Hollander chimpanzee data to illustrate the unbounded likelihood. McCullagh (1980) suggested that, as opposed to the standard two-way analysis of variance without interaction being carried out on the raw data, it should either

Table 10.1 Brown and Hollander chimpanzee data

Chimpanzee	Sign									
	1	2	3	4	5	6	7	8	9	10
1	178	60	177	36	225	345	40	2	287	14
2	78	14	80	15	10	115	10	12	129	80
3	99	18	20	25	15	54	25	10	476	55
4	297	20	195	18	24	420	40	15	372	190

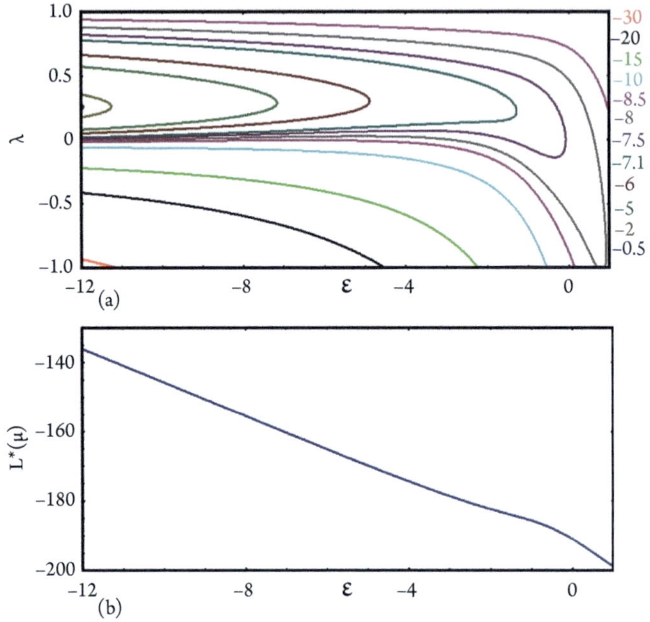

Figure 10.1 Box-Cox normal model for chimpanzee data; (a) contour plot of partially maximized log-likelihood $L^*(\mu, \lambda)$; (b) corresponding profile log-likelihood $L^*(\mu)$.

be transformed or a general linear model be fitted, because the model may result in negative fitted values. This opinion is reinforced by statistics in Atkinson's Table 9.3, which indicate that a shift parameter is desirable. Following Atkinson (1985), the possible outlier corresponding to the first chimpanzee, eighth sign has been omitted from the estimation process because of its highly influential nature.

Figure 10.1(a) is a contour plot of the partially maximized log-likelihood (10.6). To emphasize the behaviour as $\mu \to -y_1$, the log scale $\mu = -y_1(1 - 10^{\varepsilon})$ adopted by Atkinson is used for μ. The contour plot displays standard features apparent when using ML with this transformation; the log-likelihood is a constant when $\lambda = 1$, since the value of μ is irrelevant in this case, as the linear model already includes a constant, and, in addition, as $\mu \to -y_1$, parabolic contours of increasing log-likelihood are obtained. However, the problem of unboundedness is also evident here, as the plot shows the log-likelihood steadily increasing as $\mu \to -y_1$. The plot of the corresponding profile log-likelihood

$$L^*(\mu) = \max_{\lambda} L^*(\mu, \lambda) \tag{10.12}$$

in Figure 10.1(b) exhibits the unboundedness even more clearly.

Figure 10.2 shows the plots for $L_h^*(\mu)$ obtained using the modified likelihood method. In contrast to Figure 10.1(a), Figure 10.2(a) shows a clear stationary maximum, without a ridge, whose height tends to infinity. Although parameter estimates were unobtainable

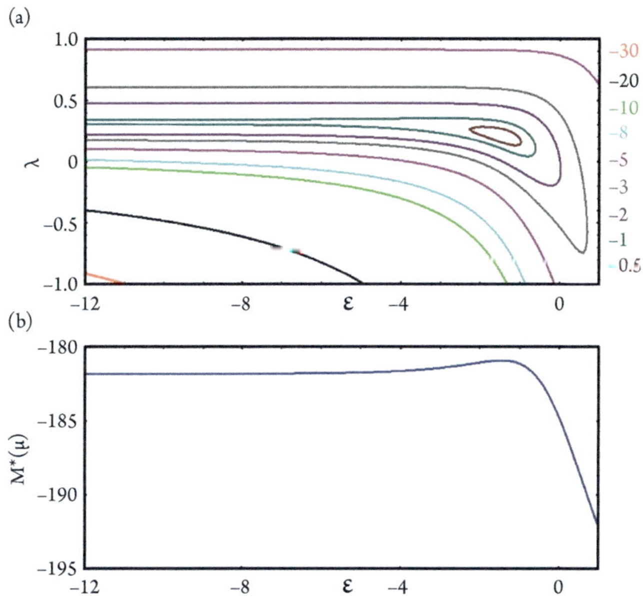

Figure 10.2 Box-Cox normal model for chimpanzee data; (a) contour plot of partially maximized modified log-likelihood $M^*(\mu, \lambda)$; (b) corresponding profile modified log-likelihood $M^*(\mu)$.

when using the standard likelihood, the stationary maximum displayed in this modified likelihood case yields the parameter estimates $\tilde{\lambda} = 0.186$ and $\tilde{\mu} = -9.64$.

The example shows that the modified likelihood approach provides a simple procedure for obtaining estimates of the parameters of the Box-Cox shifted power transformation, and successfully overcomes the underlying difficulty causing the unboundedness problem.

However, even if a fit is obtainable, we may wish to question whether the model is appropriate at all. In the next section, we examine in particular if there is any undesirable effect arising from the fact that the actual fitted distribution is a truncated normal rather than a full normal.

10.4 Consequences of Truncation

The Box-Cox shifted power transformation aims to obtain approximate normality of the transformed variable, so that standard analysis of variance techniques may be applied to the data. Yet even if the expectations satisfy $E(y(\mu, \lambda)) = \mathbf{X}\boldsymbol{\beta}$, the corresponding distribution of error may not be normal. Draper and Cox (1969) conclude that the transformation can help regularize the data even when normality is not achieved. However, this is not always the case.

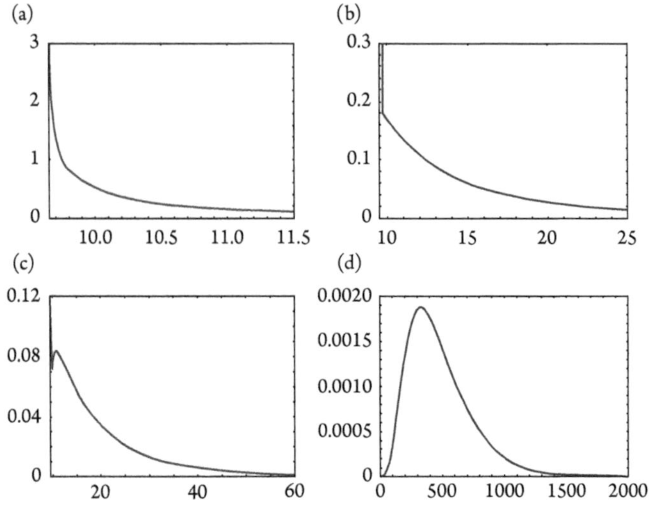

Figure 10.3 PDFs of four selected chimpanzees as given by Box-Cox model fitted to full chimpanzee data sample. (a) chimpanzee 2, sign 8; (b) chimpanzee 2, sign 2; (c) chimpanzee 3, sign 5; (d) chimpanzee 4, sign 9.

We have seen that though E in (10.8) is assumed approximately normally distributed, it is truncated at zero. This can result in the fitted distribution of Y being bimodal when $0 < \lambda < 1$, having a normal mode and also a spike as $y \to -\mu$. The spike typically has a very small associated probability and is well separated from the main mode. In some cases, both can merge into one, and the truncation can lead to curiosities in the fitted Y distributions. For instance, Figure 10.3 shows the fitted Y densities corresponding to responses of different chimpanzees in our example. It will be seen that all the different distributional forms occur. For chimpanzee 2, sign 8, the density is J-shaped; for chimpanzee 3, sign 5, the density is bimodal, whilst for chimpanzee 4, sign 9, it has essentially a single mode.

If the intention is simply to represent the data well, it may be preferable to drop the requirement of normality and consider instead an alternative model. For instance, the normal distribution could be replaced by one with positive support. It is this aspect that we now pursue.

10.5 Box-Cox Weibull Model

We give an example of an alternative model for the transformation. If E_i in (10.8) is assumed to be exponentially distributed, thus having positive density at the origin, then the original observations Y_i have the shifted Weibull distribution of eqn (2.1) or Table 6.2. That is, if the shifted power transformation has the property

$$\frac{(Y_i+\mu)^\lambda - 1}{\lambda} = E_i - \frac{1}{\lambda},$$

where $E_i \sim \text{Exp}(\delta)$, then $(Y_i+\mu)^\lambda \sim \text{Exp}(\gamma)$ where $\gamma = \lambda\delta$. It follows that Y is a shifted Weibull (a, b, c) as given in Table 6.2, where $a = -\mu$, $b = \gamma^{1/\lambda}$, $c = \lambda$. Thus Y has a shifted Weibull distribution with shape parameter λ (equal to the power parameter of the transformation). Using ML, we still have the infinite likelihood problem if $\lambda < 1$.

10.5.1 Fitting Procedure

Cheng and Iles (1987) show that the modified likelihood approach eliminates the unbounded likelihood problem when fitting the Weibull distribution, and so reinforces justification of its use here. Fitting by the ML or the modified likelihood methods can be carried out using GLIM-style methodology. If E_i in (10.8) is taken to be exponentially distributed with mean θ_i, the log-link $\eta_i = \mathbf{X}_i\boldsymbol{\beta} = \ln\theta_i$ with inverse $\theta_i = e^{\mathbf{X}_i\boldsymbol{\beta}}$ can be used to give a shifted Weibull distribution for Y_i. The log-link ensures positive means are fitted. The model is flexible in that a range of both negatively and positively skewed distributions is possible, depending on the value of λ.

Atkinson (1985) shows that for a fixed μ and λ, the MLE of $\boldsymbol{\beta}$ can be found by iterative least squares from

$$\boldsymbol{\beta}(k+1) = (\mathbf{X}^{\mathrm{T}}\mathbf{X})^{-1}\mathbf{X}^{\mathrm{T}}\mathbf{C}(k)$$

$(k = 0, 1, \ldots.)$, where $\mathbf{C} = (c_1, \ldots, c_n)$ and $c_i(k) = \eta_i(k) - 1 + q_i(\mu,\lambda)/\theta_i(k)$ where $\boldsymbol{\eta}(k) = \mathbf{X}\boldsymbol{\beta}(k)$, $\theta_i(k) = \exp(\eta_i(k))$ and $q_i(\mu,\lambda) = (y_i+\mu)^\lambda$. For the coefficient of the constant term, an initial value could be $\beta_1(0) = \ln(\sum q_i/n)$, with $\beta_i(0) = 0$ for the other coefficients. Because of the relationship with least squares, Atkinson and the references therein identify analogues of linear regression techniques which are available as analytical tools here.

$L^*(\mu,\lambda)$ and $L_h^*(\mu,\lambda)$ may be compared by examining the components of each. For this reason, we write the partially maximized log-likelihood as

$$L^*(\mu,\lambda) = \sum_{i=1}^{n} l_i(\mu,\lambda).$$

If $q_i(\mu,\lambda)$ is $\text{Exp}(\theta_i)$, then the log-likelihood for y_i must include the Jacobian $J = \lambda(y_i + \mu)^{\lambda-1}$. This gives

$$\begin{aligned} l_i(\mu,\lambda) &= \ln\left(\theta_i^{-1}\exp(-q_i(\mu,\lambda)/\theta_i)\,|J|\right) \\ &= -\ln\theta_i - q_i(\mu,\lambda)/\theta_i + \ln\lambda + (\lambda-1)\ln(y_i+\mu) \\ &= \ln\lambda - q_i(\mu,\lambda)/\theta_i - \eta_i + (\lambda-1)\ln(y_i+\mu), \end{aligned}$$

where η_i and θ_i are the converged values of the $\eta_i(k)$ and $\theta_i(k)$ iterates. $L_h^*(\mu,\lambda)$ is exactly the same as above, except in the case where $y_i = y_1$, when $l_i(\mu,\lambda)$ is given by

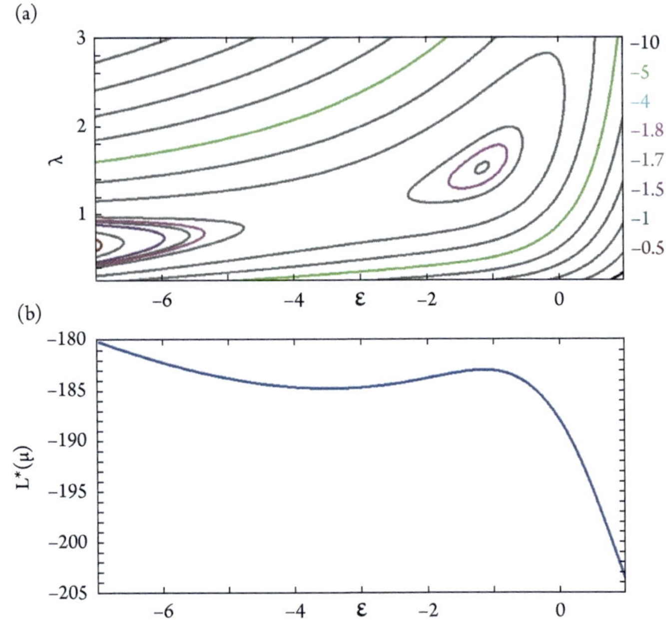

Figure 10.4 Box-Cox Weibull model for chimpanzee data; (a) contour plot of partially maximized log-likelihood $L^*(\mu, \lambda)$; (b) corresponding profile log-likelihood $L^*(\mu)$.

$$l_i(\mu, \lambda) = \ln\left(\exp(-q_i(\mu, \lambda)/\theta_i) - \exp(-q_i(\mu + h, \lambda)/\theta_i)\right).$$

10.6 Example Using Box-Cox Weibull Model

Figure 10.4 gives the plots of the Weibull model applied to the chimpanzee data using ML. Although in this case a local maximum exists, there is still unbounded behaviour as $\mu \to -y_1$. The stationary maximum yields the MLEs $\hat{\lambda} = 1.51$ and $\hat{\mu} = -9.28$. The modified likelihood results are exhibited in Figure 10.5. Here a global maximum is produced, providing a less ambiguous estimate. The estimates obtained are $\tilde{\lambda} = 1.51$ and $\tilde{\mu} = -9.29$.

We have not plotted the fitted Weibull PDFs obtained in the example for different chimpanzee/sign combinations, but they will of course all be complete PDFs with the same positively skewed shape corresponding to the power $\lambda = 1.51$.

10.7 Advantages of the Box-Cox Weibull Model

The Weibull distribution offers an alternative transformation if the aim is to describe the data well, and GLIM-style methods can be used to analyse the transformed data. If the

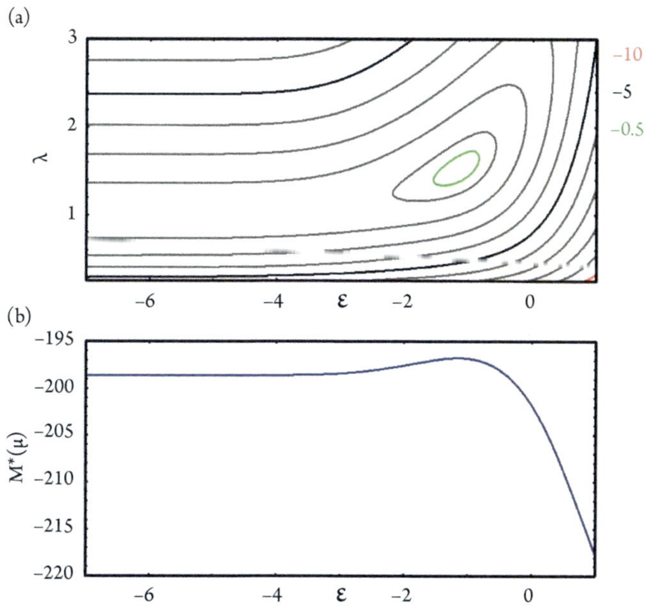

Figure 10.5 Box-Cox Weibull model for chimpanzee data; (a) contour plot of partially maximized modified log-likelihood $M^*(\mu, \lambda)$; (b) corresponding profile modified log-likelihood $M^*(\mu)$.

exponential distribution, having only one parameter, is too limited for error structure, more control could be obtained by taking E in (10.8) to be two-parameter inverse-Gaussian distributed. This would result in fits analogous to the normal distribution. As well as avoiding a bimodal density for Y, two further advantages of using a distribution with positive support also stem from the truncation. With the transformation to approximate normality, the likelihood comprises exact normals, and so does not take into account the truncation. This may bias the estimation. For a distribution with positive support, the likelihood is exact, and so the estimation procedure is more robust. In addition, both tail quantities can be estimated for a distribution with positive support, since the truncation is avoided. Royston (1993) discusses this in more detail.

11 Change-Point Models

A change-point model may be appropriate when there is a shift in the underlying parameters of a distribution. Such a model can be fitted to survival and reliability data. A varied literature exists on change-point models. One such problem was considered by Hinkley (1970), where two distinct distributions are fitted as a mixture, with one for observations less than τ, the change-point, and one for observations greater than τ, with τ unknown.

In this chapter, we discuss the change-point hazard rate model where the PDF, denoting time to failure, is

$$\begin{aligned} f(y) &= a\exp(-ay) && 0 \leq y \leq \tau \\ &= b\exp(-a\tau - b(y-\tau)) && y > \tau. \end{aligned} \tag{11.1}$$

If $a \neq b$, this has a discontinuity at $y = \tau$, with τ to be estimated.

Estimation of τ by maximum likelihood is non-standard, as the likelihood is not bounded when $\tau \to y_{(n)}$, the largest observation.

For the model of eqn (11.1), Matthews and Farewell (1982) use numerical techniques to obtain ML estimates of the parameters, and derive a likelihood ratio test for the null hypothesis $\tau = 0$. Nguyen et al. (1984) consider a consistent estimator of τ based on a mixture density, with the estimator depending on the sample mean and variance of the observations $y_{(i)} > \tau$. A very simple joint estimator $(\hat{a}, \hat{b}, \hat{\tau})$ is given by Yao (1986), where $\hat{\tau}$ is a *constrained estimator* subject to the simple constraint $\hat{\tau} \leq y_{(n-1)}$, the second-largest observation. The constraint prevents $\tau \to y_{(n)}$ so that the likelihood does not become unbounded. Yao (1986) shows that $\hat{\tau}$ is consistent, and in his Proposition (5) obtains the limiting distribution of $(\hat{a}, \hat{b}, \hat{\tau})$, where $\hat{a}$ and $\hat{b}$ are normally distributed but $\hat{\tau}$ has a more complicated limiting distribution that is in two parts, each the distribution of a different random sum. For the case where a and b are known, Yao (1987, Theorem 2) gives a very explicit formula for the CDF of the limiting distribution of $\hat{\tau}$ in terms of the normal CDF, relating it to the maximum of a two-sided Wiener process. Pham and Nguyen (1990) give an ML estimator for the parameters defined in a compact but random parameter region, showing that it is strongly consistent, and giving its

Non-Standard Parametric Statistical Inference. Russell Cheng.

asymptotic distribution. Worsley (1986) and Loader (1991) give methods for obtaining a confidence region for a change-point based on the likelihood ratio. A summary of work involving testing for the presence of a change-point is given in Smith (1989). A further estimator of τ is given by Chang et al. (1994) that covers the case of censored samples, and which is based on the difference in slopes of the cumulative hazard plot before and after the change point. As noted by the authors, the estimator is related to one considered by Matthews and Farewell (1985) based on the score process. Zhao et al. (2009) focus on censoring in situations where there are long-term survivors.

Much of the literature just summarized is mathematical, and some of the asymptotic results seem not that easy to apply. However, it is clear that estimation of the parameters can be successfully carried out using ML. It is worth pointing out that bootstrapping, the use of which is justified by Pham and Nguyen (1993), provides a simple way of obtaining distributional properties, and we will illustrate this with an example at the end of the chapter.

It is easy to obtain the profile log-likelihood using the change-point as the profiling parameter. However, the profile likelihood, in addition to being unbounded, is not smooth, with discontinuities at $\tau = y_i$ for all i. A simple procedure is to reveal the discontinuities by calculating the profile log-likelihood with τ set to values just either side of selected, or possibly every, observation. The profile likelihood plot can look somewhat disconcerting at first sight, but actually the discontinuities can make the maximum rather easy to identify.

An alternative is to use a spacings method. This removes the discontinuities and, more importantly, removes the unboundedness.

Our remaining discussion will therefore be on the more practical aspects of estimation. Though we will for completeness give computational details, the numerical implementation is actually quite straightforward, especially in the full sample case.

We begin by examining the likelihood function, including its evaluation with censored data. A spacings method is then applied to show how the unboundedness is avoided and how censoring can be handled in a rather neat way.

11.1 Infinite Likelihood Problem

Let $y_1, y_2, \ldots, y_n$ be the order statistics of a sample of size n drawn from the PDF (11.1). The likelihood of the sample, when $y_j \leq \tau < y_{j+1}$ $(j = 1, \ldots, n-1)$, is

$$Lik = \left[\prod_{i=1}^{j} a\exp(-ay_i)\right]\left[\prod_{i=j+1}^{n} b\exp(-a\tau - b(y_i - \tau))\right],$$

giving a log-likelihood of

$$L = j\ln a - a\sum_{i=1}^{j} y_i + (n-j)\ln b - (n-j)a\tau - b\sum_{i=j+1}^{n}(y_i - \tau).$$

For a fixed τ, differentiation of L with respect to a and b gives

$$\hat{a}(\tau) = \frac{j}{(\sum_{i=1}^{j} y_i) + (n-j)\tau}$$

and

$$\hat{b}(\tau) = \frac{n-j}{(\sum_{i=j+1}^{n} y_i) - (n-j)\tau}. \tag{11.2}$$

However, if $y_{n-1} \leq \tau < y_n$, then $\hat{b}(\tau) = (y_n - \tau)^{-1}$, and examination of the profile log-likelihood

$$L^*(\tau) = \max_{a,b} L(a, b, \tau)$$

shows that it becomes unbounded as $\tau \to y_n$, as the term $\ln(\hat{b}(\tau))$ becomes arbitrarily large, whilst the other terms remain finite and non-zero.

As an example, we use a data set given by Crowder et al. (1991), comprising failure times of batches of Kevlar49 fibres (Note that the data is not the same as that of the differently numbered Kevlar149 fibres of Chapter 7.) loaded to four different stresses, some of which are right-censored. The data are given in Table 11.1, where censored observations are marked (*).

These data are actually a mixture of fibres from several spools, and by obtaining residual plots Crowder et al. showed the existence of a significant spool effect which could not be ignored in the analysis. Nevertheless, the data in its present form are useful in illustrating the following change-point methodology. For instance, examination of the data, particularly those values obtained at stress 29.7 MPa, reveals change-point behaviour, where a certain proportion of the sample seems to fail over a specific range of earlier times than the remainder. Ignoring the spool effect, we therefore consider the model of (11.1) with a single change-point. Although this may be a simplistic approach for analysing this particular data set, it provides an initial approximation of mixed behaviour such as this, and is similar to that considered by Cooper (1992) for analysing observations involving Kevlar fibres collected by Howard (1988). In that case, when considering a mixture of two Weibull distributions, a threshold was determined which marked the point where the prominence of one distribution over the other was interchanged.

The presence of censored observations requires special consideration when evaluating the likelihood function. Although these data contain only right-censored observations, in the following section, the more general case of obtaining the log-likelihood for a sample containing randomly censored observations is detailed.

Table 11.1 Failure times (hours) of Kevlar49 fibres for stress levels: (a) 23.4 MPa, (b) 25.5 MPa, (c) 27.6 MPa, (d) 29.7 MPa

(a)	4000.0	5376.0	7320.0	8616.0	9120.0	14400.0
	16104.0	20231.0	20233.0	35880.0	*41000.0	*41000.0
	*41000.0	*41000.0	*41000.0	*41000.0	*41000.0	*41000.0
	*41000.0	*41000.0	*41000.0			
(b)	225.2	503.6	1087.7	1134.3	1824.3	1920.1
	2383.0	2442.5	2974.6	3708.9	4908.9	5556.0
	6271.1	7332.0	7918.7	7996.0	9240.3	9973.0
	11487.3	11727.1	13501.3	14032.0	29808.0	31008.0
(c)	19.1	24.3	69.8	71.2	136.0	199.1
	403.7	432.2	453.4	514.1	514.2	541.6
	544.9	554.2	664.5	694.1	876.7	930.4
	1254.9	1275.6	1536.8	1755.5	2046.2	6177.5
(d)	2.2	4.0	4.0	4.6	6.1	6.7
	7.9	8.3	8.5	9.1	10.2	12.5
	13.3	14.0	14.6	15.0	18.7	22.1
	45.9	55.4	61.2	87.5	98.2	101.0
	111.4	144.0	158.7	243.9	254.1	444.4
	590.4	638.2	755.2	952.2	1108.2	1148.5
	1569.3	1750.6	1802.1			

11.2 Likelihood with Randomly Censored Observations

We can account for censored observations in a very general way using the following technique. The method makes use of the Kaplan-Meier estimation process. Suppose we have a complete random sample of values $y_1^0, y_2^0, \ldots, y_n^0$ with no censoring, from a distribution $F^0(y; \boldsymbol{\theta})$. The effect of censoring certain values in this complete random sample can be viewed as there being a set of censoring variables $T_1, T_2, \ldots, T_n$ drawn from a censoring distribution $F_c(t)$, where each T_i is associated with the corresponding y_i^0, whether it is observed or not. Define

$$y_i = \min(y_i^0, T_i), \tag{11.3}$$

and set a corresponding flag δ_i, with $\delta_i = 0$ if T_i is less than y_i^0 when T_i forms a barrier and prevents the true y_i^0 from being seen. That is,

$$\begin{aligned} \delta_i = 1 \quad & \textit{if } y_i = y_i^0 \quad \text{(uncensored)} \\ 0 \quad & \textit{if } y_i = T_i \quad \text{(censored).} \end{aligned}$$

Thus, for the Kevlar data, if the fibre fails before the study ends, y_i^0 is recorded, but if the fibre is still intact at the end of the study, T_i is recorded. With the data collected at stress level 23.4 MPa, there are ten values where $y_i = y_i^0$ and eleven where $y_i = T_i$.

The distribution of Y as defined in (11.3) can then be written in terms of the complete and censored distributions. D'Agostino, and Stephens (1986) note that if T_i is independent of y_i^0, then the distribution function $F(y; \boldsymbol{\theta})$ of Y is defined as

$$F(y; \boldsymbol{\theta}) = 1 - \{1 - F^0(y; \boldsymbol{\theta})\}\{1 - F_c(y)\}. \tag{11.4}$$

If $F^0(y; \boldsymbol{\theta})$ is not specified, we can use the Kaplan-Meier estimator to estimate it. Details of this are now given.

11.2.1 Kaplan-Meier Estimate

This estimate of $F^0(y; \boldsymbol{\theta})$ for randomly censored data is analogous to the empirical distribution function (EDF) for complete data. Indeed, if no observation is censored, the following estimate becomes the EDF $F_n^0(y)$.

The pairs (y_i, δ_i) from (11.3) are placed in ascending order of the y_i. Let $R_i = j$ (the rank of y_i) if $y_i = y_{(j)}$. The estimate of $F^0(y; \boldsymbol{\theta})$ is given by

$$\begin{aligned} {}_cF_n^0(y) &= 0 && Y < y_{(1)} \\ &= 1 - \prod_{i:y_i \le Y} ((n - R_i)/(n - R_i + 1))^{\delta_i} && Y < y_{(n)} \\ &= 1 && Y > y_{(n)}. \end{aligned} \tag{11.5}$$

By writing (11.4) as

$$F(y; \boldsymbol{\theta}) = F_c(y) + F^0(y; \boldsymbol{\theta})(1 - F_c(y)),$$

we see that the likelihood function must be weighted by $(1 - F_c(y))$ to account for the randomly positioned censored observations. That is,

$$f(y;\boldsymbol{\theta}) = \frac{dF(y;\boldsymbol{\theta})}{dy} = f^0(y;\boldsymbol{\theta})(1 - F_c(y))$$

at points of continuity of the CDF. This is not a PDF because there are discrete probability masses at the censored observations.

11.2.2 Tied Observations

For the censored Kevlar fibre data set at stress level 23.4 MPa, censoring occurs at one specific T value, and so $F_c(t)$ contains a jump. There can then be ties in the actual observations y_i, although this is not a real problem, because $F(y;\boldsymbol{\theta})$ will contain a jump at precisely this point. More generally, a sample may contain randomly censored observations, where at each censored value y_i, there is a probability mass P_i. From this the likelihood function is

$$\text{Lik} = \prod_{i:\delta_i=1} f(y_i;\boldsymbol{\theta}) \prod_{k:\delta_k=0} P_k,$$

where $i + k = n$, and the log-likelihood for the sample is

$$L = \sum_{i:\delta_i=1} \ln f(y_i;\boldsymbol{\theta}) + \sum_{k:\delta_k=0} \ln P_k.$$

The probability mass P_k $(k : \delta_k = 0)$ may be defined to be the height of the jump

$$\begin{aligned} P_k &= F(y_k;\boldsymbol{\theta}) - F(y_k^-;\boldsymbol{\theta}) \\ &= F_c(y_k) + F^0(y_k;\boldsymbol{\theta})(1 - F_c(y_k)) \\ &\quad - F_c(y_k^-) - F^0(y_k^-;\boldsymbol{\theta})(1 - F_c(y_k^-)), \end{aligned}$$

where $F(y_k^-;\boldsymbol{\theta})$ is the value of the CDF immediately before y_k. Now $F^0(y_k;\boldsymbol{\theta}) = F^0(y_k^-;\boldsymbol{\theta})$ since $F^0(y;\boldsymbol{\theta})$ is a smooth, continuous function. Thus,

$$P_k = \{F_c(y_k) - F_c(y_k^-)\}\{1 - F^0(y_k;\boldsymbol{\theta})\}. \tag{11.6}$$

We apply these results to the density $f^0(y;\boldsymbol{\theta})$ of (11.1), where $\boldsymbol{\theta} = (a, b, \tau)$. Define u_j to be the number of uncensored observations out of the first j observations. Then, if $y_j \le \tau < y_{j+1}$,

$$\prod_{i:\delta_i=1} f(y_i;\boldsymbol{\theta}) = \left(\prod_{\substack{i:\delta_i=1\\ i\le j}} a\exp(-ay_i)(1-F_c(y_i))\right)\times$$
$$\left(\prod_{\substack{i:\delta_i=1\\ i\ge j+1}} b\exp(-a\tau-b(y_i-\tau))(1-F_c(y_i))\right)$$
$$= a^{u_j}b^{(u_n-u_j)}\exp\left(-a\sum_{\substack{i:\delta_i=1\\ i\le j}} y_i-\sum_{\substack{i:\delta_i=1\\ i\ge j+1}}(a\tau+b(y_i-\tau))\right)\times$$
$$\prod_{i:\delta_i=1}(1-F_c(y_i))$$
$$= \exp\left(-a\sum_{\substack{i:\delta_i=1\\ i\le j}} y_i-b\sum_{\substack{i:\delta_i=1\\ i\ge j+1}} y_i-(u_n-u_j)(a-b)\tau\right)\times$$
$$a^{u_j}b^{(u_n-u_j)}\prod_{i:\delta_i=1}(1-F_c(y_i)).$$

In addition, from (11.6) we have

$$\prod_{k:\delta_k=0} P_k = \prod_{k:\delta_k=0}(F_c(y_k)-F_c(y_k^-))(1-F^0(y_k;\boldsymbol{\theta})),$$

hence, if $y_j \le \tau < y_{j+1}$,

$$\prod_{k:\delta_k=0} P_k = \left(\prod_{\substack{k:\delta_k=0\\ k\le j}}(F_c(y_k)-F_c(y_k^-))\exp(-ay_k)\right)\times$$
$$\left(\prod_{\substack{k:\delta_k=0\\ k\ge j+1}}(F_c(y_k)-F_c(y_k^-))\exp(-(a-b)\tau-by_k)\right).$$

If c_j is defined to be the number of censored observations out of the first j observations, then this becomes

$$\prod_{k:\delta_k=0} P_k = \exp\left(-a\sum_{\substack{k:\delta_k=0\\ k\le j}} y_k-b\sum_{\substack{k:\delta_k=0\\ k\ge j+1}} y_k-(c_n-c_j)(a-b)\tau\right)\times$$
$$\prod_{k:\delta_k=0}(F_c(y_k)-F_c(y_k^-)).$$

Therefore, if $y_j \leq \tau < y_{j+1}$, the log-likelihood of a sample containing both complete observations with PDF (11.1) and randomly censored observations is

$$
\begin{aligned}
L = {} & u_j \ln a + (u_n - u_j) \ln b - a \sum_{\substack{i:\delta_i=1 \\ i \leq j}} y_i - b \sum_{\substack{i:\delta_i=1 \\ i \geq j+1}} y_i - (u_n - u_j)(a - b)\tau \\
& + \sum_{i:\delta_i=1} \ln(1 - F_c(y_i)) - a \sum_{\substack{k:\delta_k=0 \\ k \leq j}} y_k - b \sum_{\substack{k:\delta_k=0 \\ k \geq j+1}} y_k - (c_n - c_j)(a - b)\tau \\
& + \sum_{k:\delta_k=0} \ln(F_c(y_k) - F_c(y_k^-)) \\
= {} & u_j \ln a + (u_n - u_j) \ln b - a \sum_{q=1}^{j} y_q - b \sum_{q=j+1}^{n} y_q - (n - j)(a - b)\tau \\
& + \sum_{i:\delta_i=1} \ln(1 - F_c(y_i)) + \sum_{k:\delta_k=0} \ln(F_c(y_k) - F_c(y_k^-)).
\end{aligned}
$$

Differentiation of L with respect to a and b results in, for a given τ,

$$\hat{a}(\tau) = \frac{u_j}{\sum_{q=1}^{j} y_q + (n - j)\tau}$$

and

$$\hat{b}(\tau) = \frac{u_n - u_j}{\sum_{q=j+1}^{n} y_q - (n - j)\tau}.$$

Note that if $y_{n-1} \leq \tau < y_n$ and y_n is censored, $\hat{b}(\tau) = 0$, which agrees with intuition, because the censored observation y_n does not allow us to fit the second part of the model (11.1). However, if y_n is uncensored, $\hat{b}(\tau)$ is infinite as $\tau \to y_n$.

11.2.3 Numerical Example Using ML

Figure 11.1 shows the profile log-likelihood $L^*(\tau)$ for the four data sets provided by Crowder et al. (1991), with τ plotted on a log scale. Figure 11.1(a) shows the profile log-likelihood for stress level 23.4 MPa, where the 11 largest observations are censored. Here, an estimate for a possible change-point can be clearly picked out as $\hat{\tau}$ = 20233 MPa. The log-likelihood is unbounded for the other three stress levels. However, from Figure 11.1(b), we would select $\hat{\tau}$ = 28230, although another noticeable peak occurs at τ = 14032, and Figure 11.1(c) for stress level 27.6 displays a maximum at $\hat{\tau}$ = 2046.2.

Figure 11.1(d) shows the profile log-likelihood for stress level 29.7 MPa with a maximum at $\hat{\tau}$ = 22.10. Following Crowder et al., a proportional hazard plot for the sample is

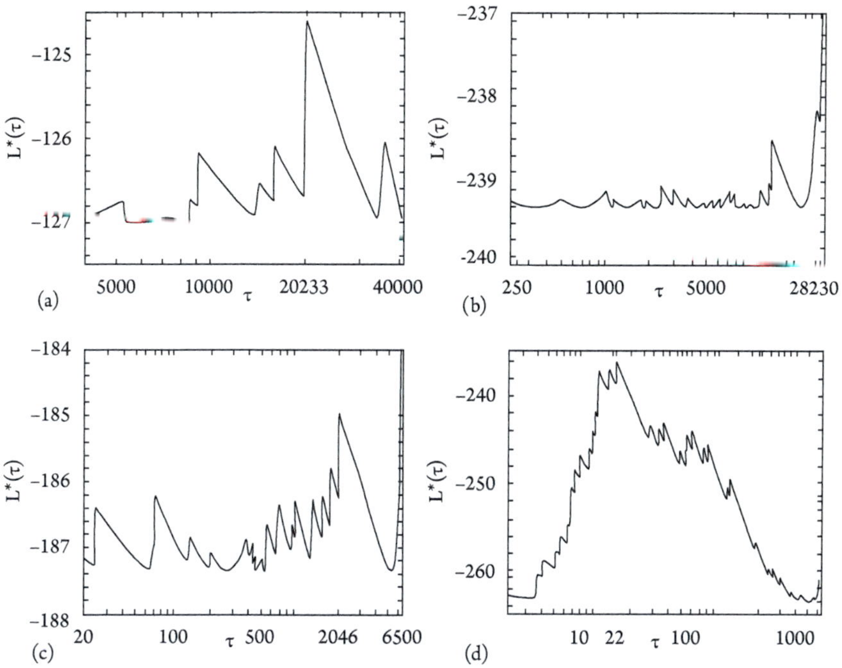

Figure 11.1 Profile log-likelihood for Kevlar49 pressure vessel data with stress levels of (*a*) 23.4 MPa, (*b*) 25.5 MPa, (*c*) 27.6 MPa, (*d*) 29.7 MPa, with τ plotted on a log scale.

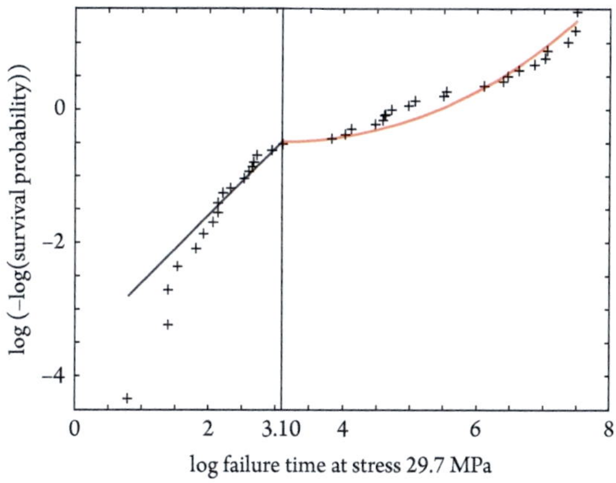

Figure 11.2 Proportional hazard plot for Kevlar49 pressure vessel data stressed at 29.7 MPa, with a vertical line indicating estimated change-point. The blue/red coloured line depicts the hazard function fitted by ML, with the blue part corresponding to where $y \le \tau$ and the red corresponding to where $y > \tau$.

displayed in Figure 11.2. Here the log of the n order statistics y_i $(i = 1, \ldots, n)$ are plotted against $\ln(-\ln(1-(i-0.5)/n))$. Examination of this figure would seem to concur with Crowder et al.'s findings that a change in distribution around this estimated value is not unreasonable for the fibres stressed at 29.7 MPa. Included in Figure 11.2 is the ML fitted distribution, namely:

$$\begin{aligned} f(y) &= 0.028\exp(-0.028y) & 0 \le y \le 22.1 \\ f(y) &= 0.00177\exp(-0.6188 - 0.00177(y-22.1)) & y > 22.1. \end{aligned} \tag{11.7}$$

The unbounded likelihood effect present here is similar to that encountered in Chapter 8. Indeed, Lawless (1982) regards the change-point parameter as an unknown threshold, noting that problems can be associated with its estimation, as in the three-parameter Weibull model with shape parameter less than unity. Again, the use of $f(y_n)dy$ for $dF(y_n)$ is inappropriate.

Smith (1986) encountered the same type of unbounded behaviour when using ML in the NEAR(2) model which, as in change-point problems, has discontinuities in the likelihood. In the case of the NEAR(2) model, the implication is that the function has many local maxima, and so usual derivative methods cannot be used. To optimize the function, Smith considered the data discretization method as suggested by Giesbrecht and Kempthorne (1976). An application of this method was also given by Matthews and Farewell (1985). By selecting an appropriate tolerance level, the desired effect of smoothing the likelihood function, whilst retaining the general shape, may be achieved. However, the method often requires examination of a number of tolerance levels for effectiveness; thus a certain subjectiveness is involved. As already explained in Section 8.3, there is an obvious flaw in the argument that the likelihood is unbounded simply because measurement error has not been taken into account. The flaw is that unboundedness will occur with *any* sample, including the sample of values measured without error (whatever this is!); the unboundedness problem would still be present.

As can be seen from the example, unboundedness is not really a problem in change-point models, providing its presence is allowed for.

We will return to this example more fully after we have considered using MPS estimation in this case, as the spacings approach is well suited to change-point models, in particular, eliminating the unboundedness problem. We also show that censored observations are easily handled.

11.3 The Spacings Function

Calculation of the spacings function involves computing the value of each area under the PDF between adjacent observations. The value of each area changes continuously with τ even though the PDF has a discontinuity at τ. The profile spacings function $H^*(\tau)$ with τ as profiling parameter is therefore a continuous function of τ. Note, however, that the derivative of $H^*(\tau)$ is discontinuous. We evaluate the spacings function, although the hybrid spacings function could be used instead.

Calculation of the spacings function involves the CDF, where for the PDF of eqn (11.1) we have

$$\begin{aligned} F(y) &= 1 - \exp(-ay) && 0 \le y \le \tau \\ &= 1 - \exp(-a\tau - b(y - \tau)) && y > \tau. \end{aligned}$$

Define $F(y_0) = 0$ and $F(y_{n+1}) = 1$. Then, if $y_j \le \tau < y_{j+1}$ $(j = 1, \ldots, n-1)$, the spacings function for the sample is

$$\begin{aligned} (n+1)H(a,b,\tau) &= F(y_1)\Big(I(j>1)\prod_{i=2}^{j}(F(y_i) - F(y_{i-1}))\Big)\times \\ &\quad \Big(I(1 \le j \le n-2)\prod_{i=j+2}^{n}(F(y_i) - F(y_{i-1}))\Big)\big(F(y_{j+1}) - F(y_j)\big) \\ &\quad \times (1 - F(y_n)) \\ &= (1 - \exp(-ay_1))\left(I(j>1)\prod_{i=2}^{j}(\exp(-ay_{i-1}) - \exp(-ay_i))\right)\times \\ &\quad \Big(I(1 \le j \le n-2)\prod_{i=j+2}^{n}(\exp(-a\tau - b(y_{i-1} - \tau)) \\ &\quad - \exp(-a\tau - b(y_i - \tau))\Big) \\ &\quad \times \big(\exp(-ay_j) - \exp(-a\tau - b(y_{j+1} - \tau))\big)\exp(-a\tau - b(y_n - \tau)). \end{aligned}$$

To obtain the profile spacings function $H^*(\tau)$, we maximize the preceding function for a fixed τ by employing a search method, such as the Nelder-Mead simplex algorithm, to obtain spacings estimates $\tilde{a}(\tau)$ and $\tilde{b}(\tau)$. Unlike ML, examination of the spacings function reveals that the unboundedness does not occur in this case, because the term $\ln(\hat{b}(\tau))$ is not present.

In order to compare the spacings method with ML in our example, we need to extend the calculations to handle censored observations.

11.3.1 Randomly Censored Observations

The spacings function is essentially a sum of $(n+1)$ terms, where each term is dependent on two consecutive observations. Each contribution to the spacings function can be calculated according to whether the adjoining values represent censored or uncensored observations.

When both y_{i-1} and y_i are uncensored, we have simply

$$\ln D_i = \ln(F(y_i; \boldsymbol{\theta}) - F(y_{i-1}; \boldsymbol{\theta})),$$

where $F(y_i;\boldsymbol{\theta})$ is given by (11.4). However, when observations are censored, there will be discrete jumps in the distribution function, and several censoring variables may occur at one point. In this case, spacings may be calculated using the following approach.

Let y_i be the observation prior to k identical censored observations $y_{i+1} = y_{i+2} = \ldots = y_{i+k} = T$, with y_{i+k+1} being the first observation after these. Define $z_i = F_y(y_i;\boldsymbol{\theta})$ and $z_{i+k+1} = F_y(y_{i+k+1};\boldsymbol{\theta})$. The remaining z_{i+j} ($j = 1,\ldots,k$) need to be placed to maximize

$$\prod_{j=1}^{k+1}(z_{i+j} - z_{i+j-1})$$

subject to $F_y(T^-;\boldsymbol{\theta}) < z_{i+j} < F_y(T^+;\boldsymbol{\theta})$ for $j = 1,\ldots,k$. We set the spacings $(z_{i+j} - z_{i+j-1})$ to be equal for $j = 2,\ldots,k$, thus assuming that each of the spacings is of size δ, with y_{i+1} and y_{i+k} positioned such that $z_{i+1} - F_y(T^-;\boldsymbol{\theta}) = F_y(T^+;\boldsymbol{\theta}) - z_{i+k} = \delta/2$. Thus,

$$\begin{aligned} k\delta &= F_y(T^+;\boldsymbol{\theta}) - F_y(T^-;\boldsymbol{\theta}) \\ &= F^0(T^+;\boldsymbol{\theta})(1 - F_c(T^+)) - F^0(T^-;\boldsymbol{\theta})(1 - F_c(T^-)) \\ &\quad + F_c(T^+) - F_c(T^-). \end{aligned}$$

Now $F^0(T^+;\boldsymbol{\theta}) = F^0(T^-;\boldsymbol{\theta}) = F^0(T;\boldsymbol{\theta})$, say. The preceding then reduces to

$$k\delta = (F_c(T^+) - F_c(T^-))(1 - F^0(T;\boldsymbol{\theta})).$$

Therefore,

$$\delta = \frac{1}{k}(F_c(T^+) - F_c(T^-))(1 - F^0(T;\boldsymbol{\theta})). \tag{11.8}$$

The $(k+1)$ terms between y_i and y_{i+k+1} contributing to the spacings function can be written as

$$\begin{aligned} (k+1)\ln \text{spacings} &= \ln\left(\int_{y_i}^{T^-} f_y(Y;\boldsymbol{\theta})dY + \delta/2\right) + (k-1)\ln\delta \\ &\quad + \ln\left(\delta/2 + \int_{T^+}^{y_{i+k+1}} f_y(Y;\boldsymbol{\theta})dY\right). \end{aligned}$$

From (11.4) we have $f_y(y;\boldsymbol{\theta}) = f^0(y;\boldsymbol{\theta})(1 - F_c(y))$, and so

$$\begin{aligned} (k+1)\ln \text{spacings} &= \ln\left((1 - F_c(y_i))\int_{y_i}^{T^-} f^0(Y;\boldsymbol{\theta})dY + \delta/2\right) \\ &\quad + (k-1)\ln\delta \\ &\quad + \ln\left(\delta/2 + (1 - F_c(T^+))\int_{T^+}^{y_{i+k+1}} f^0(Y;\boldsymbol{\theta})dY\right). \end{aligned} \tag{11.9}$$

Note that if there are two or more adjoining groups of censored observations, the method described can easily be modified to handle this.

Although the spacings $(z_{i+j} - z_{i+j-1})$ between tied, censored observations have been set equal, the values of the parameter estimates are in fact quite unaffected by the precise values of the z_{i+j} $(j = 1, \ldots, k)$. For instance, if the first I observations are uncensored and the remaining $(n - I)$ observations are censored at one point, the spacings function becomes essentially

$$(n+1)H(\boldsymbol{\theta}) = \sum_{i:\delta_i=1} \ln D_i(\boldsymbol{\theta}) + \sum_{j:\delta_j=0} \ln \delta,$$

where, from (11.8), δ =const $\times$ $(1 - F(y; \boldsymbol{\theta}))$. Thus, $\partial H(\boldsymbol{\theta})/\partial\boldsymbol{\theta}$ is unaffected by how the spacings for the censored observations are determined.

11.3.2 Numerical Example Using Spacings

Figure 11.3 shows the profile spacings function $H^*(\tau)$ for the four datasets already considered. As can be seen, the function is continuous and is finite throughout the range of τ, so that the unboundedness problem does not now occur. Again, τ is plotted on a log scale to display the behaviour more clearly.

The profile spacings function for stress level 23.4 MPa is given in Figure 11.3(a). As in the ML case, an estimate for a possible change point can be clearly picked out as $\tilde{\tau} = 20233$. For stress level 25.5, from Figure 11.3(b) we would select $\tilde{\tau} = 13944$. For stress level 27.6 MPa, Figure 11.3(c) displays a maximum at $\tilde{\tau} = 1997.8$. For stress level 29.7 MPa, Figure 11.3(d) gives a change-point estimate of $\tilde{\tau} = 20.1$.

In this last case, the fitted distribution is

$$\begin{aligned} f(y) &= 0.0293 \exp(-0.0293y) & 0 \le y \le 20.1 \\ f(y) &= 0.00168 \exp(-0.589 - 0.00168(y - 20.1)) & y > 20.1. \end{aligned} \tag{11.10}$$

11.3.3 Goodness-of-Fit

We consider two methods of testing the GoF of the estimated change-point model.

Firstly, we use the GoF test method described in Section 8.8.2 based on the Moran statistic of eqn (8.29), which relies on the fact that the spacings function, evaluated at the true parameter value, has a distribution independent of this value. However, instead of following the formal test given in that section, we use an appropriate percentage point of the distribution to calculate a critical level above which the maximized spacings function should lie. Thus, inverting the Moran GoF statistic of eqn (8.31), we have the α level test statistic

$$H_{crit}(\alpha) = -(\chi_n^2(1-\alpha) * C_2 + C_1 - \frac{k}{2})/(n+1),$$

so that if $H(\tilde{\boldsymbol{\theta}})$, the maximized spacings function value is less than $H_{crit}(\alpha)$, we would conclude that we have a significantly poor fit at level α.

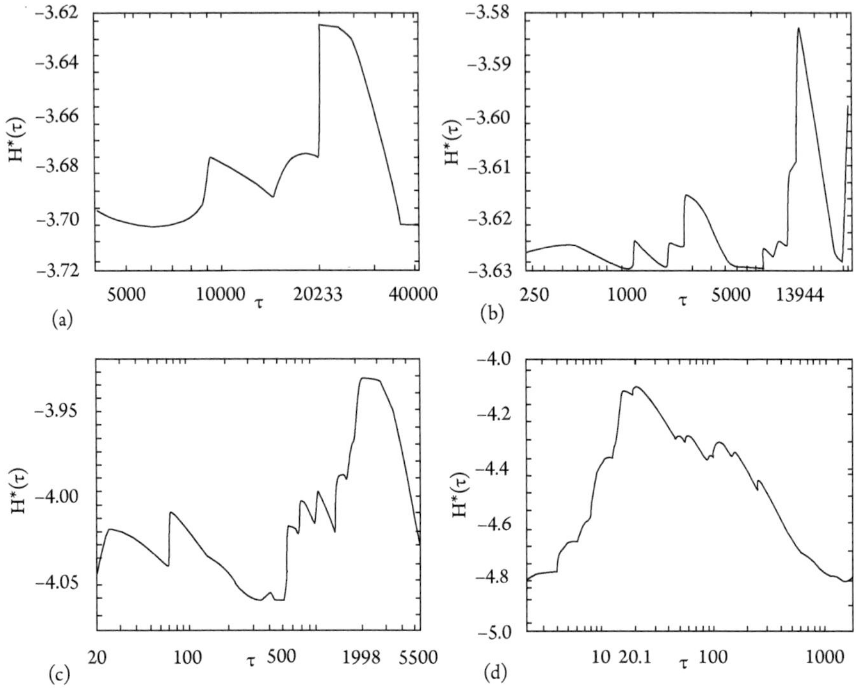

Figure 11.3 Profile spacings function for Kevlar49 pressure vessel data with stress levels of (*a*) 23.4 MPa, (*b*) 25.5 MPa, (*c*) 27.6 MPa, (*d*) 29.7 MPa, with τ plotted on a log scale.

As an example, we carry out this GoF test for the Kevlar sample stressed at 29.7 MPa. Taking $\alpha = 0.1$ gave $H_{crit}(0.1) = -4.54 < H(\tilde{\boldsymbol{\theta}}) = -4.02$, so that the fit is acceptable at level $\alpha = 0.1$.

In this example, we have a full sample. If there are censored observations, the method of estimation we have used ensures that the spacings for the censored observations are all equal. This results in 'super-uniform' spacings for these observations, and will inflate the value of the spacings function, so that the goodness-of-fit is conservative. A more accurate choice would be to place the z_{i+j} $(j = 1, \ldots, k)$ as if they were an ordered uniform sample in this interval. Therefore, in order to carry out a goodness-of-fit test, we can adjust the calculated spacings function by subtracting the contribution from the equal spacings and replacing it by uniformly sampled spacings. Thus, if y_i is the observation prior to k identical censored observations $y_{i+1} = y_{i+2} = \ldots = y_{i+k}$, with y_{i+k+1} being the first observation after these, then we replace the $(k + 1)$ ln spacings (11.9) by

$$\sum_{j=i+1}^{i+k+1} \ln(U_j - U_{j-1}),$$

where $\{U_j, j = i + 2, \ldots, i + k\}$ are ordered uniform(z_i, z_{i+k+1}) sample values, $U_i = z_i$ and $U_{i+k+1} = z_{i+k+1}$.

11.4 Bootstrapping in Change-Point Models

So far, apart from the GoF test just discussed, our analysis has been based on examination of the profile log-likelihood. As already pointed out, though the published asymptotic theory has considered the distributional properties of parameter estimates in detail, the results are not all that tractable. Bootstrapping provides a convenient tool that overcomes this. A very good theoretical justification of the use of parametric bootstrapping (BS) is given by Pham and Nguyen (1993).

In this section, we illustrate its use when fitting the change-point model (11.1) to the Kevlar49 sample stressed at 29.7 MPa. We consider both the ML fit as given in eqn (11.7) and the MPS fit as given in eqn (11.10). In each case, we generated, by parametric resampling from the fitted model, $B = 500$ BS samples all of same sample size $n = 39$ as the original. The model was then fitted to each BS sample by the same method used in fitting the model to the original data. The scatterplots of the 500 sets of fitted parameters are shown in Figure 11.4

Parameter Confidence Intervals

We used the BS method set out in flowchart (4.3) to calculate 90% percentile confidence intervals (CI) for the parameters, using the BS ML and MPS parameter estimates depicted in the scatterplots of Figure 11.4.

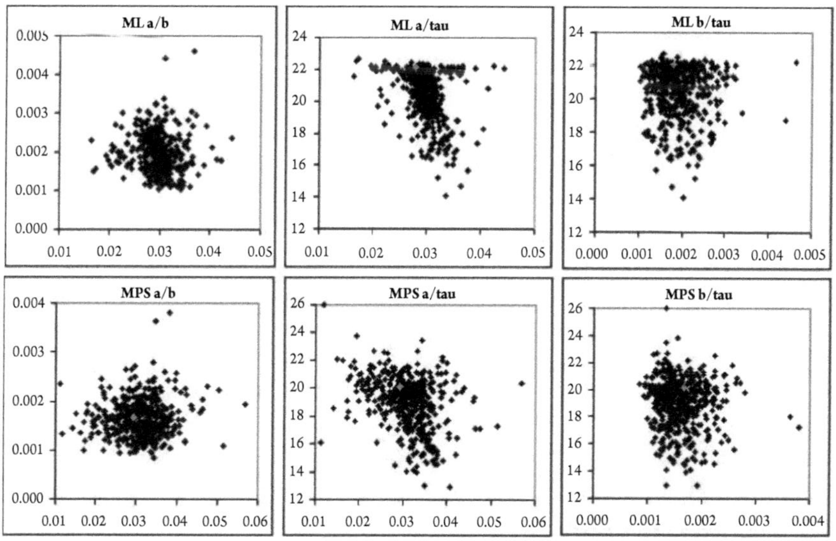

Figure 11.4 Scatterplots of 500 BS estimates of a, b, and τ for Kevlar49 pressure vessel data at stress level of 29.7 MPa, with τ plotted on a log scale. The upper row is the ML estimates, the lower row is the MPS estimates. Red dots are the estimates for the original sample.

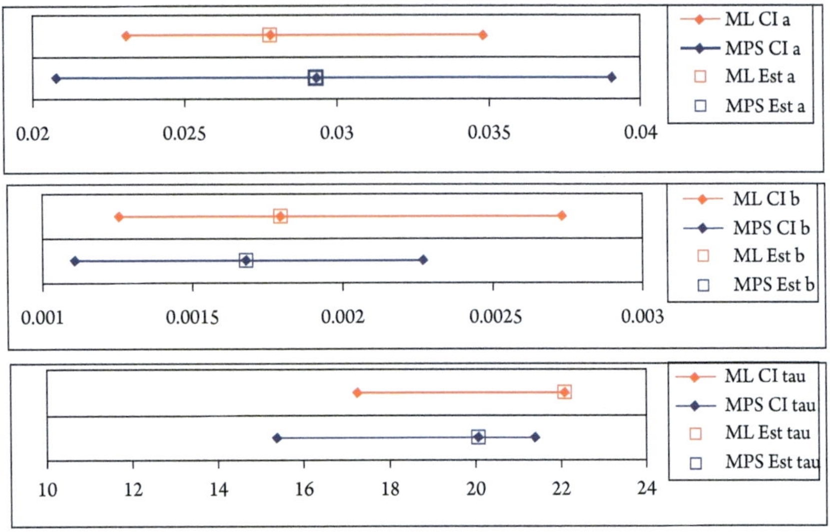

Figure 11.5 90% confidence intervals for the parameters a, b, and τ of the change-point model. Those based on the ML parameter estimates are shown in red. Those based on the MPS parameter estimates are shown in blue.

Figure 11.5 depicts the CIs. The 90% confidence intervals based on the BS ML parameter estimates are shown in red. The 90% confidence intervals based on the BS MPS parameter estimates are shown in blue.

It will be seen that the scatterplots involving τ and the CI for τ in the BS ML case are very skewed.

GoF Anderson-Darling Test

We used the BS method set out in flowchart (4.13) to provide an alternative test of goodness-of-fit, this time using the Anderson-Darling test statistic A^2. The BS parameter estimates used to generate the null hypothesis critical values were those depicted in the scatterplots of Figure 11.4.

The left-hand plots in Figure 11.6 depict the BS null hypothesis EDFs of the 500 BS Anderson-Darling statistics calculated from the BS ML (upper chart) and BS MPS (lower chart) parameter estimates. The green lines indicate the position of the A^2 test statistic calculated by ML and by MPS from the original sample. Both values indicate an acceptable fit at the 90% level.

Confidence Band for CDF

We used the method described in Section 4.3 to calculate a 90% confidence band for the CDF of the change-point model.

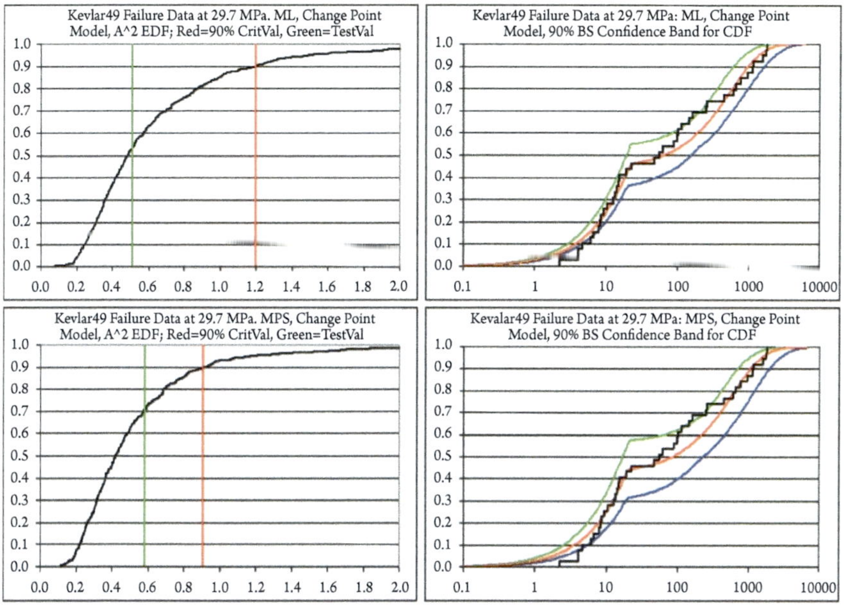

Figure 11.6 Left-hand charts: BS EDF of the A^2 GoF statistic calculated from the BS ML (upper chart) and MPS (lower chart) parameter estimates. Green lines indicate the position of the A^2 statistic calculated by ML and by MPS from the original sample. Red lines are the BS upper 10% A^2 test critical values. Right-hand charts: BS 90% confidence bands for the fitted change-point distribution. ML-upper chart, MPS-lower chart.

The right-hand charts of Figure 11.6 depict the confidence bands using the BS ML parameter estimates (upper chart) and BS MPS parameter estimates (lower chart).

It will be seen that even though the A^2 GoF test did not reject either fit, there is some evidence of lack-of-fit when y is small.

A possible remedy would be to add an overall shift parameter to the model, but we have not tried this.

11.5 Summary

In summary, though fitting change-point models is non-standard, the numerical methods that we examined appear to be straightforward to implement. Estimation of change points occurring at the upper end of the distribution might be difficult using ML, because of the unbounded likelihood problem. However, use of MPS estimation eliminates this problem.

12 The Skew Normal Distribution

12.1 Introduction

In this chapter, we consider the skew-normal distribution, a generalization of the normal that includes the normal as a special case. Estimation of the parameters in the skew normal may or may not be non-standard depending on how the distribution is parametrized. Estimation of parameters, where the parametrization might appear to be the most natural, turns out to be non-standard. The reason is that this parametrization renders the Fisher information matrix singular at the true parameter value when this corresponds to the normal special case, the singularity occurring because the log-likelihood is then particularly flat in a certain coordinate direction. Thus, standard asymptotic theory cannot reliably be used to calculate the asymptotic distribution of the estimates of all the parameters. The problem can be handled using an alternative parametrization. We discuss this non-standard problem, and show how it can be overcome.

It should be noted that the half-normal distribution is also a special case, and that this distribution can be the best fit to a data set. However, in the usual formulation, this occurs when the shape parameter, which we will be defining and representing by λ, is infinite. There is nothing untoward when this occurs, only appearing unusual because it occurs at an infinite parameter limit. We shall show that computationally it is easily handled.

We discuss only the univariate case, when the skew normal can be assumed to be a three-parameter distribution. The skew-normal distribution generalizes to a tractable multivariate form that makes it useful in situations where multivariate skewness needs modelling. There has been much work done in this area. We give references that discuss univariate and multivariate generalizations at the end of the chapter.

Non-Standard Parametric Statistical Inference. Russell Cheng.

12.2 Skew Normal Distribution

Skew-normal distributions are a family (or class) of distributions introduced by Azzalini (1985). Specifically, we shall say that Z has a *skew-normal* distribution, or that Z is $SN(\lambda)$, if it is a continuous random variable with density function

$$\phi(z; \lambda) = 2\phi(z)\Phi(\lambda z), \quad -\infty < z < \infty, \tag{12.1}$$

where $\phi(\cdot)$ and $\Phi(\cdot)$ are the standard normal density and distribution functions, respectively.

The following properties follow directly from the density.

(i) The family is parametrized by one parameter λ, with any finite value of λ, i.e. $-\infty < \lambda < \infty$, allowed. The standard normal distribution, $N(0, 1)$, is a special case obtained at the internal point $\lambda = 0$.

(ii) As $\lambda \to \infty$, $\phi(z; \lambda)$ tends to the half (i.e. folded) normal density, and to the reflected half-normal if $\lambda \to -\infty$.

(iii) If Z is a $SN(\lambda)$ random variable, then $-Z$ is a $SN(-\lambda)$ random variable.

(iv) $1 - \Phi(-z; \lambda) = \Phi(z; -\lambda)$.

(v) $\Phi(z; 1) = \{\Phi(z)\}^2$.

(vi) If Z is $SN(\lambda)$, then Z^2 is χ_1^2.

The family is mathematically quite tractable, with mean, variance, skewness, and kurtosis expressible in closed form. Writing

$$b = \sqrt{2/\pi}, \delta = \lambda/\sqrt{(1 + \lambda^2)},$$

we have

$$\begin{aligned}
E(Z) &= b\delta, \\
\text{Var}(Z) &= 1 - (b\delta)^2, \\
\gamma_1(Z) &= \frac{1}{2}(4 - \pi)\frac{|\lambda|}{\lambda}\left[\frac{(b\delta)^2}{1 - (b\delta)^2}\right]^{3/2}, \\
\gamma_2(Z) &= 2(\pi - 3)\left[\frac{(b\delta)^2}{1 - (b\delta)^2}\right]^2,
\end{aligned} \tag{12.2}$$

where γ_1 is the third standardized cumulant or skewness, and γ_2 is the fourth standardized cumulant or kurtosis. Both are zero at $\lambda = 0$, increasing, respectively, to approximately 0.9953 and 0.8692 as $\lambda \to \infty$.

The distribution function of Z, an SN random variable, is

$$\Phi(z; \lambda) = \int_{-\infty}^{z} \phi(t)\Phi(\lambda t)dt.$$

For numerical work, this can be expressed as

$$\Phi(z;\lambda) = \Phi(z) - 2T(z,\lambda), \tag{12.3}$$

where, for $z > 0$ and $\lambda > 0$,

$$T(z,\lambda) = \int_z^{\infty}\left(\int_0^{\lambda t}\phi(u)du\right)\phi(t)dt.$$

We use

$$T(-z,\lambda) = T(z,\lambda)$$

and

$$T(z,-\lambda) = -T(z,\lambda)$$

to calculate negative values of z and λ, so that (12.3) can be used for all z and λ. The function $T(z,\lambda)$ has been tabulated by Owen (1956), who gives further details of it.

Azzalini (2005) suggests the following elegant way of generating a whole class of random variates, of which *SN* is just one, when these are needed in simulation work or bootstrapping. If f_0 is a PDF symmetric about 0, and G is a CDF such that G' is a PDF symmetric about 0, then

$$f(z) = 2f_0(z)G(w(z)) \quad -\infty < z < \infty \tag{12.4}$$

is a PDF for any odd function $w(.)$.

Let $Y \sim f_0$ and $X \sim G'$, and let

$$Z = \begin{cases} Y \text{ if } X < w(Y) \\ -Y \text{ otherwise} \end{cases}. \tag{12.5}$$

From this definition, and using the fact that

$$\Pr(z < -Y < z + dz \,\&\, X > w(-z)) = \Pr(z < -Y < z + dz \,\&\, X > -w(z)),$$

because $w(\cdot)$ is an odd function, and that

$$\Pr(z < -Y < z + dz \,\&\, X > -w(z)) = \Pr(z < Y < z + dz \,\&\, X < w(z)),$$

because Y and X are both symmetrically distributed about 0, we obtain, after a little manipulation, the differential result

$$\begin{aligned} f_Z(z)dz &= \Pr(z < Z < z + dz) \\ &= 2\Pr(z < Y < z + dz \,\&\, X < w(z)) \\ &= 2f_0(z)G(w(z))dz, \end{aligned}$$

so that Z has the given density. We get the SN distribution if $G(\cdot)$ is the standard normal CDF and $w(z) = \lambda z$.

An alternative way to generate $SN(\lambda)$ distributed variates is given by Henze (1986). If we set

$$Z = \frac{\lambda}{\sqrt{1+\lambda^2}}|U| + \frac{1}{\sqrt{1+\lambda^2}}V, \tag{12.6}$$

where U and V are independent standard normal variates and $|U|$ is the magnitude of U, then $Z \sim SN(\lambda)$.

12.3 Linear Models of Z

12.3.1 Basic Linear Transformation of Z

In statistical applications, it would seem natural to use the linear transformation

$$Y = \lambda_1 + \lambda_2 Z, \quad \lambda_2 > 0, \tag{12.7}$$

where Z is a $SN(\lambda)$ random variable, so that we have a three-parameter distribution, with

$$\boldsymbol{\lambda} = (\lambda_1, \lambda_2, , \lambda)^T \tag{12.8}$$

the vector of parameters.

However, there is a problem when $\lambda = 0$. In this case, denote by $\boldsymbol{\lambda}^0$ the vector of true parameter values with $\lambda^0 = 0$, and by $\hat{\boldsymbol{\lambda}}$ its ML estimator. The standard way of deriving the asymptotic distribution of these estimators requires calculation of the Fisher information matrix, which in this case (for one observation) is

$$\mathrm{i}(\boldsymbol{\lambda}) = \begin{pmatrix} (1+\lambda^2 a_0)/\lambda_2^2 & (b\delta\frac{1+2\lambda^2}{1+\lambda^2} + \lambda^2 a_1)/\lambda_2^2 & (\frac{b}{(1+\lambda^2)^{2/3}} - \lambda a_1)/\lambda_2 \\ (b\delta\frac{1+2\lambda^2}{1+\lambda^2} + \lambda^2 a_1)/\lambda_2^2 & (2+\lambda^2 a_2)/\lambda_2^2 & -\lambda a_2/\lambda_2 \\ (\frac{b}{(1+\lambda^2)^{2/3}} - \lambda a_1)/\lambda_2 & -\lambda a_2/\lambda_2 & a_2 \end{pmatrix},$$

where

$$a_k = E\left\{Z^k\left[\frac{\phi(\lambda Z)}{\Phi(\lambda Z)}\right]^2\right\}, \quad k = 0, 1, 2.$$

Evaluating these expectations explicitly, we find, after some manipulation, that

$$a_0(\lambda) = \frac{2}{\pi}, \quad a_1(\lambda) = 0, \quad a_2(\lambda) = \frac{2}{\pi}.$$

Thus,

$$\mathrm{i}(\mathbf{0}) = \begin{pmatrix} 1/\lambda_2^2 & 0 & b/\lambda_2 \\ 0 & 2/\lambda_2^2 & 0 \\ b/\lambda_2 & 0 & \frac{2}{\pi} \end{pmatrix},$$

which is singular.

An informal explanation of the problem is given by Azzalini (1985), who points out that near $\lambda = 0$, λ has the same order of magnitude as $|\gamma_1|^{1/3}$. Though γ_1 can be estimated with the usual variance rate n^{-1}, the estimate of λ will have a variance that is a larger order of magnitude. Morever, as $E(Y) = \lambda_1 + \lambda_2 b\delta$ and $\delta \simeq \lambda$ near $\lambda = 0$, the parameter λ_1 cannot be estimated with variance rate n^{-1} either.

Despite this problem, the parameters are identifiable using ML estimation. From the definition of Y, (12.7), we have $Z = (Y - \lambda_1)/\lambda_2$, and substituting this into the PDF of Z, the log-likelihood can be written as

$$L(\boldsymbol{\lambda}|\mathbf{y}) = \sum_{i=1}^{n} \log\{2\phi[(y_i - \lambda_1)/\lambda_2]\Phi[\lambda(y_i - \lambda_1)/\lambda_2]\}, \tag{12.9}$$

with the ML estimator of $\boldsymbol{\lambda}$ obtained by maximizing this to give

$$\hat{\boldsymbol{\lambda}} = \arg\max L(\boldsymbol{\lambda}|\mathbf{y}).$$

The problem where the Fisher information matrix has a singularity at a given point has been thoroughly investigated by Rotnitzky et al. (2000). The key result for our purposes is Theorem 3 given in Rotnitzky et al. (2000), which was applied by Chiogna (2005) to the *SN* case to establish the precise asymptotic behaviour of the parameter estimators. We summarize the detailed analysis of the ML estimators given by Chiogna (2005) for the model (12.7).

Specifically, Chiogna shows for the special case where $\lambda^0 = 0$, that (correcting the typographical error in the first component expression)

$$[n^{1/2}(\hat{\lambda}_1 - \lambda_1^0 - b\hat{\lambda}_2\hat{\lambda}), n^{1/2}(\hat{\lambda}_2 - \lambda_2^0 + \frac{1}{2}b\hat{\lambda}_2\hat{\lambda}^2), n^{1/6}\hat{\lambda}]^T \tag{12.10}$$

converges to $(Z_1, Z_2, Z_3^{1/3})$, where (Z_1, Z_2, Z_3) is a normal random vector with mean zero and covariance matrix equal to the inverse of the covariance matrix of the vector $(u_{\chi}(\boldsymbol{\chi}^0, \lambda^0)^T, \frac{1}{6}\frac{d}{d\lambda}\mathrm{j}_{\lambda\lambda}^{\lambda_{II}}(\boldsymbol{\chi}^0, \lambda^0))^T$, where $\boldsymbol{\chi} = (\lambda_1, \lambda_2)^T$ and $u_{\chi}(\boldsymbol{\chi}, \lambda)$ is the 2×1 score vector for the component $\boldsymbol{\chi}$ (that is, the vector of first derivatives of the log-likelihood function with respect to $\boldsymbol{\chi}$), and $\mathrm{j}_{\lambda\lambda}^{\lambda_{II}}(\boldsymbol{\chi}, \lambda)$ is the second partial derivative with respect to λ of the one-observation log-likelihood under a two-step reparametrization $\boldsymbol{\lambda} \to \boldsymbol{\lambda}_I \to \boldsymbol{\lambda}_{II}$ specified by Rotnitzky et al. (2000), details of which are given by Chiogna (2005).

Chiogna gives the expressions

$$u_{\chi}(\chi^0, \lambda^0) = \left(\frac{z_i}{\lambda_2^0}, \frac{z_i^2 - 1}{\lambda_2^0} \right)^T$$

and

$$\frac{d}{d\lambda} j_{\lambda\lambda}^{\lambda_{II}}(\chi^0, \lambda^0) = -b(3b + 1)z_i^3 - \frac{4b^3}{\lambda_2^0} z_i$$

for one generic observation y_i where $z_i = (y_i - \lambda_1)/\lambda_2$. However, it will be seen from the expressions for the components in (12.10) that ML estimation using this parametrization yields estimators whose asymptotic distributions are not very easy to use from the practical point of view.

12.3.2 Centred Linear Transformation of Z

Azzalini (1985) suggested a neat way to overcome the estimation problems that occur using the linear transformation (12.7) by using a *centred* linear transformation

$$Y = \theta_1 + \theta_2 \frac{Z - E(Z)}{\sqrt{\text{var}(Z)}} = \theta_1 + \theta_2 \frac{Z - b\delta}{\sqrt{1 - (b\delta)^2}}, \tag{12.11}$$

and taking the parameters as

$$\boldsymbol{\theta} = (\theta_1, \theta_2, \gamma_1)^T, \tag{12.12}$$

where γ_1 is the skewness given in (12.2). In this version, the parameters are simply the mean and SD of the distribution, that is,

$$E(Y) = \theta_1 \text{ and } SD(Y) = \theta_2,$$

so that the parameters are much more readily interpretable in practical use than λ_1 and λ_2.

Equating parameters in (12.7) and (12.11) gives

$$\lambda_1 = \theta_1 - \frac{\theta_2 b\delta}{\sqrt{1 - (b\delta)^2}}, \quad \lambda_2 = \frac{\theta_2}{\sqrt{1 - (b\delta)^2}}. \tag{12.13}$$

As shown by Chiogna, taking these expressions together with γ_1 as defined (12.2) shows that the new parametrization is equivalent to making the following substitutions in the original linearization (12.7):

$$\lambda_1 = \theta_1 - \theta_2 \eta_1^{1/3}, \; \lambda_2 = \theta_2(1 + \eta_1^{2/3})^{1/2}, \; \lambda = \eta_1^{1/3}[b^2 + \eta_1^{2/3}(b^2 - 1)]^{-1/2} \tag{12.14}$$

(correcting a typing error in the formula for λ given in Chiogna), where we have written $\eta_1 = \frac{2}{4-\pi}\gamma_1$ to make the formulas slightly neater. In the numerical examples, we will use the parametrization

$$\boldsymbol{\theta}^I = (\theta_1, \theta_2, \eta_1)^T \tag{12.15}$$

rather than $\boldsymbol{\theta}$. The ML estimator of $\hat{\boldsymbol{\theta}}^I$ can be obtained directly by maximizing the log-likelihood (12.9), with λ replaced by $\boldsymbol{\theta}^I$, using the substitutions (12.14).

In this alternative parametrization, let $\boldsymbol{\varphi} = (\theta_1, \theta_2)^T$ so that $\boldsymbol{\theta}^I = (\boldsymbol{\varphi}^T, \gamma_1)^T$, and let $\boldsymbol{\theta}^{I0} = (\boldsymbol{\varphi}^{0T}, \gamma_1^0)^T$ be the true parameter values. Again, we consider the special case $\lambda^0 = 0$, so that $\gamma_1^0 = 0$. For this case, Chiogna shows that the random vector $(n^{1/2}(\hat{\theta}_1 - \theta_1^0)$, $n^{1/2}(\hat{\theta}_2 - \theta_2^0)$, $n^{1/2}\gamma_1^0)$ converges to a random normal vector with mean zero and variance equal to the inverse of the covariance matrix of the score vector $(u_\varphi(\boldsymbol{\varphi}^0, \gamma_1^0)^T$, $u_{\gamma_1}(\boldsymbol{\varphi}^0, \gamma_1^0))^T$, where $u_\varphi(\boldsymbol{\varphi}, \gamma_1)$ and $u_{\gamma_1}(\boldsymbol{\varphi}, \gamma_1)$ are the scores in the new parametrization analogous to $u_\chi(\boldsymbol{\chi}, \lambda)$ and $u_\lambda(\boldsymbol{\chi}, \lambda)$ in the original.

The information matrix in the new parametrization has the form

$$\mathrm{i}(\boldsymbol{\varphi}^0, \gamma_1^0) = \begin{pmatrix} \frac{1}{\theta_2^2} & 0 & 0 \\ 0 & \frac{2}{\theta_2^2} & 0 \\ 0 & 0 & \frac{1}{6} \end{pmatrix}$$

at $\boldsymbol{\theta}^I = \boldsymbol{\theta}^{I0}$ (where $\lambda^0 = 0$), showing that all three parameters can be estimated with variance of order n^{-1}, even in this special case. We then have, in this special case,

$$n^{1/2}\hat{\theta}_1 \xrightarrow{d} N(\theta_1^0, \theta_2^{02}) \text{ and } n^{1/2}\hat{\theta}_2 \xrightarrow{d} N(\theta_1^0, \frac{1}{2}\theta_2^{02}) \text{ as } n \to \infty.$$

This is the same result as when $\hat{\theta}_1$ and $\hat{\theta}_2$ are the ML estimators of the mean θ_1^0 and SD θ_2^0 of a $N(\theta_1^0, \theta_2^{02})$ sample. If, therefore, it is thought possible or even likely that **y** is a normally distributed sample, one can fit this centred linearization of the *SN* model using the asymptotic distribution of $\hat{\lambda}$ to test if the actual value obtained for $\hat{\lambda}$ is significantly different from zero; and if not, treating $\hat{\theta}_1$ and $\hat{\theta}_2$ as the ML estimators of a $N(\theta_1, \theta_2^2)$ sample. However, the asymptotic distribution of $\hat{\lambda}$ is then needed, without assuming that $\lambda = 0$, so that the information matrix would need evaluation, probably numerically. Salvan (1986) gives a locally most powerful invariant test of normality.

An alternative parametrization, which shows that the use of skewness as the shape parameter is not absolutely necessary, is to note that the equations (12.13) can easily be used to express the λ_1 and λ_2 original parameters in terms of other parametrizations.

For example, if we use the linearization (12.11) with parameter

$$\boldsymbol{\theta}^{II} = (\theta_1, \theta_2, \gamma), \tag{12.16}$$

where

$$\gamma = \lambda^3$$

is the third parameter, then the equivalent of (12.14) is

$$\lambda_1 = \theta_1 - \frac{\theta_2 b\delta}{\sqrt{1-(b\delta)^2}}, \lambda_2 = \frac{\theta_2}{\sqrt{1-(b\delta)^2}}, \lambda = \gamma^{1/3}, \tag{12.17}$$

where δ is treated as function of γ, namely,

$$\delta = \frac{\lambda}{\sqrt{1+\lambda^2}} = \frac{\gamma^{1/3}}{\sqrt{1+(\gamma^{1/3})^2}}.$$

The information matrix in this parametrization has the form

$$\mathrm{i}(\boldsymbol{\varphi}^0, \gamma^0) = \begin{pmatrix} \frac{1}{\theta_2^2} & 0 & 0 \\ 0 & \frac{2}{\theta_2^2} & 0 \\ 0 & 0 & \frac{1}{3}\frac{(\pi-4)^2}{\pi^3} \end{pmatrix},$$

showing that it is not singular at $\gamma^0 = 0$.

ML estimation in terms of this parametrization proceeds in exactly the same way as before, only now the substitutions (12.17) are used for $\boldsymbol{\lambda}$ in the log-likelihood $L(\boldsymbol{\lambda}|\mathbf{y})$, so that the numerical maximization is carried out directly in terms of the new parametrization $\boldsymbol{\theta}^{II}$.

12.3.3 Parametrization Invariance

It is of interest to see that the parameters of the three-parameter linear *SN* model can be selected so that they possess standard asymptotic properties, as this enables basic inference to be easily carried out. It should be noted, however, that using ML estimation, no parametrization really has any practical advantage over any other in the sense of providing an estimate of the overall distribution, or indeed of any quantity of practical interest that is a function of the parameters, that is superior to other parametrizations. This is because of the following well-known invariance property of ML estimators, already pointed out in Section 3.2. We give an alternative derivative-based demonstration here of this property.

Recall that if the original parametrization is $\boldsymbol{\lambda}$, then $\hat{\boldsymbol{\lambda}} = \arg\max L(\boldsymbol{\lambda})$, where for simplicity we have omitted the dependence of the log-likelihood $L(\boldsymbol{\lambda})$ on the sample $\mathbf{y}$. For an interior point $\hat{\boldsymbol{\lambda}}$, this is a solution of the likelihood equations, so that we have

$$\left.\frac{\partial L(\boldsymbol{\lambda})}{\partial \boldsymbol{\lambda}}\right|_{\boldsymbol{\lambda}=\hat{\boldsymbol{\lambda}}} = \mathbf{0}.$$

Suppose now we have $\boldsymbol{\theta} = \mathbf{g}(\boldsymbol{\lambda})$, a bijection transformation of $\boldsymbol{\lambda}$, so that $\boldsymbol{\theta} = \mathbf{g}(\boldsymbol{\lambda})$ and $\boldsymbol{\lambda} = \mathbf{g}^{-1}(\boldsymbol{\theta})$ are both well defined. Then

$$\frac{\partial L(\mathbf{g}^{-1}(\boldsymbol{\theta}))}{\partial \boldsymbol{\theta}} = \frac{\partial L(\boldsymbol{\lambda})}{\partial \boldsymbol{\lambda}} \cdot \frac{\partial \mathbf{g}^{-1}(\boldsymbol{\theta})}{\partial \boldsymbol{\theta}}$$

and

$$\frac{\partial L(\mathbf{g}^{-1}(\boldsymbol{\theta}))}{\partial \boldsymbol{\theta}}\Big|_{\boldsymbol{\theta}=\mathbf{g}(\hat{\boldsymbol{\lambda}})} = \frac{\partial L(\boldsymbol{\lambda})}{\partial \boldsymbol{\lambda}}\Big|_{\boldsymbol{\lambda}=\hat{\boldsymbol{\lambda}}} \cdot \frac{\partial \mathbf{g}^{-1}(\boldsymbol{\theta})}{\partial \boldsymbol{\theta}}\Big|_{\boldsymbol{\theta}=\mathbf{g}(\hat{\boldsymbol{\lambda}})} = \mathbf{0}.$$

Thus, the ML estimator of $\boldsymbol{\theta}$ is $\hat{\boldsymbol{\theta}} = \mathbf{g}(\hat{\boldsymbol{\lambda}})$, showing that, subject to the transformation between $\boldsymbol{\lambda}$ and $\boldsymbol{\theta}$ being a bijection, we can obtain the ML estimator of one set of parameters from that of any other set.

The implication of this property is that the statistical behaviour of the estimate of any quantity of interest that depends on the parameters is invariant with regard to the actual parameterization used; in other words, the behaviour will be independent of the parametrization chosen, so that we are free to choose whichever parametrization that we like for inference purposes. This freedom of choice is reinforced if one uses bootstrapping to calculate confidence intervals or carry out hypothesis and GoF tests based on ML estimation, as the invariance property shows that the results obtained from parametrizations that are bijectively equivalent will be invariant.

12.4 Half-Normal Case

As listed in point (ii) earlier, $\phi(z;\lambda)$ tends to the half-normal density when $\lambda \to \infty$. When fitting any one of the three parameter versions discussed in the previous section, this possibility is easily spotted and dealt with if we plot the profile log-likelihood

$$L^*(\lambda) = \max_{\chi} L(\boldsymbol{\chi}, \lambda)$$

with profiling parameter λ, where the maximization is with respect to $\boldsymbol{\chi} = (\lambda_1, \lambda_2)$, the other two parameters. We find that the graph of $L^*(\lambda)$ has two horizontal asymptotes as $\lambda \to \pm\infty$, with asymptotic values equal to the log-likelihoods of the two possible half-normal model fits. If the overall maximum of the graph is at either of these values, then the corresponding half-normal is the best fit.

The possibility of λ becoming infinite is considered by Sartori (2006) in the skew-normal distribution and by Álvarez and Gamero (2012) in the skew-t-distribution, a closely related distribution. These authors do not mention the fact that there is a non-degenerate limit as $\lambda \to \pm\infty$, appearing to view this limit as undesirable. Indeed, Sartori (2006) proposes an estimator for λ involving a modified score function so that λ is always finite for the skew normal. Álvarez and Gamero (2012) have shown that this score statistic also keeps λ finite in the skew-t case, provided the degrees of freedom are known and greater than or equal to 2.

Limiting the magnitude of λ may be convenient computationally, but it is not clear if there are any practical applications where it would be meaningful or desirable per se.

12.5 Log-likelihood Behaviour

We illustrate the behaviour of the log-likelihood of a sample $y = y_1, y_2, \ldots, y_n$ of size n corresponding to the different parametric models considered in the previous section.

12.5.1 FTSE Index Example

In our first example, we use the same FTSE share price data set that we considered in Section 9.7. Fitting the skew normal to this data set provides an example of where $\hat{\lambda}$, the ML estimator of λ, is finite.

We take the sample in the form as given in eqn (9.17):

$$y_i = \ln(p_i/p_{i-1}), i = 1, 2, \ldots, n,$$

where p_i is the closing FTSE100 index on day i. In the present example, we took the full data set of $n = 250$ observations as given by Cheng (2011).

We calculated the ML estimators in two ways. One used Nelder-Mead numerical optimization of the log-likelihood treated as a function all three parameters, the precise parameters depending on the form of the linearization and the quantity taken as the third parameter.

Azzalini (1985) observed that, in numerical optimization of the log-likelihood, care is needed to avoid a numerical algorithm stopping prematurely simply because the log-likelihood appears flat, as will be the case near a point of inflection. Chiogna points out, in a corollary to her analysis, that this is what happens for the distribution of Y in (12.7), and that the profile log-likelihood

$$L_P(\lambda) = \max_{\chi} L(\chi, \lambda|\mathrm{y})$$

always has a point of inflexion at $\lambda = 0$. A convenient and robust approach that avoids difficulties with the inflection point, as suggested by Azzalini, is to maximize the log-likelihood by plotting the profile likelihood to ensure that an actual maximum is located. This is the second way in which we obtained the ML estimators.

Figure 12.1 shows three profile log-likelihoods for the FTSE data. The first is where the y observations are *SN Z* variables transformed with the simple linearization (12.7) using the parameters $(\lambda_1, \lambda_2, \lambda)$ as in (12.8). The second and third are where the y observations are *SN Z* variables using the centred linearization (12.11), respectively, with parameters $\boldsymbol{\theta}^I = (\theta_1, \theta_2, \eta_1)$ as in (12.15) and parameters $\boldsymbol{\theta}^{II} = (\theta_1, \theta_2, \gamma)$ as in (12.16). It will be seen that in the first case, using the parameters λ, the profile log-likelihood is quite flat at the inflection point $\lambda = 0$, with the maximum further to the right. In the second case, using the parameters $\boldsymbol{\theta}^I$, the graph in the neighborhood of the inflection point has greater slope, so that the maximum is more pronounced. In the third case, using the parameters $\boldsymbol{\theta}^{II}$, the graph has a knee near $\gamma = 0$, but the maximum is quite distinct, as in the $\boldsymbol{\theta}^I$ case.

The invariance property of ML estimators means that the value of the maximized log-likelihood is the same in all three parametrizations, subject to rounding error or possible

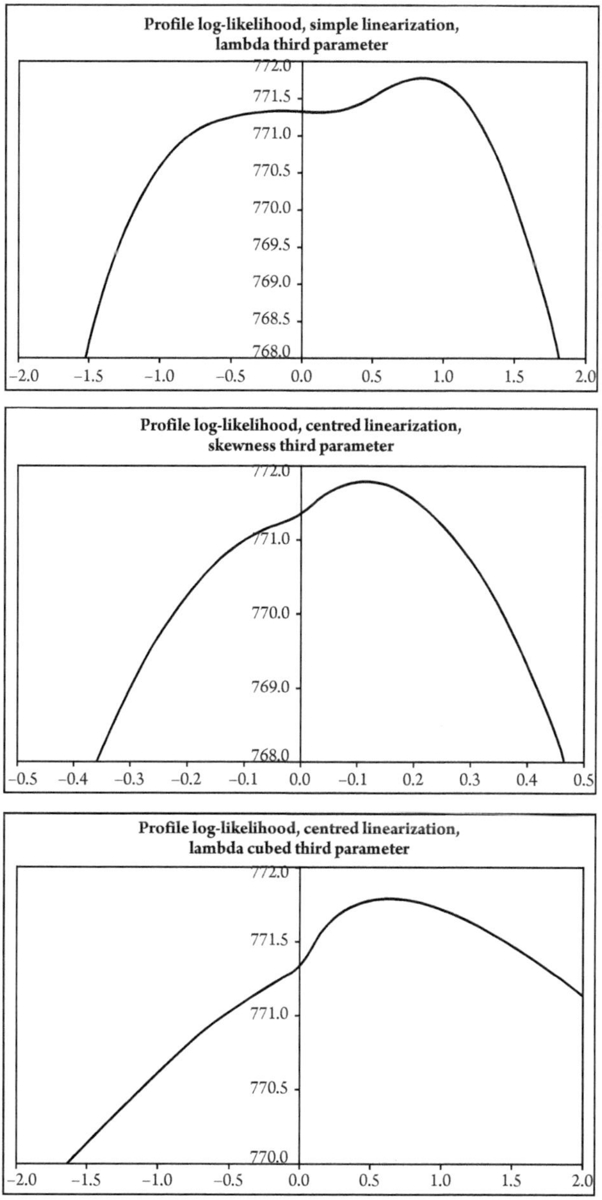

Figure 12.1 Three skew-normal profile log-likelihood plots for the FTSE data set using the λ, θ', and θ'' parametrizations.

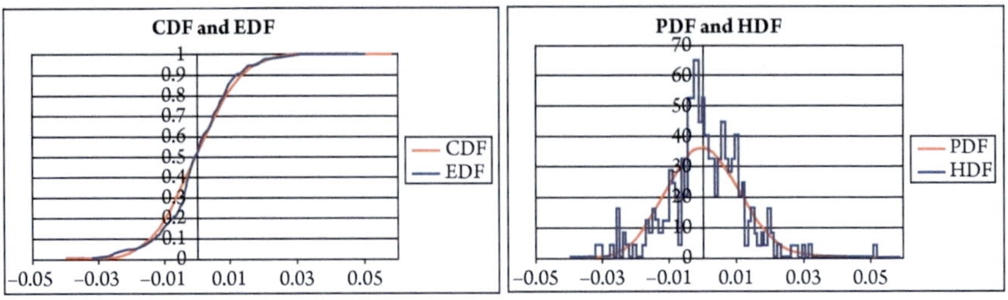

Figure 12.2 Graphs of the CDF and PDF of the linearized skew-normal distribution fitted to the FTSE data set.

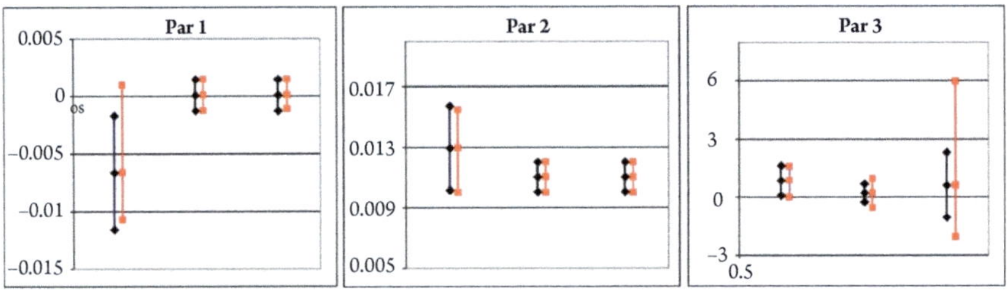

Figure 12.3 Confidence intervals for the parameters of the *SN* model fitted to the FTSE data set, using the three parametrizations. Each subgraph shows three pairs of 95% CIs for the same parameter in each of the three parametrizations. In each pair, the left-hand CI is calculated using asymptotic formula (black diamond marker lines) and the right-hand CI is calculated by bootstrapping (red square marker lines). The parametrizations are $\boldsymbol{\lambda} = (\lambda_1, \lambda_2, \lambda)$, $\boldsymbol{\theta}^I = (\theta_1, \theta_2, \eta_1)$, $\boldsymbol{\theta}^{II} = (\theta_1, \theta_2, \gamma)$, the parameters being labelled Par 1, Par 2, Par 3 in this figure.

optimization inaccuracy. Thus, whichever ML parameter estimator, be it $\hat{\boldsymbol{\lambda}}$, $\hat{\boldsymbol{\theta}}^I$, or $\hat{\boldsymbol{\theta}}^{II}$, is used in the calculation, the same fitted distribution will be obtained. The CDF and PDF of the fitted *SN* distribution using the $\boldsymbol{\theta}^I$ parametrization is shown in Figure 12.2. The CDFs and PDFs fitted using the other two parametrizations were indeed effectively identical and are not shown.

Confidence intervals were calculated using the formulas, such as eqn (3.7) given in Chapter 3 covering standard normal theory. We also calculated confidence intervals, such as the percentile CI using the parametric bootstrapping method described in Chapter 4, with $B = 500$ bootstrap replicates. This was done for all three parametrizations, and the results are summarized in Figure 12.3. It will be seen that the asymptotic and bootstrap CIs are generally quite similar, though the bootstrap CIs can be occasionally rather wider and asymmetrical, an indication perhaps that asymptotic conditions have not been reached for the given sample size. This becomes clear if we examine the behaviour of the bootstrap ML estimates of the parameters.

Figure 12.4 shows the $B = 500$ bootstrap scatterplots of parameter pairs for the three parametrizations discussed. It will be seen that in the first case, which uses the simple

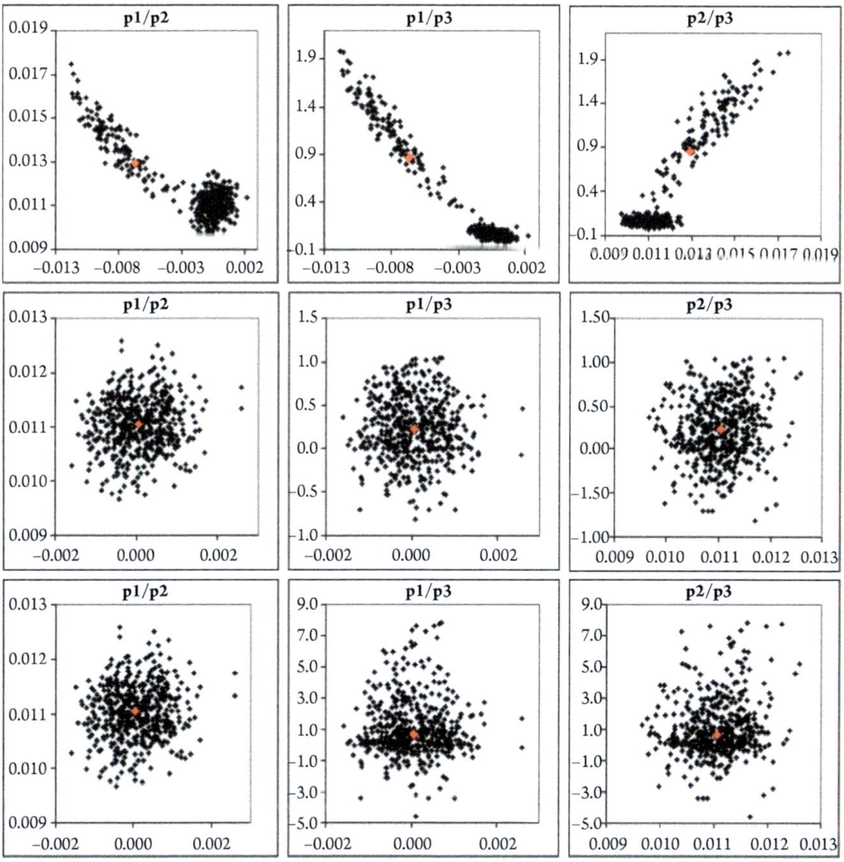

Figure 12.4 Bootstrap parameter scatterplots (B = 500) for the FTSE data set using the λ (top row), θ^{I} (middle row), and θ^{II} (bottom row) parametrizations. The labelling p1, p2, p3 corresponds to Par 1, Par 2, Par 3 in Figure 12.3.

linearization with parameter set λ, the estimators are correlated, with distributions that are distinctly non-normal, this latter reflecting the behaviour seen in the non-concave profile log-likelihood plot. However, the other two parametrizations, where the centred linearization is used, appear more satisfactory, especially where $\eta_1 = 2\gamma_1/(4 - \pi)$ is used as the third parameter. The scatter-plots show the value of bootstrapping in giving a rapid indication of whether asymptotic theory will be satisfactory or not for a given data sample.

Another use of bootstrapping is in providing a GoF test. In all three parametrizations, the value of zero for the third parameter, p_3, corresponds to a normal fit. The confidence intervals for p_3 in Figure 12.3, all include 0, except in the case of the CI calculated from asymptotic theory in the first parametrization, and even in this case the value zero is only

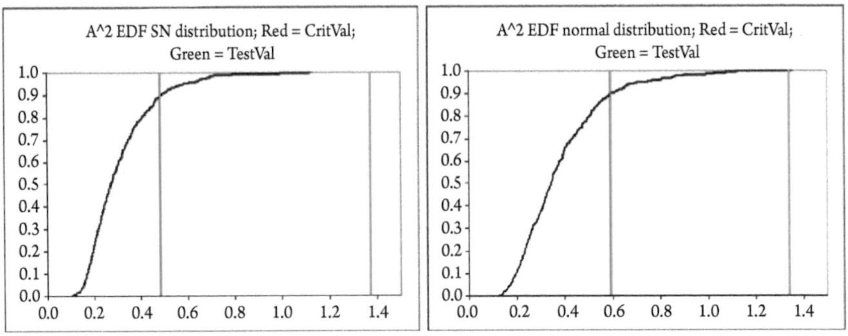

Figure 12.5 Left-hand graph: The EDF of a sample of 500 BS observations of the A^2 goodness-of-fit statistic, generated from the *SN* distribution fitted to the FTSE data. Red line: BS 90% critical value. Green line: A^2 test value. Right-hand graph: Corresponding results for fitted normal model.

just missed. Thus, the normal fit would appear not unsatisfactory. However, comparison of the fitted PDF and the sample frequency histogram in Figure 12.2 indicates a possibly poor fit. The sample appears quite symmetric. The lack-of-fit would seem to be a difference in kurtosis rather than a difference in skewness. The only symmetric distribution that can be represented in the skew-normal family is the normal, so it cannot cope with a symmetric distribution with a different kurtosis. As a formal test of GoF, the Anderson-Darling test statistic of eqn (4.12) can be bootstrapped as described in Chapter 4. This was done in this case. The left-hand graph in Figure 12.5 shows the bootstrap estimate of the distribution of A^2 under the null hypothesis that A^2 has been calculated from a sample drawn from the fitted *SN* distribution. This produced an estimate of the null hypothesis 95% critical value of $A^2_{0.95} = 0.585$. The value for the data sample is $A^2 = 1.38$, showing that the *SN* model has not provided a good fit. This result is very much in line with the findings reported in Section 9.7, where the Pearson type IV, Johnson S_U, and stable law distributions, with their more flexible kurtotic behaviour, were found to provide satisfactory fits.

Note that inclusion of zero in the CI for the parameter γ_1 in the present situation does not provide real evidence for treating the sample as normal. Examination of the other two parameter estimates in the *SN* distribution does not help, as they can be expected to behave as the estimates of a normal distribution with $\hat{\theta}_1 \simeq \hat{\mu}$ and $\hat{\theta}_2 \simeq \hat{\sigma}$, where $\hat{\mu}$ and $\hat{\sigma}$ are the ML estimates of the parameters in the normal distribution $N(\mu, \sigma^2)$. This is the case in our example, where the values are $\hat{\theta}_1 = 0.00005$, $\hat{\mu} = 0.00004$, $\hat{\theta}_2 = 0.01105$, and $\hat{\sigma} = 0.01106$.

We might expect that a straight GoF test of normality rather than skew-normality would give a similar result. This is indeed the case. Parametric bootstrap values of A^2 calculated from samples drawn from the fitted normal distribution yielded $A^2_{0.95} = 0.596$ with test value $A^2 = 1.34$, values very similar to those obtained in the *SN* case. The right-hand graph in Figure 12.5 shows the bootstrap EDF of the A^2 statistic in this case.

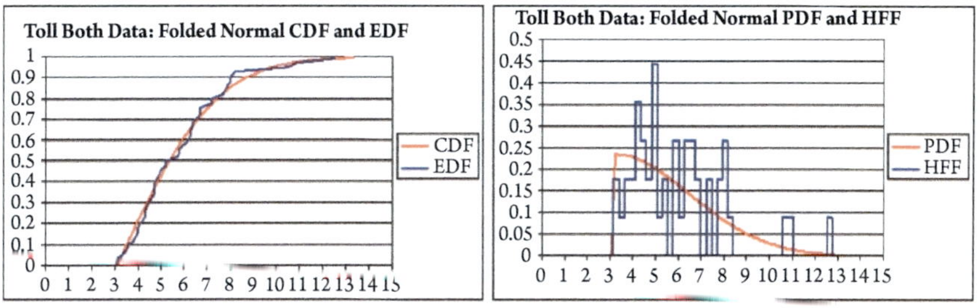

Figure 12.6 EDF and histogram of the frequency function (HFF) of the Toll Booth data set and CDF and PDF of the folded normal distribution fitted by ML estimation.

12.5.2 Toll Booth Service Times

In our second example, we consider the toll booth data set introduced in Section 3.8 and set out in Table 3.1. The data are quite skewed, with a sample skewness of 1.18, which is much more skewed than the FTSE data where the sample skewness was only 0.15. The example is a case where the estimate $\hat{\lambda}$ is not finite.

Nelder-Mead optimization was used to fit the simple linearized *SN* model with the $\boldsymbol{\lambda}$ parametrization. Search iterations had to be halted as the λ values were increasing with no sign of stopping. With $\lambda = 1,200,000$, the other parameter values were $\hat{\lambda}_1 = 3.10$ and $\hat{\lambda}_2 = 3.38$. The fitted model is negligibly different from the special case of a folded normal obtained as $\lambda \to \infty$. The CDF and PDF of this fitted *SN* distribution are depicted in Figure 12.6. Also depicted in the figure are the sample EDF and histogram frequency function (HFF).

To confirm that the folded normal is the correct choice of fit, we also calculated the profile log-likelihood, as a function of λ, and this is shown in Figure 12.7. It will be seen that there is not a finite maximum value. The graph has different horizontal asymptotes as $\lambda \to \infty$ and $\lambda \to -\infty$, with the overall maximum being approached as $\lambda \to \infty$. The same horizontal asymptotic behaviour also occurs with the FTSE profile log-likelihood plot. This is not evident in the plots of Figure 12.1, because the range of λ was kept small in that figure to focus on the maximum point. Figure 12.8 depicts the profile log-likelihood for the FTSE data as a function of λ over an extended λ range, still showing the maximum point near $\lambda = 0$, but also showing the same horizontal asymptotic behaviour as $\lambda \to \infty$ or as $\lambda \to -\infty$, evident in the profile log-likelihood plot for the toll booth data.

This asymptotic behaviour occurs with any sample, as the following analysis shows.

Using the simple linearization with $\boldsymbol{\lambda}$ as the parameter vector, the log-likelilhood is

$$L(\boldsymbol{\lambda}|\mathbf{y}) = \sum_{i=1}^{n} \ln[2\lambda_2^{-1}\phi(z_i)\Phi(\lambda z_i)], \text{ where } z_i = \frac{(y_i - \lambda_1)}{\lambda_2}.$$

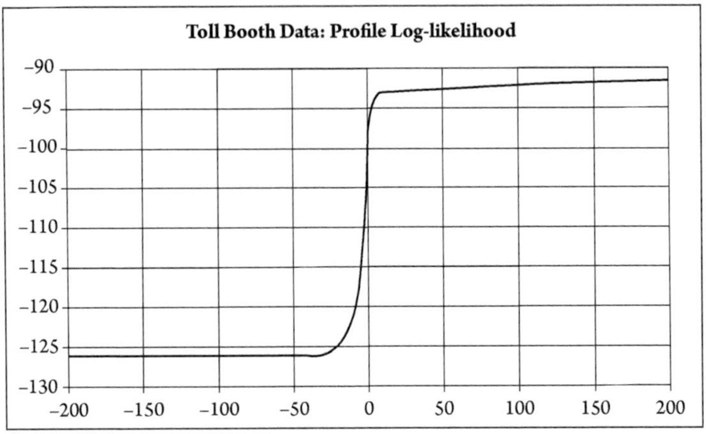

Figure 12.7 Profile log-likelihood of the λ *SN* model fitted to the toll booth data set, with λ taken as the profiling parameter.

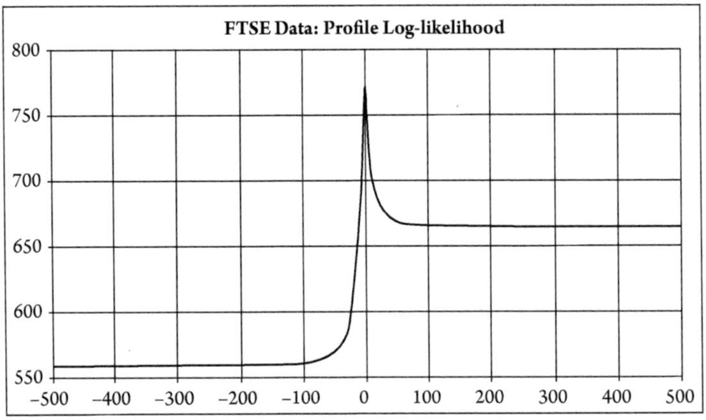

Figure 12.8 Profile log-likelihood for the FTSE data set using the λ parameterization plotted over a wide range of λ values.

Assume that the observations are ordered so that y_1 and y_n are the smallest and largest observations, respectively. Then, if λ_1 and λ_2 are fixed with $\lambda_1 < y_1$ and $\lambda_2 > 0$, the z_i are all strictly positive, so that the $\Phi(\lambda z_i)$ will all be strictly increasing functions of λ, with $\Phi(\lambda z_i) \to 1$ from below as $\lambda \to \infty$. Thus,

$$L(\boldsymbol{\lambda}|\mathbf{y}) \to \sum_{i=1}^{n} \ln[2\lambda_2^{-1}\phi(z_i)]$$

as $\lambda \to \infty$, moreover, with the convergence being monotone from below. As $\phi(z)$ is an increasing function when z decreases, the largest value of this limit as λ_1 varies subject to $\lambda_1 \le y_1$ is obtained as $\lambda_1 \to y_1$. Setting $\lambda_1 = y_1$ gives a slightly lower limit as $\lambda \to \infty$, as then $z_1 = 0$, so that $\Phi(z_1)$ remains fixed at $\Phi(z_1) = 0.5$ instead of increasing to unity; so, as a reminder of this, we write this best choice as $\hat{\lambda}_1^- = y_1^-$, where y_1^- is arbitrarily close to but less than y_1.

In the limit, as $\lambda \to \infty$, the best value for λ_2 satisfies

$$\frac{\partial}{\partial \omega} \sum_{i=1}^{n} \ln[2\lambda_2^{-1}\phi(z_i)] = 0,$$

yielding

$$\hat{\lambda}_2^- = \left[n^{-1} \sum_{i=1}^{n} (y_i - y_1^-)^2) \right]^{1/2}.$$

The limiting fit obtained with $\lambda_1 = \hat{\lambda}_1^-$, and $\lambda_2 = \hat{\lambda}_2^-$ whilst letting $\lambda \to \infty$, is the standard folded normal distribution that is skewed to the right.

The value of the maximized log-likelihood in this case is

$$L^- = \frac{n}{2} \ln(\frac{2}{\pi}) - n \ln \hat{\lambda}_2^- - \frac{n}{2}. \qquad (12.18)$$

Note that the monotone behaviour of $L(\lambda|\mathbf{y})$ as λ increases from $-\infty$ to ∞ for any fixed pair of λ_1, λ_2 in the region $R_2 = \{(\lambda_1, \lambda_2)| -\infty < \lambda_1 < y_1, 0 < \lambda_2 < \infty\}$ means that there can be no finite local maximum point in the region $R_3 = \{\boldsymbol{\lambda}| -\infty < \lambda_1 < y_1, 0 < \lambda_2 < \infty, -\infty < \lambda < \infty\}$, as the value of $L(\boldsymbol{\lambda}|\mathbf{y})$ at any finite fixed point of R_3 can always be improved by increasing the value of λ.

Similar asymptotic limiting behaviour is obtained by letting $\lambda \to -\infty$. In this case, we require $\lambda_1 > y_n$. Then

$$L(\boldsymbol{\lambda}|\mathbf{y}) \to \sum_{i=1}^{n} \ln[2\lambda_2^{-1}\phi(z_i)],$$

only now as $\lambda \to -\infty$, where convergence is monotone and from below again. The best choice for λ_1 and λ_2 in this case is

$$\lambda_1^+ = y_n^+, \quad \hat{\lambda}_2^+ = \left[n^{-1} \sum_{i=1}^{n} (y_n^+ - y_i)^2) \right]^{1/2},$$

where y_n^+ is a value arbitrarily close to but larger than y_n. In this case, we get the *mirror* folded normal as the best fit. The value of the maximized log-likelihood in this case is

$$L^+ = \frac{n}{2}\ln\left(\frac{2}{\pi}\right) - \frac{n}{2} - n\ln\hat{\lambda}_2^+. \tag{12.19}$$

The question arises as to which half-normal is the better fit. The standard positively skew version obtained with $\hat{\lambda}_1 = y_1^-$ is the better fit if $L^- > L^+$, which will be the case if the difference

$$\sum_{i=1}^{n}(y_n - y_i)^2) - \sum_{i=1}^{n}(y_i - y_1)^2) = 2n(y_n - y_1)\left(\frac{y_1 + y_n}{2} - \bar{y}\right)$$
$$> 0,$$

i.e. if the mean of the y's, $\bar{y}$, is less than the mid-range $\frac{1}{2}(y_1 + y_n)$.

In the example, we have $\bar{y} = 5.8$ and $\frac{1}{2}(y_1 + y_n) = \frac{1}{2}(3.1 + 12.5) = 7.8$, so that the standard folded normal obtained by setting $\lambda_1^- = 3.1$ and $\lambda_2^- = 3.39$ is the better fit. The values are almost identical to those obtained when carrying out the three-parameter *SN* fit.

The two maxima calculated from (12.18) and (12.19) are

$$L^- = -91.43, \quad L^+ = -125.56,$$

also in agreement with the values in Figure 12.7 obtained by numerical optimization in fitting the full *SN* distribution.

In the situation where it is desired to fit the *SN* distribution to a data sample, but there is the possibility that the data are too skew, we would suggest, in view of the discussion just given, that the *SN* distribution be fitted using numerical optimization to calculate ML estimates for the centred linearization using the $\boldsymbol{\theta}^I$ parametrization, but with limits imposed on the magnitude of the third parameter $\eta_1 (= 2\gamma_1/(4 - \pi))$ to keep λ bounded. The value of λ can be calculated from η_1 using the expression in (12.14). For example,

$$-2.318 < \eta_1 < 2.318$$

is close to the full range of possible values for η_1, but ensures

$$-104 < \lambda < 104.$$

12.6 Finite Mixtures; Multivariate Extensions

We have only discussed the basic skew-normal model. Since the pioneering papers of Azzalini (1985, 2005), there has been much interest in extensions to more general skew-symmetric models, including multivariate versions, and to finite mixtures of skew-normal distributions. We will not attempt any detailed discussion of these developments, but just point out that the non-standard problems that we have discussed in the basic skew-normal case extend to these generalizations, and all have been the subject of detailed study. We therefore simply provide references that the reader interested in such extensions may wish to consult.

Other Skew Distributions

Many other skew distributions, both univariate and multivariate, have been studied. A review is given by Azzalini (2005). See also Azzalini and Genton (2008). Zhu and Galbraith (2009) list some nine different versions. All but two have identical polynomial tail decay rates. An exception is that of Jones and Faddy (2003), which has two parameters separately controlling the two polynomial rates of tail decay; another is that of Aas and Haff (2006), where one tail has a polynomial decay rate and the other has an exponential decay rate. The version given by Zhu and Galbraith (2009) themselves has two parameters, each controlling the rate of decay in one tail, with an additional third parameter controlling the skewness in the central part of the distribution. Though the model does not satisfy regularity conditions for ML estimation, the authors do establish consistency, asymptotic normality, and efficiency giving an explicit expression for the asymptotic variance.

Multivariate forms are described, for example, by Azzalini and Dalla Valle (1996), Azzalini and Capitanio (2003), and Gupta (2003). Branco and Dey (2001, 2002) have considered multivariate skew-elliptical distributions.

Centred Parametrizations

These have been discussed by Arellano-Valle and Azzalini (2008, 2013) in the multivariate skew-normal and skew-t cases, respectively.

Perturbation Representation

Azzalini (2005) points out that the formula (12.4) for the PDF of a skew distributed random variable and its stochastic representation (12.5) can be viewed as that of a random variable from a 'basis' distribution with PDF f_0, that is then subject to a perturbation stemming from a random variable drawn from a different distribution. We might consider (12.5) as a particular instance of a *mixing distribution*. We define another form of such mixing distributions in Chapter 13, which we call *randomized-parameter distributions*, where the random perturbation is of a parameter of f_0, the PDF of the basis distribution. Using such models is attractive in situations where greater flexibility is needed in representing tail behaviour. In the case of skewed distributions with PDF (12.4), there is also a benefit from the relative computational ease of estimating their parameters using the EM algorithm. This is discussed by da Silva Ferreira et al. (2011). Variates from such mixing distributions are often easy to generate, using, for example, the stochastic representation

(12.5). Henze (1986) gives some alternative methods of generating skew-normals, one of which we have already given in eqn (12.6).

Finite Mixtures

Finite multivariate mixtures of skew-normals have been investigated by Cabral, Lachos, and Prates (2012), who use an EM-type algorithm for parameter estimation. A Bayesian approach to fitting finite mixtures of multivariate skew-normal distributions is given by Cabral, Lachos, and Madruga (2012). We shall discuss finite mixture models in Chapters 17 and 18. Though we do not discuss explicit skew-normal finite mixtures in those chapters, indeterminacy is a prominent non-standard problem in finite mixtures, so it may be useful to consult these chapters before looking at the references just given.

13 Randomized-Parameter Models

In this chapter, we examine a means of introducing more flexibility into the tails of a given distribution, which we will call the *base* distribution, by regarding one of its parameters as being random rather than fixed. In use, the randomized parameter does not have to be treated in a special way, because the effect of the randomization can be handled by a numerical integration that turns the model into a standard parametric model containing just one more parameter than the original base model.

We shall call the final model a *randomized-parameter* model, with the name emphasizing how it is obtained from the original base model. We will consider only the specific case where the base model has two parameters and the randomized-parameter model has three.

It is emphasized that despite the name, the randomized-parameter model is just an ordinary parametric model with all its parameters treated in the standard way, so that they are deterministic but with unknown values which have to be estimated. The effect of randomizing a parameter in the base model is converted simply into the additional flexibility obtained in having the extra parameter.

A randomized-parameter model is a continuous mixture model that tends to have broader tails than the original base distribution. Although it will usually provide a more accurate fit than the base model, the latter may, on occasion, be the best fit to a sample. We consider likelihood ratio (LR) and GoF tests for deciding whether the base or randomized-parameter model is the better fit. We give a numerical example where we apply the A^2 Anderson-Darling GoF test. As an additional aspect to the chapter, the data are grouped in the example. EDF GoF tests like that using A^2 cannot then normally be used even when critical values are tabulated, as they assume continuously variable observations. Grouping changes the critical value, usually quite dramatically. We show that use of bootstrapping overcomes this difficulty.

Non-Standard Parametric Statistical Inference. Russell Cheng.

13.1 Increasing Distribution Flexibility

To set the scene, we start with a brief summary of different ways of extending parametric densities by the inclusion of further parameters.

Two-parameter skewed distributions are often used in life-testing and in other reliability studies (Bain, 1978; Cohen and Whitten, 1988). Such two-parameter distributions include the gamma, inverse-Gaussian, lognormal, and Weibull models. Although all of these can provide a reasonable fit for small samples, as the sample size increases, clear differences appear between models, and the fit is often no longer adequate. Extra parameters may then be needed to achieve a better fit. Various methods exist to provide more flexibility in a model; we consider three such methods.

13.1.1 Threshold and Location-Scale Models

One method is where a shifted threshold parameter is incorporated into a two-parameter model with CDF $F(y; \boldsymbol{\theta})$ to obtain a three-parameter distribution of the form $F(y - \alpha; \boldsymbol{\theta})$. The method results in an overall translation of the position of a distribution. This has been discussed in detail in Chapters 6 and 8.

The shifted threshold augmentation can be taken a step further when the distribution $F(y; \boldsymbol{\theta})$ does not include explicit scale or location parameters, so that the components of $\boldsymbol{\theta}$ are essentially just shape parameters. We can apply a location-scale transform so that the distribution is $F[(y - \alpha)/\beta; \boldsymbol{\theta}]$. The stable law distribution is of this form, as also are the Pearson and Johnson systems. These two systems include where $\boldsymbol{\theta}$ has just one component as well as where $\boldsymbol{\theta}$ is two-dimensional. The added flexibility in tail behaviour provided by the stable law, Pearson Type IV, and Johnson S_U distributions compared with the normal distribution for fitting to data with long tails has already been illustrated in Section 9.7 with a numerical example involving financial data. Other examples that we have discussed, though not particularly with regard to tail behaviour, include the Burr XII distribution and many of the distributions considered in Chapter 6.

13.1.2 Power Transforms

Another method is to introduce a power transform to the two-parameter model $F(y; \boldsymbol{\theta})$. This will affect the overall shape of the distribution and result in a probability distribution of the form $F(y^{\alpha}; \boldsymbol{\theta})$. Stacy (1962) obtained a generalization of the gamma distribution by applying a Weibull shape parameter as an exponent in the exponential factor of the gamma distribution. Cohen (1969) considered the application with the distributions interchanged to achieve the same result. By having two shape parameters, the generalized gamma distribution has more flexibility than the Weibull or gamma distribution alone. Jørgensen (1982) states that a power generalization of a distribution F is the class of distributions of $Y^{1/\lambda}$ for $\lambda \neq 0$ where Y follows the distribution F, and employs this approach to obtain generalizations of the inverse-Gaussian distribution. Another example is the method discussed in Chapter 10 involving a family of power transforms introduced by Box and Cox (1964).

Most recently, the skew-normal distribution discussed in Chapter 12 and its generalizations and extensions to different types of skew Student's t-distribution has been the subject of extensive study. The effect on tail behaviour is one aspect that is now receiving attention. The paper of Zhu and Galbraith (2009) targets control of tail behaviour specifically.

13.1.3 Randomized Parameters

The approaches just mentioned introduce additional flexibility into a model either by altering the location and range of the original distribution or by changing its shape. However, in reliability studies, the tail behaviour is often of particular concern, and a distribution may give a reasonable overall fit but be unsatisfactory in the tails. In some cases, it may be more important to understand the tail behaviour of a distribution than to fit the entire distribution, in order to predict extreme values. One way of obtaining more flexibility for tail-fitting is the method suggested by Hougaard (1986), where a scaling factor is incorporated into a distribution that specifically damps the upper tail. We adopt a different approach, which affects the tails indirectly by randomizing one of the parameters of the distribution. The distribution is of the form $F(y; \theta_i Z, \boldsymbol{\theta}_{\setminus i})$, where the ith component θ_i is replaced by $\theta_i Z$, where Z is a random variable with a continuous distribution of its own. The other parameters are unchanged, and have been denoted by $\boldsymbol{\theta}_{\setminus i}$, the $\boldsymbol{\theta}$ vector with the ith component omitted.

We shall call the distribution of Z the *mixing distribution* and denote its PDF by $h(z; a)$, where a is an additional parameter different from $\boldsymbol{\theta}$.

It is important to note that in a Y sample, $y_1, y_2, \ldots, y_n$, each y_j is sampled from $F(y; \theta_i z_j, \boldsymbol{\theta}_{\setminus i})$ with the $z_1, z_2, \ldots, z_n$ a random sample from the mixing distribution, so that a different z_j is used in generating each y_j. All but one of the parameters of the original distribution retain their original interpretation, with the remaining parameter being randomly 'perturbed' in a way that depends on the mixing parameter a. This random element will typically spread the observations more than in a sample drawn from the original distribution, giving fatter tails, yet retaining the overall shape. Thus the added randomization has a straightforward interpretation, which is appealing in applications such as life-span models; for example, the probability of surviving up to a certain time might be a function not only of age but also of other factors which may vary between individuals.

A key point is that the model does not have to be analysed in the form $F(y; \theta_i Z, \boldsymbol{\theta}_{\setminus i})$, where Z is regarded as random. The precise effect of the randomization can be obtained by the concatenating integral operation, which gives the distribution of Y explicitly as

$$G(y, \alpha, \boldsymbol{\theta}) = \int F(y; \theta_i z, \boldsymbol{\theta}_{\setminus i}) h(z; a) dz.$$

We shall call $G(y; a, \boldsymbol{\theta})$ the *randomized-parameter* model. The main focus of this chapter is how to generate and use such randomized-parameter models.

Many examples of randomized-parameter models can be found in the literature. Johnson, Kotz, and Balakrishnan (1994, 1995), who term such models *compound* or simply *mixture* models, review a number involving various distributions, including beta, chi-squared, gamma, Laplace, normal, Poisson, Student's t, and Weibull.

To be specific, we will from now on focus on the case where the original distribution has just two parameters.

An elementary example is given by Stuart and Ord (1987), where a normal distribution with a gamma mixing distribution results in a variable which, when standardized, is t-distributed. Harris and Singpurwalla (1968) derive an unconditional life-time distribution by treating a parameter of a failure distribution as being random with a known distribution. These are obtained by assuming a random hazard function; for instance, the scale parameter of an exponential time-to-failure distribution is treated as having a gamma distribution. Another compound distribution considered by Harris and Singpurwalla is derived by treating the scale parameter of a Weibull distribution as being gamma-distributed, obtaining the Burr Type XII distribution. This well-known model has already been discussed in Section 6.1.2. Johnson, Kotz, and Balakrishnan (1994, Chapter 21) give a summary of how the distribution is obtained by treating a powered-up scale parameter of a Weibull distribution as having the gamma density. Dubey (1968) also obtains the Burr distribution by using a two-parameter Weibull distribution with a mixing parameter which is gamma-distributed, and refers to it as a compound Weibull or Weibull-gamma model. Elandt-Johnson (1976) makes use of Dubey's results to illustrate the way that a compound distribution can be represented in a form involving the moment generating function (mgf) of the mixing distribution, where the Burr distribution is obtained from the mgf of the gamma distribution. A summary of the relationship between different types of Burr distribution and other distributions is given by Tadikamalla (1980). The importance of the Burr distribution in failure models, and to represent biological and clinical data, is well recognized (Wingo, 1983; Nigm, 1988).

An application where heterogeneity between individuals in population-based mortality studies is modelled by an inverse-Gaussian distributed frailty quantity has been considered by Hougaard (1984). Here the gamma distribution is used with the inverse-Gaussian model as the mixing distribution. Use of the inverse-Gaussian distribution for mixing often gives workable generalizations, but, for most of the models we shall be considering, gives a generalization involving a modified Bessel function of the third kind.

Distributions used in biology and reliability tend to involve exponential factors, and using the gamma density for mixing is often tractable. Indeed, we obtain three-parameter generalizations of the four aforementioned two-parameter distributions by using the gamma model as the mixing distribution.

This chapter is mainly a collation exercise in which we collect together and examine particular examples scattered in the literature where a three-parameter generalization is obtained by randomizing one of the parameters of a two-parameter distribution. We shall focus on giving details of the resulting tail behaviour.

13.1.4 Hyperpriors

In Bayesian analysis of a parametric distribution, the parameters have prior distributions which themselves also have parameters. Thus, for example, in a Bayesian approach to fitting a normal distribution $N(\mu, \sigma^2)$, μ will have a prior, so that we would regard μ as being random in the Bayesian sense with $\mu \sim N(\theta, \tau^2)$ but with θ, τ^2 specified values. However, if this was thought not sufficient, added flexibility can be obtained by assuming, say, that θ is also random with its own *hyperprior*, so that $\theta \sim H(\alpha, \beta)$, where H is some distribution with its own *hyperparameters*, α, β in this example. In a typical Markov chain Monte-Carlo (MCMC) analysis, the objective is to estimate the posterior distribution of parameters. For example, in our normal example, we may want the posterior distribution of μ. This requires sampling μ values from its prior distribution, the $N(\theta, \tau^2)$ in our example. However, if θ has a hyperprior distribution, this would need to be sampled first, with the sampled value θ', say, then used to form the prior $N(\theta', \tau^2)$ that we sample to obtain our sampled μ value. It is clear that the introduction of a hyperprior is exactly the same process as the generalization of a base prior distribution into a randomized-parameter prior distribution.

In Section 17.3.1 and Section 18.2.3, we shall discuss an example that occurs in the Bayesian analysis of finite mixture distributions.

13.2 Randomized Parameter Procedure

For convenience, we summarize our definition of a randomized-parameter model. Let Y be a continuous random variable from a two-parameter *base* model with pdf $g(y; \lambda, \mu)$, depending on the parameters λ and μ. Then replace λ by λZ, where Z is a continuous random variable with mixing density $h(z; \alpha)$, depending on a parameter α. Then $E_Z[g(y; \lambda Z, \mu)]$ gives the PDF

$$f(y; \lambda, \mu, \alpha) = \int_z g(y; \lambda z, \mu)\, h(z; \alpha)\, dz \tag{13.1}$$

of the *randomized-parameter distribution* of Y.

We use the gamma, standardized to have mean unity, as the mixing density. Thus Z has the gamma $(1/\alpha, \alpha)$ distribution with PDF

$$h(z; \alpha) = \frac{\alpha^{-1/\alpha}}{\Gamma(1/\alpha)} z^{1/\alpha - 1} \exp\left(-\frac{z}{\alpha}\right). \tag{13.2}$$

13.3 Examples of Three-Parameter Generalizations

13.3.1 Normal Base Distribution

Consider the normal distribution $N(\mu, \sigma^2)$ with mean μ and SD σ. We obtain Student's t-distribution if we use the above gamma mixing model, (13.2), and apply it to a normal random variable Y parametrized as $N(\mu, \lambda^{-1})$, so that $\lambda = \sigma^{-2}$. Here the randomized base model, conditional on $Z = z$, is

$$g(y; \lambda z, \mu) = \left(\frac{\lambda z}{2\pi}\right)^{1/2} \exp\left(-\frac{\lambda z(y-\mu)^2}{2}\right),$$

and using this in (13.1), the integral is easily evaluated to give the unconditional randomized-parameter density

$$f(y; \lambda, \mu, \alpha) = \left(\frac{\lambda\alpha}{2\pi}\right)^{1/2} \frac{\Gamma(\alpha^{-1}+\frac{1}{2})}{\Gamma(\alpha^{-1})} \left(1 + \frac{\lambda\alpha(y-\mu)^2}{2}\right)^{-(\alpha^{-1}+1/2)}.$$

If we standardize by letting $t = (y-\mu)\sqrt{\lambda}$, then

$$f(t; \alpha) = \left(\frac{\alpha}{2\pi}\right)^{1/2} \frac{\Gamma(\alpha^{-1}+\frac{1}{2})}{\Gamma(\alpha^{-1})} \left(1 + \frac{\alpha t^2}{2}\right)^{-(\alpha^{-1}+1/2)}$$

is the t-distribution with $2/\alpha$ degrees of freedom.

Note that in this case it is the inverse variance that has been randomized.

We examine the effect of the randomization on the tail behaviour. The tails of the normal distribution are $O(\exp(-\lambda y^2/2 + \lambda\mu y))$, whereas the tails of Student's t-distribution are $O(|\, y\,|^{-1-2/\alpha})$. Thus the three-parameter randomized-parameter model has fatter tails than the two-parameter base model, as the normal tails tend to zero more rapidly.

13.3.2 Lognormal Base Distribution

Likewise, consider the lognormal $LN(\mu, \sigma^2)$ basic model

$$f_{LN}(y) = \frac{1}{\sqrt{2\pi}\sigma y} \exp\left\{-\frac{1}{2\sigma^2}[\ln y - \mu]^2\right\},\ a < y,$$

which is the PDF of eqn (9.14) with $a = 0$ and $\delta = \sigma^{-1}$. Then if we suppose $Y \sim LN(\mu, \lambda^{-1})$ so that $\lambda = \sigma^{-2}$, as in the normal case, and randomize λ using the gamma density of eqn (13.2), this gives a similar result. We obtain the three-parameter randomized-parameter model with PDF

$$f(y; \lambda, \mu, \alpha) = \left(\frac{1}{y}\right) \left(\frac{\lambda\alpha}{2\pi}\right)^{1/2} \frac{\Gamma(\alpha^{-1}+\frac{1}{2})}{\Gamma(\alpha^{-1})} \left(1 + \frac{\lambda\alpha(\ln y-\mu)^2}{2}\right)^{-(\alpha^{-1}+1/2)}. \quad (13.3)$$

This is referred to as the 'log-t' distribution by Johnson, Kotz, and Balakrishnan (1995, eqn 28.73).

The tails of the two-parameter lognormal are $O(y^{\lambda\mu-1}y^{-\lambda \ln y/2})$, whilst the tails of the log-t are $O(y^{-1}\,|\ln y\,|^{-1-2/\alpha})$. Note that the term y^{-1} dominates as $y \to 0$, and so $f(.)$ is infinite near the origin. However, the fitted density is unlikely to assign much probability to y being near to zero unless the sample is very skewed. The upper tail of the log-t distribution is much broader than the basic lognormal model.

13.3.3 Weibull Base Distribution

The method is applicable to other basic models with long tails. Let $Y \sim Weibull(\mu, \sigma)$, with PDF as given in Table 6.2, where we have set $a = 0$ and replaced b and c by σ and μ, respectively, to conform with the notation used in this chapter. Then $Y \sim Weibull(\mu, \lambda^{-1/\mu})$ has PDF

$$f_{Weibull}(x, \mu, \lambda^{-1/\mu}) = \mu\lambda y^{\mu-1} \exp(-\lambda y^{\mu}).$$

Replacing λ by λZ, where Z has the gamma mixing distribution of eqn (13.2), and using eqn (13.1), we obtain the three-parameter Burr XII distribution

$$f(y; \lambda, \mu, \alpha) = \lambda\mu y^{\mu-1}(1 + \lambda\alpha y^{\mu})^{-(1+\alpha^{-1})},$$

which we considered in Sections 6.1.2 and 6.1.3 under a different parametrization and with an added fourth threshold parameter. To avoid confusion, it is the three-parameter Burr XII form that appears in the remainder of this chapter.

The effect on the tails of replacing λ by λZ is to broaden the upper tail from $O(y^{\mu-1} \exp(-\lambda y^{\mu}))$ of the Weibull to $O(y^{-1-\mu/\alpha})$ of the Burr XII. The lower tail remains unchanged as $O(y^{\mu-1})$, where the distribution is J-shaped if $\mu < 1$.

13.3.4 Inverse-Gaussian Base Distribution

A less well-known model is obtained using the inverse-Gaussian (μ, λ) with distribution

$$g(y; \lambda, \mu) = \left(\frac{\lambda}{2\pi y^3}\right)^{1/2} \exp\left(-\frac{\lambda(y-\mu)^2}{2\mu^2 y}\right) \tag{13.4}$$

as the base model. Applying the gamma mixing to the λ parameter, we get the three-parameter generalization with PDF

$$f(y; \lambda, \mu, \alpha) = \left(\frac{\lambda}{2\pi y^3}\right)^{1/2} \alpha^{-1/\alpha} \frac{\Gamma(\alpha^{-1}+1/2)}{\Gamma(\alpha^{-1})} \left(\frac{1}{\alpha} + \frac{\lambda(y-\mu)^2}{2\mu^2 y}\right)^{-(\alpha^{-1}+1/2)}, \tag{13.5}$$

which we call the 'inverse-Gaussian-t' distribution. The upper tail is broadened from $O(y^{-3/2} \exp(-\lambda y/2\mu^2))$ to $O(y^{-\alpha^{-1}-2})$ and the lower tail is broadened from $O(y^{-3/2} \exp(-\lambda/2y))$ to $O(y^{\alpha^{-1}-1})$.

In all the examples, if $\alpha \to 0$ then the mixing density degenerates into a discrete atom at unity, so that the randomized-parameter density reduces to the original two-dimensional base model. Under the restriction $\alpha \geq 0$, the two-parameter base model is not degenerate but is however a boundary model.

As a check, we consider the inverse-Gaussian case.

If we let α become small, corresponding to degeneracy of the gamma density, we can apply Stirling's formula for large α^{-1} to obtain

$$\Gamma(\alpha^{-1}) \simeq \sqrt{2\pi}\alpha^{1/2-1/\alpha}\exp(-\alpha^{-1})$$

and

$$\Gamma\left(\frac{1}{\alpha}+\frac{1}{2}\right) \simeq \sqrt{2\pi}\left(\frac{1}{\alpha}+\frac{1}{2}\right)^{1/\alpha}\exp\left(-\left(\frac{1}{\alpha}+\frac{1}{2}\right)\right).$$

Substituting these expressions into (13.5), we have

$$f(y;\lambda,\mu,\alpha) = \left(\frac{\lambda}{2\pi y^3}\right)^{1/2}\alpha^{-1/\alpha}\frac{(\alpha^{-1}+1/2)^{1/\alpha}}{\alpha^{-1/\alpha+1/2}}e^{-1/2}\left(\frac{1}{\alpha}+\frac{\lambda(y-\mu)^2}{2\mu^2 y}\right)^{-(\alpha^{-1}+1/2)}.$$

Then, as $\alpha \to 0$, this density becomes

$$f(y;\lambda,\mu,0) = \left(\frac{\lambda}{2\pi y^3}\right)^{1/2}\exp\left(-\frac{\lambda(y-\mu)^2}{2\mu^2 y}\right),$$

which is precisely the inverse-Gaussian model of (13.4).

The two-parameter base models of the other three-parameter generalizations can be found in a similar way.

13.4 Embedded Models

Consider now the case where the gamma (α^{-1},α) distribution is used as the mixing density with a base model which is also gamma, but with the parametrization $(\mu,(\lambda\mu)^{-1})$, that is, with PDF

$$g(y;\lambda,\mu) = \frac{(\lambda\mu)^{\mu}}{\Gamma(\mu)}y^{\mu-1}\exp(-\lambda\mu y),$$

so that μ and $(\lambda\mu)^{-1}$ are the respective power and scale parameters in the conventional parametrization. Here, the base model has upper tail that is $O(y^{\mu-1}\exp(-\lambda\mu y))$ and the lower tail $O(y^{\mu-1})$. The resulting three-parameter randomized-parameter model is the Pearson Type VI $(\mu,\alpha^{-1},(\lambda\mu\alpha)^{-1})$ distribution with PDF

$$f(y;\lambda,\mu,\alpha) = \frac{(\lambda\mu\alpha)^{\mu}y^{\mu-1}}{B(\mu,\alpha^{-1})(1+\lambda\mu\alpha y)^{\mu+1/\alpha}}, \tag{13.6}$$

where the beta function is defined as

$$B(\mu,\alpha^{-1}) = \frac{\Gamma(\mu)\Gamma(\alpha^{-1})}{\Gamma(\mu+\alpha^{-1})}.$$

The upper tail of the basic model is now broadened to $O(y^{-1-1/\alpha})$ whilst the distribution is still J-shaped if $\mu < 1$. If $\alpha \to 0$, then the gamma mixing density becomes degenerate, and we recover the embedded two-parameter gamma $(\mu, (\lambda\mu)^{-1})$ distribution.

Note, however, that this two-parameter base distribution is parametrized in such a way that the shape and scale are linked to allow for another limiting form. The mean of the distribution is λ^{-1}, and the variance $(\lambda^2\mu)^{-1}$, hence, if $\mu \to \infty$, the two-parameter base model becomes a discrete atom located at $y = \lambda^1$. In this case, the three-parameter Pearson Type VI distribution of eqn (13.6) reduces to the non-degenerate two-parameter distribution with PDF

$$g(y; \lambda, \alpha) = \frac{(\lambda\alpha)^{-1/\alpha}}{\Gamma(\alpha^{-1})} \quad y^{-(1+1/\alpha)} \exp(-(\lambda\alpha y)^{-1}),$$

which is the inverted-gamma or Pearson Type V $(\alpha^{-1}, (\lambda\alpha)^{-1})$ distribution, with α^{-1} the power parameter and $(\lambda\alpha)^{-1}$ the scale parameter. The upper tail of this distribution is $O(y^{-(1+\alpha^{-1})})$, whilst the behaviour of the lower tail is $O(y^{-(1+\alpha^{-1})} \exp(-(\lambda\alpha y)^{-1}))$, which is thinner than the lower tail of the three-parameter randomized-parameter model.

Thus, the three-parameter Pearson Type VI distribution contains both the gamma and the inverted-gamma distributions as embedded two-parameter limits; something well known and which we have already encountered in examining the Pearson family in its own right in Chapter 9.

Our parametrization is in the form where the generalization method can be directly applied to give the gamma limit.

When Y is the inverted-gamma distribution, the conditional density of Y given $Z = z$ is

$$g(y; \frac{z}{\lambda}, \alpha) = \left(\frac{z}{\lambda\alpha}\right)^{1/\alpha} \frac{y^{-(1+1/\alpha)}}{\Gamma(\alpha^{-1})} \exp\left(-\frac{z}{\lambda\alpha y}\right).$$

Thus, this model is obtained when μ^{-1} and λ^{-1} play the roles of α and λ in our original parametrization.

A summary of the three-parameter generalizations of the two-parameter distributions discussed is given next. It can be seen that the inverse-Gaussian t-distribution has the thinnest upper tail, whilst the log-t has the broadest.

Summary of three-parameter densities $f(y; \lambda, \mu, \alpha)$ obtained by applying mixing to z in the two-parameter basic densities $g(y; \lambda z, \mu)$

$N(\mu, \lambda^{-1}) \to$ Student's t $(\upsilon = 2\alpha^{-1})$:

$$g(y;\lambda z, \mu) = \left(\frac{\lambda z}{2\pi}\right)^{1/2} \exp\left(-\lambda z\frac{(y-\mu)^2}{2}\right)$$

$$f(y;\lambda,\mu,\alpha) = \left(\frac{\lambda\alpha}{2\pi}\right)^{1/2} \frac{\Gamma(\alpha^{-1}+1/2)}{\Gamma(\alpha^{-1})} \left(1+\lambda\alpha\frac{(y-\mu)^2}{2}\right)^{-\alpha^{-1}-1/2}$$

$LN\ (\mu, \lambda^{-1}) \rightarrow$ log-t (λ, μ, α):

$$g(y;\lambda z,\mu) = \left(\frac{1}{y}\right)\left(\frac{\lambda z}{2\pi}\right)^{1/2} \exp\left(-\lambda z\frac{(\ln y-\mu)^2}{2}\right)$$

$$f(y;\lambda,\mu,\alpha) = \left(\frac{1}{y}\right)\left(\frac{\lambda\alpha}{2\pi}\right)^{1/2} \frac{\Gamma(\alpha^{-1}+1/2)}{\Gamma(\alpha^{-1})} \left(1+\lambda\alpha\frac{(\ln y-\mu)^2}{2}\right)^{-\alpha^{-1}-1/2}$$

Weibull $(\mu, \lambda^{-1/\mu}) \rightarrow$ BurrXII (λ, μ, α):

$$g(y;\lambda z,\mu) = \mu y^{\mu-1}\lambda z \exp\left(-\lambda z y^{\mu}\right)$$

$$f(y;\lambda,\mu,\alpha) = \lambda\mu y^{\mu-1}(1+\lambda\alpha y^{\mu})^{-1-\alpha^{-1}}$$

inverse-Gaussian $(\mu, \lambda) \rightarrow$ inverse-Gaussian-t (λ, μ, α):

$$g(y;\lambda z,\mu) = \left(\frac{\lambda z}{2\pi y^3}\right)^{1/2} \exp\left(-\frac{\lambda z(y-\mu)^2}{2\mu^2 y}\right)$$

$$f(y;\lambda,\mu,\alpha) = \left(\frac{\lambda}{2\pi y^3}\right)^{1/2} \alpha^{-1/\alpha}\frac{\Gamma(\alpha^{-1}+1/2)}{\Gamma(\alpha^{-1})} \left(\frac{1}{\alpha}+\frac{\lambda(y-\mu)^2}{2\mu^2 y}\right)^{-\alpha^{-1}-1/2}$$

gamma $(\mu, (\lambda\mu)^{-1})$, inverted-gamma $(\alpha^{-1}, (\lambda\alpha)^{-1}) \rightarrow$
Pearson Type VI $(\mu, \alpha^{-1}, (\lambda\mu\alpha)^{-1})$:

$$g(y;\lambda z,\mu) = \left(\frac{(\lambda\mu z)^{\mu}}{\Gamma(\mu)} y^{\mu-1}\exp(-\lambda\mu z y)\right),$$

$$g(y;\frac{z}{\lambda},\alpha) = \left(\frac{z}{\lambda\alpha}\right)^{\alpha^{-1}} \frac{y^{-1-\alpha^{-1}}}{\Gamma(\alpha^{-1})} \exp\left(-\frac{z}{\lambda\alpha y}\right)$$

$$f(y;\lambda,\mu,\alpha) = \left(\frac{(\lambda\mu\alpha)^{\mu} y^{\mu-1}\Gamma(\mu+\alpha^{-1})}{\Gamma(\mu)\Gamma(\alpha^{-1})(1+\lambda\mu\alpha y)^{\mu+\alpha^{-1}}}\right)$$

The parametrization used in the previous formulas retains the λ, μ, α parameters used in formulating the randomized-parameter models. When fitting a distribution, whether one of the original two-parameter base models or the corresponding three-parameter randomized-parameter model, there is no need to adhere to these parametrizations. As most of the models are well known, it may be convenient in numerical calculations to use an existing package that is to hand, but where the parametrization may be different, so that a conversion is needed if estimates are required of the λ, μ, α parameter values.

13.5 Score Statistic Test for the Base Model

When estimating the parameters of the randomized-parameter models by ML, the possibility of the third parameter being precisely zero has to be allowed for. This corresponds to the original model being the best fit to the data. A simple method is to use the score statistic.

Apart from the inverted-gamma case, all of the three-parameter randomized-parameter models discussed in this chapter have a log-likelihood that can be expanded as a Taylor series about $\alpha = 0$, that is,

$$L(\lambda, \mu, \alpha) = L_0(\lambda, \mu) + L_1(\lambda, \mu)\alpha + R,$$

where $L_0(\lambda, \mu)$ is the log-likelihood of the basic model $g(y; \lambda, \mu)$, $L_1(\lambda, \mu)$ is the score statistic $\partial L/\partial\alpha\ |_{\alpha=0}$, and R is a remainder term which is $O_p(\alpha^2)$ as $\alpha \to 0$. From this, we see that as $\alpha \to 0$, the log-likelihood tends to that of the base model, which is a regular special case. As we have previously noted, the restriction that $\alpha \geq 0$ is not an issue if we simply make the test one-sided. We can use the same method as used previously to test if this regular special case is the best fit. Let $(\hat{\lambda}(\alpha), \hat{\mu}(\alpha))$ and $(\hat{\lambda}, \hat{\mu})$ be the values of (λ, μ) maximizing, respectively, $L(\lambda, \mu, \alpha)$ at a given α, and $L_0(\lambda, \mu)$. Cheng and Iles (1990) show that as $\alpha \to 0$,

$$(\hat{\lambda}(\alpha), \hat{\mu}(\alpha)) \to (\hat{\lambda}, \hat{\mu}), \tag{13.7}$$

and that the profile log-likelihood

$$L^*(\alpha) = \max_{\lambda,\mu} L(\lambda, \mu, \alpha)$$

can be written as

$$L^*(\alpha) = L_0(\hat{\lambda}, \hat{\mu}) + L_1(\hat{\lambda}, \hat{\mu})\alpha + O_p(\alpha^2). \tag{13.8}$$

As seen in Chapter 6, the sign of $L_1(\hat{\lambda}, \hat{\mu})$ indicates how $L^*(\alpha)$ is approached as $\alpha \to 0$. If $L_1(\hat{\lambda}, \hat{\mu}) < 0$, then $L^*(\alpha)$ increases as $\alpha \to 0$ and there is, at least, a local maximum at $(\lambda, \mu, \alpha) = (\hat{\lambda}, \hat{\mu}, 0)$. In this case, the basic two-parameter model is the best fit. Conversely, if $L_1(\hat{\lambda}, \hat{\mu}) > 0$, then the full three-parameter randomized-parameter model is the best fit. Hence once again, the sign of $L_1(\hat{\lambda}, \hat{\mu})$ can be used as an informal test to determine which of the two models to fit.

Table 13.1 L_1 statistics for testing the adequacy of the two-parameter base model

Base Distribution	$\hat{L}_1$
$N(\mu,\lambda^{-1})$	$\frac{n}{8}\left(\hat{\lambda}^2\sum\frac{(y_i-\hat{\mu})^4}{n}-3\right)$
$LN(\mu,\lambda^{-1})$	$\frac{n}{8}\left(\hat{\lambda}^2\sum\frac{(\ln y_i-\hat{\mu})^4}{n}-3\right)$
$Weib(\mu,\lambda^{-1/\mu})$	$\frac{n}{2}\left(\sum\frac{(y_i^{\hat{\mu}}\hat{\lambda}-1)^2}{n}-1\right)$
$InvGaussian(\mu,\lambda)$	$\frac{n}{8}\left(\frac{1}{n}\sum\left(\hat{\lambda}\frac{(y_i-\hat{\mu})^2}{y_i\hat{\mu}^2}-1\right)^2-2\right)$
$Gamma(\mu,(\lambda\mu)^{-1})$	$\frac{n}{2}\left(\frac{1}{n}\sum(\hat{\lambda}\hat{\mu}y_i-\hat{\mu})^2-\hat{\mu}\right)$
$PTV(\alpha^{-1},(\lambda\alpha)^{-1})$	$\frac{n}{2}\left(\frac{1}{n}\sum\left(\frac{1}{\hat{\lambda}\hat{\alpha}y_i}-\frac{1}{\hat{\alpha}}\right)^2-\frac{1}{\hat{\alpha}}\right)$

13.5.1 Interpretation of the Test Statistic

Table 13.1 lists the $\hat{L}_1$ statistic for testing the adequacy of the embedded model. Note that these are not the $T_s^{\frac{1}{2}}$ score test statistics which require standardization with the variance multiplier if asymptotic theory is invoked to obtain null hypothesis critical values. However, if bootstrapping were used to construct critical test values, this can be carried out using $\hat{L}_1$ directly as the test statistic.

For all of the distributions in the table, the $\hat{L}_1$ statistic has a simple interpretation in terms of the dispersion of the model. For example, if, for the normal distribution, we replace λ and μ by their MLEs in

$$L_1(\lambda,\mu) = -\frac{n}{8} - \frac{\lambda}{4}\sum(y_i-\mu)^2 + \frac{\lambda^2}{8}\sum(y_i-\mu)^4,$$

we obtain the statistic

$$L_1(\hat{\lambda},\hat{\mu}) = \frac{n}{8}\left(\hat{\lambda}^2\sum\frac{(y_i-\hat{\mu})^4}{n}-3\right).$$

The criterion $L_1(\hat{\lambda},\hat{\mu}) > 0$ reduces to the estimated kurtosis, being greater than 3 in this case. This agrees with intuition, because this corresponds to the tails being thicker than those of the normal distribution, like those of the t-distribution. The score statistic for the lognormal/log-t generalization is of a similar form.

For the Weibull distribution, the score statistic is found to be

$$L_1(\hat{\lambda},\hat{\mu}) = \frac{n}{2}\left(\sum\frac{(y_i^{\hat{\mu}}\hat{\lambda}-1)^2}{n}-1\right). \tag{13.9}$$

We know $Y^{\mu}\lambda$ is exponentially distributed with unit mean and variance, so the quantity can be regarded as being the difference between the variance of the sample and the expected variance if Y is Weibull $(\mu, \lambda^{-1/\mu})$. If the data are spread out, suggesting that a third parameter may be required in the fit, the variance will be larger than expected, and so $L_1(\hat{\lambda}, \hat{\mu})$ in (13.9) will be positive. This corresponds to the three-parameter Burr XII being the preferred model.

The test statistic for the inverse-Gaussian distribution is again a comparison of variance. In this case,

$$L_1(\hat{\lambda}, \hat{\mu}) = \frac{n}{8}\left(\frac{1}{n}\sum\left(\hat{\lambda}\frac{(y_i - \hat{\mu})^2}{y_i\hat{\mu}^2} - 1\right)^2 - 2\right). \tag{13.10}$$

Shuster (1968) shows that if Y is inverse-Gaussian (μ, λ), then $\lambda(Y - \mu)^2/(Y\mu^2)$ is $\chi^2(1)$ with mean unity and variance equal to 2. From (13.10), if the variance of the sample is greater than the expected variance when Y is inverse-Gaussian (μ, λ), $L_1(\hat{\lambda}, \hat{\mu})$ will be positive and the three-parameter inverse-Gaussian t-distribution will be selected.

The three-parameter Pearson Type VI distribution is a special case, as it contains both the gamma and the inverted-gamma distributions as two-parameter limits. Consequently, it is necessary to check the fits of both embedded models here. The parametrization in Table 13.1 is in the form where the score statistic results can be directly applied with the gamma limit, giving

$$L_1(\hat{\lambda}, \hat{\mu}) = \frac{n}{2}\left(\frac{1}{n}\sum(\hat{\lambda}\hat{\mu}y_i - \hat{\mu})^2 - \hat{\mu}\right). \tag{13.11}$$

If Y is gamma $(\mu, (\lambda\mu)^{-1})$, then $\lambda\mu Y$ is gamma $(\mu, 1)$ with mean and variance equal to μ. Thus, once again, the criterion is based on a comparison of variance; if the variance of the sample is larger than would be expected of a sample from this gamma distribution, the three-parameter generalization is selected to fit the data.

Note that the inverted-gamma distribution is obtained with μ^{-1} and λ^{-1} in the Pearson Type VI distribution playing the roles of α and λ in (13.7) and (13.8), with the limit therefore being obtained as $\mu \to \infty$. The test criterion is based on the sign of

$$L_1(\hat{\lambda}, \hat{\alpha}) = \frac{n}{2}\left(\frac{1}{n}\sum\left(\frac{1}{\hat{\lambda}\hat{\alpha}y_i} - \frac{1}{\hat{\alpha}}\right)^2 - \frac{1}{\hat{\alpha}}\right). \tag{13.12}$$

13.5.2 Example of a Formal Test

For all the examples given in the previous section, an elementary estimation procedure is to begin by fitting the base model to obtain MLEs of its parameters, these being $(\hat{\lambda}, \hat{\mu})$ in all cases (except the inverted-gamma, for which the parameters are $(\hat{\lambda}, \hat{\alpha})$, and then use the sign of $L_1(\hat{\lambda}, \hat{\mu})$ ($L_1(\hat{\lambda}, \hat{\alpha})$ in the inverted-gamma case) to determine whether the three-parameter generalization would give a better fit to the sample.

Only if this is so would all three parameters of the randomized-parameter model then be estimated. Confidence intervals, if required, can then be calculated using standard ML methodology.

In a more formal test of the significance of the fit of the model, the standard score statistic

$$T_S^{\frac{1}{2}} = [\mathrm{I}^{\alpha\alpha}(0,\hat{\boldsymbol{\lambda}})]^{1/2}[L_1(0,\hat{\lambda},\hat{\mu})] \tag{13.13}$$

as given in eqn (6.33) is particularly convenient, as under the null hypothesis $H_0 : \alpha = 0$, it has a standard normal distribution aymptotically. However, evaluation of the variance term $[\mathrm{I}^{\alpha\alpha}(0,\hat{\boldsymbol{\lambda}})]^{1/2}$ is required. The method for doing this was discussed at length in Chapter 6, with several examples given in Section 6.1.4, including how to orthogonalize the information matrix to simplify the calculation. We shall therefore only give one example here.

We consider the log-t case where the PDF is as given in eqn (13.3). It will be notationally slightly simpler to write $\lambda = \sigma^{-2}$. Then the one-observation log-likelihood has the form $L = L_0 + L_1\alpha + L_2\alpha^2 + O_p(\alpha^3)$, where

$$\begin{aligned}
L_0 &= -\ln y + \frac{1}{2}\ln\lambda - \frac{1}{2}\ln(2\pi) - \frac{\lambda}{2}(\ln y - \mu)^2, \\
L_1 &= \left(-\frac{1}{8} + \frac{\lambda^2}{8}(\ln y - \mu)^4 - \frac{\lambda}{4}(\ln y - \mu)^2\right), \\
L_2 &= \left(-\frac{\lambda^3}{24}(\ln y - \mu)^6 + \frac{\lambda^2}{16}(\ln y - \mu)^4\right).
\end{aligned}$$

The term L_0 shows that we have the lognormal special case obtained when $\alpha = 0$. The one-observation information is given by eqn (6.4), and a straightforward calculation of this and its inverse gives

$$\mathrm{I}(0,\mu,\lambda) = \begin{bmatrix} \frac{7}{8} & 0 & \lambda^{1/2} \\ 0 & \lambda & 0 \\ \lambda^{1/2} & 0 & 2\lambda \end{bmatrix}, \mathrm{I}^{-1}(0,\mu,\lambda) = \begin{bmatrix} \frac{8}{3} & 0 & -\frac{4}{3\lambda^{1/2}} \\ 0 & \frac{1}{\lambda} & 0 \\ -\frac{4}{3\lambda^{1/2}} & 0 & \frac{7}{6\lambda} \end{bmatrix}.$$

Thus, the variance of $\hat{\alpha}$ under the assumption $\alpha = 0$ is $\mathrm{I}^{\alpha\alpha}(0,\mu,\sigma) = 8/(3n)$. Using the value of $\hat{L}_1$ for the lognormal model from Table 13.1, we have

$$T_S^{\frac{1}{2}} = \sqrt{\frac{8}{3n}}\hat{L}_1 = \sqrt{\frac{n}{24}}\left(\hat{\lambda}^2\sum\frac{(y_i - \hat{\mu})^4}{n} - 3\right), \tag{13.14}$$

a result matching that for the sample kurtosis obtained in Fisher (1930) by combinatorial means. We shall illustrate use of this in the numerical example of the next section.

13.5.3 Numerical Example

An example of the application of randomized-parameter models and their selection criteria is now given. We consider the fitting of three-parameter mixing models in estimating the risk of a pregnancy resulting in the birth of a child with Down's syndrome, where it is known that increased risk can be indicated by both high maternal age and low maternal serum alpha-fetoprotein (AFP) level. Cuckle, Wald, and Thompson (1987) show how the overall risk can be determined from these age-specific and AFP-specific factors. From this, a screening procedure can be carried out which is based on age and AFP level, where pregnancies above a chosen overall risk level are referred for further tests. Cuckle et al. (Table 6) provide percentiles of the distribution of AFP levels, so that for a given overall risk level and maternal age, the corresponding percentile is used as a cut-off in the screening procedure. Thus an accurate estimate of the distribution of AFP levels is needed, and often a tail value is required. Cuckle et al. give details of the calculation using estimates of risk based on the lognormal model.

The distribution of AFP levels has been recorded in many surveys. We use a sample collected by the obstetrics unit at the Royal Gwent Hospital, taken at gestational age 18 weeks, for which we thank Dr M. Penny. The data are given in Table 13.2 where, with zero values omitted, the sample size is 641.

The lognormal distribution and its log-t generalization are fitted to these data. For comparison, and to illustrate the effect that mixing has on the tails of the two-parameter distributions, we also consider fitting the Weibull, gamma, and inverse-Gaussian models with their three-parameter generalizations. In all cases, the score statistic is positive, indicating that the three-parameter models provide the better fits. Table 13.3 gives the score statistic values obtained.

For a formal test, we need to standardize the value of $\hat{L}_1$ to the form $T_S^{\frac{1}{2}}$ as given in eqn (13.13). We illustrated calculation of this in the case of the lognormal giving the value of $T_S^{\frac{1}{2}}$ in eqn (13.14). For our AFP data, this gives

$$T_s^{\frac{1}{2}} = \sqrt{\frac{8}{3n}\hat{L}_1} = \sqrt{\frac{8}{3 \times 641}31.15} \simeq 2.01. \tag{13.15}$$

This has a p-value of approximately 0.02, so that using the one-sided score test, the lognormal fit would be rejected at the 10% and 5% level, though not at the 1% level.

The fits of both the basic distributions and their generalizations are shown in the following figures. Since the tail fits are of particular interest, the tail areas are plotted using the -logsurvivor scale. By accentuating the tails of the distribution, this method of plotting gives more emphasis to the critical parts of the fit. Following D'Agostino, and Stephens (1986), the lower tail fits are illustrated by the plots of the first $n/2$ order statistics on a vertical scale of $-\ln F(y)$, whilst the upper tail fits are illustrated by the plots of the last $n/2$ order statistics on a vertical scale of $-\ln(1 - F(y))$. The standard plotting position $F_n(y) = (i - 1/2)/n$ $(i = 1, \ldots, n)$ is used to represent the empirical distribution function in the figures.

Table 13.2 Royal Gwent Hospital maternal serum AFP data, collected at gestational age 18 weeks

Obsvn	11	13	15	16	17	18	19	20
Freq	2	2	1	2	4	4	9	8
Obsvn	21	22	23	24	25	26	27	28
Freq	8	8	12	16	10	12	11	16
Obsvn	29	30	31	32	33	34	35	36
Freq	22	13	21	18	13	17	15	19
Obsvn	37	38	39	40	41	42	43	44
Freq	17	18	18	20	18	18	20	14
Obsvn	45	46	47	48	49	50	51	52
Freq	20	11	11	14	15	11	8	4
Obsvn	53	54	55	56	57	58	59	60
Freq	9	8	11	9	3	5	6	6
Obsvn	61	62	63	64	65	66	67	68
Freq	3	2	2	5	4	4	5	1
Obsvn	69	70	71	72	73	74	75	76
Freq	2	3	3	5	3	3	1	3
Obsvn	77	79	80	82	83	84	86	88
Freq	7	2	4	2	2	1	1	1
Obsvn	89	90	91	92	96	103	107	110
Freq	1	1	2	1	1	1	1	1
Obsvn	113	115	125	140	149	151		
Freq	1	1	1	1	1	1		

The parameter estimates obtained when fitting the basic models and their three-parameter generalizations are listed in Table 13.4.

Figure 13.1 gives the fits of the two-parameter lognormal and its three-parameter generalization. Although the log-t distribution is slightly better in the upper tail, there appears to be little difference between the two fits, which is reflected in the similarity of the parameter estimates in Table 13.4. The fits of the two-parameter Weibull and three-parameter Burr XII are exhibited in Figure 13.2. Here, there is a marked

Table 13.3 Score statistic values of two-parameter base distributions fitted to the AFP data

2-parameter distribution	$\hat{L}_1$
lognormal	31.15
Weibull	309.7
inverse Gaussian	39.79
Pearson Type V	230.3
gamma	369.6

Table 13.4 ML parameter estimates of five two-parameter base models and corresponding randomized three-parameter models

Distribution	$\hat{\lambda}$	$\hat{\mu}$	$\hat{\alpha}$
lognormal	2.514	3.676	
log-t	2.685	3.676	0.1241
Weibull	9.260×10^{-5}	2.396	
Burr XII	5.693×10^{-8}	4.548	1.049
inverse-Gaussian	248.4	42.80	
inverse-Gaussian-t	292.7	42.62	0.1524
Pearson Type V	0.02739		0.1529
gamma	0.02336	6.365	
Pearson Type VI	0.02574	16.65	0.09166

difference in the fits obtained, with the Burr XII model performing fairly well in both tails, whereas the Weibull model proves rather unsatisfactory. Some improvement in fit, albeit less great, is also obtained with the three-parameter inverse-Gaussian-t over the two-parameter inverse-Gaussian, shown in Figure 13.3. Figure 13.4 displays the fits obtained from both the two-parameter inverted-gamma (PTV) and gamma distributions and their three-parameter PTVI generalization. In both the lower and upper tails, an improved fit is obtained with the PTVI distribution.

Overall, the three-parameter models generally tend to give a noticeably improved fit, particularly in the upper tails.

As estimates of tail AFP levels are of special interest, Table 13.5 gives some examples of selected percentiles for the models, together with the empirical percentile values.

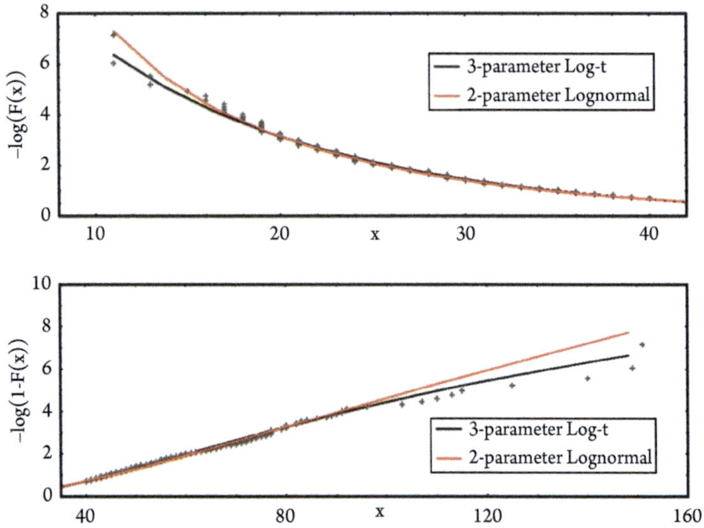

Figure 13.1 Lower tail (upper chart) and upper tail (lower chart) fits to maternal serum AFP data, given by the three-parameter log-t and two-parameter lognormal models.

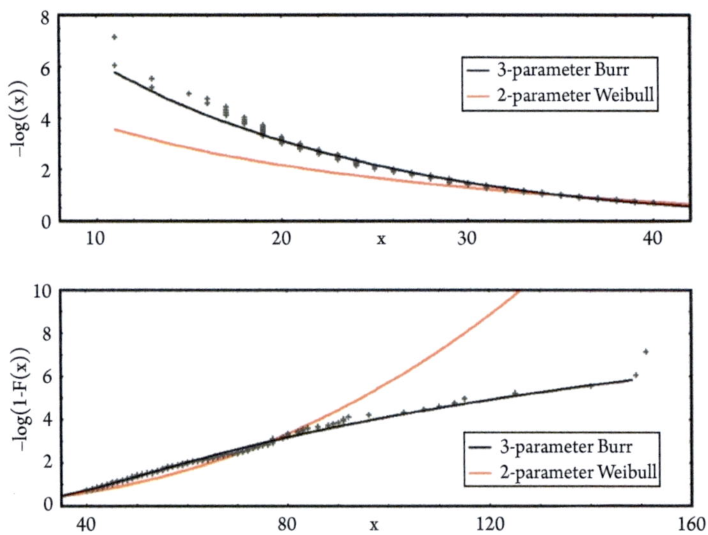

Figure 13.2 Lower tail (upper chart) and upper tail (lower chart) fits to maternal serum AFP data, given by the three-parameter Burr and two-parameter Weibull models.

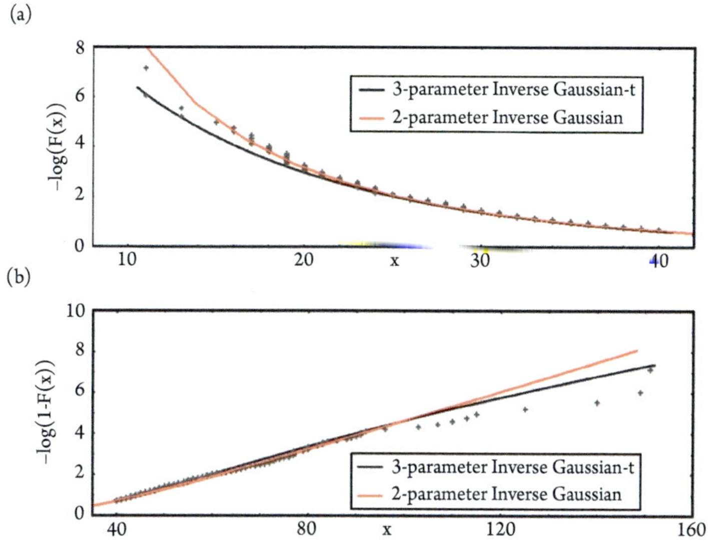

Figure 13.3 Lower tail (upper chart) and upper tail (lower chart) fits to maternal serum AFP data, given by the three-parameter inverse-Gaussian-t and two-parameter inverse-Gaussian models.

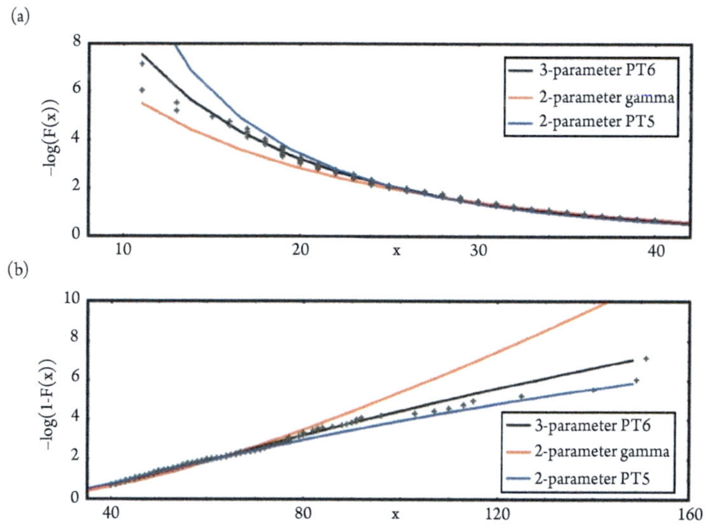

Figure 13.4 Lower tail (upper chart) and upper tail (lower chart) fits to maternal serum AFP data, given by the three-parameter Pearson Type VI, two-parameter Pearson Type V, and two-parameter gamma models.

Table 13.5 Selected fitted percentiles and corresponding empirical values

Percentage Point Distribution	0.5	1.0	5.0	10.0	90.0	95.0	99.0	99.5
empirical	13.0	16.0	20.1	24.0	66.8	77.0	111.8	137.0
lognormal	14.2	15.7	20.5	23.7	65.7	76.0	99.6	110.0
log-t	13.3	15.1	20.6	24.0	65.0	75.6	103.3	117.1
Weibull	5.3	7.1	14.0	18.9	68.4	76.3	91.3	96.8
Burr XII	12.2	14.3	20.5	24.2	64.6	76.6	111.9	131.4
inverse-Gaussian	14.6	16.0	20.6	23.6	66.3	76.5	99.4	109.1
inverse-Gaussian-t	13.2	15.0	20.4	23.8	64.8	75.0	100.2	112.1
Pearson Type V	15.9	17.2	21.3	24.0	67.2	80.3	115.0	132.5
gamma	11.5	13.3	19.2	23.0	65.5	74.0	91.8	98.9
Pearson Type VI	14.5	15.9	20.8	23.9	65.8	76.6	102.7	114.7

The fitted percentiles are found to reinforce the contention of better fit, with considerable improvement in fit achieved in the upper tail by using the three-parameter generalizations of the lognormal and Weibull. A more formal test of adequacy of fit is now considered.

13.5.4 Goodness-of-Fit

We consider a formal GoF test using the Anderson-Darling (A-D) test statistic A^2 of eqn (4.12). As the test is designed to be sensitive to lack of fit in the tails, this seems a particularly appropriate GoF test to use in the present context.

The AFP example allows us to examine a complication when using an EDF test statistic that arises because the observations in the data are not given individually, but are grouped into cell frequencies. It is well known that this will have an effect, sometimes substantial, on the distribution of test statistics such as A^2, making their value larger than when observations have been individually observed and accurately recorded. Thus the GoF test cannot be carried out with tables calculated under the assumption that the observations are continuous and are subject to negligible rounding error.

However, if we use bootstrapping, then we can reproduce the grouping of observations in the bootstrap process, so that the critical values will correctly include the effect of grouping.

An interesting aspect of bootstrapping grouped observations is that there is no need to sample individual observations from the continuous fitted distribution $F(y, \hat{\boldsymbol{\theta}})$ first, followed by grouping. It is much more efficient to calculate the appropriate cell boundaries of the groups first. Thus, if there are $m + 1$ group intervals (η_i, η_{i+1}), $i = 0, 1, 2, \ldots, m$, where $\eta_0 = -\infty$, $\eta_{m+1} = \infty$, then we only need calculate the CDF values $F(\eta_i, \hat{\boldsymbol{\theta}})$, $i = 1, 2, \ldots, m$, which immediately give the cell probabilities

$$p_0 = F(\eta_1, \hat{\boldsymbol{\theta}}),\ p_i = F(\eta_{i+1}, \hat{\boldsymbol{\theta}}) - F(\eta_i, \hat{\boldsymbol{\theta}}),\ i = 1, 2, \ldots m-1,\ p_m = 1 - F(\eta_m, \hat{\boldsymbol{\theta}}).$$

A BS sample is then obtained simply as a sample of discretely distributed observations from the multinomial distribution with these p_i as cell probabilities.

An ideal way of carrying out this discrete multinomial sampling is to use the well-known aliasing method given by Walker (1977). This requires only the easy pre-calculation of an array of cutoff probabilities $q(i)$ and an array of corresponding alias integers $j(i)$, $i = 0, 1, \ldots, m$, each of dimension $(m + 1)$. These arrays are calculated just once. A grouped sample of observations is then obtained by sampling each individual observation in just two steps:

1. Generate I, an integer uniformly distributed over $0, 1, \ldots, m$, and an independent $U \sim U(0, 1)$.
2. If $U \leq q(I)$, return $N = I$, otherwise return $N = j(I)$.

By way of illustration, we carried out BS GoF tests for five of the models described previously. The calculated Anderson-Darling statistics for the fits obtained for the AFP data set are listed in Table 13.6, in the A^2 column, together their p-values and the 90% critical values, $A_{0.1}^{*2}$ as estimated from the EDF of 500 BS A_j^{*2}, $j = 1, 2, \ldots, 500$, calculated from BS samples generated in the way just described.

The BS A^2 GoF test statistic p-value = 0.13 of the two-parameter lognormal model fitted by ML shows that the fit is satisfactory at the upper 10% level, but with the GoF test value a little marginal. In this example, the score test was more stringent with value 2.01 as given in eqn (13.15) giving a p-value = 0.02, so that the lognormal model is rejected at the 5% level, though it would be accepted at the 1% level. The other three two-parameter

Table 13.6 Calculated Anderson-Darling statistics of models fitted to AFP18 data set

Distribution	A^2	p-val	$A_{0.1}^{*2}$	$A^{(C)2}$	$A_{0.1}^{(C)2}$
lognormal	**0.70**	**0.13**	**0.75**	0.70	0.631
Weibull	11.98	0	1.04	12.08	0.637
Burr XII	**0.44**	**0.34**	**0.58**		
gamma	2.57	0	0.77	2.57	0.635
Pearson Type V	1.51	0	0.78		
Pearson Type VI	**0.58**	**0.28**	**0.76**		

models, the Weibull, gamma, and inverted gamma (PT V) are all firmly rejected by the GoF test, with p-values = 0.0 in all cases. However, the two three-parameter models, Burr XII and PT VI, are very satisfactory. This confirms the view obtained from the graphical plots that the fits of the three-parameter randomized-parameter models tend to be much more accurate than those of the base two-parameter distributions.

Upper 10% critical values of A^2 are available in D'Agostino, and Stephens (1986) for three of the models we consider: the lognormal, Weibull, and gamma. The upper 10% critical values for these three models are listed as $A_{0.1}^{(C)2}$ in Table 13.6. We have added the superscript (C) to indicate that, as already remarked, the values are only appropriate for non-grouped data. We can compare these $A_{0.1}^{(C)2}$ values with the corresponding $A_{0.1}^{*2}$ values obtained by bootstrapping with the BS samples grouped in the same way as in the original data set. We find $A_{0.1}^{(C)2} < A_{0.1}^{*2}$ in all three cases, with the differences all noticeable.

Also, the lognormal and Weibull tables given by D'Agostino, and Stephens (1986) require an additional adjustment to the A^2 value. In the normal/lognormal case, which is covered in D'Agostino and Stephens Table 4.7, the test statistic is $A^{(C)2} = A^2(1 + 0.75/n + 2.25/n^2)$. The sample size of $n = 641$ means that there is little difference between A^2 and $A^{(C)2}$ in this example. In the Weibull case, which is covered in Table 4.17, the test statistic is $A^{(C)2} = A^2(1 + 0.2/\sqrt{n})$. The gamma case is covered in Table 4.21 with no adjustment needed, so that $A^{(C)2} = A^2$. The appropriate $A^{(C)2}$ values in our example are also listed in Table 13.6. Note that the gamma critical value also depends on the power parameter μ, which was $\hat{\mu} \simeq 6$ in our fitted model.

Based on the tabulated $A_{0.1}^{(C)2}$ values, we would reject all three fits, including the lognormal. Though marginal, the smallness of the tabulated critical value of $A_{0.1}^{(C)2} = 0.631$ has changed the result compared with the GoF version of the test, where $A_{0.1}^{*2} = 0.73$ has meant that formally, at least, the lognormal would not be rejected.

14 Indeterminacy

This chapter and the following one are concerned with the problem of indeterminacy and its consequences, particularly in model selection. In this chapter, we define the problem and describe a well-known approach suggested by Davies (1977, 1987, 2002). This approach is technically interesting, but is somewhat complicated and can be difficult to apply in practical problems. In the following chapter, we suggest a simple alternative approach, applying it to the problem of fitting nonlinear regression models in situations where indeterminacy arises. In this case, we show that indeterminacy is linked with embeddedness which was discussed in Chapter 5.

We stress that we have not investigated optimality properties of our suggested alternative. However, as will become evident, the approach follows the stepwise method of identifying important factors in linear regression, where the emphasis is on a simple but flexible approach that is easy to follow and to implement, with optimality not critically important. Our approach is offered very much in that spirit.

The reader more interested in the simple approach might wish to omit discussion of the Davies method in this chapter on first reading.

14.1 The Indeterminate Parameters Problem

When we discussed the embedded problem in Chapter 5, we focused on the situation where it arises in fitting a p-parameter model with parameters denoted by $\boldsymbol{\theta} = (\theta_1, \ldots, \theta_p) \in \boldsymbol{\Theta}$, when the best fit is not an interior point, but a boundary point. In discussing the indeterminate parameters problem, we focus on this same scenario in which a subvector $\boldsymbol{\varphi}$, comprising $l < p$ of the parameters, is allowed to tend to a special value not dependent on the data, typically on the boundary of $\boldsymbol{\Theta}$. With no loss of generality, we can assume this special value to be $\boldsymbol{\varphi}=\mathbf{0}$. In a regular problem, there are then $p - l$ $(= r$, say) remaining parameter values that need to be fixed.

The *indeterminate parameters problem* is when q of the remaining r parameters, where $0 < q \leq r$, simply vanish and no longer appear anywhere in the model when $\boldsymbol{\varphi} = \mathbf{0}$. If $\boldsymbol{\psi}$ is the subvector of parameters that vanish in this way, we shall say that $\boldsymbol{\psi}$ is *indeterminate* (when $\boldsymbol{\varphi}=\mathbf{0}$), denoting this by

Non-Standard Parametric Statistical Inference. Russell Cheng.

$$\boldsymbol{\varphi} \succ \boldsymbol{\psi}. \tag{14.1}$$

Strictly speaking, when $\boldsymbol{\varphi} \succ \boldsymbol{\psi}$, we should say that $\boldsymbol{\psi}$ is *potentially* or *conditionally indeterminate*, only actually becoming indeterminate when $\boldsymbol{\varphi} = \mathbf{0}$. However, this exactness of expression is somewhat clumsy to use repetitively, so when there is no confusion, we will usually just refer to $\boldsymbol{\psi}$ as being indeterminate, with the understanding that it is conditional on another parameter $\boldsymbol{\varphi}$ taking a special value, usually $\boldsymbol{\varphi} = \mathbf{0}$.

The simplest case is where setting just one parameter (so that $l = 1$) to zero results in another parameter being eliminated from the model. Thus, if we set $\theta_i = 0$ in the log-likelihood $L(\boldsymbol{\theta}; \mathbf{y})$, and as a direct result of this all terms involving θ_j disappear, we write

$$\theta_i \succ \theta_j \tag{14.2}$$

to mean θ_j is made indeterminate by θ_i being zero.

Each such indeterminacy condition is best regarded as being a flag that toggles a fundamental change in the actual structure of the model triggered by one particular parameter, θ_i in condition (14.2), taking the particular value zero.

The formulation (14.2) is a little more general than would appear at first sight, as it covers other identifiability problems. For example, let $f(y; \boldsymbol{\theta}) = \theta_1 + \theta_2 \exp(\theta_3 y)$. Then if $\theta_3 = 0$, θ_1 and θ_2 are observable only as $\theta_1 + \theta_2$. If we set $\psi_1 = \theta_1$, $\psi_2 = \theta_1 + \theta_2$, $\psi_3 = \theta_3$ then we have

$$f(y; \boldsymbol{\psi}) = \psi_1 + (\psi_2 - \psi_1) \exp(\psi_3 y), \tag{14.3}$$

and the indeterminacy is reduced to $\psi_3 \succ \psi_1$. This particular model is considered in more detail in Section 15.6.2. We will encounter this second form of indeterminacy frequently, for example, in the double exponential regression model of eqn (15.13), to be discussed in Section 15.5.1.

It should be pointed out that indeterminacy is referred to in other ways. Thus, Davies (1977) refers to it as the situation where hypothesis testing involves a nuisance parameter that is present only under the alternative. Garel (2005) refers to the problem as one of non-identifiability of parameters. For simplicity, we refer to the problem as one of indeterminacy, and not continually refer to these alternative terminologies.

14.1.1 Two-Component Normal Mixture

A typical example is given by Smith (1989), who cites the well-known case of a two-component mixture model with PDF

$$f(y; \xi, \theta) = (2\pi)^{-1/2}[(1 - \xi) \exp(-y^2/2) + \xi \exp(-(y - \theta)^2/2], \tag{14.4}$$

$0 < \xi < 1$, where we have both $\xi \succ \theta$ and $\theta \succ \xi$.

The model (14.4) has been the subject of extensive study. Even though the main interest is the theoretical complication that the model gives rise to despite its apparent simplicity, the problem nevertheless does occur in practice. An application of this model slightly generalized to a two-component mixture of a general continuous distribution, as given in eqn (2.5), is considered by Chen, Ponomareva, and Tamer (2014).

Hartigan (1985) considers the likelihood ratio test of $H_0 : \xi = 0$ versus $H_1 : \xi \neq 0$. The log-likelihood of (14.4) is given by

$$L = -\frac{n}{2}\ln(2\pi) - \frac{\sum y_i^2}{2} + \sum \ln\left(1 - \xi + \xi \exp(y_i\theta - \theta^2/2)\right). \tag{14.5}$$

Note that under the null hypothesis, the distribution does not depend on θ. For fixed θ, the log-likelihood ratio is

$$\begin{aligned}\lambda_n(\xi,\theta) &= L(\mathbf{y};\xi,\theta) - L(\mathbf{y};0,\theta) \\ &= \sum_{i=1}^{n} \ln\left(1 - \xi + \xi \exp(y_i\theta - \theta^2/2)\right).\end{aligned} \tag{14.6}$$

If the problem were regular, it would follow from standard asymptotic theory that

$$\Lambda = \sup_{\xi\in[0,1],\,|\theta|<\infty} \lambda_n(\xi,\theta)$$

is asymptotically $\chi_2^2/2$ distributed. However, due to the indeterminacy present both when $\xi = 0$ and when $\theta = 0$, the situation is not regular in this example.

Hartigan argued that the aforementioned test is not only non-standard, but that Λ tends to ∞ in probability as $n \to \infty$, however, Hartigan does not give full details in his proof. The implication is that, for large n, the null hypothesis would very likely be rejected when it is, in fact, true. Hartigan conjectured that divergence is slow at a rate of $\ln\ln n$.

Numerical simulations have been carried out by Thode et al. (1988), whose results seem to indicate that for n up to 1000, approximating the distribution of Λ by that of a chi-squared χ_ν^2 where $1 \leq \nu \leq 2$ cannot be ruled out when the null is true.

Results produced by Böhning and Dietz (1995) of a simulation based on the mixture model with $n = 10,000$ and 5000 replications gave an almost complete agreement with $\Lambda \sim \chi_1^2$ when the null is true. Böhning and Dietz (1995) also tested a second form of null $H_0' : \xi = 1$. The resulting null distribution of Λ in this case includes an atom at $\Lambda = 0$ with probability $p \simeq 0.44$, and a continuous component more like a χ_ν^2 with $\nu < 1$. These results indicate that divergence is extremely slow.

Bickel and Chernoff (1993) showed that Λ diverges at least as fast as $\ln\ln n$. More recently, a definitive result has been given by Liu and Shao (2004), who proved, in a tour de force of statistical analysis, that 2Λ has a limiting extreme value (EVMax, see eqn (6.24)) distribution, namely,

$$\lim_{n\to\infty} \Pr\{2\Lambda - \ln\ln n + \ln(2\pi^2) \leq x\} = \exp(-e^{-x/2}),$$

confirming Hartigan's conjecture.

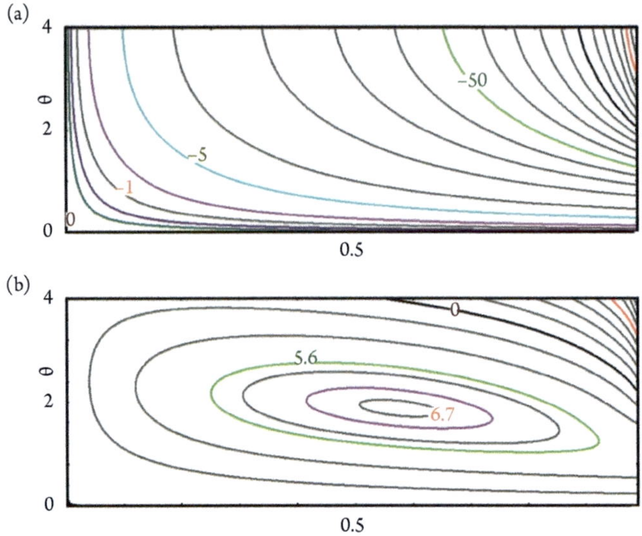

Figure 14.1 Contour plot of log-likelihood ratio LR ratio $\lambda_n(\xi,\theta)$ given in eqn (14.6) with observations sampled (a) under null, and (b) under specified alternative.

The result of Liu and Shao (2004) clarifies the behaviour of 2Λ under the null. Hall and Stewart (2005) have considered the power of the test, showing that it is less than in standard situations, using a system of local alternative hypotheses to quantify the extent to which the power is reduced.

The related problem, where the mixture is of exponential distributions rather than the normal, has been considered by Böhning et al. (1994).

Figure 14.1(a) illustrates the behaviour of the LR ratio $\lambda_n(\xi,\theta)$ given in eqn (14.6), showing a contour plot when the observations are sampled under the null; that is, when $Y \sim N(0,1)$. Here, $\lambda_n(\xi,\theta)$ is zero at $\xi = 0$ and $\theta = 0$, and remains flat in the neighborhood of these limits, decreasing when ξ and θ both move away from zero, so that $\Lambda_n = 0$ in this case. In contrast, Figure 14.1(b) is a contour plot of $\lambda_n(\xi,\theta)$ when the observations are sampled from (14.4) with $\xi = 0.5$ and $\theta = 1.5$. Here, λ_n behaves in a standard way, with regular contours obtained.

The problem is a general one, for example, arising if one attempts to fit a mixture model with k components to data drawn from a mixture with $k_0 < k$ components. The true distribution is then represented not by a single point but by a boundary region of the space defining the k-component model being fitted. In practical terms, Feng and McCulloch (1996) show that though a supposedly standard application of the likelihood ratio test may be unsatisfactory, this arises only because the true density cannot be represented by a single fixed parameter value even if comes from the boundary region representing the true model, as the likelihood ratio will not then behave in a standard way. However, Feng and McCulloch (1996) show that the position is not irretrievable, as the ML estimates will converge to *some point* in the subset representing the true distribution.

The conditions given by Feng and McCulloch (1996) under which this will occur assume complicated conditions, including high-order differentiabilty. Cheng and Liu (2001) show that the result still holds under the much weaker conditions, essentially as given by Wald (1949).

Cheng and Liu (2001) also point out that it is generally possible to define the parameter space so that individual points of the space correspond to distinct distributions, but that such a reparameterization is not usually simple. As pointed out by Garel (2005), such a reparametrization raises issues concerning the differentiability of the PDF, with the second derivative becoming unbounded under the null. Ghosh and Sen (1985) suggest overcoming the problem instead by imposing a 'separation condition', namely, $|\theta| \geq \varepsilon > 0$, where ε is a given constant, when the null becomes simply $H_0 : \xi = 0$ or $\xi = 1$. Many authors, not listed here but cited in Garel (2005), have attempted to remove the separation condition with varying levels of satisfactoriness. Garel (2005) himself gives a set of five assumptions involving the second derivative of the PDF which, if satisfied, allows the test of the null hypothesis to be conducted with two separate regular tests. The assumptions seem quite technical, and we do not discuss this approach further here but refer the reader to Garel (2005) for details.

In the remainder of this chapter, we consider the well-known method proposed by Davies (1977, 1987, 2002), and for comparison suggest a simple alternative stepwise method which acts somewhat like the separation condition approach, but without the need to invoke such a condition explicitly.

14.2 Gaussian Process Approach

14.2.1 Davies' Method

Davies (1977, 1987, 2002) proposed a method of dealing with the problem of hypothesis testing when a nuisance parameter is present only under the alternative, using a Gaussian stochastic process approach. We describe the original basic version only.

Consider the test

$$H_0 : \xi = 0 \text{ versus } H_1 : \xi > 0, \tag{14.7}$$

where the n observations have density such as (14.4), with $\xi \succ \theta$. The approach suggested by Davies is to base the test on a locally optimal test statistic which would be used if θ were known, and treat it as a stochastic process, $S_n(\theta)$, say, dependent on the nuisance parameter θ. Possible choices for $S_n(\theta)$ would be the Wald test statistic

$$S_n(\theta) = n^{1/2}\gamma(\theta)\hat{\xi}_n(\theta),$$

or the signed square root of the likelihood ratio

$$S_n(\theta) = (2\ln \lambda_n(\theta))^{1/2}\, sgn\, \hat{\xi}_n(\theta),$$

or the standardized score statistic

$$S_n(\theta) = n^{-1/2} \sum_{i=1}^{n} \left[\frac{\partial}{\partial \xi} (\ln f(y_i; \xi, \theta)) / \gamma(\theta) \right]_{\xi=0},$$

where

$$\gamma^2(\theta) = \mathrm{Var}\left(\left[\frac{\partial}{\partial \xi} (\ln f(y_i; \xi, \theta)) \right]_{\xi=0} \right).$$

Each of these quantities has asymptotically a $N(0, 1)$ distribution. For a fixed θ, $S_n(\theta)$ is then a regular test statistic of (14.7).

Under suitable regularity conditions, the stochastic process $S_n(\theta)$ converges weakly to a Gaussian process $S(\theta)$ in distribution as $n \to \infty$. Under the null hypothesis, if θ is limited to lying in the range $[L, U]$, then

$$T = \sup_{L \leq \theta \leq U} S(\theta) \tag{14.8}$$

can be used as a test statistic. Under the assumption of appropriate smoothness conditions, Davies considers the test with critical region

$$T > c. \tag{14.9}$$

More generally, the critical value is represented by a continuously differentiable curve, $c(\theta)$. Based on an approach used by Cramér and Leadbetter (1967), a formula is obtained for the expected number of 'up-crossings' of zero by $S(\theta) - c(\theta)$ under the null hypothesis. This gives an approximation to the distribution of T as

$$\begin{aligned} P(T - c(\theta) > 0) \leq \quad & \Phi(-c(L)) + (2\pi)^{-1} \textstyle\int_L^U \exp\left(-c^2(\theta)/2\right) \\ & \times \left(-\rho_{11}(\theta)\right)^{1/2} \varphi\left(c_1(\theta)/\left(-\rho_{11}(\theta)\right)^{1/2}\right) d\theta \end{aligned} \tag{14.10}$$

where $\Phi(.)$ is the cdf of the standard normal distribution,

$$\rho_{11}(\theta) = \left[\frac{\partial^2 \rho(\theta_1, \theta_2)}{\partial \theta_1^2} \right]_{\theta_1 = \theta_2 = \theta}, \tag{14.11}$$

with $\rho(\theta_1, \theta_2)$ the covariance function of $S(\theta)$, $c_1(\theta) = \partial c(\theta)/\partial\theta$ and

$$\varphi(x) = \exp\left(\frac{-x^2}{2}\right) - x \int_x^{\infty} \exp\left(\frac{-t^2}{2}\right) dt.$$

The expression is simplified if the boundary is taken as a constant c, applicable in situations where the range of θ is restricted. In this case, the bound on the significance level of the test (14.8) is given by

$$P(T - c > 0) \leq \Phi(-c) + (2\pi)^{-1} \exp\left(-c^2/2\right) \int_L^U \left(-\rho_{11}(\theta)\right)^{1/2} d\theta. \tag{14.12}$$

14.2.2 A Mixture Model Example

We apply Davies' method to the mixture model (14.4), taking $S_n(\theta)$ to be the score statistic. From (14.5),

$$\frac{\partial L}{\partial \xi} = \sum_{i=1}^{n} \frac{-1 + \exp(y_i\theta - \theta^2/2)}{1 - \xi + \xi \exp(y_i\theta - \theta^2/2)},$$

and so we have

$$S_n(\theta) = \frac{\sum Z_i(\theta)}{\sqrt{n}\gamma(\theta)}, \tag{14.13}$$

where $Z_i(\theta) = \exp(y_i\theta - \theta^2/2) - 1$ and

$$\gamma(\theta) = \text{S.D.}\left[Z_i(\theta)\right] = \left(E\left[Z_i^2(\theta)\right] - E^2\left[Z_i(\theta)\right]\right)^{1/2},$$

with expectations evaluated under the null hypothesis. Thus, if y_i is distributed as a standard normal variable with PDF $\phi(.)$, then

$$\begin{aligned} E\left[Z_i(\theta)\right] &= E\left[\exp(y_i\theta - \theta^2/2) - 1\right] \\ &= \textstyle\int_{-\infty}^{\infty} (\exp(y\theta - \theta^2/2) - 1)\phi(y)dy \\ &= \textstyle\int_{-\infty}^{\infty} (\exp(-(y - \theta)^2/2)dy - 1 \\ &= 0 \end{aligned}$$

and

$$\begin{aligned} E\left[Z_i^2(\theta)\right] &= \textstyle\int_{-\infty}^{\infty} (\exp(2y\theta - \theta^2) - 2\exp(y\theta - \theta^2/2) + 1)\phi(y)dy \\ &= (2\pi)^{-1/2} \textstyle\int_{-\infty}^{\infty} (\exp(-(y - 2\theta)^2/2 + \theta^2) - 2\exp(-(y - \theta)^2/2) \\ &\quad + \exp(-y^2/2))dy \\ &= (2\pi)^{-1/2} \exp(\theta^2) \textstyle\int_{-\infty}^{\infty} \exp(-(y - 2\theta)^2/2)dy - 1 \\ &= \exp(\theta^2) - 1, \end{aligned}$$

giving

$$\gamma(\theta) = (\exp(\theta^2) - 1)^{1/2}.$$

This can be expanded as

$$\gamma(\theta) = \left(1 + \theta^2 + \frac{\theta^4}{2} - 1 + O(\theta^6)\right)^{1/2},$$

and similarly

$$Z_i(\theta) = \exp(-\theta^2/2)(1 + \theta y_i + \theta^2 y_i^2/2 + O(\theta^3 y_i^3)) - 1,$$

so that $\gamma(\theta) \simeq \theta$ and $Z_i(\theta) \simeq \theta y_i$, if $\theta \to 0$. Therefore,

$$S_n(\theta) = \frac{\sum Z_i(\theta)}{\sqrt{n}\gamma(\theta)} \simeq \frac{\theta \sum y_i}{\sqrt{n}\theta},$$

and

$$\mathrm{Var}\left[S_n(\theta)\right] \simeq \frac{1}{n}\left(n\mathrm{Var}\left[y_i\right]\right) = 1.$$

Hence

$$S_n(\theta) = \frac{\sum(\exp(y_i\theta - \theta^2/2) - 1)}{n^{1/2}(\exp(\theta^2) - 1)^{1/2}} \tag{14.14}$$

has a $N(0, 1)$ distribution.

In order to calculate the boundary, c, for a given significance level α, we require $\rho_{11}(\theta)$ from (14.11). We have

$$\mathrm{cov}(S(\theta_1), S(\theta_2)) = \mathrm{cov}\left(\frac{\sum_{i=1}^n Z_i(\theta_1)}{\sqrt{n}\gamma(\theta_1)}, \frac{\sum_{j=1}^n Z_j(\theta_2)}{\sqrt{n}\gamma(\theta_2)}\right).$$

Since the observations are independent,

$$\mathrm{cov}\left(\frac{Z_i(\theta_1)}{\sqrt{n}\gamma(\theta_1)}, \frac{Z_j(\theta_2)}{\sqrt{n}\gamma(\theta_2)}\right) = \frac{E\left[Z_i(\theta_1)Z_i(\theta_2)\right]}{n\gamma(\theta_1)\gamma(\theta_2)} \qquad \text{if } i = j$$

and zero otherwise. Letting $r_{ij} = (\exp(\theta_i\theta_j) - 1)^{1/2}$, we then have

$$\mathrm{cov}(S(\theta_1), S(\theta_2)) = \frac{nE\left[Z_i(\theta_1)Z_i(\theta_2)\right]}{n\gamma(\theta_1)\gamma(\theta_2)} = \frac{(r_{12})^2}{(r_{11})(r_{22})}$$

and

$$\frac{\partial\rho(\theta_1, \theta_2)}{\partial\theta_1} = \frac{\theta_2(r_{11})(r_{22})\exp(\theta_1\theta_2) - \theta_1(r_{12})^2\exp(\theta_1^2)(r_{11})^{-1}(r_{22})}{(r_{11})^2(r_{22})^2},$$

so that

$$\rho_{11}(\theta) = \left[\frac{\partial^2\rho(\theta_1, \theta_2)}{\partial\theta_1^2}\right]_{\theta_1=\theta_2=\theta} = \frac{-\exp(\theta^2)(\exp(\theta^2) - \theta^2 - 1)}{(\exp(\theta^2) - 1)^2}. \tag{14.15}$$

The upper bound (14.12) can only be used if $\int_L^U(-\rho_{11}(\theta))^{1/2}d\theta$ is convergent. Expression (14.15) can be written as

$$\rho_{11}(\theta) \simeq -\exp(\theta^2)(1+\theta^2+(\theta^4/2)-\theta^2-1+O(\theta^6))(1+\theta^2-1+O(\theta^4))^{-2}$$
$$= -\exp(\theta^2)((\theta^4/2)+O(\theta^6))/(\theta^4+O(\theta^6)).$$

Then, $\rho_{11}(\theta) \simeq -1/2$ if $\theta \to 0$ and $\rho_{11}(\theta) \simeq -1$ if $\theta \to \infty$. Thus, $\rho_{11}(\theta)$ is finite for $\theta > 0$.

A simulation was carried out on this example to test the procedure, generating 1000 samples of size 10 from (14.4) with differing combinations of ξ and θ. For each sample, $S_n(\theta)$ was evaluated for a range of θ values between $L = 0$ and $U = 4$, and T was obtained using (14.8). This value of T was then compared with the boundary c, where c had to be calculated, for a given significance level α, from

$$\alpha = \Phi(-c) + (2\pi)^{-1}\exp(-c^2/2)\int_0^4 (-\rho_{11}(\theta))^{1/2}d\theta. \tag{14.16}$$

Using (14.15), this reduces to

$$\alpha = \Phi(-c) + 3.6371(2\pi)^{-1}\exp(-c^2/2).$$

Taking a significance level of 0.1, $c \simeq 2.005$.

Figure 14.2(a) is a typical plot of $S_n(\theta)$ for a sample obtained from (14.4) under the null hypothesis. It can be seen that T is well below the boundary value, and so H_0 is not rejected in this case. Figure 14.2(b) is a typical plot of $S_n(\theta)$ for a sample obtained with

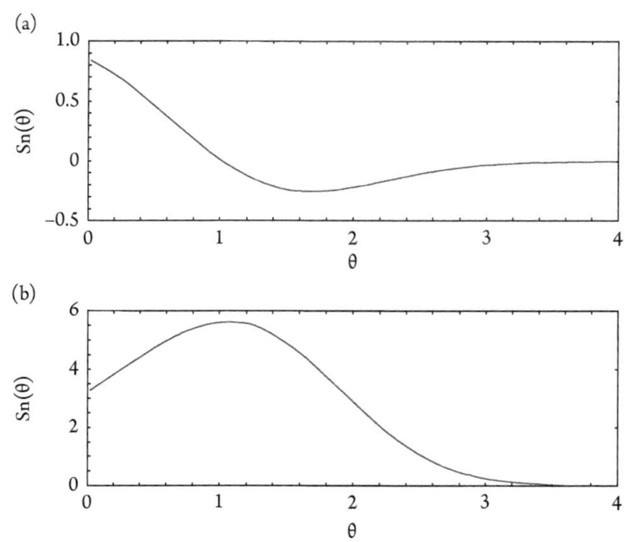

Figure 14.2 Typical plot of score statistic (14.14) with observations sampled (a) under null, and (b) under specified alternative.

$\xi = 0.5$ and $\theta = 1.5$. Here, $T = \sup_{0\le\theta\le4} S(\theta)$ is well above the boundary value of 2.005, and so H_0 is correctly rejected in this instance. The proportion of times out of each 1000 samples that $H_0 : \xi = 0$ was rejected using this method with a significance level of 0.1 was recorded, and Table 14.1 in the following section gives the results obtained. From the table, we see that the test is conservative when $\xi = 0$.

Although Davies' method can overcome the problem of hypothesis testing when indeterminacy is present, its implementation is quite an involved process. In practice, the method also requires rather large samples, as Davies notes that the probabilities calculated are likely to be sensitive to deviations of $S(\theta)$ from normality. In addition, the example just given assumes the nuisance parameter θ to lie in the closed interval $[L, U]$. The theory can be applied for an open interval, but only if (14.16) converges. Berman (1986) reports other difficulties which may occasionally arise with the approach, concerning convergence properties and situations where the regularity conditions are not satisfied.

Davies (1987) extended the method to testing the hypothesis that a vector is zero against the alternative that at least one component is non-zero. In this case, $S(\theta)$ has a chi-squared distribution as opposed to the normal distribution previously fitted. Davies (2002) extended the method further still to a linear model with unknown residual variance. These extensions have not been pursued here.

14.3 Test of Sample Mean

Implementation of Davies' approach can be elaborate, and so sometimes a simpler, less sophisticated technique may be preferable. Whilst Davies' approach tests H_0 against a range of particular alternatives H_1, a very simple alternative method would be to view the problem as purely one of goodness-of-fit, by testing whether the model under H_0 is adequate, without explicitly specifying the alternative. In the previously discussed mixture model example, this involves testing if the sample is $N(0, 1)$. If $V = \sum_1^n Y_i/\sqrt{n}$, the null hypothesis is rejected if $V > Z_\alpha$, where Z_α is the $100(1 - \alpha)$ percentile of the standard normal distribution, determined by the significance level chosen. The proportion of times H_0 is rejected using this test with $\alpha = 0.1$ is also given in Table 14.1. In this case, neither method appears to be totally preferable to the other, although the V statistic requires considerably less computational effort.

Consider the effect of applying the Davies approach in this example with the roles of ξ and θ interchanged. In the score statistic (14.13), we would have

$$Z_i(\xi) = \frac{\partial}{\partial\theta}(\ln f(y_i; \xi, \theta))\bigg|_{\theta=0} = \xi y_i$$

and

$$\gamma^2(\xi) = \mathrm{Var}\left[Z_i(\xi)\right] = \xi^2.$$

Table 14.1 Proportion of times H_0 is rejected for mixture model (14.4) with $\alpha = 0.1$, number of samples = 1000, sample size = 10, and for $\theta \in [0, 4]$. **T** : Davies' test, **V** : test of sample mean

ξ	θ	T	V
0.0	–	0.061	0.090
0.5	0.5	0.212	0.314
0.5	1.0	0.486	0.591
0.5	1.5	0.773	0.824
0.5	2.0	0.925	0.910

Thus

$$S_n(\xi) = \frac{\sum_{i=1}^{n} Y_i}{\sqrt{n}},$$

which is precisely the test of sample mean. This was observed by Titterington et al. (1985).

If θ is not assumed to lie in a finite interval in the mixture example, the boundary should no longer be a constant c but an appropriately defined continuously differentiable curve, $c(\theta)$. For example, if $\theta \in [0, \infty)$, we can consider the boundary $c(\theta) = a + b\theta$. Then, from (14.10), we have

$$\alpha = \Phi(-a) + (2\pi)^{-1} \int_0^{\infty} \exp(-(a + b\theta)^2/2)(-\rho_{11}(\theta))^{1/2}\varphi(b/(-\rho_{11}(\theta))^{1/2})d\theta. \tag{14.17}$$

This allows the range of θ values used to be unrestricted, yet still maintains the required significance level. The intercept, a, could be fixed at the value of c obtained when the range of θ was restricted, and then the gradient, b, set to maintain the significance level. That is, as U (the upper limit of the range of allowable θ) increases, so b should increase so that fewer early test statistic values exceed the boundary. The value of b can be found from (14.17) using an iterative procedure such as the bisection method.

As $U \to \infty$, the value of b should tend to a finite limit. A quick simulation, setting $a = 2.005 = c$ from the previous example where $U = 4$ gives $b = 5.66 \times 10^{-7} \simeq 0$ (as expected when $U = 4$). Table 14.2 shows that as $U \to \infty$, if $a = 2.005$, then b does indeed tend to a finite limit.

Table 14.2 Values of b for the boundary $c(\theta) = a + b\theta, \theta \in [0, U]$, when $a = 2.005$, for $U \to \infty$

U	b
7	0.06980
9	0.08300
25	0.09431
35	0.09434
50	0.09434
100	0.09434

14.4 Indeterminacy in Nonlinear Regression

14.4.1 Regression Example

Many examples of indeterminacy exist in nonlinear regression. Consider an example where there are n sets of observations, with k observations in each set, where for each set i the response is given by the regression equation

$$y_{ij} = \frac{\xi(t_j^\theta - 1)}{\theta} + \epsilon_{ij}, \; j = 1, 2, \ldots, k, \; i = 1, 2, \ldots, n. \tag{14.18}$$

In the numerical example to follow, we take $n = 10, k = 5$, with $\mathbf{t} = (0.1, 0.3, 0.5, 0.7, 0.9)$, and $\epsilon_{ij} \sim N(0, 1)$. In this model, $\xi \succ \theta$, and so the test of $H_0 : \xi = 0$ against $H_1 : \xi > 0$ is not regular, as θ vanishes under the null hypothesis. We apply Davies' method to this problem.

We shall write

$$a_j(\theta) = \frac{t_j^\theta - 1}{\theta}, \; j = 1, 2, \ldots, k.$$

14.4.2 Davies' Method

From (14.18) we have that $Y_{ij} \sim N(\xi a_j(\theta), 1)$, so that

$$f(y_{ij}; \theta) = \frac{1}{\sqrt{2\pi}} \exp\left(-\frac{[y_{ij} - \xi a_j(\theta)]^2}{2}\right)$$

and

$$\ln f(y_{ij}; \theta) = -[\ln(2\pi)]/2 - [y_{ij} - \xi a_j(\theta)]^2/2,$$

giving

$$\frac{\partial \ln f}{\partial \xi} = y_{ij}\xi a_j(\theta) - \xi a_j^2(\theta).$$

Let

$$Z_i(\theta) = \sum_{j=1}^{k} y_{ij}a_j(\theta),\ i = 1, 2, \ldots, n.$$

Under the null hypothesis, the y_{ij} are mutually independent and standard normally distributed so that

$$E\left[y_{ij}a_j(\theta)\right] = E\left[y_{ij}\right] a_j(\theta) = 0$$

and

$$\text{Var}\left[y_{ij}a_j(\theta)\right] = \text{Var}\left[y_{ij}\right] a_j^2(\theta) = a_j^2(\theta).$$

Hence,

$$E\left[Z_i(\theta)\right] = 0,$$

and

$$\text{Var}\left[Z_i(\theta)\right] = \sum_{j=1}^{k} a_j^2(\theta) = \gamma^2(\theta), \text{ say.}$$

The score statistic for all n sets of observations taken together is

$$S_n(\theta) = \frac{\sum_{i=1}^{n} Z_i(\theta)}{\sqrt{n}\gamma(\theta)}.$$

As before,

$$\rho(\theta_1, \theta_2) = \text{cov}(S(\theta_1), S(\theta_2)) = \frac{E\left[Z_i(\theta_1)Z_i(\theta_2)\right]}{\gamma(\theta_1)\gamma(\theta_2)}.$$

Writing

$$a_j = a_j(\theta_1),\ b_j = a_j(\theta_2),$$

we have

$$E\left[Z_i(\theta_1)Z_i(\theta_2)\right] = \sum_{j=1}^{k} a_j b_j,$$

and

$$\rho(\theta_1, \theta_2) = \frac{\sum_{j=1}^{k} a_j b_j}{\gamma(\theta_1)\gamma(\theta_2)}.$$

These expressions can be approximated by taking t as $U[0, 1]$. Then,

$$\begin{aligned}\sum_{j=1}^{k} a_j b_j &= k\int_0^1 (t^{\theta_1} - 1)(t^{\theta_2} - 1)\theta_1^{-1}\theta_2^{-1} dt \\ &= k\theta_1^{-1}\theta_2^{-1}\int_0^1 (t^{\theta_1+\theta_2} - t^{\theta_1} - t^{\theta_2} + 1)dt \\ &= k\theta_1^{-1}\theta_2^{-1}[(\theta_1 + \theta_2 + 1)^{-1} - (\theta_1 + 1)^{-1} - (\theta_2 + 1)^{-1} + 1].\end{aligned}$$

Similarly,

$$\sum_{j=1}^{k} a_j^2(\theta) = k\theta^{-2}[(2\theta + 1)^{-1} - 2(\theta + 1)^{-1} + 1] = 2k(2\theta + 1)^{-1}(\theta + 1)^{-1}.$$

Therefore,

$$\rho(\theta_1, \theta_2) = \frac{\theta_1 + \theta_2 + 2(2\theta_1 + 1)^{1/2}(2\theta_2 + 1)^{1/2}}{2(\theta_1 + 1)^{1/2}(\theta_2 + 1)^{1/2}(\theta_1 + \theta_2 + 1)}.$$

This gives

$$\rho_{11}(\theta) = \left[\frac{\partial^2 \rho(\theta_1, \theta_2)}{\partial\theta_1^2}\right]_{\theta_1=\theta_2=\theta} = -\frac{(4\theta + 3)}{4(2\theta + 1)^2(\theta + 1)^2}.$$

Recall that the upper bound (14.12) can only be used if (14.16) is convergent. When $\theta = 0$, $\rho_{11}(\theta) = -0.75$. When $\theta \to \infty$, $\rho_{11}(\theta) \to 0$. We have

$$\begin{aligned}\int_0^\infty (-\rho_{11}(\theta))^{1/2} d\theta &= \int_0^\infty (4\theta + 3)^{1/2}(4\theta^2 + 6\theta + 2)^{-1} d\theta \\ &= (\ln(2 + \sqrt{3}))/2 + (\pi/6) \\ &= 1.18207.\end{aligned}$$

Thus (14.16) is finite as required.

Davies' method was applied as in the previous example, using the boundary $c(\theta) = a + b\theta$. A simulation was carried out of 2000 replicates of $n = 10$ sets of $k = 5$ observations, using $\alpha = 0.1$. The intercept a was fixed at 1.618, the value of the constant boundary c that would be obtained if the range of θ were restricted to $[0, 4]$. Then, from (14.17), we have $b = 0.02659$. For each replicate, $S_n(\theta)$ was evaluated for a range of θ values between $L = 0$ and $U = 9$, and again T was obtained using (14.8). Table 14.3 lists the proportion of times out of each 2000 replicates that $H_0 : \xi = 0$ was rejected.

Table 14.3 Proportion of times H_0 is rejected for nonlinear regression model (14.18) with $\alpha = 0.1$, number of replicates = 2000, number of sets of k observations = 10, $k = 5$, and for $\theta \in [0, 9]$. **T** : Davies' test, **V** : test of sample mean, **V′** : test of weighted sample mean

ξ	θ	T	V	V′
0.00	–	0.065	0.109	0.101
0.50	0.25	0.970	0.926	0.977
0.50	0.50	0.905	0.862	0.920
0.50	1.00	0.723	0.709	0.736
1.00	0.50	1.000	1.000	1.000
1.00	1.00	0.996	0.987	0.991
1.00	1.50	0.945	0.938	0.938

14.4.3 Sample Mean Method

The alternative method, based on the sample mean, was also used in this example. Here our test statistic is

$$V = \frac{\sum_{i=1}^{n}\sum_{j=1}^{k} y_{ij}}{\sqrt{nk}}.$$

However, for observations sampled under the alternative hypothesis, the response y_{ij} in (14.18) will always be negative, and so in this case we reject the null hypothesis if $-V > Z_\alpha$, where Z_α is the $100(1-\alpha)$ percentile of the standard normal distribution. The results using this test are also given in the table.

The results obtained using V are consistent with what would be expected on examination of the regression equation. From (14.18), we can see that as $\theta \to \infty$, $(t_j^\theta - 1)/\theta \to 0$ since $t_j < 1\ \forall\ j$. Then, $y_{ij} \to \epsilon_{ij}$ which is $N(0, 1)$. Hence, the null hypothesis is not rejected very often in this case. Conversely, as θ decreases, $(t_j^\theta - 1)/\theta$ increases, and so it is easier to reject the null hypothesis. The value of ξ may be selected to counterbalance this loss of power in the former case. From the table, we see that the test is more powerful in instances when $\xi > \theta$.

14.4.4 Test of Weighted Sample Mean

An advantage of the test based on the sample mean is that the alternative does not have to be explicitly specified. However, if there is reason to favour a particular alternative, then the test may be improved by utilizing this extra information. In the previous example, we can modify our test based on the sample mean by considering the regression equation (14.18). Under the null, $E\left[y_{ij}\right] = 0$. Responses obtained when t_j ($j = 1, \ldots, 5$) is small will show the greatest departure from the null, with the differences decreasing as t_j increases. Thus, if the responses are weighted so that the larger responses contribute more to the sample mean than the smaller ones, such as

$$y'_{ij} = y_{ij}t_j^{-1},$$

then this should provide a more sensitive test. In this case,

$$E\left[\sum_{i=1}^{n}\sum_{j=1}^{k} y'_{ij}\right] = E\left[\sum_{i=1}^{n}\sum_{j=1}^{k} \frac{y_{ij}}{t_j}\right] = 0$$

as before, but

$$\mathrm{Var}\left[\sum_{i=1}^{n}\sum_{j=1}^{k} y'_{ij}\right] = \mathrm{Var}\left[\sum_{i=1}^{n}\sum_{j=1}^{k} \frac{y_{ij}}{t_j}\right] = n\sum_{j=1}^{k}\left(\frac{1}{t_j^2}\right).$$

Thus, our test statistic is now

$$V' = \left(\sum_{i=1}^{n}\sum_{j=1}^{k} \frac{y_{ij}}{t_j}\right)\left(n\sum_{j=1}^{k}\left(\frac{1}{t_j^2}\right)\right)^{-1/2}.$$

Table 14.3 also gives the proportion of times H_0 is rejected using this test of weighted sample mean, V'. We see that the test performs well when the observations are sampled under H_0, and is also noticeably more sensitive in detecting slight departures from the null.

Using a goodness-of-fit method, the less simple model would only be fitted if H_0 is rejected. In the next chapter, we extend this approach more systematically to model building in nonlinear regression, focusing on how this is affected by indeterminacy and the existence of embedded models.

15 Nested Nonlinear Regression Models

15.1 Model Building

In this chapter, we consider the problem of choosing that nonlinear regression model which best fits a given data sample from a family of nested nonlinear models.

In our approach, we assume that there is an overriding *full model* which contains all the structure and parameters that it might be important to include in our model if it is to adequately explain the characteristics present in the data. Our aim is to systematically search through the *nested submodels* of the full model, as defined in Section 3.3, to find one that provides an adequate representation, but which is the simplest in the sense of containing the fewest parameters. Even in the linear case, there is already a wealth of detail that can be discussed, involving issues such as model adequacy and the effects of aliasing factors not allowed for in the model. Wu and Hamada (2000) provide a comprehensive discussion, particularly when fitting the linear model, including backwards, forwards, and stepwise methods that systematically add or remove individual factors from the model. Here, we focus on nonlinear models where embedded models and indeterminate parameters can arise, discussing how such problems can be handled.

For ease of reference, the general notation and terminology we use in discussing nested models follows that of Cheng and Traylor (1995).

We will assume that our data has the form (x_i, y_i), $i = 1, 2, \ldots, n$, with

$$y_i = \eta(x_i, \boldsymbol{\theta}) + \varepsilon_i, \tag{15.1}$$

where the ε_i are independent $N(0, \sigma^2)$ variables and $\boldsymbol{\theta} = (\theta_1, \theta_2, \ldots, \theta_p)$ is a vector of p unknown parameters not including σ^2. Here, $\eta(\cdot, \boldsymbol{\theta})$ is the regression function of the full model. A submodel in which only $q \leq p$ components of $\boldsymbol{\theta}$ are included will be called a submodel of order q, with its regression function denoted by η_q.

Non-Standard Parametric Statistical Inference. Russell Cheng.

Suppose that the true parameter value $\boldsymbol{\theta}^0$ lies in the neighbourhood of $\boldsymbol{\theta} = \mathbf{0}$ and that the log-likelihood L of the observations satisfies the standard conditions given in Chapter 3, with L having a Taylor series expansion

$$L(\boldsymbol{\theta}) = L(\boldsymbol{\theta}^0) + (\partial L(\boldsymbol{\theta})/\partial\boldsymbol{\theta})|_{\boldsymbol{\theta}^0}^T(\boldsymbol{\theta} - \boldsymbol{\theta}^0) \\ + \frac{1}{2}(\boldsymbol{\theta} - \boldsymbol{\theta}^0)^T(\partial^2 L(\boldsymbol{\theta})/\partial\boldsymbol{\theta}^2)|_{\boldsymbol{\theta}^0}(\boldsymbol{\theta} - \boldsymbol{\theta}^0) + R(\boldsymbol{\theta}, \boldsymbol{\theta}^0),$$

where, as $n \to \infty$,

$$n^{-1/2}(\partial L(\boldsymbol{\theta})/\partial\boldsymbol{\theta})|_{\boldsymbol{\theta}^0} \sim N(0, \mathrm{i}(\boldsymbol{\theta}^0)),\ n^{-1}\left[-\partial^2 L(\boldsymbol{\theta})/\partial\boldsymbol{\theta}^2|_{\boldsymbol{\theta}^0}\right] \to \mathrm{i}(\boldsymbol{\theta}^0),$$

and R is asymptotically negligible in probability; $\mathrm{i}(\boldsymbol{\theta}^0)$ being the one-observation information matrix, assumed positive definite. Thus,

$$n^{1/2}(\hat{\boldsymbol{\theta}} - \boldsymbol{\theta}^0) \sim N(0, \mathrm{i}^{-1}(\boldsymbol{\theta}^0))$$

asymptotically as $n \to \infty$, where $\hat{\boldsymbol{\theta}}$ is the MLE.

Let M_q be the set of submodels of order q, $0 \leq q \leq p$. We call $\eta_q \in M_q$ a *direct submodel* of $\eta_{q+1} \in M_{q+1}$ if it can be obtained by setting one of the parameters in η_{q+1}, θ_i, say, to 0.

A necessary condition for η_q to be a direct submodel of η_{q+1} is that, for any θ_i of η_{q+1}, no other θ_j of η_{q+1} is made indeterminate by setting $\theta_i = 0$.

Suppose $\eta_q \in M_q$ is a direct submodel of $\eta_{q+1} \in M_{q+1}$ obtained by setting one of the parameters in η_{q+1}, θ_i, say, to 0. Let $\hat{L}_q$ and $\hat{L}_{q+1}$ be the maximized likelihoods obtained by fitting η_q and η_{q+1}. Then, fitting η_q is clearly equivalent to fitting η_{q+1} under the null hypothesis $H_0 : \theta_i = 0$. The likelihood ratio (LR) test of eqn (3.9) in this case is

$$T = 2(\ln \hat{L}_{q+1} - \ln \hat{L}_q), \tag{15.2}$$

and from eqn (3.13) this has the χ^2 distribution with just one degree of freedom, as the null hypothesis involves just the one component θ_i. We can therefore use T to give a regular test of $H_0 : \eta = \eta_q$ versus $H_1 : \eta = \eta_{q+1}$, with

$$H_0 : \theta_i = 0 \text{ rejected at significance level } \alpha \text{ if } T > \chi_1^2(1 - \alpha), \tag{15.3}$$

where $\chi_1^2(1 - \alpha)$ is the upper α quantile of the χ^2 distribution with one degree of freedom.

We define a *nested lattice* to be a directed, connected acyclic graph whose nodes are submodels of a full model, whose arrow links $\eta_q \to \eta_{q+1}$ are all *direct links* with η_q, a direct submodel of η_{q+1}. A *full nested lattice* (of a full model), is the nested lattice that includes all possible submodels and all possible direct links. For a full nested lattice, the submodel $\eta = 0$, which will be denoted by η_0, is the only *source* node. The full model is the only *sink* node.

We illustrate this terminology by first considering the linear model in a little more detail. We will then indicate how embeddedness and indeterminacy need to be accounted for in the nonlinear case.

15.2 The Linear Model

In a linear model, the dependent variable of interest is a (scalar) continuous random variable, denoted by Y, that is linearly dependent on P explanatory variables $X_j, j = 1, 2, \ldots, P$. The model selection problem usually focuses on which explanatory variable should be included in the fitted model. However, it will be simpler for our purposes to focus on parameters rather than explanatory variables.

Consider the following simple example in which our full model is the linear model

$$Y = \mu + aX_1 + bX_2 + \varepsilon, \tag{15.4}$$

where μ, a, and b are the three parameters in the regression function. There are $2^3 = 8$ submodels which we can set out in a nested lattice as depicted in Figure 15.1. Thus, the full model $\eta = \mu + aX_1 + bX_2$ is the only end node which we depict at the top in our figure, and the *null model* $\eta_0 = 0$ is the start node which we depict at the bottom. Moving (up) along any link in the figure between two models represents the addition of a further parameter to the lower model leading to the higher model; thus, $\mu \to \mu + ax_1$ represents starting with the model that is just a constant $\eta = \mu$ and then adding the linear term ax_1 to give the linear model $\eta = \mu + ax_1$ involving just the first explanatory variable X_1.

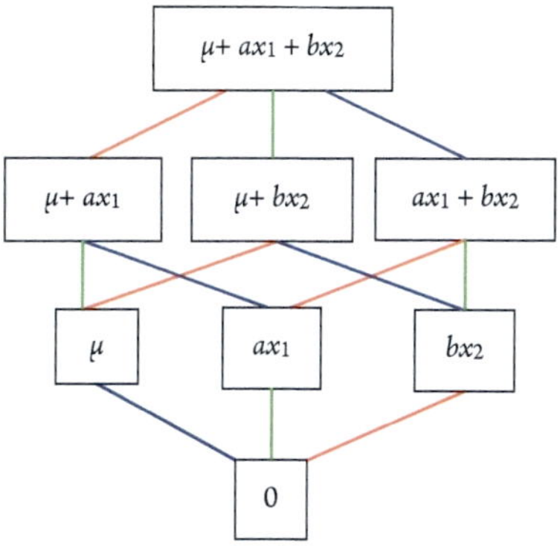

Figure 15.1 Example of nested linear regression model

Consider now the fitting of a submodel to a given data set. We estimate parameters using maximum likelihood. For a linear model such as in eqn (15.4), we can fit any submodel directly (provided the design matrix associated with the observations is nonsingular). However, as our objective is to obtain a good fit using a parsimonious submodel, we will use the forward stepwise regression fitting method (see Wu and Hamada, 2000), where we construct a *fitted* nested sublattice with individual terms added one at a time. Thus, in our example, the individual terms μ, ax_1, and bx_2 can be added one at a time in any order to the model, with the model refitted to the data after each term is included. Figure 15.1 depicts this with a blue link denoting the addition of μ, a red the addition of ax_1, and a green the addition of bx_2. In this example, and indeed for any other linear model, we have the following important

Linear Model Property

If the errors in the sample observations are independently and identically distributed (IID) normal random variables, then, when fitting the parameters forwards stepwise,

(i) the steps represented by the links, as in Figure 15.1, are all *regular* in the sense that a standard likelihood ratio (LR) test can be carried out as each additional parameter is included to see if it is significantly different from zero.

(ii) Moreover, whether a test is significant or not at any step, further parameters can continue to be included stepwise.

(iii) If the experimental design is orthogonal, the estimates and LR test results will always be the same for any given parameter irrespective of the step at which the parameter is fitted. Thus, we can fit the full model and carry out tests on the individual parameters based on any one sequence of steps.

Perhaps it is because of the ubiquity of use of the linear model and the last property that leads to the expectation that some similar property must hold in fitting nested nonlinear regression, and in particular that all parameters in a full model, once this is defined, can somehow be fitted, or at least must somehow be accounted for, even in situations when they are indeterminate. This is not the case in nonlinear models, which we consider in the next section.

15.3 Indeterminacy in Nested Models

In fitting a nested lattice, we call a directed path from η_0 to η_q *regular* and the nodes *regular* if

(i) all the submodels from η_0 to η_q inclusive can be consistently estimated, and

(ii) the LR tests corresponding to all the links on the path are regular.

Note that the second condition does not require that all the parameters fitted at each step along the path have to be found significantly different from zero.

For the linear model, any directed path from η_0 to any η_q will be regular, as additional terms can be added irrespective of whether fitted parameters along the path are significantly different from zero or not.

In fitting a nonlinear regression model using a lattice in which indeterminacy can occur, our aim is still, as in the linear case, to use a stepwise forward construction, building up a model by adding one parameter at a time. However, the order in which parameters are selected for inclusion is now important. Suppose we have two parameters $a \succ b$, so that b vanishes when $a = 0$. This is the situation discussed by Hartigan (1985) in which the LR test is inconsistent if the value of a is not known and it is possibly zero. We take the view that if $a \succ b$ and *we know that* $a = 0$, then it is meaningless to attempt to fit b. However b can be meaningfully fitted if we have already fitted *a and shown it to be significantly different from zero*. Our strategy is therefore to construct a nested lattice that is *conditionally regular* in that, for every condition $a \succ b$ that exists in the model, b is only considered for inclusion if a has already been fitted and found to be significantly different from zero. Thus, there is a key difference from the linear case. For a model that contains indeterminacy conditions, then

If the model is to be fitted by stepwise forward selection, the parameters cannot be selected for consideration in a totally arbitrary order.

A very simple example is the two-parameter model

$$\eta = b\exp(cx). \tag{15.5}$$

The parameter b controls the term $\exp(cx)$ that it multiplies, acting like an indicator of whether the model is η_0 or not. Consider the forward substitution lattice $\eta_0 \to b \to b\exp(cx)$. The first link $\eta_0 \to b$ represents fitting a constant b, and the associated LR test of $H_0^{(1)} : b = 0$ is regular. The second link represents fitting the model $b\exp(cx)$ where $b \succ c$, so that the associated LR test has null $H_0^{(2)} : c = 0$ that is regular only if $b \neq 0$. The parameters therefore have to be fitted in the order given in the lattice to enable the LR tests to be used to check that indeterminacy does not occur in the fitting process.

A three-parameter example is given by Seber and Wild (2003), who report the loss of identifiability in the model

$$\eta = a + b\exp(cx), \tag{15.6}$$

where setting $c = 0$ results in $(a + b)$ being only identifiable together. We write the model in the equivalent form

$$\eta = a + (d - a)\exp(cx), \tag{15.7}$$

where $d = a + b$, so that a vanishes if $c = 0$, i.e. we have the indeterminacy $c \succ a$.

As $\eta = b\exp(cx)$ is a submodel of (15.6), we also have the indeterminacy $b \succ c$. A satisfactory way of handling both is provided by the lattice shown in Figure 15.2(a), where the full nested lattice is the single directed path $0 \to b \to b\exp(cx) \to a + b\exp(cx)$.

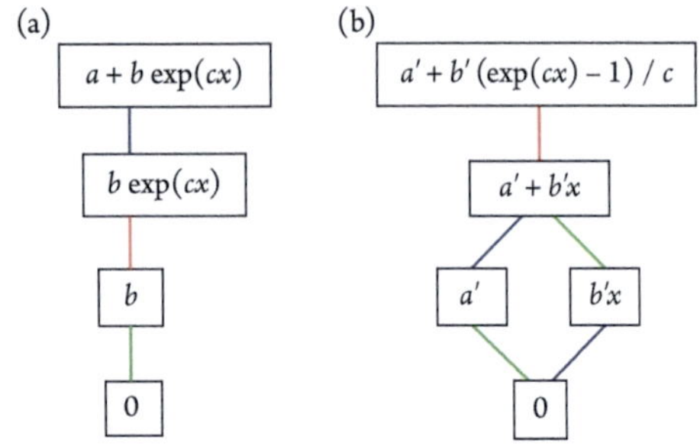

Figure 15.2 Two versions of nested exponential regression model of eqn (15.6) given by Seber and Wild (2003).

The first step fits $\eta = b$ together with the LR test of $H_0^{(1)} : b = 0$. This is a regular test. However, in the second step, $b \to b\exp(cx)$ involves fitting the model $\eta = b\exp(cx)$, where $b \succ c$. This is where the first LR test is important, as if the null of that test $H_0^{(1)} : b = 0$ is rejected, then c will not be indeterminate. The estimates can then be consistently estimated, and in particular with the LR test of $H_0^{(2)} : c = 0$ then consistent. The final step, $b\exp(cx) \to a + b\exp(cx)$ involves the previously mentioned indeterminacy $c \succ a$, as essentially identified by Seber and Wild. This shows that the second LR test is also important, as we require that $H_0^{(2)} : c = 0$ be rejected if the full model is to be successfully fitted.

Seber and Wild tried to handle the $c \succ a$ indeterminacy in a different way, using a reparametrization of the model to

$$\eta = a' + b'(\exp(cx) - 1)/c, \tag{15.8}$$

so that if $c \to 0$ the model becomes

$$\eta = a' + b'x. \tag{15.9}$$

Figure 15.2(b) gives the lattice for the reparametrized model (15.8). Note that though this eliminates the $c \succ a$ indeterminacy (in the original parametrization), it does *not* resolve the $b \succ c$ indeterminacy. Seber and Wild encountered this problem also, observing that the model becomes insensitive to the value of c when $b \approx 0$. The problem here is the indeterminacy $b \succ c$, which shows that if $b \approx 0$, then c becomes essentially meaningless. In Figure 15.2(b), the parameters are a', b', and c. The final link $a' + b'x \to a' + b'(\exp(cx) - 1)/c$ is regular only if $b' \neq 0$, so the LR test of $H_0 : b' = 0$ has to be rejected before this final step can be taken. This test is carried out when b' is fitted, which can be either the first or second step, as a' and b' can be fitted in either order.

Returning to the $c \succ a$ indeterminacy, the reader will recognize (15.9) to be an example of an embedded model of (15.6). In the next two sections, we show that the form of this $c \succ a$ indeterminacy occurs quite generally, where (i) indeterminacy will always occur when a model contains an embedded model which is the best fit, but that (ii) an indeterminacy arising in this way can always be handled simply by reparametrizing the full model so that the embedded model becomes a regular special case.

15.3.1 Link with Embedded Models

Consider the situation where a full model contains an embedded model obtained by letting $\theta_i \to 0, \theta_j \to \infty$, with $\theta_i\theta_j$ constant. Cheng and Traylor (1995) show that such an embedded model will cause a problem of indeterminacy if it is the best fit.

We write the parameters as $\boldsymbol{\theta} = (\theta_i, \theta_j, \boldsymbol{\theta}')$ so that written in full, the log-likelihood is $L(x; \theta_i, \theta_j, \boldsymbol{\theta}')$. For simplicity, we write the log-likelihood as $L(x; \theta_i, \theta_j)$ when we are focusing on its dependence on θ_i and θ_j only. Suppose that when $\theta_i \to 0, \theta_j \to \infty$ with $\theta_j' = \theta_i\theta_j$ fixed, then $L(x; \theta_i, \theta_j) \to L_0(x; \theta_i\theta_j) = L_0(x; \theta_j')$, where L_0 is the log-likelihood of the embedded model. That is, L is of the form

$$L(x; \theta_i, \theta_j) = L_0(x; \theta_i\theta_j) + L_1(x; \theta_i, \theta_j),$$

where $L_1 \to 0$ as $\theta_i \to 0$ and $\theta_j \to \infty$, with $\theta_i\theta_j = \theta_j'$ fixed. Equivalently, $L_1(x; \theta_i, \theta_j'/\theta_i) \to 0$ as $\theta_i \to 0$ and with θ_j' fixed. Suppose now that

$$L_1(x; \theta_i, \theta_j'/\theta_i) = \sum_k c_k \theta_i^{\alpha_k} \left(\frac{\theta_j'}{\theta_i}\right)^{\beta_k}.$$

This tends to zero as $\theta_i \to 0$ with θ_j' fixed. This implies $\alpha_k > \beta_k$, and so

$$L_1(x; \theta_i, \theta_j'/\theta_i) = \sum_k c_k \theta_i^{\gamma_k} (\theta_j')^{\beta_k},$$

where $\gamma_k = \alpha_k - \beta_k > 0 \; \forall k$. Then

$$L_1(x; \theta_i, \theta_j'/\theta_i) = \sum_k c_k \theta_i^{\gamma_k} \theta_i^{\beta_k} \theta_j^{\beta_k},$$

and so each term involving θ_j disappears due to the presence of θ_i. Thus, if an embedded model is obtained by letting $\theta_i \to 0, \theta_j \to \infty$ with $\theta_i\theta_j = \theta_j'$ constant, then $\theta_i \succ \theta_j$.

When an indeterminacy occurs in this way through the existence of an embedded model, then instability in parameter estimates will occur if the embedded model is the best fit. However, the problem is easily resolved simply by reparametrizing the full model so that the embedded model becomes a standard special case. We show in the next section that this automatically removes the indeterminacy.

15.4 Removable Indeterminacies

In the previous section, we showed that an indeterminacy will arise if we try to fit a sub-model that is an embedded model of the full model. The indeterminacy can, however, be removed simply by a reparametrization that makes the embedded model into a regular model.

Suppose $\theta_i \succ \theta_j$, and that L—or, in the regression case, η—can be expanded as a Taylor series about $\boldsymbol{\theta} = \mathbf{0}$. Then, in this expansion, any term containing θ_j must also contain θ_i. Write the kth term in which θ_j appears as

$$\ldots a_k \theta_i^{i_k} \theta_j^{j_k} \ldots = a_k (\theta_i^{i_k/j_k} \theta_j)^{j_k} \ldots, \tag{15.10}$$

and let the smallest value obtainable for the ratio i_k/j_k in the expansion be called r. Then, if $i_k/j_k \geq r > 0 \; \forall k$, with equality for some k, we can replace θ_i and θ_j in η by θ_i' and θ_j', where

$$\begin{aligned} \theta_i' &= \theta_i \\ \theta_j' &= \theta_i^r \theta_j. \end{aligned} \tag{15.11}$$

Use of the relationship

$$\theta_j = \frac{\theta_j'}{\theta_i'^r} \tag{15.12}$$

in (15.10) where the ratio is r leaves at least one term containing θ_j' and not θ_i'. Hence, setting $\theta_i' = 0$ does not now make θ_j' indeterminate. Note that it is important that the ratio $i_k/j_k > 0$ for all terms involving both θ_i and θ_j in the original expansion, because substitution of (15.12) would otherwise lead to a non-convergent sequence when we set $\theta_i' = 0$.

As already mentioned, we can interpret the foregoing analysis as one where the indeterminacy is caused because we are trying to fit an embedded model of the original parametrization. In the original parametrization, θ_j vanishes when $\theta_i = 0$. However, in the reparametrization (15.11), we have $\theta_j' = \theta_i^r \theta_j$ instead of θ_j, and can keep θ_j' fixed as $\theta_i' = \theta_i \to 0$, but only if $\theta_j \to \infty$ in the original parametrization, with this unstable parameter limit corresponding to the embedded model. The effect of the reparametrization is to turn the embedded model into a regular model. This is precisely what has occurred in the example of eqn (15.7), where

$$\begin{aligned} \eta &= a + (d - a)\exp(cx) \\ &= d\exp(cx) + a(1 - \exp(cx)), \end{aligned}$$

so that a vanishes when $c = 0$, i.e. $c \succ a$. Reparametrizing to

$$\eta = d\exp(cx) + a'(\exp(cx) - 1)/c$$

(which is equivalent to (15.8)), we can set $\theta_i = \theta_i' = c, \theta_j = a, \theta_j' = a'$, with the defining relation $\theta_j' = \theta_i^r \theta_j$ in (15.11) taking the form

$$a' = \theta_j' = \theta_i^r \theta_j = ca, \text{ where } r = 1.$$

Neither a' nor d become indeterminate as $c \to 0$, as we obtain the linear model $\eta = d + a'x$ of (15.9), showing this to be an embedded model in the original parametrization.

In the next section, we consider three explicit examples in more detail.

15.5 Three Examples

The first example in this section illustrates the situation where the regression contains more than one term, with each term involving a different set of parameters and where indeterminacy arises not only amongst the parameters just within a set, but also between the parameters in different sets.

The other two are numerical examples of fitting a nonlinear regression model using the forward stepwise method outlined in Section 15.1.

15.5.1 A Double Exponential Model

In this example, we consider the double-exponential model

$$\eta = \alpha + \beta_1 \exp(\gamma_1 x) + \beta_2 \exp(\gamma_2 x), \tag{15.13}$$

focusing on models in the neighbourhood $\boldsymbol{\theta} = (\alpha, \beta_1, \beta_2, \gamma_1, \gamma_2) = \mathbf{0}$. This is a more complicated model than we have discussed so far.

The simplest case is what we have encountered previously in examples like (15.8). Thus, if $\gamma_1 = 0$, then α and β_1 are observable only as $(\alpha + \beta_1)$. The model can be reparametrized using $a' = \alpha + \beta_1$ to give

$$\eta = \alpha' \exp(\gamma_1 x) + \alpha(1 - \exp(\gamma_1 x)) + \beta_2 \exp(\gamma_2 x).$$

Now $\gamma_1 \succ \alpha$ so let $\beta' = \alpha\gamma_1$, giving

$$\eta = \alpha' \exp(\gamma_1 x) + \beta' \left(\frac{1 - \exp(\gamma_1 x)}{\gamma_1}\right) + \beta_2 \exp(\gamma_2 x).$$

Then, if $\gamma_1 \to 0$, the submodel

$$\eta = \alpha' + \beta' x + \beta_2 \exp(\gamma_2 x) \tag{15.14}$$

is obtained.

A different problem occurs if $\gamma_1 - \gamma_2 \to 0$, as β_1 and β_2 are then observable only as $(\beta_1 + \beta_2)$. Taking $\beta' = \beta_1 + \beta_2$ and $\delta = \gamma_1 - \gamma_2$ in the full model (15.13) gives the reparametrized form

$$\eta = \alpha + \beta' \exp((\gamma_1 - \delta)x) + \beta_1 \exp(\gamma_1 x)(1 - \exp(-\delta x)).$$

Now $\delta \succ \beta_1$, so let $\beta'' = \beta_1 \delta$ to give

$$\eta = \alpha + \beta' \exp((\gamma_1 - \delta)x) + \beta'' \exp(\gamma_1 x)\left(\frac{1 - \exp(-\delta x)}{\delta}\right).$$

Then, $\delta \to 0$ gives the submodel

$$\eta = \alpha + (\beta' + \beta'' x)\exp(\gamma_1 x). \tag{15.15}$$

Consider now the submodel just obtained in eqn (15.14), which we can write as

$$\eta = \alpha + \beta_1 x + \beta_2 \exp(\gamma_2 x).$$

This has an embedded model in its own right. We can obtain this by again identifying potentially indeterminate parameters, however, as an alternative, we use a Taylor series approach. We have

$$\alpha + \beta_1 x + \beta_2 e^{\gamma_2 x} = \alpha + \beta_1 x + \beta_2 + \beta_2\gamma_2 x + \frac{1}{2}\beta_2\gamma_2^2 x^2 + O(\gamma_2^3).$$

If, therefore, we set

$$\alpha = \alpha' - 2\frac{\beta_2'}{\gamma_2^2},\ \beta_1 = \beta_1' - \frac{2\beta_2'}{\gamma_2},\ \beta_2 = \frac{2\beta_2'}{\gamma_2^2},$$

where α', β_1', and β_2' are regarded as arbitrary but fixed as $\gamma_2 \to 0$, then we can invert the expressions to get

$$\alpha' = \alpha + 2\frac{\beta_2'}{\gamma_2^2},\ \beta_1' = (\beta_1 + 2\frac{\beta_2'}{\gamma_2}),\ \beta_2' = \frac{1}{2}\beta_2\gamma_2^2.$$

We then have

$$\begin{aligned}
&\alpha + \beta_1 x + \beta_2 + \beta_2\gamma_2 x + \frac{1}{2}\beta_2\gamma_2^2 x^2 + O(\gamma_2^3)\\
&= \alpha + 2\frac{\beta_2'}{\gamma_2^2} + (\beta_1 + 2\frac{\beta_2'}{\gamma_2})x + \beta_2' x^2 + O(\gamma_2^3)\\
&= \alpha' + \beta_1' x + \beta_2' x^2 + O(\gamma_2^3),
\end{aligned}$$

so that we get the submodel

$$\eta = \alpha' + \beta_1' x + \beta_2' x^2$$

when $\gamma_2 \to 0$.

Using similar arguments for other cases, we identify the entire lattice for the double-exponential model, and this is shown in Figure 15.3.

It must be emphasized that for simplicity we have written each submodel in the lattice in a simple canonical way. A consequence is that the meaning of any given parameter symbol is not necessarily the same in different models. This includes models $\eta_q \to \eta_{q+1}$ connected by a direct link, where η_q is an embedded model of η_{q+1}. For example, one of the top links is shown as

$$\alpha + (\beta_1 + \beta_2 x)e^{\gamma_1 x} \to \alpha + \beta_1 e^{\gamma_1 x} + \beta_2 e^{\gamma_2 x}, \tag{15.16}$$

where the right-hand model is the full double exponential model of eqn (15.13). In the discussion just given, we showed that we can find a reparametrization of the full model so that the stable four-parameter submodel $\alpha + (\beta' + \beta'' x)e^{\gamma_1 x}$, as given in eqn (15.15), is obtained when $\gamma_2 \to \gamma_1$. This model clearly has the same form as the left-hand model in the link (15.16). However, for simplicity of presentation in Figure 15.3, we have just used the same symbols β_1 and β_2 in both models, though the derivation of (15.15) shows we actually have $\beta' = (\beta_1 + \beta_2)$ and $\beta'' = \beta_1(\gamma_1 - \gamma_2)$, and these relationships must hold in comparing the two fitted models.

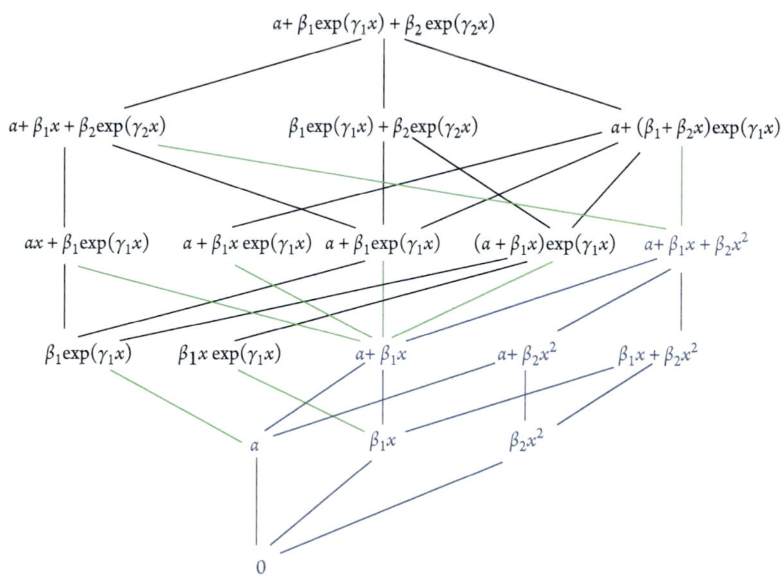

Figure 15.3 Full nested lattice of the double exponential regression model of eqn (15.13).

Note, however, that the LR test associated with a link depends only on the maximized log-likelihood values of the two linked models, and not on the precise parametrization used. The parametrization given in the figure can therefore always be used in fitting a particular submodel.

15.5.2 Morgan-Mercer-Flodin Model

Dudzinski and Mykytowycz (1961) give a data set linking the age of a rabbit to eye lens thickness. The data have been considered by Ratkowsky (1983), who discusses fitting a Morgan-Mercer-Flodin (MMF) model to the 18-point data set given in Table 15.1 below. The observations are as in eqn (15.1), where we take the variant

$$\eta = \frac{a + bx^{1+d}}{1 + cx^{1+d}} \tag{15.17}$$

as our full model with the regression parameters $\boldsymbol{\theta} = (a, b, c, d)$. These parameters together with the error variance σ^2 are estimated by ML. We focus on models with parameter values in an open neighbourhood including $\boldsymbol{\theta} = \mathbf{0}$ representing models close to a ratio of linear expressions. Figure 15.4 shows the lattice corresponding to eqn (15.17), where submodels of the full model are obtained in a very obvious way by setting subsets of the parameters equal to zero. The figure includes the conditionally regular LR test values as obtained for the eye lens data, using the stepwise fitting procedure of Section 15.1, with the LR test of eqn (15.2) and null hypothesis H_0 of eqn (15.3) used at each step; with each parameter, θ_i, being added only when all parameters θ_j, for which the condition

Table 15.1 18-point reduced data set given by Ratkowsky (1983). The dry weight of eye lens (Y) in milligrams is given as a function of age (X) in days

X	Y	X	Y
15	22.75	218	173.03
29	40.55	227	173.73
50	63.47	246	176.13
64	79.09	300	186.09
75	86.10	317	216.41
91	101.70	357	195.31
125	134.90	535	209.70
147	152.20	660	231.00
183	153.22	768	232.12

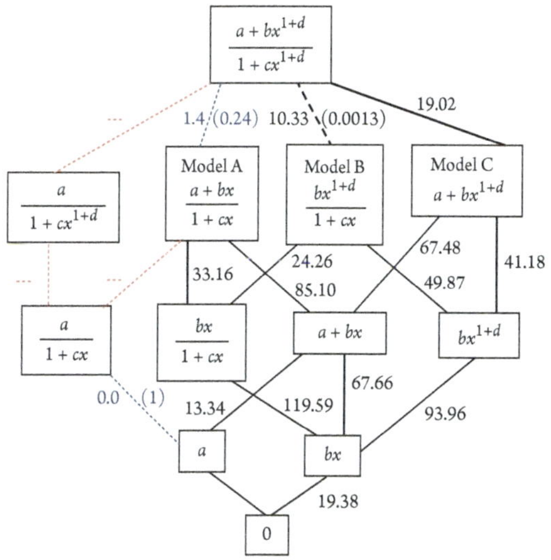

Figure 15.4 Lattice of submodels of the Morgan-Mercer-Flodin model (15.17) fitted to eye lens data, showing LR test value for each link. Each number is the chi-squared LR test value with its p-value in brackets (not shown if $p < 0.001$). The link where the test value is 0 indicates c is not significantly different from zero, with further steps along that path unreliable or redundant, and which are therefore not made.

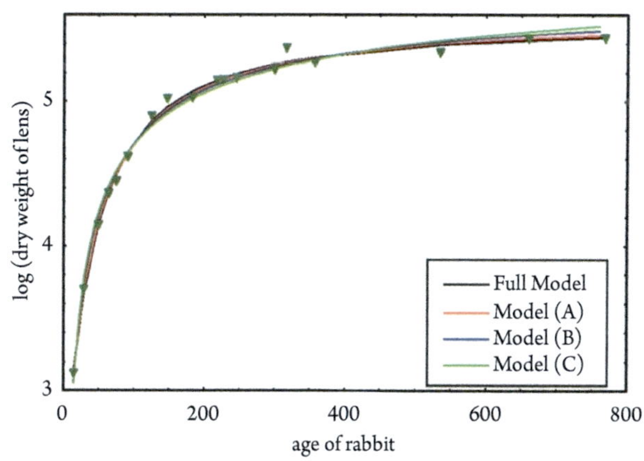

Figure 15.5 Fits of Morgan-Mercer-Flodin models to the rabbit eye lens thickness data.

$\theta_j \succ \theta_i$ holds, have already been fitted and tested to be significantly different from zero. From Figure 15.4, we see that the best model is $\eta = (a + bx)/(1 + cx)$, in accordance with Ratkowsky's findings. Figure 15.5 shows the fitted regression lines corresponding to the full model and this selected submodel.

The usual limitation of a stepwise approach should be noted. The LR tests are not independent, so calculation of an overall confidence level of the fitted model by multiplication of the confidence levels of individual tests is only approximate.

The formulation so far allows only an exploration of parameter values in the neighbourhood of $\boldsymbol{\theta} = \mathbf{0}$. However, it may be that the parameter values well away from $\boldsymbol{\theta} = \mathbf{0}$ cannot be ruled out. In essence, we need to allow the possibility of an embedded model being a good fit. For instance, in the MMF model (15.17), letting $a, b, c \to \infty$ subject to $a/c = \alpha, b/c = \gamma$ finite gives

$$\lim_{a,b,c\to\infty} \frac{a + bx^{1+d}}{1 + cx^{1+d}} = \frac{\alpha}{x^{1+d}} + \gamma.$$

Similarly, letting $a, c \to \infty$ in the model $\eta = a/(1 + cx^{1+d})$ subject to $a/c = \alpha$ finite gives

$$\lim_{a,c\to\infty} \frac{a}{1 + cx^{1+d}} = \frac{\alpha}{x^{1+d}},$$

and so on. These are embedded models of the original parametrization, and can be added to the lattice and treated in the same way as any other model of the lattice. However, to retain numerical stability when making any test, a parametrization should be used where both the null and the alternative models are stable special cases. For example, in the MMF model, a and a/x are both direct submodels of $a/(1 + cx)$, but a/x is embedded and is obtained at parameter values about a point well away from $\boldsymbol{\theta} = \mathbf{0}$, and so the alternative parametrization $a/(c' + x)$ is needed to make it into a stable special case. Augmenting models in this way by, in effect, amalgamating two lattices, each of which is essentially an embedded version of the other, is discussed by Cheng and Traylor (1995). Note, however, that because the two lattices do not correspond to parameter values in the same neighbourhood, no single parametrization can be used for which all submodels are stable special cases. However, tests of any of the models in the lattice can be carried out as long as a stable parametrization is used. Therefore, in the link $a/x \to a/(1 + cx)$, it is required that the test be made in the form of the stable parametrization $a/x \to a/(c' + x)$.

Different parametrizations of a model may be of interest, depending on the context of the application. Continuing with the MMF model, a further variant is the regression function

$$\eta = \frac{a + bx^d}{1 + cx^d}, \tag{15.18}$$

a generalization of the Box-Cox model focusing on models where the power of x is near $d = 0$. The model suffers from the indeterminacies $d \succ c$ and $d \succ b$, so a regular lattice for models near $\boldsymbol{\theta} = \mathbf{0}$ cannot be directly constructed. If x^d is replaced, as in the Box-Cox model, by $(x^d - 1)/d$, then this indeterminacy is removed, and a regular lattice can be constructed. Note that an embedded model in the original parametrization is obtained when $d \to 0$. Hence, the lattice obtained will be exactly that of Figure 15.4 if occurrences of x^{1+d} are replaced by $(x^d - 1)/d$ and if individual terms involving x are replaced by $\ln x$. Although such a reparametrization is no longer about $a = b = c = 0$ in the original parametrization, this is not unsatisfactory if the concern is with models involving d near to zero.

15.5.3 Weibull Regression Model

In this last example, we consider the Weibull regression model

$$\eta = \alpha + \beta \exp(\gamma x^{\delta}). \tag{15.19}$$

We include this example to illustrate the point that the meaning of a particular parameter symbol may change between submodels.

The model contains several potentially indeterminate cases. If $\gamma \to 0$, then α and β only appear as $\alpha + \beta$. However, writing $\alpha' = \alpha + \beta$ and $\beta' = \beta\gamma$, the model (15.19) becomes

$$\eta = \alpha' + \beta' \left(\frac{\exp(\gamma x^{\delta}) - 1}{\gamma} \right), \tag{15.20}$$

so that if $\gamma \to 0$, we obtain the special case

$$\eta = \alpha' + \beta' x^{\delta}. \tag{15.21}$$

The model of eqn (15.19) can also be obtained by taking $\beta'' = \beta \exp(\gamma)$ and $\gamma' = \gamma\delta$ in (15.19), giving

$$\eta = \alpha + \beta'' \exp\left(\gamma' \frac{(x^{\delta} - 1)}{\delta} \right), \tag{15.22}$$

so that if $\delta \to 0$, we obtain the special case

$$\eta = \alpha + \beta'' x^{\gamma'}, \tag{15.23}$$

which is equivalent to (15.21), but where $\alpha \neq \alpha'$, $\beta' \neq \beta''$ and $\delta \neq \gamma'$.

Thus, both reparametrizations (15.20) and (15.22) have the same model (15.21) or (15.23) as a regular submodel, but which is embedded in the original parametrization (15.19). This has an important numerical implication in that if we do not reparametrize, but the embedded model is the best fit, or even if it is just close to it, then numerically we can end with very different estimates of the parameter values, depending on how the fitted model is being approached by the fitting process. Our numerical example will illustrate this.

The model (15.21)/(15.23) itself has an embedded model. Setting $\beta'' = \beta'\delta$ and $\alpha'' = \alpha' + (\beta''/\delta)$ in (15.21), and letting $\delta \to 0$, we obtain

$$\eta = \alpha'' + \beta'' \ln x.$$

Table 15.2 Ballistic current (Y) in picoAmps crossing a gold/gallium arsenide interface, as a function of energy (X) in eVolts

X	Y	X	Y	X	Y	X	Y
0.438	132	0.755	133	1.093	134	1.410	224
0.458	132	0.795	132	1.113	138	1.450	241
0.497	131	0.815	134	1.152	140	1.470	302
0.517	131	0.855	134	1.172	145	1.510	342
0.557	132	0.874	134	1.212	148	1.530	465
0.577	131	0.914	134	1.232	154	1.569	550
0.616	132	0.934	133	1.271	160	1.589	720
0.636	132	0.974	134	1.291	166	1.629	814
0.676	132	0.994	134	1.331	171	1.649	1092
0.696	132	1.033	132	1.351	186		
0.735	133	1.053	132	1.391	199		

Finally, (15.19) has the regular submodel

$$\eta = \beta \exp(\gamma x^{\delta}). \tag{15.24}$$

But this is simply (15.22) with $\alpha = 0$; so, using (15.23) with $\alpha = 0$, we have that (15.24) has $\beta'' x^{\gamma'}$ as an embedded model.

Use of the model (15.19) is illustrated by fitting it to the data set of Table 15.2, given by Traylor (1994), where the measurements are of two passes of the ballistic current, measured in picoAmps, crossing a gold/gallium-arsenide interface as a function of kinetic energy (measured in eV). The current increases exponentially once the energy is above a threshold (the Schottky barrier height), and so gives a measure of the electronic structure of the interface. Though we have not modelled this threshold energy level explicitly, the three parameters of the exponential term give sufficient flexibility to represent the current flow quite accurately. The nested lattice resulting from the enumeration of all the submodels of the full model (15.19) is given in Figure 15.6.

Although there is a systematic difference between the two sets of measurements, the results of the two passes are combined simply to illustrate the fits of models in the lattice.

The Weibull regression model represents the data approximately as a background level together with an exponential term. The three parameters of the exponential term give sufficient flexibility to represent the current flow quite accurately.

Table 15.3 gives the parameter ML estimates of five fits.

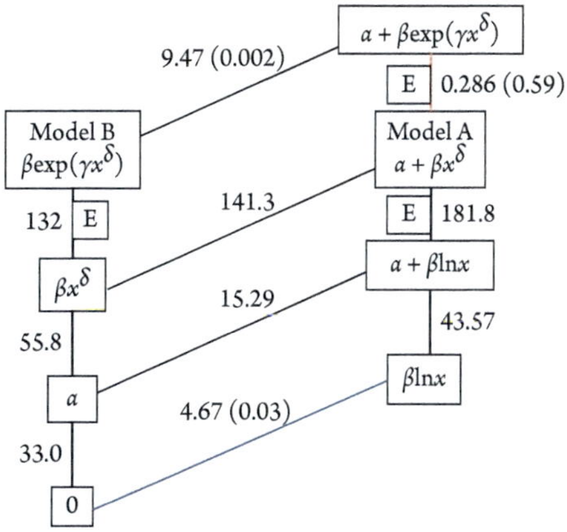

Figure 15.6 LR test results for Weibull regression model lattice when fitted to ballistic current data. For each link tagged with an 'E', the model with one fewer parameter is an embedded model of the other; the parameter symbols do not have the same meaning in the two models. Each number is the chi-squared value of the LR test for the link with the p-value in brackets (with cases where $p < 0.0001$ not shown).

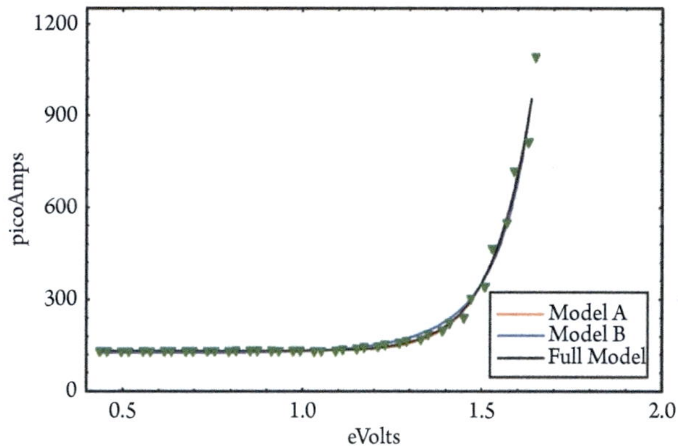

Figure 15.7 Fit of two three-parameter power models, models (A) and (B), to ballistic current data. The fit of the full Weibull model is also shown and is visually almost identical to model (A).

The first three fits, cases (a), (b), and (c) in Table 15.3, are, respectively, the original four-parameter model of eqn (15.19) and the two four-parameter reparametrizations as in eqn (15.20), case (b), and as in eqn (15.22), case (c). All three contain the three-parameter power model of eqn (15.21) or (15.23) as a special case, but where it is embedded in case (a). All three full four-parameter models give fits that are almost identical, with maximized log-likelihood values in the range $\hat{L} = -184.54$ to -184.52. The fits are shown as one curve in Figure 15.7, as all three are visually indistinguishable.

Table 15.3 Fit of regression models (8.6) and (8.8) to ballistic current data. The standard deviations are given in brackets

(a) Full model	$\eta = \alpha + \beta \exp(\gamma x^\delta)$
$\hat{\alpha} = -3844$ (1590)	$\hat{\beta} = 3977$ (1589)
$\hat{\gamma} = 0.000163$ (6.44×10^{-5})	$\hat{\delta} = 14.30$ (0.380)
$\hat{\sigma} = 19.58$ (2.30)	$\hat{L} = -184.519$
(b) Alternative full model	$\eta = \alpha + \beta' \exp(\gamma(x^\delta - 1)/\delta)$
$\hat{\alpha} = 132.9$ (4.44)	$\hat{\beta}' = 0.988$ (1.63)
$\hat{\gamma} = 11.97$ (6.75)	$\hat{\delta} = 0.518$ (1.25)
$\hat{\sigma} = 19.59$ (2.14)	$\hat{L} = -184.544$
(c) Alternative full model	$\eta = \alpha + \beta' \exp(\gamma'(x^\delta - 1)/\delta)$
$\hat{\alpha}' = -3810$ (11289)	$\hat{\beta}' = 3944$ (11291)
$\hat{\gamma}' = 0.00252$ (0.0090)	$\hat{\delta} = 15.29$ (2.26)
$\hat{\sigma} = 19.58$ (2.14)	$\hat{L} = -184.519$
(d) 3-par power model A	$\eta = \alpha + \beta' x^{\gamma'}$
$\hat{\alpha} = 133.9$ (3.64)	$\hat{\beta}' = 0.4746$ (0.11)
$\hat{\gamma}' = 15.12$ (0.48)	
$\hat{\sigma} = 19.64$ (2.14)	$\hat{L} = 184.662$
(e) 3-par model B	$\eta = \beta' \exp(\gamma' x^\delta)$
$\hat{\beta}' = 126.9$ (4.37)	$\hat{\gamma}' = 0.043$ (0.0076)
$\hat{\delta} = 7.82$ (0.34)	
$\hat{\sigma} = 21.91$ (2.39)	$\hat{L} = -189.251$

However, examination of the fitted parameter values shows that two of the fits are very similar, whilst the other is rather different. The underlying reason is that the best full model fit is actually almost identical to the fit obtained with the three-parameter power model of eqn (15.21) or eqn (15.23). The fit of this model is given in Table 15.3 as case (d), where the model is called model A. Figure 15.7 shows that the model A fit is visually almost identical to the full model fit.

The parameter estimates of these fits are as follows.

Case (a) is the fit of the original model of eqn (15.19), with parameter estimates $\hat{\alpha} = -3844$, $\hat{\beta} = 3977$, $\hat{\delta} = 14.3$, and a small $\hat{\gamma} = 0.000163$. These values are actually unstable, as will become clear when we look at the other fits.

Case (b) is where we take the full model as in (15.22), with parameter estimates $\hat{\alpha} = 132.9$, $\hat{\beta}'' = 0.988$, $\hat{\gamma}' = 11.97$, where $\hat{\delta} = 0.518$ is relatively small, so that the model is actually close the power model version of eqn (15.23).

Case (c) is where we fit the full model as in eqn (15.20), giving estimates $\hat{\alpha}' = -3810$, $\hat{\beta}' = 3944$, $\hat{\delta} = 15.29$, and $\hat{\gamma} = 0.0025$, so that this fitted model matches that of the power version of eqn (15.21). However, the fit is actually essentially identical to the full model of case (a), showing that the maximization process has not converged to the stable power model version of eqn (15.23), but instead to that of eqn (15.21), as occurred in case (a).

Cases (b) and (c) therefore serve as a warning that where reparametrization to remove an embedded model can be done in two different ways, it will not be clear which will be numerically stable. For instance, if we had started with case (a) and then simply used resulting estimates to calculate the parameters of the case (c) model of eqn (15.20), we would have obtained $\hat{\alpha}' = \hat{\beta} - \hat{\alpha} = 3977 - 3844 = 133$, $\hat{\beta}' = \hat{\beta}\hat{\gamma} = 3977 \times 0.000163 = 0.648$, $\hat{\delta} = 14.3$, $\hat{\gamma} \simeq 0$, which correspond much more closely to the case (b) estimates.

The case (d) fit of the model $\eta = \alpha + \beta' x^{\gamma'}$ gave estimates $\hat{\alpha} = 133.9$, $\hat{\beta}' = 0.4746$, and $\hat{\gamma}' = 19.64$, comparable to the case (b) estimates.

Summarizing: the large estimates of $\hat{\alpha}$ and $\hat{\beta}$ and their large standard deviations in cases (a) and (c) indicate that neither fit is satisfactory. Cases (b) and (d) are satisfactory.

Case (e) gives the parameter estimates for

$$\eta = \beta \exp(\gamma x^{\delta}),$$

the other three-parameter model in the lattice. It is labelled as model B in Table 15.3(e) and Figure 15.7. Though stable, the fit is not so satisfactory in this case.

Figure 15.6 also includes the LR test values obtained using forward stepwise regression fitting, corroborating the discussion just given. The LR test of the link between the power model of eqn (15.21) and the full model of eqn (15.19) is not at all significant, showing that the power model would be selected as the most parsimonious model giving an adequate fit.

These examples give instances where the indeterminate parameters problem, and therefore the embedded model problem, can be handled by reparametrization. The directed lattice approach is general and simple, which may appeal to a practitioner. However, if $\theta_i \succ \theta_j$ and $\theta_j \succ \theta_i$, and neither is removable, then a regular lattice cannot be constructed. A straightforward way of accommodating such a situation within the framework of regular one-step tests is to insert an *intermediate* model between the models, where both θ_i and θ_j are missing and where both θ_i and θ_j are present. It is this method that we now explore.

15.6 Intermediate Models

We begin with an example. The regression model

$$\eta = \alpha(\exp(\beta x) - 1)$$

suffers from the indeterminacies $\alpha \succ \beta$ and $\beta \succ \alpha$. Therefore, the situation $\eta = 0$ can occur if $\alpha = 0$ or $\beta = 0$ when the regular test of $\eta = \alpha(\exp(\beta x) - 1)$ versus $\eta = 0$ cannot be made using ML.

In general, a regular lattice for η cannot be constructed if $\theta_i \succ \theta_j$ and $\theta_j \succ \theta_i$. Unless one or other indeterminacy is removable, then a regular lattice cannot be obtained. In the previous example, the indeterminacy $\beta \succ \alpha$ happens to be removable. However, we consider instead an alternative approach. This is simply to introduce a restricted version of the full model, obtained by setting β to a particular value; we call this an intermediate model. The approach is then to test H_0 against this restricted version of H_1, for which the test is regular. For instance, if we take $\beta = 1$ for our intermediate model, we obtain the lattice $0 \to \alpha(\exp(x) - 1) \to \alpha(\exp(\beta x) - 1)$, and a series of regular tests may now be used. Note that this approach effectively constitutes one specific cross-section through Davies' method discussed in Section 14.2.1, and so the two methods are locally equivalent here. Whereas Davies' method required calculation of the test statistic for a range of parameter values, the intermediate model approach selects only one of these parameter values to test whether the inclusion of the parameter α is worthwhile.

Cheng and Traylor (1995) give a general method for introducing an intermediate model by taking a function $\psi = \psi(\theta_1, \theta_2)$ as a parameter with the property $\psi(0, 0) = 0$. If ψ can be inverted to give $\theta_2 = \theta_2(\theta_1, \psi)$, with $\theta_2(\theta_1, \psi) \neq 0$ when $\psi \neq 0$, then the lattice

$$\eta(\theta_1 = 0) \to \eta(\theta_1, \theta_2(\theta_1, 0)) \to \eta(\theta_1, \theta_2(\theta_1, \psi)) \tag{15.25}$$

can be taken to be regular. The usefulness of this approach will depend on the power of the two tests. This is dependent on how the intermediate model is positioned. We have not investigated ways of positioning the intermediate model in general, but simply consider a specific example.

15.6.1 Mixture Model Example

We apply the intermediate model approach to the mixture model $f(y; \xi, \theta) = (1 - \xi)\phi(y) + \xi\phi(y - \theta)$ of eqn (14.4), where $\phi(y) = (2\pi)^{-1/2}\exp(-y^2/2)$ is the PDF of the standard normal distribution. If $\xi = \theta_1$ and $\theta = \theta_2$, then setting $\psi = \theta - \xi$ gives $\theta = \psi + \xi$, and so (15.25) becomes

$$\begin{aligned} H_0 &: f = \phi(y) \\ H_1 &: f = (1 - \theta_1)\phi(y) + \theta_1\phi(y - \theta_1) \\ H_2 &: f = (1 - \theta_1)\phi(y) + \theta_1\phi(y - \theta_1 - \psi). \end{aligned}$$

Note that the test of H_0 versus H_1 is not regular in the full sense. Under H_1, the series expansion of L about $\boldsymbol{\theta} = \mathbf{0}$ is

$$L = S_1\theta_1^2 + \frac{1}{2}(S_2 - n)\theta_1^3 + \frac{1}{6}(S_3 - 3S_2 - 3S_1)\theta_1^4 + \ldots \tag{15.26}$$

where $S_j = \sum_1^n y_i^j$, and where θ_1 is non-negative (as it is exactly the mixing proportion ξ). As in Example 9.12 of Cox and Hinkley (1974), the log-likelihood has no linear term in θ_1, the situation discussed in Section 12.3.1. In the present case, as mentioned by Cox and Hinkley (1974), we can resolve the problem by considering estimation of θ_1^2 instead.

We assume that the higher-order terms in (15.26) are negligible, and so by taking the leading three terms we obtain

$$\frac{\partial L}{\partial \theta_1} = 2S_1\theta_1 + \frac{3}{2}(S_2 - n)\theta_1^2 + \frac{2}{3}(S_3 - 3S_2 - 3S_1)\theta_1^3. \tag{15.27}$$

If $S_1 < 0$, then $\partial L/\partial\theta_1 < 0$ near $\theta_1 = 0$, implying that a maximum occurs at $\hat{\theta}_1^2 = 0$. However, if $S_1 > 0$, then we need to examine (15.27) in more detail. From (15.27) we have

$$\frac{\partial L}{\partial \theta_1} = \theta_1\left[2S_1 + \frac{3}{2}(S_2 - n)\theta_1 + \frac{2}{3}(S_3 - 3S_2 - 3S_1)\theta_1^2\right],$$

and setting $\partial L/\partial\theta_1 = 0$ gives

$$\hat{\theta}_1^2 = \left[\frac{-3(S_2 - n)/2 \pm \sqrt{9(S_2 - n)^2/4 - 16S_1(S_3 - 3S_2 - 3S_1)/3}}{4(S_3 - 3S_2 - 3S_1)/3}\right]^2. \tag{15.28}$$

Under the null hypothesis, Y is $N(0, 1)$. Therefore, $E[S_1] = E[S_3] = 0$, and $E[S_2] = nE[y^2] = n$, with $\mathrm{Var}[S_j] = O(n)$. That is, S_1 and S_3 are $O(\sqrt{n})$, whilst S_2 is $n + O(\sqrt{n})$. Selecting the dominant terms in the expression gives the asymptotic result

$$\hat{\theta}_1^2 \simeq \left(\frac{\sqrt{16S_1S_2}}{4S_2}\right)^2 = \frac{S_1}{S_2}.$$

Thus, asymptotically $\hat{\theta}_1^2 = 0$ if $S_1 < 0$, whilst if $S_1 > 0$ $\hat{\theta}_1^2 = S_1/S_2$, which is $O(n^{-1/2})$.

From this analysis, we have that $\hat{\theta}_1^2 = 0$ with probability $\frac{1}{2}$, and is a positive folded $N(0, n^{-1})$ variable with probability $\frac{1}{2}$. Although $\hat{\theta}_1^2 \simeq S_1/S_2$, a more accurate estimate would be obtained from (15.28), selecting the negative square root, since θ_1 is non-negative. Then the test for the rejection of the null hypothesis would be based on $w > c$, where $w = \hat{\theta}_1^2 n^{1/2}$ and where c is the $100(1 - \alpha)$ standard normal percentile. Table 15.4 gives the proportion of times that the null hypothesis is rejected on the basis of this test. As can be seen, the test performs quite satisfactorily in this example, with the results comparing favourably to those given in Table 14.1 in Chapter 14.

15.6.2 Example of Methods Combined

In this section, we illustrate the use of the intermediate model approach in conjunction with occurrences of embedded models. As an example, we return to the regression

Table 15.4 Proportion of times H_0 is rejected for mixture model of eqn (14.4) with $\alpha = 0.1$, number of samples = 1000, sample size = 10, using an intermediate model

ξ	θ	P(reject)
0.0	–	0.094
0.5	0.5	0.318
0.5	1.0	0.618
0.5	1.5	0.859
0.5	2.0	0.949

function of eqn (14.3), which we write here as

$$\eta_3 = \psi_1(1 - \exp(\psi_3 x)) + \psi_2 \exp(\psi_3 x), \tag{15.29}$$

and which we shall now regard as the full model. To identify the submodels, we could begin by setting $\psi_1 = 0$ in the model, and then set $\psi_3 = 0$ in this submodel to obtain the regular path in the lattice $\eta_0 \to \psi_2 \to \psi_2 \exp(\psi_3 x) \to \eta_3$.

A second route would be to set $\psi_2 = 0$ in the full model to obtain the submodel

$$\eta_2 = \psi_1(1 - \exp(\psi_3 x)). \tag{15.30}$$

A problem now arises, because $\psi_1 \succ \psi_3$ and $\psi_3 \succ \psi_1$, and so a regular test of $H_0 : \eta_0$ versus $H_1 : \eta_2 = \psi_1(1 - \exp(\psi_3 x))$ cannot be constructed. In this case, we now have two obvious ways of overcoming the problem. Firstly, the indeterminacy $\psi_3 \succ \psi_1$ is removable using the method used in the example of Section 15.4: writing $\theta_1' = \psi_1\psi_3$ and replacing ψ_1 by θ_1'/ψ_3, we get the model $\eta = \theta_1' x$, as $\psi_3 \to 0$. We then have the regular path $\eta_0 \to \theta_1' x \to \theta_1'(1 - \exp(\psi_3 x))/\psi_3$. The model $\eta = \theta_1' x$ is exactly the model that would be obtained if we considered the embedded model approach directly. However, in the original parametrization (15.30), $\theta_1' x$ is an embedded model which is obtainable only as another parameter becomes infinite.

An alternative method would be to introduce a simple intermediate model. For instance, selecting $\psi_3 = 1$, say, in (15.30) gives the intermediate model $\eta = \psi_1(1 - \exp(x))$, and the path $\eta_0 \to \psi_1(1 - \exp(x)) \to \psi_1(1 - \exp(\psi_3 x))$ comprises a sequence of regular steps. Similarly, setting $\psi_1 = 1$, say, in (15.30) would also lead to a regular construction. Whether the intermediate model is obtained by setting ψ_3 or ψ_1 equal to some particular constant depends on the preferences of the practitioner.

15.7 Non-nested Models

The focus of this chapter has been on model building with nested models. A different problem is where we are fitting two models to a data sample, where the models are essentially quite different from each other. The models are therefore non-nested and the problem is non-standard, as the usual standard likelihood-based methods of comparing model fits are not applicable. Suppose, therefore, that $f_0(y; \boldsymbol{\theta}_0)$ and $f_1(y; \boldsymbol{\theta}_1)$ are the PDFs of the two models, and that f_0 and f_1 have different functional forms, with the parameter vectors $\boldsymbol{\theta}_0$ and $\boldsymbol{\theta}_1$ on which each depends not related and possibly of different dimensions. In two seminal papers, Cox (1961, 1962) examines this situation, pointing out two ways of formulating and analysing the problem. One way is where the problem is treated as one of significance testing, in which the null hypothesis H_0 is that f_0 is the correct model, whilst the alternative H_1 is that f_1 is the correct model, this being treated only as a general alternative for which high power is needed. In this first way the two models are therefore not treated symmetrically. The second way, which does treat the two models symmetrically, is where one simply selects the better model, using some appropriate criterion to make this choice.

Cox (1962) gives a method for handling each problem. Our main interest is in the second problem, where Cox (1962) suggests a method for selecting the best model. Our method of using an intermediate model to examine models in a stepwise way in the nested case is somewhat similar. However, for completeness, we shall also outline the significance testing method given by Cox (1962), and we do this first.

Let $\hat{\boldsymbol{\theta}}_0, \hat{\boldsymbol{\theta}}_1$ be the maximum likelihood estimators of the parameters in the two models when each is considered separately, and let $L_0(\hat{\boldsymbol{\theta}}_0)$ and $L_1(\hat{\boldsymbol{\theta}}_1)$ be the corresponding respective maximized log-likelihoods. To test the null hypothesis H_0 that f_0 is the true model, Cox (1962) proposes the following modified likelihood ratio as the significance test statistic:

$$T_0 = \{L_0(\hat{\boldsymbol{\theta}}_0) - L_1(\hat{\boldsymbol{\theta}}_1)\} - E_{\hat{\boldsymbol{\theta}}_0}\{L_0(\hat{\boldsymbol{\theta}}_0) - L_1(\hat{\boldsymbol{\theta}}_1)\}.$$

This compares the observed difference between the two log-likelihoods and an estimate of the expected difference under H_0. Cox shows that under H_0, T_0 is asymptotically normal with mean zero and variance $\mathrm{Var}(T_0)$, where the form of the latter is in general quite complicated. However, Cox (1962) derives consistent estimators for $\mathrm{Var}(T_0)$, giving formulas that enable them to be calculated. A simple example is given by Cox (1962), where the null is the lognormal model with PDF $f_0 = y^{-1}(2\pi\lambda)^{-1/2}\exp\{-(\ln y - \mu)^2/(2\lambda)\}$ and the alternative is the exponential distribution with PDF $f_1 = \beta^{-1}\exp(-y/\beta)$. Unfortunately, as pointed out by Atkinson (1970), the formula for T_0 given by Cox (1962, eqn 20) contains an error, though in the applications the correct form is used. The correct form is tractably explicit and is given in Atkinson (1970), eqns (50) and (51), with asymptotic variance as in eqn (52). In general, consistent estimators of $\mathrm{Var}(T_0)$ are not so easy to obtain. As mentioned in Royston and

Thompson (1995), the asymptotic normal approximation is generally not very accurate unless the sample size n is large.

Further work along the lines set out by Cox (1962) has been carried out by a number of authors. We do not discuss this further here, as this work is well reviewed by Pace and Salvan (1990), who develop a unified setting covering many pairs of distributions from the generalized exponential family. We refer the reader to Pace and Salvan (1990) for details. However, Pace and Salvan do make use of two ideas that we shall discuss further.

One idea is to construct what Pace and Salvan (1990) call an *encompassing* family with PDF, $\bar{f}$, say, which includes both the f_0 and f_1 models as special cases. Pace and Salvan also use the term 'embedding' family, but to avoid confusion with how embedding is used in this book, we will not use embedding in their sense. Though, loosely speaking, f_0 and f_1 are to be regarded as quite distinct, Pace and Salvan (1990) are able to define, quite precisely in multiparameter exponential families, the extent to which f_0 and f_1 can intersect, with duplicated parameters. In particular, they give conditions on this intersection in their Proposition 3, under which a uniformly most powerful similar (UMPS) test will exist that will reject H_0 for a given test statistic t' conditional on sufficient statistics s and on $y_{(1)}$ and $y_{(n)}$.

This use of a *conditional test* is the second idea used by Pace and Salvan (1990). Such a conditional UMPS test is attractive on two counts. Firstly, where they do exist, they are easily obtainable. Secondly, Pace and Salvan (1990) give simulation results for when f_0 is the lognormal distribution and f_1 is the gamma distribution, showing the superiority of such a conditional test compared with unconditional tests based on the asymptotic distribution of normalized log-likelihoods.

The second method discussed by Cox (1962) and which is examined in further detail by Atkinson (1970) uses an approach somewhat like that considered by Pace and Salvan (1990), where the models f_0 and f_1 are extended to an encompassing model. In the Cox (1962) and Atkinson (1970) approach, this is achieved by means of an additional mixing parameter, λ. Let

$$g(y; \boldsymbol{\theta}_0, \boldsymbol{\theta}_1, \lambda) \propto (f_0(y; \boldsymbol{\theta}_0))^{1-\lambda}(f_1(y; \boldsymbol{\theta}_1))^{\lambda}, \ 0 < \lambda < 1, \tag{15.31}$$

be a PDF formed from $f_0(y; \boldsymbol{\theta}_0)$ and $f_1(y; \boldsymbol{\theta}_1)$. The null hypothesis $H_0 : g = f_0$ corresponds to $\lambda = 0$, when $\boldsymbol{\theta}_1$ is indeterminate (that is, $\lambda \succ \boldsymbol{\theta}_1$). To test H_0, Cox proposed a score statistic evaluated at $\hat{\boldsymbol{\theta}}_0$, with $\lambda = 0$. If we take the extended parameter space as corresponding to points $(\boldsymbol{\theta}_0, \boldsymbol{\theta}_1)$, we can regard moving from the model $f_0(y; \boldsymbol{\theta}_0)$ to the model $f_1(y; \boldsymbol{\theta}_1)$ as corresponding to moving along a path with points $(\boldsymbol{\theta}_0(\lambda), \boldsymbol{\theta}_1(\lambda))$ parametrized by λ, starting from $(\boldsymbol{\theta}_0, \mathbf{0})$ and ending at $(\mathbf{0}, \boldsymbol{\theta}_1)$ as λ moves from $\lambda = 0$ to $\lambda = 1$. The models corresponding to $(\boldsymbol{\theta}_0(\lambda), \boldsymbol{\theta}_1(\lambda))$ can be regarded as intermediate models. Introduction of a mixing parameter is similar to the intermediate model approach that we considered in the nested model case, where the structure of the first step test in (15.25) is identical. Testing the hypothesis that $\lambda = 0$ or $\lambda = 1$ is the same as testing for departures from one model in the direction of the other; $\lambda = 0.5$ implies both models fit the data equally well (or badly).

Atkinson (1970) considers obtaining inferences about λ, and develops a test statistic which is shown to be asymptotically equivalent to that of Cox. Davidson and MacKinnon (1981) extend this type of approach by introducing new procedures which may be applied to test against several alternative models simultaneously. Godfrey and Pesaran (1983) examine their behaviour with small samples, and suggest two alternative tests based on an adjustment of Cox-type statistics.

Focusing on the role played by λ has been exploited by Royston and Thompson (1995), who describe an interesting way for carrying out the significance test of H_0 when two non-nested regression models are involved. For simplicity, we refer to this paper as R&T in this section. R&T utilize an alternative conditioning approach that is tractable and which can lead to greater precision in estimating λ. The method is a development of one originally considered by Davidson and MacKinnon (1981) and refined by MacKinnon (1983). The MLEs $\hat{\boldsymbol{\theta}}_0$ and $\hat{\boldsymbol{\theta}}_1$ of the parameters $\boldsymbol{\theta}_0$ and $\boldsymbol{\theta}_1$ of the model of eqn (15.31) are first obtained separately by maximizing $L_0(\boldsymbol{\theta}_0)$ and $L_1(\boldsymbol{\theta}_1)$. The conditional log-likelihood $L_g(\hat{\boldsymbol{\theta}}_0, \hat{\boldsymbol{\theta}}_1, \lambda)$, with the $\boldsymbol{\theta}$ parameters held at their MLE values, is then maximized with respect to λ only. The resulting estimate $\hat{\lambda}(\hat{\boldsymbol{\theta}}_0, \hat{\boldsymbol{\theta}}_1)$, where we have indicated its dependence on the estimates $\hat{\boldsymbol{\theta}}_0, \hat{\boldsymbol{\theta}}_1$, is not the true MLE of λ. However, if we can obtain a consistent estimator $v(\hat{\boldsymbol{\theta}}_0, \hat{\boldsymbol{\theta}}_1)$ of $\mathrm{Var}\{\hat{\lambda}(\hat{\boldsymbol{\theta}}_0, \hat{\boldsymbol{\theta}}_1)\}$, this would allow a confidence interval (CI) to be calculated for λ, and if the CI does not include the value $\lambda = 0$, then H_0 is rejected.

R&T applied this method to two non-nested regression models

$$y = \eta_0(x, \boldsymbol{\theta}_0) + \varepsilon_0 \text{ and } y = \eta_1(x, \boldsymbol{\theta}_1) + \varepsilon_1,$$

obtaining the conditional estimator $\hat{\lambda}(\hat{\boldsymbol{\theta}}_0, \hat{\boldsymbol{\theta}}_1)$ for the unified model

$$y = (1 - \lambda)\eta_0(x, \boldsymbol{\theta}_0) + \lambda\eta_1(x, \boldsymbol{\theta}_1) + \varepsilon_0.$$

Moreover, R&T then obtain a consistent variance estimator $v(\hat{\boldsymbol{\theta}}_0, \hat{\boldsymbol{\theta}}_1)$ of $\mathrm{Var}\,\{\hat{\lambda}(\hat{\boldsymbol{\theta}}_0, \hat{\boldsymbol{\theta}}_1)\}$ in this case. Rather than calculating a CI for λ, R&T formalize the approach as a significance test using a test statistic C, which they specify. The calculations required are set out in detail by R&T, and for the most part are straightforward, being based on linear regression through the original. If approximate estimates indicate that the bias of $\hat{\lambda}$ is high, then it is probably preferable to use the C^* statistic suggested by MacKinnon (1983).

R&T give two interesting numerical examples involving fitting non-nested regression models to practical data sets. One concerns the development of reference centiles where the fetal mandible length, y, is regressed against length of gestation, x. The other occurred in a study of glucose turnover in humans, where y, the level of deuterium enrichment in blood, as administered by injection, is measured as a function of time, x, for which a nonlinear regression fit was required. R&T report some success with their approach, with the relative simplicity of the calculations making the method attractive compared with 'Cox-type' tests.

16 Bootstrapping Linear Models

16.1 Linear Model Building: A BS Approach

This chapter discusses the problem of explanatory variable selection in linear regression. The problem is straightforward and completely standard in principle, so technically falls outside the remit of this book. However, in practice, the problem is less straightforward when the experimental design is non-orthogonal and the number of explanatory variables is large. We therefore include this chapter for two reasons.

Firstly, the topic fits in well with our discussion in the previous chapter on model selection in a nested nonlinear family. The multilinear model is a special case that in principle does not have the complications occurring in the nonlinear case. However, when the experimental design is large and non-orthogonal, efficient identification of important explanatory variables, even when linear, is not so straightforward. Stepwise regression methods are often advocated and used, see, for example, Wu and Hamada (2000). These have a rationale that is akin to our lattice approach, so that discussion of how to implement such stepwise methods is not totally out of place here, even though the model is linear.

Secondly, we shall consider how BS resampling can be used in a non-standard way to fit the linear model. We aim to show that our approach is simple and effective, with distinct advantages over standard sequential methods that are often advocated and employed. Given the emphasis we have placed on BS methods in this book, this provides a second reason for including consideration of the linear model.

The basic idea discussed in this chapter was originally suggested by Cheng (2008) and then in more detail in Cheng (2009). Here, we outline the methodology described in the latter reference, in which bootstrapping is given a central role in the explanatory variable selection process, and, for convenience in referencing, we use the same terminology. Our numerical example is, however, different from those discussed in the references, allowing discussion of issues not covered in those articles.

Non-Standard Parametric Statistical Inference. Russell Cheng.

Once models have been obtained that are expected to be a good explanation of the data, bootstrapping has a second use in providing a natural way to gauge the adequacy of selected models and for choosing amongst several, possibly many, competing models. We also discuss how this can be done.

It is of interest to note that such methods are now beginning to be recognized as very appropriate for model selection in areas like simulation. A good example is given by Fishman (2006, §2.8).

Our approach is easily explained once we have formally defined the linear model, which we do next.

16.1.1 Fitting the Full Linear Model

The linear model and estimation of its parameters are well known, see, for example, Searle (1971), but we set out the details here for ease of reference in our subsequent discussion.

We suppose the variable of interest, Y, is a (scalar) continuous random variable and that it is linearly dependent on P explanatory variables (or factors) X_j, $j = 1, 2, \ldots, P$. We consider the (full) linear model

$$\begin{bmatrix} Y_1 \\ Y_2 \\ \vdots \\ Y_n \end{bmatrix} = \begin{bmatrix} 1 & X_{12} & X_{13} & \ldots & X_{1P} \\ 1 & X_{22} & X_{23} & \ldots & X_{2P} \\ \vdots & \vdots & \vdots & \ddots & \vdots \\ 1 & X_{n2} & X_{n3} & \ldots & X_{nP} \end{bmatrix} \begin{bmatrix} b_1 \\ b_2 \\ \vdots \\ b_P \end{bmatrix} + \begin{bmatrix} \varepsilon_1 \\ \varepsilon_2 \\ \vdots \\ \varepsilon_n \end{bmatrix}, \tag{16.1}$$

where Y_i, $i = 1, 2, \ldots, n$, are the observations; X_{ij} are the explanatory variable values, also recorded; b_j, $j = 1, 2, \ldots, P$, are the unknown coefficients corresponding to each of the P explanatory variables; and ε_i, $i = 1, 2, \ldots, n$, are random errors. We have taken $X_{i1} = 1$, $i = 1, 2, \ldots, n$, so that b_1 corresponds to a general scalar constant. We thus treat the constant as a coefficient, so that, as far as the model selection and fitting process is concerned, we do not treat it differently from the other coefficients. In what follows, when we refer to a 'factor' or explanatory variable, it is to be understood that this general constant is included.

We shall assume that the ε_i, $i = 1, 2, \ldots, n$, are identically distributed with mean zero and variance

$$\mathrm{Var}(\varepsilon) = \sigma^2. \tag{16.2}$$

To simplify our discussion, we will assume that the random errors are normally distributed.

We shall, where convenient, write (16.1) in the alternative matrix form

$$\mathbf{Y} = \mathbf{X}\mathbf{b} + \boldsymbol{\varepsilon}. \tag{16.3}$$

Equation (16.1) is the full model in which all explanatory variables are included. Each possible *model* will be specified as

$$m = \{j_1, j_2, \ldots, j_p\}, \tag{16.4}$$

where

$$j_1 < j_2 < \cdots < j_p, \text{ with } p \leq P,$$

are just those and only those factors appearing in the model. We can therefore write the observations corresponding to this model as

$$\begin{bmatrix} Y_1 \\ Y_2 \\ \vdots \\ Y_n \end{bmatrix} = \begin{bmatrix} X_{1j_1} & X_{1j_2} & \ldots & X_{1j_P} \\ X_{2j_1} & X_{2j_2} & \ldots & X_{2j_P} \\ \vdots & \vdots & \ddots & \vdots \\ X_{nj_1} & X_{nj_2} & \ldots & X_{nj_P} \end{bmatrix} \begin{bmatrix} b_{j_1} \\ b_{j_2} \\ \vdots \\ b_{j_P} \end{bmatrix} + \begin{bmatrix} \varepsilon_1 \\ \varepsilon_2 \\ \vdots \\ \varepsilon_n \end{bmatrix}, \tag{16.5}$$

or in the matrix form

$$\mathbf{Y} = \mathbf{X}(m)\mathbf{b}(m) + \boldsymbol{\varepsilon}. \tag{16.6}$$

Where necessary, we shall write

$$p(m) = p \tag{16.7}$$

for the number of coefficients specified in the model m. Also, we will denote the full model by M, so that $p(M) = P$.

When we fit the model m, we shall use the least squares estimates (see Searle, 1971, for example)

$$\hat{\mathbf{b}}(m) = \left[\mathbf{X}^{\mathrm{T}}(m)\mathbf{X}(m)\right]^{-1}\mathbf{X}^{\mathrm{T}}(m)\mathbf{Y} \tag{16.8}$$

for the unknown coefficient values, these also being the ML estimates under the normal error assumption, and

$$\begin{aligned} \hat{\sigma}^2(m) &= [n - p(m)]^{-1}\sum_{i=1}^{n}[Y_i - \hat{Y}_i(m)]^2 \\ &= [n - p(m)]^{-1}[\mathbf{Y} - \mathbf{X}(m)\hat{\mathbf{b}}(m)]^{\mathrm{T}}[\mathbf{Y} - \mathbf{X}(m)\hat{\mathbf{b}}(m)] \end{aligned} \tag{16.9}$$

for the unbiased estimate of the variance of the ε_i, where

$$\hat{Y}_i(m) = \mathbf{X}_i(m)\hat{\mathbf{b}}(m) = \sum_{j\in m} X_{ij}\hat{b}_j(m), \tag{16.10}$$

with $\mathbf{X}_i(m)$ the ith row of the matrix $\mathbf{X}(m)$.

16.1.2 Model Selection Problem

Assume now that the coefficients of the *full model* have been estimated as can be done using the formulas of the previous section, providing $n > p(m)$. The *model selection problem* of interest is where we wish to identify *simpler* models in which some of the explanatory variables are omitted because they are unimportant. To avoid confusion, we shall use the term 'full model' to indicate when all P explanatory variables are included. The term 'model' will in general mean one that includes only a subset of the full set explanatory variables. Also, it is well to make a distinction between the terms 'model' and 'fitted model'. The term 'model' merely refers to where the explanatory variables appearing in the model have been specified, but where the associated coefficients do not have particular values; the term 'fitted model' will be used when the coefficients in the model have been set to fitted values.

There are a total of 2^P distinct subsets of the explanatory variables, so that this is the number of distinct models that might be considered.

Many authors exclude the null model $\mathbf{y} = \boldsymbol{\varepsilon}$, so that the total number of distinct models is then $2^P - 1$. However, in this chapter, we will include it, so that the total number of possible models is 2^P.

Ideally, we would wish to find a best model without having to examine and compare all 2^P models.

Wu and Hamada (2000) have discussed this problem at length. They considered the very well-known backward, forward, and stepwise explanatory variable selection methods. These are all sequential, with explanatory variables considered one at a time for possible inclusion, or elimination, including possible backtracking. Nevertheless, simply because of the order in which explanatory variables are considered, it is possible with non-orthogonally designed experiments to end up with a selected model that does not include all those explanatory variables that are important.

Our suggested BS sampling method aims to avoid the uncertainty of a stepwise procedure, but where we examine *just a subset* of the 2^P models. However, in examining only a subset, we wish to ensure that good, or potentially competitive, submodels are not excluded. We do this by constructing a subset of *promising submodels* that is far smaller in number than the 2^P, but which is still likely to include most, if not all, competing models. Before describing how to do this, we discuss a basic approach, where all 2^P models are considered.

16.1.3 'Unbiased Min p' Method for Selecting a Model

Cheng (2009) gives two methods for selecting a 'best' model. Here, we describe just one, the other being just a special case.

We shall define a 'best' model using a criterion based on the C_p statistic proposed by Mallows (1973):

$$C_p(m) = [n - p(m)]\hat{\sigma}^2(m)/\hat{\sigma}^2(M) + 2p(m) - n\,. \tag{16.11}$$

This estimates the expected prediction error, taking into account the variance and bias of the fitted model, and is asymptotically equivalent (see Nishii, 1984) to the Akaike Information Criterion, see Akaike (1973, 1974), which for the linear model is $AIC(m) = n\log[\hat{\sigma}^2(m)] + 2p(m)$; see also Shibata (1981). Mallows (1973) shows that if the model m (with p factors) has no bias, then the expected value of C_p is close to p, that is,

$$C_p \simeq p. \tag{16.12}$$

If not all important factors are included, the expected value of C_p will be larger than p. A possible selection method would be to calculate $C_p(m)$ for each model m of the 2^P possible of (16.1), then select as the best model m for which $C_p(m)$ is minimum. Mallows (1995) points out that if m^+ is a model containing one factor additional to those already in a model m, then the extra factor would be worth including if

$$C_{p+1}(m^+) - C_p(m) = 2 - (S_1/\hat{\sigma}^2(M)) < 0,$$

where S_1 is the 1-degree-of-freedom (df) sum of squares due to the additional factor. We can therefore use a test that includes the factor if

$$t^2 = S_1/\hat{\sigma}^2(M) > 2. \tag{16.13}$$

For an orthogonal design, the sum of squares S_1 corresponding to any given factor is the same irrespective of which factors occur in the model. The t^2 values for each of the factors can therefore all be obtained *simultaneously* simply by fitting the full model, and the minimum C_p and corresponding 'best' model obtained by including just those factors whose t^2 satisfy (16.13).

For the non-orthogonal case, we could still fit the full model, and for each factor j calculate the so called t-value of its fitted coefficient $\hat{b}_j$:

$$t_j = \hat{b}_j/s_j\,, \tag{16.14}$$

where $s_j = \sqrt{d_j\hat{\sigma}^2(M)}$ is the estimated standard deviation of $\hat{b}_j$, with d_j the jth entry in the main diagonal of the dispersion matrix, i.e.

$$d_j = [(\mathbf{X}^{\mathrm{T}}\mathbf{X})^{-1}]_{jj}. \tag{16.15}$$

A simple criterion for selecting a model is to include only those factors j for which

$$|t_j| > a, \tag{16.16}$$

where a is a chosen critical level.

We consider how a value for a might be selected.

If the true value of b_j is $b_j = 0$, then under the normal error assumption, t_j has Student's t-distribution with $n - P$ degrees of freedom. If we therefore denote the complementary distribution function for the absolute value $|t_j|$ by $\bar{T}_{n-P}(\cdot)$, then the probability of success of the test (16.16), under the assumption that $b_j = 0$, is

$$\pi_a = \Pr\{|t_j| > a\} = \bar{T}_{n-P}(a).$$

Given the p-value of the estimate $\hat{b}_j$, namely, $\bar{T}_{n-P}(|t_j|)$, the factor j is retained if

$$\bar{T}_{n-P}(|t_j|) < \pi_a\,. \tag{16.17}$$

The factor selection approach based on C_p, where the test (16.13) is used, is the special case of the test (16.16), where $a = \sqrt{2}$, so that the critical p-value in (16.17) is then $\pi_a = 0.1573$ when $n - P$ is large.

Note, however, that if the number of parameters P is large and the (unknown) true values of many coefficients are at or near zero, the selection test (16.17) would include nearly 16% of such negligible coefficients in the model.

The effect of varying a can be seen more fully by considering the asymptotic probability that a factor with coefficient of size $b_j = bs_j$ will be selected and seeing how this varies with b. If we assume that $n - P$ is large, then s_j can be treated as being a known constant, so that $\hat{b}_j \sim N(bs_j, s_j^2)$. The probability we would include the factor is then

$$\begin{aligned} \Pr\{\text{Factor } j \text{ is included in model}\} &= 1 - \Pr\{-as_j < \hat{b} < as_j\} \\ &= 1 - \Pr\{-a - b < (\hat{b} - bs_j)/s_j < a - b\} \\ &= 1 - \Phi(a - b) + \Phi(-a - b), \end{aligned} \tag{16.18}$$

where $\Phi(\cdot)$ is the standard normal distribution function. Figure 16.1 shows how this probability varies as a function of b for different selected a, showing that values larger than $a = \sqrt{2}$ in (16.13), such as $a = \sqrt{6}$ or $a = 3$, might be more appropriate in exploratory

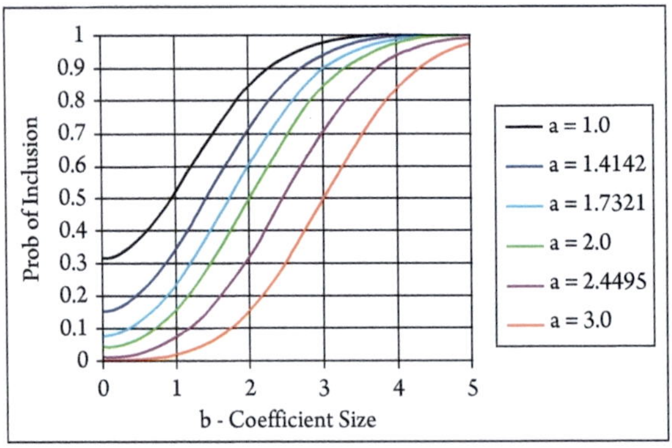

Figure 16.1 Asymptotic probability that a factor with coefficient size b is included in a model using the t test with critical level a.

studies, where we are only interested in identifying significantly large b, and would prefer the probability of retaining a zero coefficient to be much smaller than 16%, though we would wish to avoid having $C_p > p$.

We therefore define the following basic selection method:

'Unbiased Min p' Model Selection Method

(i) Find the smallest p for which there are models m satisfying $C_p(m) \leq p$ and let

$$p_0 = \min\{p : C_p(m) \leq p\}. \tag{16.19}$$

(ii) Amongst all such models m, with $p(m) = p_0$, find the one for which $C_p(m)$ is minimum.

A visually very satisfying way to identify this model is to plot C_p versus p for all possible models, and examine the lower envelope of this scatterplot of points. For the orthogonal case, where there are a large number of factors with coefficient values uniformly distributed in the neighbourhood of zero with density λ, Mallows (1995) has shown that the scatterplot has a lower boundary that is the (convex) cubic polynomial in p

$$C_p - P \approx \frac{(P-p)^3}{12\lambda^2} - 2(P-p), \tag{16.20}$$

and that this boundary intersects the line $C_p = p$ at $P - p = 2\sqrt{3}\lambda$. Figure 16.2 depicts the scatterplot for our example that we shall be discussing in Section 16.2.4, and the boundary, which, though not particularly well approximated by a cubic in this case, is nevertheless readily distinguishable, with its intersection with the line $C_p = p$ quite discernible.

Our selection method selects a point near this intersection by taking the smallest p, p_0 as in (16.19), for which there are points of the scatterplot below the line $C_p = p$, and then selecting from amongst those models with $p = p_0$, the one with minimum C_p.

In the orthogonal case, models at, or near, this intersection point will include essentially just those factors for which (16.16) is satisfied with $a = \sqrt{3}$, which is equivalent to using (16.17) to include just those factors whose estimated coefficients have p-value less than $\pi_a = 0.083$.

The condition (16.12) that $C_p \approx p$ obtains when the model contains no bias, so that the model is completely appropriate whilst having the smallest p possible. For this reason, we call (16.19) the 'unbiased min p' method.

The basic method described so far is not practical, as there is a *dimensionality problem* if one were to attempt fitting all 2^P possible models when P is large. For example, with just 20 explanatory variables there would already be 1,048,576 models.

Our suggested way round this problem is to identify a set of *promising models* using bootstrap resampling. The number of models in this set is easily controlled and so can be made much smaller than 2^P. We show, however, that it will almost certainly contain many good candidate models. It is thus satisfactory to select a 'best' model from this subset.

We discuss this bootstrap approach in the next section.

16.2 Bootstrap Analysis

We shall use bootstrapping for two distinct purposes. Firstly, as already mentioned in the previous section, it can be used for identifying a set of promising models. However, we shall also use bootstrapping to deal with the following second problem.

Once a model has been selected as being the best fit to a data set, we have the problem of determining what might be termed the *quality of the selected model*. For example, there may be several models with values of $C_p(m)$ close to that of the best, so that we may not be sure which model really is the best. This question would be answered if we had many (independent but identically distributed) data samples and not just the one original sample, as we could determine the best model for each sample and see if the same model is being selected as being best for all the samples. In practice, what we can expect is that different models will be selected as being the best fit for the different samples, but the variation in choice will enable the selected models to be compared. BS resampling enables such additional data samples to be generated.

We first outline how BS samples are generated in the next subsection, before going on to describe our two distinct uses of bootstrapping.

16.2.1 Bootstrap Samples

Cheng (2009) discusses two ways of generating BS samples that asymptotically have the same form as (16.1). Here, we consider just one of them using *parametric bootstrapping*, where it is assumed that the random errors ε_i, $i = 1, 2, \ldots, n$, in (16.1) are normally distributed and independent. We first fit the full model M to the original data sample, obtaining $\hat{Y}_i(M)$ as in eqn (16.10) and $\hat{\sigma}^2(M)$ as in eqn (16.9) (both with $m = M$). The BS sample then takes the form

$$Y_i^* = \hat{Y}_i(M) + e_i^*, \; i = 1, 2, \ldots, n, \tag{16.21}$$

where the e_i^*, $i = 1, 2, \ldots, n$, are a random sample from the fitted normal distribution, i.e.

$$e_i^* \sim N(0, \hat{\sigma}^2(M)), \quad i = 1, 2, \ldots, n. \tag{16.22}$$

We write $\hat{\mathbf{b}}^*$ and $\hat{\sigma}^{*2}$ for the estimates (16.8) and (16.9) obtained from fitting the model (16.1) to the BS observations (16.21).

The justification for bootstrapping is provided by Freedman (1981, Theorem 2.2). Assume that (16.1) and (16.2) hold and that $\mathbf{X}(n)$ is not random, with

$$\frac{1}{n}\mathbf{X}^{\mathrm{T}}(n)\mathbf{X}(n) \rightarrow \mathbf{V} \tag{16.23}$$

as $n \to \infty$. Then

$$\sqrt{n}\{\hat{\mathbf{b}}^*(n) - \hat{\mathbf{b}}(n)\} \sim N(\mathbf{0}, \sigma^2\mathbf{V}^{-1}) \tag{16.24}$$

and

$$\hat{\sigma}^*(n) \to \sigma. \tag{16.25}$$

This result assumes that P is fixed as $n \to \infty$, which we shall assume in what follows.

16.2.2 BS Generation of a Set of Promising Models

The 'unbiased min p' method of selecting a best model requires consideration only of those models that are near p_0, as defined in (16.19) or in the C_p/p scatterplot . Our first use of bootstrapping is therefore to generate a set of *promising models* located in this region. The number of models in this set does not need to be anywhere near 2^P, but it does need to be large enough to enable the lower boundary (16.20) to be clearly identified, at least near its intersection with $C_p = p$. Ideally, it needs to contain all the models with scatterplot points near this intersection point. Cheng (2009) gives two methods; we describe the more general one here.

'Many Models per Sample' Generation of Promising Models by Bootstrapping

Step (1) Fit the full model to the original data and use this fitted full model to generate B BS samples, each of the form (16.21).

Step (2) For each BS sample:

(i) Fit the full model, M, to the sample and determine, as defined in (16.14), the t-value, t_j, of each of the fitted coefficients, $\hat{b}_j, j = 1, 2, \ldots, P$.

(ii) Order the coefficients by their $|t_j|$ values:

$$|t_{j_1}| \geq |t_{j_2}| \geq \cdots \geq |t_{j_P}|, \tag{16.26}$$

so that $\hat{b}_{j_1}$ is the most significant.

(iii) Set a critical t-value, a (we used $a = \sqrt{3}$ as before), and include all the following models in the promising set S:

$$\begin{aligned} m_1 &= \{j_1\} \\ m_2 &= \{j_1, j_2\} \\ &\vdots \\ m_k &= \{j_1, j_2, \ldots, j_k\}, \end{aligned} \tag{16.27}$$

where the last factor j_k satisfies

$$\left|t_{j_k}\right| \geq a > \left|t_{j_{k+1}}\right|. \tag{16.28}$$

Thus, the model m_i is the one where the i most significant factors have been retained, with a cutoff that only factors with t-level greater than a are allowed in a model. The last model, m_k in (16.27), is the one that includes just those coefficients with $|t|$-value a or greater, this being the sole model selected in the 'one model per sample' method suggested by Cheng (2009).

16.2.3 Selecting the Best Model and Assessing its Quality

To select the best model, we simply use the 'unbiased min p' method, but apply it just to the set of promising models rather than the full set of all 2^p models.

Our second use of bootstrapping is to study the quality of the selected model. This is most easily done by adding the following steps to the method used to generate a set of promising models.

BS Quality Assessment of Selected Best Model

Step (3) For each of the B BS samples, fit the set S of promising models, and calculate the C_p value for each model, selecting as the best model for this sample that which minimizes C_p. Cheng (2009) suggests the restriction that only models where $p \leq p_0$ are considered, but we do not do so here. Denote by $S_0 \subseteq S$, the models of S selected as the best for at least one of the BS samples.

Step (4) Display the models of S_0, ranked in order of the frequency with which they are selected as being the best model in the B BS samples, displaying these frequencies as well.

Step (5) Display the empirical distribution functions of the C_p values of a selected number of those models in S_0 most frequently selected as being the best.

Let $\alpha(m)$ be the probability that model m will be selected as the best model in the sense of minimizing C_p. Step (3) estimates these probabilities by fitting all the models in S. Out of all 2^P models, those that are not a good fit will have very little probability of being included in S, because of how it is constructed. Hence, they would not be considered for possible inclusion in S_0. However, every model has a *positive* probability of being included in S. Thus, asymptotically, as $B \to \infty$, S must tend to the full set of all models, so we will converge to the situation where every model is considered for possible selection as the best. Hence, for each m, $\alpha(m)$ can reasonably be estimated from the frequency with which m is selected as being the best model in Step (3).

Step (4) simply highlights those models that have been most frequently selected as being the best fit.

Step (5) assesses the behaviour of the C_p values of those models that have been most frequently selected as being the best fit. For such a model to be satisfactory, one would expect the distribution of its C_p value, over the BS samples, to be concentrated mainly in the region where $C_p \leq p$.

16.2.4 Asphalt Binder Free Surface Energy Example

In our example, we illustrate calculation of a subset of promising models, S, as set out in Section 16.2.2, and the quality assessment analysis of Section 16.2.3. We will also consider another fitting issue not considered in Cheng (2009).

We consider the data sample given by Wei et al. (2014), who examined how a key property of asphalt, the asphalt binder surface free energy, $Y = (sfe)$ was affected by its chemical composition as measured in terms of 12 explanatory X_i variables. Specifically, these were four fractions: saturates (sat), aromatics (aro), resins (s), and asphaltenes (asp), and also wax content (wax) and seven elemental contents: carbon (C), hydrogen (H), oxygen (O), nitrogen (N), sulphur (S), nickel (Ni), and vanadium (V). Thus, including a constant $(const)$, the full linear model has 13 parameters. The abbreviations shown in brackets are used in our figures. We shall for simplicity designate a general parameter as b_i, but for ease of reference as $b(sat)$, for example, when referring to the parameter corresponding to a specific explanatory variable.

Wei et al. (2014, Table 1) give the experimental results for $n = 23$ different asphalts, which they analysed using linear and multilinear models involving both the original and transformed explanatory X_i values. We concentrate on the multilinear fit using the original X_i's with the model precisely as in eqn (16.1), including a constant term with $X_1 = 1$. Wei et al. (2014) give the results of the analysis of the full model in their Section 3.3.4, with the fitted model given in their eqn (4) with a multiple regression coefficient of $R = 0.972$ that they were 'excited to see'. Satisfyingly, our fitted full model gave the same fitted model coefficients (subject to rounding error) as given by Wei et al. (2014) and the same R value, corroborating their results.

We now look for a 'best' model using the method of Sections 16.2.2 and 16.2.3.

In Step (1) of the analysis, we generated $B = 500$ BS samples of the original data.

In Step (2), we fitted the full model to each BS sample, and then selected fitted submodels using $a = \sqrt{3}$ so that $\pi_a = 0.083$ in the selection process. This produced a set, S, of 370 promising models in our example. The C_p values of these models when fitted to the *original* data are plotted against p in Figure 16.2. Applying (16.19) gave $p_0 = 3$. The C_p/p points corresponding to the top four models are highlighted in Figure 16.2, with the best model coloured in red and the other three in order of decreasing redness. Only *asp*, *wax*, and H are included as explanatory variables in the best model. Fitting it to the original sample gave the estimated coefficients

$$b(asp) = -0.48 \qquad b(wax) = -1.25 \qquad b(H) = 2.79, \tag{16.29}$$

so this would constitute our best fitted model.

In the second stage, as described in Section 16.2.3, the models of S were fitted to each BS sample, and the model with the minimum C_p was selected as being the best model for

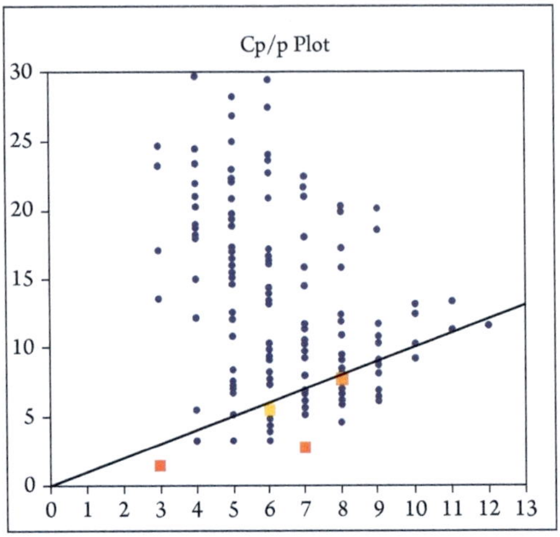

Figure 16.2 13-parameter asphalt model. C_p versus p plot of 370 promising models (not all included in the frame shown) found for the asphalt free surface energy data including only linear component factors, using the 'many models per sample' BS method. The C_p values are those obtained when the promising models are fitted to the original sample. The (p, C_p) point position of the top model is coloured red, with next best three ordered in decreasing redness.

that BS sample. This yielded a subset $S_0 \subset S$ of 85 distinct models, each of which was the best fit for at least one of the 500 BS samples.

We have taken the opportunity to illustrate how results might be displayed by reproducing in Figure 16.3 the actual spreadsheet tabular output, which enables colour to be used to convey and enhance the results. The 25 models selected most frequently as being the best are displayed in the main body of Figure 16.3. The right-hand column (headed MFrq for model frequency) gives the frequency each of these models was the best fitted model for some given BS sample, with the models placed in rank order in the table according to this frequency.

The model most frequently found to be the best fitted model was actually that of eqn (16.29), which was selected as the best in 65 of the $B = 500$ BS samples. This provides corroboration that this model is a good choice.

Together, the top 25 models were selected as being the best in 350 of the BS samples; that is, 70% of the time.

The second and third most right-hand columns in Figure 16.3 give the C_p values obtained by fitting each of these models to the *original* data set together with corresponding p values. The C_p/p plot is not shown, but is similar to that of Figure 16.2 in the region of the line $C_p = p$, though with fewer points above the line $C_p = p$ than in that figure, as is to be expected.

The top row of numbers are the p-values of each estimated b coefficient value, $\hat{b}_i$. The second row is the frequency that each b parameter of the associated factor appears

	const	sat	aro	res	asp	wax	C	H	O	N	S	Ni	V			
P-Val	0.46	0.66	0.78	0.92	0.94	0.08	0.53	0.16	0.25	0.87	0.33	0.36	0.37			
ParFrq	47.4	45.4	68.2	63.2	63.6	84.4	38.4	67.4	53.2	20.4	44	28.6	29.4	p	Cp	MFrq
1	0	0	0	0	−0.48	−1.25	0	2.79	0	0	0	0	0	3	1.5	65
2	−34.27	0	0.30	0.24	0	−1.35	0	3.51	2.31	0	−0.51	0	0	7	2.8	25
3	0	−1.17	−0.53	−0.72	−0.95	0	0.99	0	3.94	0	0	0.06	0.00	8	7.9	18
4	−46.57	0	0.37	0.38	0	−1.09	0	3.63	2.28	0	0	0	0	6	5.5	17
5	−29.70	0	0.29	0.29	0	−1.44	0	3.12	0	0	−0.51	0	0	6	4.2	16
6	−14.33	0	0.38	0.42	0	−0.49	0	0	0	5.64	0	0	0	5	7.5	14
7	60.04	0.29	0.46	0.45	0	−1.46	0	3.54	0	0	0	0	0.00	7	5.3	13
8	0	0	0.29	0.24	0.00	−1.30	−0.33	3.02	2.68	0	−0.77	0	0	8	5.9	12
9	0	−1.01	−0.47	−0.63	−0.89	−0.49	0.93	0	2.72	0	0	0.04	0.00	9	8.2	12
10	0	−0.68	−0.17	−0.32	−0.45	−0.85	0.36	1.90	3.46	0.23	−0.59	0.03	0.00	12	11.6	12
11	0	−0.21	0	0	−0.48	−1.14	0.36	0	0	0	0	0	0	4	5.5	12
12	0	−0.72	0	0	0	0	0.26	0	5.93	0	−1.11	0.04	0.00	6	8.2	12
13	0	−0.53	0	−0.14	−0.57	0	0.36	0	3.71	0	0	0.05	0.00	7	8.0	11
14	−42.79	0	0.35	0.31	0.14	−1.34	0	3.69	2.81	0	−0.66	0	0	8	4.6	11
15	0	0	0.22	0.28	0	−1.32	0	0.64	0	0	−0.75	0	0	5	7.6	10
16	−66.67	0.22	0.55	0.52	0.33	−1.40	0	4.06	2.65	0	−0.58	0	0	9	6.4	10
17	−16.23	0	0.17	0	−0.33	−1.19	0	3.31	0.09	4.10	−0.08	0	0	8	6.2	10
18	−39.17	0	0.34	0.27	0.05	−1.18	0	3.48	1.79	2.32	−0.47	0	0	9	6.1	10
19	0	−0.75	−0.36	−0.44	−0.78	−0.87	0.80	0	0	0	0	0.03	0.00	8	9.1	10
20	−49.81	0.18	0.45	0.37	0.00	−1.10	0	3.50	0.48	3.61	0	0	0	9	6.9	10
21	0	−1.14	−0.43	−0.62	-0.74	0	0.87	0	4.77	0	−0.38	0.06	0.00	9	9.1	8
22	−33.64	0	0.21	0	0	−1.76	0	4.89	3.20	0	−0.93	0	0	6	7.4	8
23	−70.74	0.46	0.63	0.63	0.17	−1.49	0	3.83	0	0	0	0	0	7	5.2	8
24	−52.05	0.31	0.48	0.46	0	−1.45	0	3.53	0	0	0	0	0	6	3.4	8
25	0	−0.65	0	0	0	0	0.26	0	5.91	0	−1.23	0.03	0	5	10.9	8

Figure 16.3 The 13-parameter asphalt model. The top 25 of the final selection of 85 'best' models ranked according to the frequency the models were found to be the best fit for some given sample in the 500 BS samples. Each row gives the estimated b_i coefficients in the model fit to the *original* sample. Cells in blue indicate explanatory variables not included in the model. The top row are the p-values of each fitted b_i in the full model fit to the original sample. The second row gives the number of times, expressed as a percentage, that each b_i is included in the 500 best model fits to the BS samples.

in all 500 best fit models of the BS samples; so this frequency includes models not listed in the top 25 given in the table. Each frequency is expressed as a percentage.

Comparing the frequency of selection of each b_i coefficient with its corresponding p-value reveals an unsatisfactory feature of the model.

One would expect a b_i coefficient with a small p-value to be important, so that its frequency of appearance in the 500 best fitted models would be correspondingly high. Figure 16.4 is a plot of this frequency of appearance of each b_i against its p-value in the full model fit to the original data. One would expect the general trend of the points to indicate the inverse relationship between the frequency and p-value by being negative, but, as is clear from the figure, the four fractional variables *sat*, *aro*, *res*, and *asp* all appear disproportionately often in the best fits for the BS samples, contradicting the very high p-value of the estimated b_i.

We have not investigated the reasons fully, but have instead carried out a second analysis where the *square* of each of the four fractional variables is also included as an

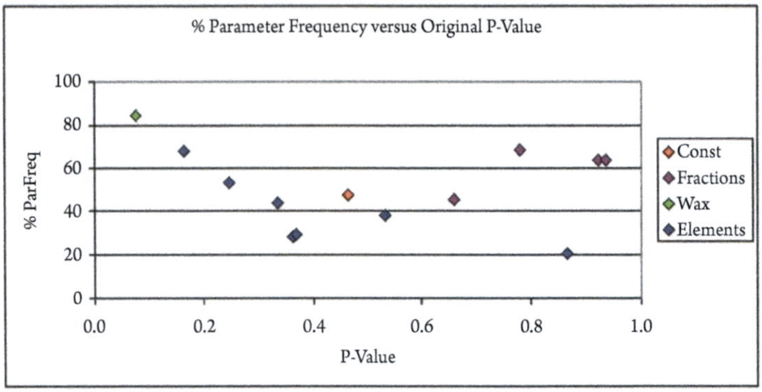

Figure 16.4 13-parameter asphalt model. Plot of the number of times (expressed as a percentage frequency) that each parameter is included in the 500 best model fits to the BS samples versus the *p*-value of the parameter estimate in the full model fit to the original data.

	const	sat	aro	res	asp	sat2	aro2	res2	asp2	wax	C	H	O	N	S	Ni	V			
P-Val	0.48	0.18	0.03	0.06	0.91	0.03	0.08	0.01	0.39	0.50	0.65	0.10	0.12	0.46	0.59	0.14	0.88			
b Frq	50.4	71.4	96.6	94.4	30.4	95	83	100	53.2	42.2	44.2	84	83.4	40.2	34.4	78	40.2	p	Cp	MFrq
1	0	-1.66	1.86	-1.92	0	0.08	-0.03	0.02	-0.01	-0.45	0	2.58	5.19	0	0	0.05	-0.001	12	8.3	29
2	0	-1.23	1.48	-1.92	0	0.07	-0.02	0.03	-0.01	0	0	2.27	5.50	0	0	0.05	0	10	7.9	26
3	32.70	-1.41	1.95	-2.17	0	0.09	-0.02	0.03	-0.01	0	-0.49	2.32	5.70	0	0.09	0.05	0	13	10.2	21
4	0	0	1.50	-1.24	0	0.02	-0.01	0.02	0	0	-0.45	2.60	4.44	0	0	0.04	0	9	8.7	21
5	59.91	-2.19	1.66	-2.16	-0.79	0.09	-0.03	0.02	0	0	0	0	5.60	0	0	0.06	-0.002	11	10.0	20
6	44.14	-1.69	2.02	-2.52	0	0.10	-0.03	0.03	-0.01	0	-0.46	2.35	4.25	2.75	0.31	0.04	0	14	11.5	15
7	7.18	-1.08	1.96	-1.45	0	0.08	-0.02	0.02	0	-0.28	-0.52	2.71	5.21	0.35	-0.02	0.04	0	14	12.3	14
8	0	-1.27	1.40	-1.82	0	0.07	-0.02	0.02	-0.01	-0.40	0	2.66	4.71	0	0	0.04	0	11	8.3	14
9	46.74	-1.69	1.99	-2.49	-0.09	0.10	-0.03	0.03	-0.01	0	-0.45	2.25	4.18	2.78	0.31	0.04	0.000	16	15.5	13
10	0	0	1.15	0	0.76	0.07	0	0.02	0	-0.94	-1.26	4.13	4.53	0	-0.63	0	0.003	10	14.1	12
11	39.52	0	1.83	-1.51	0	0.06	-0.02	0.03	0	0	-1.18	3.19	0	6.72	0.64	0	0.004	11	12.2	12
12	33.85	-1.60	1.80	-2.10	0	0.09	-0.02	0.03	-0.01	-0.30	-0.40	2.54	4.75	1.42	0	0.04	0.000	15	13.4	11
13	0	0	0.48	-0.97	0	0.02	0	0.02	0	0	-0.28	2.52	4.35	0	0	0.03	0	8	10.4	10
14	0	-1.53	2.11	-2.23	0	0.08	-0.03	0.03	-0.02	-0.56	0	3.08	0	6.31	0.78	0.02	-0.001	13	13.7	10
15	37.14	-1.63	1.94	-2.32	0	0.09	-0.03	0.03	-0.01	-0.24	-0.42	2.61	3.89	2.64	0.25	0.03	0	15	13.1	9
16	0	-1.30	1.55	-1.98	0	0.07	-0.02	0.03	-0.01	-0.36	0	2.67	4.53	0	0.17	0.04	0	12	10.1	9
17	45.90	-2.27	1.96	-2.90	-0.26	0.10	-0.03	0.03	-0.02	0	0	1.73	4.07	2.96	0.42	0.05	-0.001	15	14.0	8
18	0	0	0.73	-1.29	-0.23	0	-0.01	0.02	0	0	0	1.91	3.76	0	0	0.04	0	8	10.9	8
19	-24.69	-0.91	1.55	-1.40	0	0.06	-0.02	0.02	0	0	0	2.32	5.48	0	0	0.04	0	10	9.3	8
20	0	-1.22	1.54	-1.76	0.64	0.08	-0.02	0.02	-0.02	-0.98	-0.34	4.27	2.17	3.98	0	0	0	13	14.6	8
21	24.55	-2.03	1.75	-2.35	0	0.09	-0.03	0.03	-0.02	-0.42	0	2.38	4.82	1.17	0	0.04	-0.002	14	11.8	8
22	53.34	0	0.59	-1.22	0	0.04	0	0.02	0	-0.28	-1.02	3.25	2.40	3.16	0	0	0.003	11	11.9	8
23	18.15	-1.92	1.79	-2.27	0	0.09	-0.03	0.03	-0.02	-0.41	0	2.38	5.31	0	0	0.05	-0.002	13	10.0	7
24	68.08	-2.36	1.69	-2.43	-0.68	0.10	-0.03	0.02	-0.01	0	0	0	5.69	0	0	0.06	-0.002	12	11.8	7
25	13.00	-1.26	1.32	-1.77	-0.36	0.06	-0.02	0.02	0	0	0	1.77	5.28	0	0	0.05	0	11	9.9	7

Figure 16.5 Spreadsheet output for the 17-parameter asphalt model. The top 25 of the final selection of 113 'best' models ranked according to the frequency the models were found to be the best fit for some given sample in the 500 BS samples. Each row gives the estimated b_i coefficients in the model fit to the *original* sample. Cells in blue indicate explanatory variables not included in the model. The top row are the *p*-values of each fitted b_i in the full model fit to the original sample. The second row gives the number of times, expressed as a percentage, that each b_i is included in the 500 best model fits to the BS samples.

explanatory variable. These are designated by *sat2*, *aro2*, *res2*, and *asp2*. Their inclusion therefore increases the flexibility in the way the four fractions can influence the surface free energy, $Y = sfe$. The full model fit was

$$\begin{aligned}\widehat{sfe} = &\ 33.1 - 1.75sat + 0.0964aro + 1.97res - 0.0267asp\\ &-2.41sat2 + 0.0296aro2 + 0.0614res2 - 0.0171asp2\\ &-0.296wax - 0.329C + 2.66H + 3.92O\\ &+2.59N + 0.257S + 0.0346Ni - 0.000393V\end{aligned}$$

with an R value of 0.993.

A BS analysis was then carried out in exactly the same way as for the original model, with $B = 500$ BS samples of the data set generated and a set S of promising models obtained from these. In this case, there were 814 models in S. These were each fitted to the 500 BS samples and a best fit obtained for each sample. There were 113 distinct 'best'models, these comprising the set S_0. The results are shown in Figure 16.5, Figure 16.6, and Figure 16.7.

Figure 16.5 reproduces the spreadsheet output showing the top 25 best models ranked in terms of how often each was chosen to be the best model for a BS sample. The C_p/p

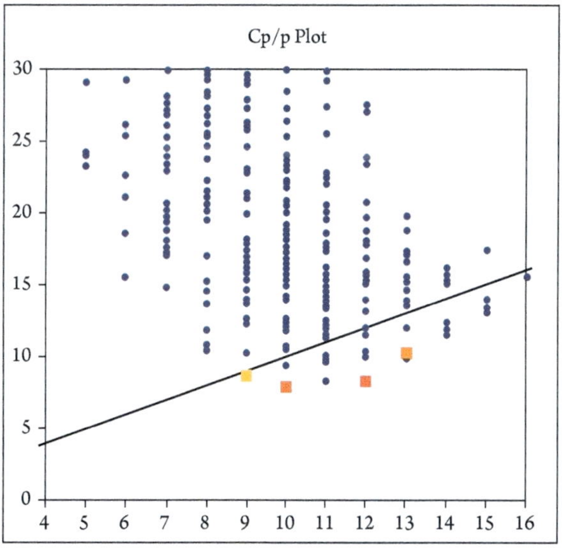

Figure 16.6 17-parameter asphalt model. C_p versus p plot of 814 promising models (not all included in the frame shown) found for the asphalt free surface energy data, including four selected squared component factors as well as all linear components, using the 'many models per sample' BS method. The C_p values are those obtained when the promising models are fitted to the original sample. The (p, C_p) position of the best model is shown in red, with the next three ranked downwards in decreasing redness.

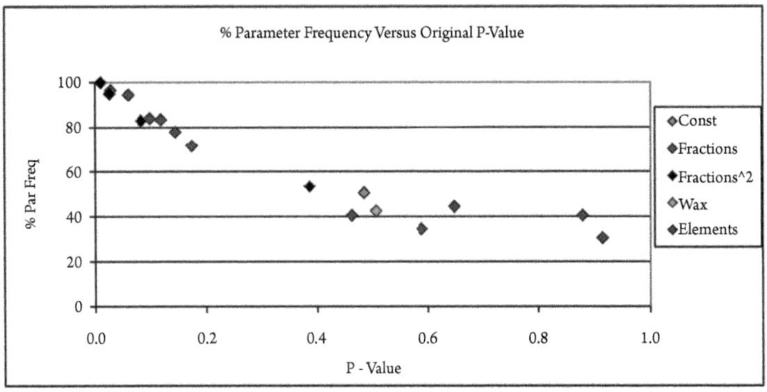

Figure 16.7 17-parameter asphalt model. Plot of the number of times (expressed as a percentage frequency) that each parameter is included in the 500 best model fits to the BS samples versus the *p*-value of the parameter estimate in the full model fit to the original data.

values of the top four models tabulated in the spreadsheet output are shown in red in the C_p/p plot given in Figure 16.6, indicating they all have $C_p < p$. The right-hand column in the spreadsheet gives the frequency each model was selected as being the best; the top four models taken together were chosen 97 times. The top 25 models taken together were selected as best in 315 of the 500 BS samples; that is, 63% of the time. The fitted b_i values for each model are shown in each row. It will be seen that *sat, aro, res, sat*2, *aro*2, and *res*2 are all regularly chosen, as are the elements *H, O*, and *Ni*; so these seem to be the important explanatory variables.

Figure 16.7 depicts the plot of the frequency of inclusion of each parameter b_i versus the p-value of the estimated b_i in the full model fit to the original sample, showing the inverse relationship clearly. Moreover, the grouping of the points for *sat, aro, res, sat*2, *aro*2, *res*2, *H, O*, and *Ni*, all with small p-value and high frequency of inclusion, is now what we would expect.

Comparing the 13- and 17-parameter models, it will be seen that the inclusion of squared explanatory variables has led to a much more satisfactory fit being achieved. As mentioned, we have not examined the data fully, but the fact that inclusion of quadratic terms has led to a much more satisfactory fit would suggest that there are significant nonlinearities present in the data not captured by a simple linear model.

16.3 Conclusions

We have discussed a non-standard way of using bootstrapping to analyse the selection and fitting of linear models in multiple regression. The bootstrapping is used for two purposes.

First, it is used to produce B parametric BS samples of the original data set. The full model is fitted to each BS sample, and these fits are used to produce promising sub-models using a 'many models per sample' method. This guarantees a maximum of BP models, though duplication means the number of distinct promising models is usually significantly smaller.

The way the promising models is constructed means that they are likely to have small fitted C_p values, as is borne out in the numerical examples, so that though far fewer than the full set of possible submodels, the set of promising models will usually contain models that are a good fit to the original data. Selection of the overall best model can therefore focus on choosing a model from the set of promising models.

The bootstrapping also allows an assessment to be made of the quality of such a model selected as being the best. This is done by examining if the selected model will also be the best fit for other samples obtained by bootstrapping so that the BS samples have the same statistical properties as the original data. Such information is not available using a standard best subset analysis or a stepwise regression analysis.

17 Finite Mixture Models

17.1 Introduction

As mentioned in the Acknowledgements, this chapter and the final chapter are based on joint ongoing work with Dr Christine Currie of Southampton University, so that, apart from the Hidalgo stamp issues example discussed in the final chapter, she is essentially a co-author.

Finite mixture models have been the subject of extensive study. Maximum likelihood and Bayesian Markov chain Monte Carlo (MCMC) estimation are the two main methods used in fitting finite mixture models to data, and we review both. We cover only the univariate case. A good general reference is McLachlan and Peel (2000), covering both methods.

Finite mixture models are an extremely flexible family which should be potentially very useful in applications. However, despite extensive study, perhaps because of it, the practitioner is confronted by what can appear to be a confusing picture with a number of pitfalls to be negotiated. We have already discussed a special case of the two-component normal mixture model in Section 14.1.1, which illustrated some of the non-standard difficulties that occur. Wasserman, a notable researcher of mixture models, gives an amusing informal indication, in Wasserman (2012), of the kind of problems that can arise when trying to fit a mixture model to data, with the advice that 'mixtures, like tequila, are inherently evil and should be avoided at all costs' http:// normaldeviate.wordpress.com/ 2012/08/04/ mixture-models-the-twilight-zone-of-statistics/. We are perhaps not quite so pessimistic!

In this chapter, we review the ML and two Bayesian MCMC estimation methods. Our conclusion is that neither the ML nor the MCMC approaches are entirely satisfactory on their own, particularly in identifying what are known as *overfitted models* (to be defined below). However, an alternative Bayesian approach with a focus more like that of ML is possible, which makes identification and estimation much easier and more transparent. The method replaces ML by maximum a posteriori (MAP) point estimation conditionally on k, the number of components, over a range of k values. We show that importance sampling (IS), which is seldom discussed in finite mixture modelling, can then be used to calculate the posterior distribution of k. This is quite straightforward to do, avoiding the

Non-Standard Parametric Statistical Inference. Russell Cheng.

complications involved with use of MCMC. We call this approach the MAPIS method. The chapter is set out as follows.

(i) As there are a number of different but interrelated aspects to bear in mind when discussing the methods, we shall give our discussion some structure by first setting out the model formally and summarizing immediately what we consider to be the strengths and weaknesses of the ML and Bayesian MCMC methods. We then outline our suggested MAPIS method.

(ii) We will then examine each of the ML, Bayesian MCMC and MAPIS approaches in more detail to justify our comments more fully. We do not claim that MAPIS is a complete panacea that overcomes all problems of a ML or a Bayesian MCMC approach, but nevertheless suggest that its use avoids the main difficulties which can occur with use of any of the other approaches on their own.

(iii) We illustrate our discussion with three examples.

17.2 The Finite Mixture Model

We consider the fitting of a univariate finite mixture model to a random sample $\mathbf{y} = (y_1, y_2, \ldots, y_n)^T$ drawn from a distribution with probability density function (PDF) that is a weighted sum of a finite number of component continuous PDFs:

$$f(y|\boldsymbol{\psi}(k), \mathbf{w}(k), k) = \sum_{j=1}^{k} w_j(k) g(y|\boldsymbol{\psi}_j(k)), \tag{17.1}$$

where each component density $g(\cdot|\boldsymbol{\psi}(k))$ has the same functional form depending on a vector $\boldsymbol{\psi}(k)$ of parameters, but with different values $\boldsymbol{\psi}_j(k)$ for each component j. We consider only the case where the component densities each just have two parameters, with the mean μ_j and standard deviation (SD) σ_j as the parameters. There is a focus in the literature on the normal distribution, and for presentational clarity, this is the component distribution discussed in detail in this chapter. But our discussion applies equally to many other cases. We will outline in Section 18.2.1 in the next chapter how our discussion applies to the lognormal, extreme value (EVMin and EVMax), Weibull, gamma, and inverse Gaussian (IG) cases.

The vector of weights $\mathbf{w}(k) = (w_1(k), w_2(k), \ldots, w_k(k))$ is also treated as a parameter. This must belong to the simplex

$$\boldsymbol{\Delta}_k \equiv \{\mathbf{w}(k) | \, 0 \le w_j(k) \le 1, \, j = 1, 2, \ldots, k; \, \sum_{j=1}^{k} w_j(k) = 1\}.$$

In general, one would expect to have $-\infty < \mu_j < \infty$ and $0 < \sigma_j < \infty$. Thus, for k components,

$$\boldsymbol{\psi}(k) \in \boldsymbol{\Psi}_k \equiv \boldsymbol{\Psi} \times \boldsymbol{\Psi} \times \ldots \times \boldsymbol{\Psi},$$

where

$$\boldsymbol{\Psi} = \Re \times \Re^{+}.$$

All the quantities k, $(\boldsymbol{\psi}_j(k), w_j(k))$, for $j = 1, 2, \ldots, k$ are assumed unknown. We will consider only the situation where we can specify a given finite $k_{\max}$ with the unknown $k \leq k_{\max}$. In the theoretical study of one of the Bayesian MCMC methods that we consider, $k_{\max}$ is allowed to tend to infinity, but for practical numerical work, we will always assume that a $k_{\max} < \infty$ can be specified. We focus on estimation of k, $(\boldsymbol{\psi}_j(k), w_j(k))$ under the following assumption:

Assumption A0

The sample $y = (y_1, y_2, \ldots, y_n)$ is drawn from the model (17.1) where the number of components k_0 is finite and fixed and the corresponding component parameters have specific fixed values: $(\boldsymbol{\psi}_0(k_0), \mathbf{w}_0(k_0))$. However, k_0, $(\boldsymbol{\psi}_0(k_0), \mathbf{w}_0(k_0))$ are not known.

In the theoretical analysis of fitting such a finite mixture model to data, this is the usual assumption, made whether the ML or a Bayesian method of estimation is being studied. Note, however, that though the ML and Bayesian methods both aim is to estimate k_0, and $(\boldsymbol{\psi}_0(k_0), \mathbf{w}_0(k_0))$, the two methods do this in rather different ways. We will examine both more fully, but begin straightaway with a summary of the pros and cons of each approach.

17.2.1 MLE Estimation

In MLE, we focus on obtaining $\hat{k}$ and $(\hat{\boldsymbol{\psi}}(\hat{k}), \hat{\mathbf{w}}(\hat{k}))$, *point estimates* of k_0, $(\boldsymbol{\psi}_0(k_0), \mathbf{w}_0(k_0))$. The problem is non-standard on two counts:

(D1): If $\sigma_j \to 0$, the corresponding component becomes a discrete atom, so that the model is no longer strictly continuous.

(D2): If $w_j \to 0$, then the jth component vanishes, with both μ_j and σ_j becoming indeterminate.

Problem (D2) is the more problematic, as it makes non-standard the estimation of k. If we attempt to fit a model where $k > k_0$, which is called *overfitting* by Rousseau and Mengersen (2011), then $k - k_0$ components do not exist and their parameters are indeterminate. In Section 14.1.1, we have already considered a two-component normal that includes the (one component) standard normal as a special case, reviewing the extensive literature even this ostensibly simple example has generated. A slightly more general variant of this model was given in Chapter 2 with PDF as in eqn (2.5). This model was discussed by Chen, Ponomareva, and Tamer (2014) in a real practical context and, as will be seen from their discussion, resolving whether the data comes from one or two components is not straightforward.

In overfitting, the true model is represented by an entire boundary subregion of the parameter space and not by a single point. However, as discussed by Redner (1981), Feng and McCulloch (1996), and Cheng and Liu (2001), though the parameter estimates are unstable, the ML estimator, barring problem (D1), will converge to the true model. The distributional properties of quantities like the parameter estimates and even the maximized log-likelihood are non-standard, but can still be estimated using bootstrapping. Thus statistically viable methods of checking goodness-of-fit can be used in estimating k.

We will discuss MLE more fully later, but for clarity summarize our view of MLE here immediately.

(A1) ML estimation is much easier to implement and less subjective compared with Bayesian estimation, where selection of a prior distribution has to be done with great care.

(A2) ML estimation gives point estimates of component parameter values ($\boldsymbol{\psi}(k)$, $\mathbf{w}(k)$) conditional on k. Still conditional on k, confidence intervals can then be easily found.

(A3) Problem (D1), where components become discrete atoms, can occur. In practice, this is not a great issue, as where it might happen can be anticipated; for example, if extreme clustering of data observations occurs, this can be easily noted.

(A4) In estimating k_0, the behaviour of the MLE is non standard when $k > k_0$.

Of the four points made, the last is the most problematic. We discuss it in more detail straightaway in the next section.

17.2.2 Estimation of k under ML

McLachlan and Peel (2000) give a very thorough review of methods for estimating k when using ML, see particularly Sections 6.3-6.11. On the theoretical side, the distribution of the log-likelihood is non-standard, because regularity conditions are not satisfied in overfitted models. As a result, accurate distributional results for estimators of k are difficult to obtain theoretically, as already discussed in Section 14.1.1. McLachlan and Peel (2000, Sections 6.4 and 6.5), describe the extensive work done on the problem, from which unfortunately no clear definitive simple method really emerges.

More recent approaches based on cluster methods have been suggested by Tibshirani et al. (2001) and by Tadesse et al. (2005).

With ML, methods using information criteria are simple, though not fully underpinned by rigorous asymptotic theory. An early suggestion is to use the Akaike information criterion (AIC),

$$AIC_k(\hat{\boldsymbol{\theta}}(k)) = 2L_k(\hat{\boldsymbol{\psi}}(k), \hat{\mathbf{w}}(k) - 2p_k, \tag{17.2}$$

or, with an $O(n^{-1})$ correction,

$$AICC_k(\hat{\boldsymbol{\theta}}(k)) = AIC_k(\hat{\boldsymbol{\theta}}(k)) - 2p_k(p_k + 1)/(n - p_k - 1). \tag{17.3}$$

This was introduced by Akaike (1973, 1974) for standard problems. However, there is now much evidence, which we have witnessed ourselves, that use of AIC in fitting finite mixture models leads to k_0 being overestimated.

The Bayesian information criterion (BIC) introduced by Schwarz (1978),

$$BIC_k(\hat{\boldsymbol{\theta}}(k)) = 2L_k(\hat{\boldsymbol{\psi}}(k), \hat{\mathbf{w}}(k)) - p_k \ln(n), \tag{17.4}$$

performs better, as reported, for example, by Dasgupta and Raftery (1998).

The theoretical result given by Leroux (1992, Theorem 4) shows that estimating k by maximizing the AIC or BIC will ensure that the true k_0 will in the limit as $n \to 0$ not be underestimated. This is to be expected, as standard regularity conditions will apply for k up to and including k_0, so that it can be expected that $AICC_k(\hat{\boldsymbol{\theta}}(k))$ or $BIC_k(\hat{\boldsymbol{\theta}}(k))$ will increase with k up to this point. The problem is what happens once k increases past k_0. Feng and McCulloch (1996) and Cheng and Liu (2001) show that if one overfits by fitting a model where $k > k_0$, though the MLE becomes unstable, it will converge in probability to the subset of points in the full parameter space that represent the true model f_0. The implication of this, and indeed in the proof given by Wald (1949) of the consistency of the MLE, is that the maximized log-likelihood will tend to the expected value of the log-likelihood evaluated at the true parameter point, as k increases up to k_0, and then remain essentially unchanged if k is increased past k_0, so that $AICC_k$ or BIC_k will decrease linearly with k once $k > k_0$, with BIC_k doing so much more quickly than $AICC_k$. This gives an informal justification of choosing the k which maximizes either $AICC_k(\hat{\boldsymbol{\theta}}(k))$ or $BIC_k(\hat{\boldsymbol{\theta}}(k))$ as the ML estimator, $\hat{k}$ of k_0, though the argument is incomplete, as it leaves open the possibility that $\hat{k}$ will be biased too small.

More generally, Naik et al. (2007) have considered extension of AIC to a *mixture regression criterion* (MRC) for the simultaneous determination of the number of regression variables as well as components in finite mixture regression models. We do not discuss such generalizations here, as they are technically significantly more demanding than what we wish to consider, but the results reported by Naik et al. (2007) are encouraging and should be consulted by the interested reader.

Biernacki et al. (2000) discuss a classification-based likelihood criterion that can be used to improve BIC. A summary is given by McLachlan and Peel (2000, Section 6.10.3), who give simulation results showing the improvement obtained using the ICL-BIC method—a BIC-type of approximation to the integrated classification likelihood criterion (ILC) proposed by Biernacki et al. (2000). The ICL-BIC adds a further term to the BIC formula that is Bayesian, involving the posterior distribution of the indicators L_i to be defined and appearing later in eqn (17.5). We have not investigated this method for use in the ML context, because, as will be discussed when we discuss prior distributions in Section 17.3.1, the L_i are not needed if the finite mixture is simply being used as a representation of a non-standard density, which is the version of the Bayesian model we will be considering. In any case, if one is using a Bayesian approach, examination of the posterior distribution of k seems a more transparent way to select k compared with use of an information measure.

Overall, the BIC given in eqn (17.4) seems the easiest to apply whilst being sufficiently reliable. A numerical example is given in detail in the next chapter in Section 18.1.3, where it is compared with AIC.

A possible computationally intensive alternative is to use a bootstrap goodness-of-fit (BS GoF) test for each individual k. Such a test is regular when $k \leq k_0$ and, as no direct comparisons between different k are involved, BS GoF tests can be carried out sequentially, increasing k stepwise and stopping immediately the lack-of-fit is not statistically significant. Thus, with controllable level of confidence, we will stop with $k \leq k_0$, so avoiding the overfitting problem entirely. In fact, relying on the result of Feng and McCulloch (1996) and Cheng and Liu (2001) that when $k > k_0$, the model $\hat{f}$ by ML is nevertheless consistent for f_0, using BS GoF testing should not be unsatisfactory for $k > k_0$ as well.

We have studied this approach briefly, using as the GoF statistic T, the Anderson-Darling statistic A^2 of eqn (4.12), the Cramér-von Mises statistic of eqn (4.11), and the drift statistic $\Delta_0(m)$ of eqn (4.17). With T equal to any of these statistics, its value $T_k(\mathbf{y})$ for each k is easily calculated, as only the fitted CDF values F_k corresponding to the mixtures PDF (17.1) evaluated at $(\hat{\boldsymbol{\psi}}(k), \hat{\mathbf{w}}(k))$ or $(\tilde{\boldsymbol{\psi}}(k), \tilde{\mathbf{w}}(k))$ are required. The $T_k(\mathbf{y})$ values are not directly comparable on their own. However, we can estimate the null distribution of T_k from the EDF of a sample $T_k(\mathbf{y}^{*(j)})$, $j = 1, 2, \ldots, B$, of bootstrap values, where each BS sample of observations $\mathbf{y}^{*(j)}$ is generated from the null distribution $F_k(\hat{\boldsymbol{\psi}}(k), \hat{\mathbf{w}}(k))$ or $F_k(\tilde{\boldsymbol{\psi}}(k), \tilde{\mathbf{w}}(k))$.

The main downside of bootstrapping is the computational intensiveness of the calculation.

A numerical example of the GoF approach is given in detail in the next chapter in Section 18.1.3.

17.2.3 Two Bayesian Approaches

Given the great flexibility of finite mixtures, the Bayesian approach arguably allows a more comprehensive examination than ML of complicated data samples. Though we have not finished discussing ML, the bulk of this chapter is a discussion of Bayesian issues, and we summarize these here.

In Bayesian estimation, the focus is not on point estimation, but in obtaining the posterior distribution of k, and $(\boldsymbol{\psi}, \mathbf{w})$. This is obtained from the data $\mathbf{y}$ in conjunction with a specified prior distribution. As the number of components of $\boldsymbol{\psi}$ and $\mathbf{w}$ varies with k, both the prior and posterior are most simply specified by first giving the distribution of k on its own, namely, its marginal distribution, and then giving the conditional joint distribution, given k, of $\boldsymbol{\psi}(k)$ and $\mathbf{w}(k)$ as k varies. We write the posterior distribution of k as $\pi(k|\mathbf{y})$ and the conditional distribution of $(\boldsymbol{\psi}, \mathbf{w})$ given k as $\pi(\boldsymbol{\psi}(k), \mathbf{w}(k)|k, \mathbf{y})$, so that the product of this latter distribution with $\pi(k|\mathbf{y})$ gives the complete joint posterior distribution of k, $(\boldsymbol{\psi}, \mathbf{w})$.

We will consider in the main the random jump Markov chain Monte Carlo (RJMCMC) method, as this is well established and popular. A good introduction is given by Richardson and Green (1997). We do, however, albeit rather more briefly, consider another method, proposed by Ishwaran and co-authors, see Ishwaran and Zarepour (2000), Ishwaran et al. (2001), Ishwaran and James (2002), and Ishwaran and Zarepour (2002), which also uses MCMC simulation, but to approximate a Dirichlet process, with different versions given in these papers. As they are all quite similar, we will refer to them generically as the *approximate Dirichlet process* (ADP) method.

We will discuss both the RJMCMC and ADP methods in more detail shortly, but again for clarity, we summarize our comments on them straightaway here.

(B1) The most attractive feature of the Bayesian approach, at least in principle, is that it explicitly estimates $\pi(k|\mathbf{y})$, the posterior distribution of k. Ironically, this actually then enables a good point estimate of k_0 to be easily obtained, with the posterior distribution of k providing an immediate and appealing quantitative assessment of how well k_0 has been estimated; something MLE does not achieve so well, despite its emphasis on point estimation.

(B2) However, using the MCMC approach, estimates of the component parameters and weights, even in the simple form $(\breve{\boldsymbol{\psi}}(k), \breve{\mathbf{w}}(k))$ conditionally on k, is difficult. The reason, to be discussed, is closely connected with problem (B5) given later.

(B3) MCMC simulation does not directly provide an estimate of the supposed true distribution with PDF

$$f_0 = f(y|\boldsymbol{\psi}_0(k_0), \mathbf{w}_0(k_0), k_0)$$

under Assumption A0, where f is the PDF (17.1). Richardson and Green (1997) call such an estimate of f_0 the *predictive density estimate*, giving two methods for obtaining this from the MCMC simulation run, but neither method is satisfactory, as the authors themselves point out. The ADP approach is similarly unsatisfactory on this count.

(B4) The choice of prior for w_j, the weight parameters, is particularly important if the posterior is to be consistent. We shall consider the Dirichlet prior used in the RJMCMC and ADP methods with a common shape parameter δ, discussing this more fully later, as the value of δ not surprisingly has a significant effect on the posterior distribution. Though a default value allows for an initial analysis, usually a range of values will give meaningful results, so some variation should always be considered in any full analysis. Though we consider only the Dirichlet prior distribution for w, this comment obviously applies to any prior used.

The influence of this choice of δ strongly affects estimation of the posterior of k, as described in the next comment.

(B5) The posterior distribution of k over-favours high k values.
How this can occur is shown by Theorem 1 in Rousseau and Mengersen (2011), which reveals two asymptotic characteristic properties of the posterior distribution of k and $(\boldsymbol{\psi}, \mathbf{w})$ in overfitted models. We discuss the theorem more fully later, but, in a nutshell, it shows that the posterior distributions of the component parameters

depend critically on the choice of the shape parameter δ in the prior for the weight parameter $\mathbf{w}$. If δ is small and $k > k_0$, then there will be $(k - k_0)$ weights near zero in the posterior PDF f. If δ is large and $k > k_0$, there will be $(k - k_0)$ components having (μ_j, σ_j) that simply duplicate the (μ_j, σ_j) of other components. If these two situations are not picked up, the estimate of k_0 will be biased too large.

The implication of the theorem is that output of the MCMC simulation run requires postprocessing to remove components which are only apparent, but that should be omitted and not counted if a correct interpretation of results is to be arrived at. A significant difficulty is that this adjustment has to be done in a way that will work when k_0 is not known.

The theorem of Rousseau and Mengersen is an asymptotic result. In the case of finite n, the problem can appear in a way that is different to that given in the theorem. What appears to happen is seen if one examines the results of an MCMC simulation, when f_0 is known. We find that the two forms of behaviour cannot be fully eliminated by choice of the δ value alone, and that instead there is an awkward computational effect, discussed in the next comment, which can be regarded as an alternative numerical manifestation of the two problems identified in the Rousseau and Mengersen theorem.

(B6) The conditional posterior distributions given k of the individual component parameters $\mu_j, \sigma_j, w_j, j = 1, 2, \ldots, k$, are typically multimodal when $k > k_0$.

In the numerical example given in Section 17.6.2, we examine the form of the MCMC simulation output, finding that visits of the Markov chain to a given state k, where $k > k_0$, will 'hunt' between existing real components, with visits to several real components combining to count as visits to non-existent components. This can give rise to either of the two results in the Rousseau and Mengersen theorem. Even without having to appeal to the theorem of Rousseau and Mengersen (2011), problem (B6) shows that, when $k > k_0$, the way that MCMC simulation run assigns observations to individual components is too simplistic, so that the components assumed to have been been observed at each step in the run do not always correspond to specific and genuine components in the simple way assumed when counting components.

17.2.4 MAPIS Method

Our main focus in the remainder of the chapter will be to consider an alternative but still Bayesian approach that overcomes the issues raised in (B5) and (B6). We call the method MAPIS to reflect the fact that it involves two stages. In the first, point estimates of component parameters are obtained by the maximum a posteriori estimation (MAP) method, and, in the second, posterior component parameter distributions are calculated using importance sampling (IS). This produces estimated components that are more easily related to sample features when $k > k_0$, and yields a posterior distribution of k not over-favouring high k values.

We summarize the characteristics of the MAPIS approach as follows.

(C1) MAPIS, being Bayesian, shares the characteristic (B1) of the RJMCMC and ADP methods in allowing a posterior distribution for k to be calculated.

It is worth noting that obtaining posterior distributions is numerically challenging using MCMC, because sampling is correlated and from a parameter space with subspaces of different dimensions as k varies. The RJMCMC and ADP have ingenious but elaborate mechanisms to handle this problem. The difficulty of using MCMC when the number of components k is unknown seems to have led some researchers to believe that importance sampling cannot be used. This is not the case.

Use of importance sampling is quite standard, and can be implemented using a *rejection/acceptance* method to sample from an *importance sampling* (IS) distribution with a support that spans *all* the spaces of different dimensions involved. Because the individual observations in IS are mutually independent, a mechanism like reversible jumps is not therefore needed as in the RJMCMC method. In IS, estimation of the posterior distribution of k is therefore completely straightforward.

(C2) With MAP estimation, we obtain point estimates $(\tilde{\boldsymbol{\psi}}(k), \tilde{\mathbf{w}}(k))$ for $k = 1, 2, \ldots, k_{\max}$. As will be shown, unlike the MCMC approach, the difficulty of multimodality in posterior distributions of component parameters and weights does not occur.

(C3) The MAPIS method does not suffer from problem (B3). An explicit estimate (the posterior predictive density), $\tilde{f}_0$, of $f(y|\boldsymbol{\psi}_0(k_0), \mathbf{w}(k_0), k_0)$, the PDF (17.1) under Assumption A0, is easily obtained as

$$\tilde{f}_0 = f(y|\tilde{\boldsymbol{\psi}}(\tilde{k}), \tilde{\mathbf{w}}(\tilde{k}), \tilde{k}) \text{ where } \tilde{k} = \arg \max_{k=1,2,\ldots,k_{\max}} \tilde{\pi}(k|\mathbf{y}).$$

(C4) The choice of the shape parameter δ in the prior for the weights w_j can be more sensitive than in the MCMC case. This extends somewhat to the shape parameter g to be defined in the prior for the SD component parameters σ_j. The reason is that the problems (D1) and (D2), which can occur using ML, will also occur in MAP if the posterior of a component parameter σ or weight w is J-shaped. We will discuss how to choose values of the shape parameters g and δ of the priors for σ and w to avoid this.

(C5) Jasra et al. (2005) do not recommend MAP estimation, pointing out that, in selecting a single point estimate, insufficient weight is attached to alternative estimates that might be worth considering. This is true, but arguably not as serious as might at first appear. If we write $\tilde{\boldsymbol{\psi}}(k), \tilde{\mathbf{w}}(k)$, $k = 1, 2, \ldots, k_{\max}$, for the MAP estimate of the component parameters and weights conditional on the value of k, then each $\tilde{\boldsymbol{\psi}}(k), \tilde{\mathbf{w}}(k)$ represents the best estimate obtainable for that given k, with alternative estimates for that given k being considered to be inferior. A comparison between just these $\tilde{\boldsymbol{\psi}}(k), \tilde{\mathbf{w}}(k)$ therefore gives a succinct evaluation of what would be a good choice for k, reducing choice of a best overall k to the simplest comparison possible made between the *best* estimates of each given k. This avoids considering what might be an otherwise unhelpfully large number of alternatives for each k.

More specifically, to estimate the posterior distribution of k accurately, importance sampling has to cover the entire parameter space which has a dimension that varies with k. The full IS distribution comprises distinct parts, one for each k, with each based on the posterior distribution of $\tilde{\boldsymbol{\psi}}(k), \tilde{\mathbf{w}}(k)$ in the neighbourhood of the MAP estimate value $\tilde{\boldsymbol{\psi}}(k), \tilde{\mathbf{w}}(k)$ itself. There is the possibility that the total posterior probability in this neighbourhood under-represents the posterior distribution over the parameter space corresponding to this k. However, what we can expect is for the points sampled for each given k to adequately represent the posterior probability in the neighbourhood of $\tilde{\boldsymbol{\psi}}(k), \tilde{\mathbf{w}}(k)$. Thus, the estimated overall probabilities corresponding to each k will give a meaningful comparison of the posterior probability associated with each of the specific MAP estimators $\tilde{\boldsymbol{\psi}}(k), \tilde{\mathbf{w}}(k)$, with any error occurring only because of insufficient sampling of points $\boldsymbol{\psi}(k), \mathbf{w}(k)$ that we would consider to be not so worth considering compared with those sampled in the neighborhood of $\tilde{\boldsymbol{\psi}}(k), \tilde{\mathbf{w}}(k)$. Overall, we would expect the MAPIS approach, if it is inaccurate, to be only parsimoniously so in not giving sufficient weight to alternatives that are actually poorer fits than those considered the best.

In our more detailed discussion later, we will give examples illustrating this, contrasting it with when problem (B5) arises using MCMC. We leave the reader to decide which results are the more informative.

The remainder of the chapter is as follows.

Section 17.3 describes in more detail the underlying Bayesian model used by RJMCMC, ADP, and MAPIS. Also, the RJMCMC and ADP methods are described more fully.

Section 17.4 describes the MAPIS method fully. We show that a predictive density estimate is easily obtained using MAPIS.

Section 17.5 discusses the problem of calculating a *predictive density estimate*, that is, a point estimate for the PDF of the finite mixture model, when k has to be estimated.

Section 17.6 highlights *overfitted models* and the theorem of Rousseau and Mengersen (2011), showing how this gives rise to problem (B5).

Detailed numerical examples are given in the next chapter in Section 18.1 comparing the three Bayesian methods in the main, but also including results using ML.

17.3 Bayesian Hierarchical Model

In this section, we consider the Bayesian finite mixture model used in the RJMCMC, ADP, and MAPIS methods. The structure of the model is defined by the priors, which we discuss first.

17.3.1 Priors

Priors are defined for each fixed k, and in both the RJMCMC and ADP methods, they take the form

$$
\begin{aligned}
(\boldsymbol{\psi}_j(k)) &\sim H,\ j = 1,2,\ldots,k, \\
\mathbf{w}(k) &\sim \text{Dirichlet}(\delta_1,\delta_2,\ldots,\delta_k), \\
(L_i|\mathbf{w}(k)) &\sim \text{Multinomial}(\{1,2,\ldots,k\},\mathbf{w}(k)), i = 1,2,\ldots,n, \\
(Y_i|L_i,\boldsymbol{\psi}(k)) &\sim N(\mu_{L_i},\sigma_{L_i}),\ i = 1,2,\ldots,n,
\end{aligned}
\tag{17.5}
$$

where we use '$\sim$' to denote sampling from a given distribution. Thus, as k runs through the possible values $k = 1, 2, 3, \ldots$, the priors take on a hierarchical form, giving the overall model the name *Bayesian hierarchical model.*

The distribution H is the prior for all of the $\boldsymbol{\psi}_j(k) = (\mu_j(k), \sigma_j(k)),\ j = 1, 2, \ldots, k$. Normal priors are usually assumed for the $\mu_j(k)$ parameters when using MCMC methods, and they will also be assumed in MAPIS.

In RJMCMC, the prior distribution taken for SD parameters σ_j of all the components has the form $\sigma_j^{-2} \sim \Gamma(\alpha,\beta)$, with β having a hyperprior distribution $\beta \sim \Gamma(g,h)$. Actually, β can be eliminated if its prior is assumed to be independent of the sigma σ_j^{-2} prior by treating β as a randomized parameter, as discussed in Chapter 13, and integrating it out using its hyperprior distribution $\Gamma(g,h)$ in a similar way to eqn (13.1). Doing this, the distribution of σ_j then just has the PDF

$$
\pi(\sigma_j) = 2\frac{\Gamma(\alpha+g)}{\Gamma(\alpha)\Gamma(g)}h^g\left(1+h\sigma_j^2\right)^{-\alpha-g}\sigma_j^{2g-1},\quad \sigma > 0. \tag{17.6}
$$

Further details are given in Section 18.2.3.

In the MAPIS method, we use (17.6) as our prior for σ_j. Use of this version of the prior for σ is not particularly advantageous in MCMC simulations, as drawing sample values of β and σ^{-2} is straightforward. However, the explicit form of (17.6) makes it easier to assess how parameter values will affect the density, and it is clear that the results will be quite sensitive to the choice of g. For example, the density $\pi(\sigma)$ is J-shaped if $g < 0.5$, but has a strictly positive mode if $g > 0.5$. However, the posterior depends on the component densities as well as the prior, and to definitely avoid the posterior density becoming infinite in the normal component case, we need

$$
g > 1.
$$

The prior used in RJMCMC and ADP for $\mathbf{w}(k)$ is the Dirichlet distribution of order k

$$
f(\mathbf{w}(k)) = \frac{1}{B(\delta)}\Pi_{j=1}^k w_j^{\delta_j-1}, \tag{17.7}
$$

where

$$
B(\delta) = \Pi_{j=1}^k\Gamma(\delta_j)/\Gamma(\Sigma_{j=1}^k\delta_j).
$$

We will use a common value for the δ_j, so that $\delta_j = \delta$ for all j. To ensure that the posterior weight densities do not become infinite requires

$$\delta > 1.$$

The choice of δ_j, $j = 1, 2, \ldots, k$, is very important, a fact highlighted by the result of Rousseau and Mengersen (2011), and we shall be discussing this in the next section and the rest of the chapter.

The quantities L_i are latent variables indicating the component that a particular observation has been drawn from and is assumed to be unknown. Their estimation is included in the RJMCMC and ADP methods. However, they are not required if the finite mixture is simply being used as a representation of a non-standard density. This is the viewpoint adopted in the MAPIS method, so that L_i estimation is not needed.

The difference between the RJMCMC and ADP methods lies in how variation in k is handled, and in particular on how its posterior distribution is estimated. We discuss this next.

17.3.2 The Posterior Distribution of k

In the RJMCMC method, it is assumed that a finite bound k_{max} exists for k_0, i.e. $k_0 < k_{max} < \infty$, with k_{max} given, and this can then be used as an upper bound on the value k can take. In this case, a simple prior for k_0 is the uniform prior

$$\pi(k) = 1/k_{max}, \;\; k = 1, 2, \ldots, k_{max}. \tag{17.8}$$

As described in Richardson and Green (1997), the RJMCMC algorithm uses a modified Metropolis-Hastings sampling method in which k, that is, the dimension of the state space, can change at each step transition of the MC. Probability-balance problems with such a 'jump' change in k are avoided by making the jumps 'reversible' so that the acceptance probability of the jump change satisfies eqn (7) in Richardson and Green (1997). Changes in the μ, σ, w parameters are made by Gibbs sampling. We will follow Richardson and Green (1997), and refer to the calculations at each step of the simulation as a *sweep*.

If the total number of sweeps is m, and m_k is the number of sweeps where the value k is observed, then the posterior distribution of k is estimated by

$$\breve{\pi}(k|\mathbf{y}) = m_k / \sum_{k=1}^{k_{\max}} m_k = m_k/m, k = 1, 2, \ldots, k_{\max}\,. \tag{17.9}$$

We shall call

$$\breve{k} = \arg\max_k \breve{\pi}(k|\mathbf{y}) \tag{17.10}$$

the *most likely* k (as given by RJMCMC). We will consider corresponding versions of this 'most likely k' for the ADP and MAPIS methods as well.

The RJMCMC approach has been much used in applications. An implementation, Nmix, is downloadable from http://www.stats.bris.ac.uk/peter/Nmix/, and described in Richardson and Green (1997). Nmix is a very flexible package, with many options. Detailed output is also available down to the level of recording parameter values at every sweep of the MCMC simulation. This enables postprocessing, making it possible to calculate modified estimates of posterior distributions like the reduced-k estimator of the posterior distribution of k, given in equation (17.29). We will discuss this estimator in Section 17.6.1.

The ADP method we have followed is essentially that given by Ishwaran et al. (2001). As in RJMCMC, it uses MCMC simulation of the Bayesian model (17.5), with Gibbs sampling in the sweeps of the MCMC simulation. However, the key difference from RJMCMC is that this is only carried out at *one* value of k, $k = N$, with N to be chosen. Theorem 6 in Ishwaran and James (2002) shows that providing

$$\delta_j = \delta/N \tag{17.11}$$

for some fixed $\delta > 0$, then the H prior for the component parameters $\boldsymbol{\psi}(N)$ and the Dirichlet prior for the weights $\mathbf{w}(N)$ of (17.5) can be viewed together as a truncated Dirichlet process that tends to a Dirichlet process as $N \to \infty$. Moreover, the posterior will be consistent (in the sense that the posterior distribution of the mixture density f becomes concentrated around the true mixture density f_0) as the sample size $n \to \infty$, provided $N \to \infty$, but with $\log N/n \to 0$, as $n \to \infty$. This last condition shows that N should not be taken too large.

The value of δ can be set, see Ishwaran and James (2002), to a constant value of (= 1, say) or can have its own prior $\delta \sim \Gamma(\eta_1, \eta_2)$. In our implementation, we set δ to a constant.

A major problem with the ADP approach is that it does not directly estimate any posterior probability distribution. In particular, the MCMC simulation does not give a direct estimate of the posterior probability of k. Postprocessing of the sweeps is needed. Ishwaran et al. (2001) propose an *effective dimensions* method, where a correction is made to the value of $k = N$ obtained at each sweep i by examining the allocation variables $L_j^{(i)}$, $j = 1, 2, \ldots, n$, in sweep i and counting only those component subscripts k' $(= 1, 2, \ldots, N)$ as corresponding to an actual component if there is at least one $L_j^{(i)} = k'$. The weights of the components included in the count have to be adjusted to sum to one, so that

$$\underline{w}_k^{(i)} = \bar{I}_{m_k^{(i)}}(0) w_k^{(i)} \Big/ \sum_{l=1}^{N} \bar{I}_{m_l^{(i)}}(0) w_l^{(i)}, \; k = 1, 2, \ldots, N, \;\; i = 1, 2, \ldots,$$

where $m_k^{(i)} = \#(L_j^{(i)} = k)$ and $\bar{I}_m(0) = 1$ if $m \neq 0$, and $\bar{I}_m(0) = 0$ otherwise. The *effective component count* at sweep i is

$$\underline{k}_+^{(i)} = \#\text{ of } m_k^{(i)}\text{s}, \, k = 1, 2, \ldots, N, \text{ that are non-zero.}$$

The estimate of $\pi(k|\mathbf{y})$, $k = 1, 2, \ldots, N$, and the most likely k are then

$$\breve{\pi}_+(k|\mathbf{y}) = m^{-1}\sum_{i=1}^{m} I_k(\underline{k}_+^{(i)}),\ k = 1, 2, \ldots, N \tag{17.12}$$

and

$$\breve{k}_+ = \arg\max_k \breve{\pi}_+(k|\mathbf{y}). \tag{17.13}$$

In our experiments with ADP, we actually used the Nmix implementation to carry out the MCMC simulation, as Nmix has an option to run the Markov chain at a fixed $k = N$. We chose N relatively small compared to n. The numerical results to be discussed later were obtained in this way.

We will describe the MAPIS method in detail in the next section. Here we just note immediately that, for the posterior distribution of k, MAPIS provides an estimator similar to $\breve{\pi}(k|\mathbf{y}) = m_k / \sum m_k$ in (17.9), except that the m_k are integrals that can be evaluated *independently* by importance sampling so that reversible jumps are not needed.

17.4 MAPIS Method

In this section, we describe our proposed MAPIS method for estimating the finite mixture model (17.1) under assumption A0 and the assumption that we have a finite upper bound k_{max} for k_0, i.e. $k_0 < k_{max} < \infty$. The MAP and IS stages of the method divide conveniently. We consider the MAP stage first.

17.4.1 MAP Estimation

Given k, the MAP estimator $[\tilde{\boldsymbol{\psi}}(k), \tilde{\mathbf{w}}(k)]$ comprises the parameter values which maximize the posterior distribution conditional on k, this latter being given by the standard formula

$$\pi(\boldsymbol{\theta}(k), k|\mathbf{y}) = \frac{f[\mathbf{y}|\boldsymbol{\theta}(k), k]\pi[\boldsymbol{\theta}(k), k]}{\sum_{l=1}^{k_{\max}} J(l|\mathbf{y})},$$

where

$$J(k|\mathbf{y}) = \int_{\boldsymbol{\Theta}(k)} f[\mathbf{y}|\boldsymbol{\theta}(k), k]\pi[\boldsymbol{\theta}(k), k]d\boldsymbol{\theta}(k) \tag{17.14}$$

and

$$\boldsymbol{\theta}(k) = [\boldsymbol{\psi}(k), \mathbf{w}(k)] \in \boldsymbol{\Theta}(k).$$

Calculation of $[\tilde{\boldsymbol{\psi}}(k), \tilde{\mathbf{w}}(k)]$, which is conditional on k, is simplified by noting that in the maximization of $\pi(\boldsymbol{\theta}(k), k|\mathbf{y})$, we can omit the denominator $\sum_{l=1}^{K} J(l|\mathbf{y})$ as it is a summation over all k. The maximum posterior (MAP) estimator, conditional on each k, $k = 1, 2, \ldots, k_{\max}$, is therefore

$$[\tilde{\boldsymbol{\psi}}(k), \tilde{\mathbf{w}}(k)] = \arg \max_{\boldsymbol{\psi}(k), \mathbf{w}(k)} \{f[\mathbf{y}|\boldsymbol{\psi}(k), \mathbf{w}(k), k]\pi[\boldsymbol{\psi}(k), \mathbf{w}(k), k]\} \tag{17.15}$$

for $k = 1, 2, \ldots, k_{\max}$.

We do not agree with the reservations of Jasra et al. (2005) over use of MAP. These reservations seem questionable, as they do not address the problem of estimating the posterior probability of overfitted models to be considered in Section 17.6, where we show that MAPIS provides a method of estimating posterior probabilities that avoids problems (B5) and (B6) described in the Introduction.

A consequence of MAP estimation is that a 'best' overall k is immediately obtained if we define 'best' as the k at which the Bayesian information criterion (BIC) given in eqn (17.4) is maximized. The overall best k is

$$\tilde{k}_{BIC} = \arg \max_{k=1,2,\ldots,M} BIC_k(\tilde{\boldsymbol{\theta}}(k)). \tag{17.16}$$

Note that this will usually be not quite the same as the ML version $\hat{k}_{BIC}$, where $\hat{\boldsymbol{\theta}}(k)$ is used instead of $\tilde{\boldsymbol{\theta}}(k)$ in the calculation of BIC_k.

The penalized ML estimation method recommended by Ishwaran et al. (2001) and Ishwaran and James (2002) for estimating k_0, as an adjunct to their ADP approach, is essentially equivalent to using $\tilde{k}_{BIC}$.

In Section 17.4.3, we describe how importance sampling can be used to estimate the $J(k|\mathbf{y})$ integral of eqn (17.14).

Though not required for calculating point estimates, the integrals $J(k|\mathbf{y})$ are needed for calculating the posterior probability distribution of k from

$$\tilde{\pi}(k|\mathbf{y}) = \frac{J(k|\mathbf{y})}{\sum_{l=1}^{k_{\max}} J(l|\mathbf{y})}, \qquad k = 1, 2, \ldots, k_{\max}. \tag{17.17}$$

This yields the 'most likely' k as

$$\tilde{k} = \arg \max_{k} \tilde{\pi}(k|\mathbf{y}), \tag{17.18}$$

which is the MAPIS analogue of $\breve{k}$ given in (17.10). In the context of our discussion, we prefer to use $\tilde{k}$ rather than $\tilde{k}_{BIC}$.

17.4.2 Numerical MAP

We obtain the MAP estimates numerically by maximizing the posterior probability using the Nelder and Mead (1965) optimization routine . The same procedure is also used for ML, in which case it is the log-likelihood that is optimized. We have already given our reasons for using Nelder-Mead in Section 3.7, but add a few additional comments in the present context.

The basic version of the Nelder-Mead method is for unconstrained optimization. We dealt with positivity constraints on $\psi_i(k)$ and $w_i(k)$ simply by setting a parameter to half its current value whenever the basic Nelder-Mead algorithm proposes a negative parameter value for the next step. We can ensure the sum of the weights remains equal to unity by not treating the weight of the last component as being a parameter of the Nelder-Mead search, but directly setting its value at each step of the search so that the weights sum to one. If, at any step and with the last weight omitted, the sum of the remaining weights is greater than unity, then all these remaining weights are rescaled so that they sum to nearly unity and the last weight is given a near-zero value. In our implementation, a warning flag is raised if the routine exits with a supposed optimum, but with an SD or weight near zero.

In our implementation, the Nelder-Mead routine is used to minimize the negative posterior distribution, doing this sequentially for increasing $k = 1, 2, \ldots, k_{\max}$, with the optimal $\tilde{\psi}(k), \tilde{\mathbf{w}}(k)$ for the k-component fit modified to provide the starting point for $k + 1$. We describe first the basic sequential approach. The final version is actually more elaborate, and will be explained separately later.

A starting value, as required by the algorithm, is only needed for the case $k = 1$. With the parametrization used, an obvious starting point for this case is $\mu_0 = \bar{y}$ and $\sigma_0 = s$, the respective sample mean and sample standard deviation of the data $\mathbf{y}$. The starting parameters for the model with $k + 1$ components are then determined from the best estimates for the model with k components. The first k components of the $k + 1$ model are set to be identical to those of the k-component model, but with reduced weights to allow some weight to be given to the $(k + 1)$th component. The $(k + 1)$th component is then chosen based on the discrepancies between the sample and the fitted k-component model, with its associated parameters μ_{k+1}, σ_{k+1}, and w_{k+1} selected so that adding the component to the mixutre will reduce the maximum overall discrepancy.

Specifically, let y_i, $i = 1, \ldots, n$, be the observations, and let $F_k(y)$ be the cumulative density function (CDF) of the fitted k -component model. We define D_i, $i = 1, \ldots, n$, to be the difference between the empirical distribution function (EDF) and the fitted model with k components, such that

$$D_i = \frac{i - 0.5}{n} - F_k(y_i).$$

We call

$$p_0 = \max\{D_j - D_i | 1 \leq i < j \leq n\} \tag{17.19}$$

the *maximum discrepancy*, and suppose that this maximum is obtained at $i = i_0, j = j_0$. Also, let

$$p_1 = \max\{D_j - D_i | 1 \leq i < j \leq n \text{ and } (i, j < i_0 \text{ or } i, j > j_0)\},$$

where this secondary maximum occurs at $i = i_1, j = j_1$. The $(k + 1)$th component is then given the mean and SD

$$\mu_{k+1} = (y_{i_0} + y_{j_0})/2, \ \sigma_{k+1} = (y_{j_0} - y_{i_0})/2,$$

and the weight of the $(k + 1)$th component is set to be p_0, while the weights of the remaining k components are multiplied by a factor $(1 - p_0)$. It is readily verified that this procedure will reduce the maximum discrepancy p_0, though there is some possibility that other, smaller differences could be increased. However, in extensive experimentation, not reported here, we found the procedure very reliable in producing acceptable optimizations over all k.

To provide some protection against missing the global optimum at the next k , the Nelder-Mead algorithm can applied a second time using the alternative starting point corresponding to the parameters i_1, j_1, and p_1.

It is fairly obvious that this basic version of Nelder-Mead will tend to favour components with large SDs when k is small. In complex mixtures, there may be some underlying components with large SDs but where these are masked by a spread of several components with smaller SDs. To avoid the procedure getting stuck with components with large SDs, we also used a variation which did not use the alternative starting point corresponding to i_1, j_1, and p_1, but instead carried out the optimization in three steps at each k. The first is what has been already described in setting up a new component and in carrying out the initial optimization including the new component. However, we do not then immediately change k, but reoptimize using the current best solution, only with all SDs to a common value. We then reoptimize a third time by allowing the SDs to freely vary again. We only go on to the next k after this third optimization. The overall process does not therefore preclude components with large SDs from being fitted, but does allow such a component to be replaced by several components with smaller SDs as k is increased.

17.4.3 Importance Sampling

We follow Geweke (1989), who gives a method of estimating a posterior distribution using IS that can be used in our case to calculate the individual posterior integrals $J(k|\mathbf{y})$ given in (17.14). The key point is that these can be evaluated using IS, independently of one another, so there is no need to use a method like reversible jumps to coordinate moves between different k.

In IS, samples are drawn from an *importance sampling (IS) distribution*, also called *candidate distribution*. The closer the candidate distribution is to the distribution being integrated over (in this case, the posterior distribution), the more efficient the importance sampling. The candidate distribution can be defined separately for each k as

$$c_k[\boldsymbol{\psi}(k), \mathbf{w}(k)],\ \boldsymbol{\psi}(k) \in \boldsymbol{\Psi}(k), \mathbf{w}(k) \in \boldsymbol{\Delta}_k\ k = 1, 2, \ldots, k_{\max}, \tag{17.20}$$

where $c_k[\boldsymbol{\psi}(k), \mathbf{w}(k)]$ for each k is a continuous density scaled so that

$$\int_{\boldsymbol{\Delta}(k)} \int_{\boldsymbol{\Psi}(k)} c_k[\boldsymbol{\psi}(k)] d\boldsymbol{\psi}\, d\mathbf{w}(k) = k_{\max}^{-1}.$$

We follow Geweke in our choice of $c_k(\cdot)$, using a multivariate Student t-distribution with variance matrix calculated from the negative of the inverse of the Hessian of the log posterior density evaluated at $[\tilde{\boldsymbol{\psi}}(k), \tilde{\mathbf{w}}(k)]$. To be unambiguous, we shall write the candidate distribution specifically as $\tilde{c}_k(\cdot)$ to indicate when it has has been obtained using MAP estimators in this way. In what follows, some care is needed to distinguish the parameters $[\boldsymbol{\psi}(k), \mathbf{w}(k)]$ as they appear in the mixture PDF, the MAP estimators $[\tilde{\boldsymbol{\psi}}(k), \tilde{\mathbf{w}}(k)]$, and the parameters treated as variates generated by the importance sampling, which we shall denote by $[\boldsymbol{\psi}^*(k), \mathbf{w}^*(k)]$. A typical parameter point obtained in this way has the form

$$\boldsymbol{\theta}^* = \begin{pmatrix} \boldsymbol{\psi}^* \\ \mathbf{w}^* \end{pmatrix} = \begin{pmatrix} \tilde{\boldsymbol{\psi}} \\ \tilde{\mathbf{w}} \end{pmatrix} + \boldsymbol{\theta}_0^*, \tag{17.21}$$

where

$$\boldsymbol{\theta}_0^* = \tilde{\mathbf{P}}_1 \tilde{\mathbf{R}} z_\nu^*,$$

with $\mathbf{z}_\nu^*$ a vector of $\nu = 3k - 1$ independent Student-t variates, each normalized to have mean zero and variance unity. We do not really need the asterisk in the case of $\mathbf{z}_\nu$, but we have added it just to emphasize that it is the source of the randomness in the IS samples. (All quantities should carry a k subscript, but for simplicity this is omitted.) The matrices $\tilde{\mathbf{P}}_1$ and $\tilde{\mathbf{R}}$ can be calculated explicitly from the eigenvectors and eigenvalues, respectively, of the Hessian matrix $\tilde{H} = H(\tilde{\boldsymbol{\psi}}(k), \tilde{\mathbf{w}}(k), k)$ of second derivatives of $L(\boldsymbol{\psi}(k), \mathbf{w}(k), k) = \ln(p[\mathbf{y}|\boldsymbol{\theta}(k), k]$ (the negative inverse of which is $\text{Var}(\boldsymbol{\theta}_0)$); the tildes are again a reminder that the matrices been calculated using $[\tilde{\boldsymbol{\psi}}(k), \tilde{\mathbf{w}}(k)]$. As the derivation is rather technical, this is not given here but in Section 18.2.4 of the next chapter. Moreover, the construction given there ensures that the component weights satisfy $\sum_{i=1}^{k} w_i = 1$. This means that $\text{Var}(\boldsymbol{\theta}_0)$ is singular (as is the Hessian). In the importance sampling calculations, we need an explicit expression for $c_k(\cdot)$. This is most easily obtained as follows. Let $\boldsymbol{\omega}$ be the $(k-1)$-dimensional vector the *reduced set of weights* formed from the first $(k-1)$ components of $\mathbf{w}$ and write $\boldsymbol{\phi}=(\boldsymbol{\psi}, \boldsymbol{\omega})$ for the vector of component distribution parameters and this reduced set of weights. We have

$$\boldsymbol{\phi}^* = \begin{pmatrix} \boldsymbol{\psi}^* \\ \boldsymbol{\omega}^* \end{pmatrix} = \begin{pmatrix} \tilde{\boldsymbol{\psi}} \\ \tilde{\boldsymbol{\omega}} \end{pmatrix} + \tilde{\mathbf{M}} z_\nu^*,$$

where $\tilde{\mathbf{M}}$ is the matrix $\tilde{\mathbf{P}}_1 \tilde{\mathbf{R}}$ but with the last row omitted. This equation is a nonsingular linear transform of $\mathbf{z}_\nu^*$ to $\boldsymbol{\phi}^*$. The Jacobian of the transformation is $|\partial[\boldsymbol{\psi}, \boldsymbol{\omega}]/\partial(\mathbf{z}_\nu)|_{\boldsymbol{\phi}=\tilde{\boldsymbol{\phi}}} = \det(\tilde{\mathbf{M}})$ so that the PDF of $\boldsymbol{\phi}^*$, when it is evaluated in the importance sampling, is

$$\tilde{f}(\boldsymbol{\phi}^*) = \left|\frac{\partial \mathbf{z}_\nu}{\partial(\boldsymbol{\psi},\boldsymbol{\omega})}\right|_{\boldsymbol{\phi}=\tilde{\boldsymbol{\phi}}} g_\nu(\mathbf{z}_\nu^*) = \left|\frac{\partial(\boldsymbol{\psi},\boldsymbol{\omega})}{\partial(\mathbf{z}_\nu)}\right|^{-1}_{\boldsymbol{\phi}=\tilde{\boldsymbol{\phi}}} g_{\mathbf{z}_\nu}(\mathbf{z}_\nu^*) = [\det(\tilde{\mathbf{M}})]^{-1} g_\nu(\mathbf{z}_\nu^*), \quad (17.22)$$

where g_ν is the PDF of $\mathbf{z}_\nu$.

Use of (18.19) to generate IS variates does not guarantee that parameters which should be positive necessarily are positive, nor that all weights necessarily satisfy $0 < w_j < 1$. This is easily handled by rejecting any $\boldsymbol{\theta}^*$ sample where *any* such constraint which should be satisfied is not. This restricts the support of the IS distribution to precisely the region where *all* parameter constraints are satisfied. The IS sampling is therefore an acceptance/rejection procedure. Given k, the IS distribution actually sampled is modified from (17.22) to

$$\tilde{c}_k^R[\boldsymbol{\psi}(k), \boldsymbol{\omega}(k)] = [\det(\tilde{\mathbf{M}}(k))]^{-1} g_\nu(\mathbf{z}_\nu)/\tilde{R}(k), \quad (17.23)$$

where we have now included dependency on k explicitly, and $\tilde{R}(k)$ is an estimate of $R(k)$, the probability that a parameter point sampled from (17.22) is accepted (because it falls in the support of the k-component form of the mixture model being fitted). A simple estimate of $R(k)$ is easily obtained from the IS sampling as

$$\tilde{R}(k) = \frac{(\#\text{ of replications sampled from (17.22) for the given } k \text{ and accepted})}{m_k}, \quad (17.24)$$

where $m_k = (\#$ of replications sampled from (17.22) for the given $k)$.

This gives ϕ^*. The last weight is simply obtained as $w_k^* = 1 - \sum_{j=1}^{k-1} w_j^*$.

To cope with the possibility that the resulting IS distribution is a poor representation of the posterior distribution, Geweke (1989) suggests adjustments of the IS distribution in each direction of each parameter axis. We have not implemented this more elaborate version, but report results using just the Student's t-distribution. We would expect results to be satisfactory when $k = k_0$, but for k different from k_0, it is likely to introduce a bias in estimating $\pi(k|\mathbf{y})$, making it smaller than the true value. Thus, our method will produce an estimate of the posterior distribution of k that is likely to be more concentrated about k_0 than with an MCMC method.

Let $f(\cdot)$ be the mixture PDF of eqn (17.1). The importance sampling procedure with sample size I is as follows.

IS1. Draw I values of $k : k_i, i = 1, 2, \ldots, I$, independently and uniformly distributed over $1, 2, \ldots, k_{\max}$.

IS2. Draw values $[\psi^*(k_i), \mathbf{w}^*(k_i)]$ from the distribution with density $\tilde{c}_{k_i}[\boldsymbol{\psi}(k_i), \mathbf{w}(k_i)]$, as in eqn (17.23), for $i = 1, 2, \ldots, I$.

This produces a sequence of independent and identically distributed random variables $(\boldsymbol{\psi}_i^*(k_i), \mathbf{w}_i^*(k_i), k_i), i = 1, 2, \ldots, I$.

For each k, record the acceptance probabilities $\tilde{R}(k)$ of eqn (17.24).

IS3. From $(\boldsymbol{\psi}_i^*(k_i), \mathbf{w}_i^*(k_i), k_i)$, $i = 1, 2, \ldots, I$, calculate the importance sampling ratios

$$\rho[\boldsymbol{\psi}_i^*(k_i), \mathbf{w}_i^*(k_i), k_i] = \frac{p[\mathbf{y}|\boldsymbol{\psi}_i^*(k_i), \mathbf{w}_i^*(k_i), k_i]}{\tilde{c}_{k_i}^R[\boldsymbol{\psi}_i^*(k_i), \mathbf{w}_i^*(k_i)]} \text{ for } i = 1, 2, \ldots, I, \quad (17.25)$$

with $p[\mathbf{y}|\boldsymbol{\psi}_i(k_i), \mathbf{w}_i(k_i), k_i] = f[\mathbf{y}|\boldsymbol{\psi}_i(k_i), \mathbf{w}_i(k_i), k_i]\pi[\boldsymbol{\psi}_i(k_i), \mathbf{w}_i(k_i), k_i]$ and $\tilde{c}_k^R[\boldsymbol{\psi}(k), \boldsymbol{\omega}(k)]$ as in eqn (17.23).

IS4. Estimate $\pi(k|\mathbf{y})$ by

$$\tilde{\pi}(k|\mathbf{y}) = \frac{\sum_{k_i=k} \rho[\boldsymbol{\psi}_i^*(k_i), \mathbf{w}_i^*(k_i), k_i]}{\sum_{i=1}^{m} \rho[\boldsymbol{\psi}_i^*(k_i), \mathbf{w}_i^*(k_i), k_i]}, \quad k = 1, 2, \ldots, k_{\max}, \quad (17.26)$$

where, as both the prior for k and the importance sampling of k are uniform, we have no need to calculate the normalizing integrals of the posterior distribution over the $[\boldsymbol{\psi}(k), \mathbf{w}(k)]$ space explicitly. We can estimate the most likely k using $\tilde{k}$ as given in (17.18).

17.5 Predictive Density Estimation

We now consider estimation of f_0, the PDF of the finite mixture model itself. This is a point estimate and is called a *predictive density estimate* in Richardson and Green (1997), who give two methods for its calculation using RJMCMC. Neither seems entirely satisfactory, but for different reasons.

(i) In the first method, which we call the *averaged density method,* for each given value of k, the mixture density is calculated at all those steps, i, of the MCMC simulation where the given k value is obtained, and the predictive density conditional on k is then estimated by the average of these values. Thus,

$$f(y|k) = n_k^{-1} \sum_{i:k^{(i)}=k} f(y|\boldsymbol{\psi}(k^{(i)}), \mathbf{w}(k^{(i)}), k^{(i)}), \quad (17.27)$$

where $k^{(i)}$, $\boldsymbol{\psi}(k^{(i)})$, and $\mathbf{w}(k^{(i)})$ are the values of k, $\boldsymbol{\psi}(k)$, and $\mathbf{w}(k)$ sampled at step i of the MCMC run, and n_k is the number of observations where $k^{(i)} = k$. This can be taken a step further by averaging across all values of k, such that we carry out the summation in (17.27) over all i without conditioning on $k^{(i)} = k$, to obtain an 'overall' unconditional density estimate.

The estimate (17.27) produces a fit for f_0 that is usually very satisfactory. However, it does not give a point estimate of k or estimates of the parameters of the components conditional on k. Indeed, Richardson and Green warn that averaged density estimates 'do not themselves have the shape of a finite mixture distribution', suggesting only that they be used for providing 'complementary evidence on which to draw when assessing the number of components'.

Ishwaran and co-authors show that the posterior distribution of f_0 as derived by their method is asymptotically consistent, so that (17.27) does converge to f_0 in the limit. The calculated predictive densities conditional on k in the numerical examples that we give in the next chapter are typical of what we found more generally, becoming similar very quickly with increasing k, so that convergence to the unconditional predictive density is rapid as k increases.

(ii) In the second method, suitable 'plug-in' estimates, $\breve{\boldsymbol{\psi}}(k)$ and $\breve{\mathbf{w}}(k)$, are used, and $f(\cdot|\breve{\boldsymbol{\psi}}(k), \breve{\mathbf{w}}(k))$ is taken as the predictive density estimate given k. An obvious plug-in is

$$\breve{\boldsymbol{\psi}}(k) = n_k^{-1} \sum_{i:k^{(i)}=k} \boldsymbol{\psi}(k^{(i)}) \text{ and } \breve{\mathbf{w}}(k) = n_k^{-1} \sum_{i:k^{(i)}=k} \mathbf{w}(k^{(i)}), \tag{17.28}$$

where each parameter of the k-component mixture is its average value over just those observations of the Markov chain where $k^{(i)} = k$. We follow Richardson and Green calling this the *averaged MC parameters method.* With this method, therefore, we do have point estimates for the parameters.

However, as Richardson and Green point out, the method tends to give a predictive density that is too smooth. The problem is particularly serious when k corresponds to an overfitted model, as parameter posterior distributions are then multimodal in general. This makes the parameter estimates (17.28) unreliable, as the distribution calculated at the average parameter value will not then be a good representation of the parameter distribution. The resulting predictive estimator of the finite mixture conditional on k is not only oversmooth but, when k is much larger than k_0, will not even correspond to the sample in a plausible or sensible way.

The ADP method, being MCMC-based, has the same characteristics as RJMCMC when it comes to predictive density estimation.

Using MAPIS, a predictive density estimate is immediately available using (17.15) for the parameter estimates in (17.1), either with $k = \tilde{k}_{BIC}$ as in (17.16) or $k = \tilde{k}$ as in (17.18).

17.6 Overfitted Models in MCMC

This section refers frequently to the main theorem given by Rousseau and Mengersen (2011). For ease of reference, we refer to this paper simply as R&M in this section. Following R&M, we call a k-component model an *overfitted* model if $k > k_0$.

We first summarize the theorem, which is on the asymptotic behaviour of the posterior distribution in overfitted models. We then illustrate what happens when RJMCMC and MAPIS are applied to an elementary example where assumption A0 holds, with f_0 known, illustrating how the R&M theorem might be applied in this case.

17.6.1 Theorem by Rousseau and Mengersen

The theorem in R&M provides theoretical insight into how the bias in $\breve{\pi}(k|\mathbf{y})$ will occur in an overfitted model. This suggests how the bias can be reduced.

The theorem covers the situation where each component has, in the notation of R&M, d components. In our case, where $d = 2$, the theorem can be summarized as follows:

Assume A0 with the true $\boldsymbol{\psi}_0 \in \Psi$ a compact region, and consider the δ_j in the Dirichlet distribution of Bayesian model (17.5). Then, for an overfitted model with k components, where $k > k_0$ and where each component has two parameters:

(a) If $\delta_j < 1$, the expectation of the sum of the weights of the extra $(k_0 - k)$ components is asymptotically $O_p(n^{-1/2})$ as $n \to \infty$.

(b) If $\delta_j > 1$, Pr{Expectation of the sum of the weights of the true components $< (1 - \varepsilon)\} \to 1$ as $n \to \infty$. This means that some extra components will not tend to zero, and as the posterior distribution is consistent as $n \to \infty$, it follows that some individual true components will be represented by a combination of more than one fitted component, these components having a common (μ, σ) value, but possibly different weights. This representation is unstable in the sense that its precise form is unpredictable.

(c) The case $\delta_j = 1$ is not covered by Theorem 1 of R&M.

Part (b) indicates that, with $\delta_j > 1$, there is a non-negligible probability when overfitting occurs that more than k_0 components will be allocated significant weight. The implication is that estimates of the posterior probability $\breve{\pi}(k|\mathbf{y})$ for $k > k_0$ will be biased high. However, some of the components may simply be aliases of other components in the way described in the previous section, and so should be merged in some way.

Part (a) of the theorem implies that setting $\delta_j < 1$ is preferable, as the weights of redundant components in an overfitted model will tend asymptotically to zero. This justifies omitting components with small weights in either the RJMCMC or the ADP method. Without needing to consider the allocation variables L_j, an estimator of the posterior distribution of k similar to $\breve{\pi}_+(k|\mathbf{y})$ given in (17.12) can therefore be obtained as follows. At every sweep of the Markov chain in either method, we exclude all components j with a weight $w_j(k)$ that is small, for example, where $w_j(k) < w_{\text{crit}}$ for some suitably chosen w_{crit}. The weights of those components retained are rescaled so that they sum to unity. Thus in sweep i,

$$\mathbf{w}^{(i)}(k) = I_{[w_{crit},1]}(\mathbf{w}^{(i)}(k))\frac{\mathbf{w}^{(i)}(k)}{\sum_{l=1}^{k} I_{[w_{crit},1]}(\mathbf{w}^{(i)}(l))},$$

where $I_{[a,b]}(x)$ is the indicator function with $I_{[a,b]}(x) = 1$ if $x \in [a, b]$, $I_{[a,b]}(x) = 0$ otherwise. The component count of the sweep is reduced from k to $\underline{k}$, the number of components not excluded. Clearly $\underline{k} \leq k$, and this has the general effect of shifting the estimated posterior probability distribution of k to the left, with $\breve{\pi}(k|\mathbf{y})$ typically being reduced when $k > k_0$. The posterior distribution of k and corresponding most likely estimate of k are

$$\breve{\pi}_{RK}(k|\mathbf{y}) = m^{-1}\sum_{i=1}^{m} I_k(\underline{k}^{(i)}),\ k = 1, 2, \ldots, K \text{ and } \breve{k}_{RK} = \arg\max_k \breve{\pi}_{RK}(k|\mathbf{y}), \qquad (17.29)$$

where $\underset{\sim}{k}^{(i)}$ is the number of components retained at sweep i and $K = k_{max}$ in the RMCMC case and $K = N$ in the ADP case.

Though (17.29) can be used for either RJMCMC or ADP, so that it is a (posterior) reduced-k distribution in both cases, we will, to avoid confusion, use the notation $\breve{\pi}_{RK}(k|\mathbf{y})$ only when using RJMCMC, and use the notation $\breve{\pi}_{ADP}(k|\mathbf{y})$ for the ADP case. We illustrate their use in Section 18.1.1.

For the case $\delta_j = 1$, not covered by Rousseau and Mengersen's theorem, Ishwaran and co-authors show that using their method, if k_{max} is chosen so that $n/k_{max} \to 0$ as $n \to \infty$, the posterior is inconsistent. It is not clear if the default value $\delta_j = 1$ used by Richardson and Green is advisable or not, as under the assumption that k_0 is finite, k_{max} will remain finite in the RJMCMC method, so that n/k_{max} will not tend to zero as $n \to \infty$.

In practice, when n is finite, it is not evident that the choice of δ will give results that are as clear-cut as given in Rousseau and Mengersen's theorem, and whether both forms of behaviour of the estimated components might be possible, especially if δ is chosen near $\delta = 1$.

We give a numerical example to illustrate what typically happens in practice, and discuss how Rousseau and Mengersen's theorem might be applied in practice.

17.6.2 A Numerical Example

In an overfitted model, problem (B6) mentioned in Section 17.2.3 occurs with the RJMCMC and the ADP methods when the estimated individual posterior distributions of the components of $\boldsymbol{\psi}$ and $\mathbf{w}$ will typically be multimodal.

This characteristic was recognized by Richardson and Green, who actually identified two difficulties: (i) label switching and (ii) genuine multimodality, though they do not offer a definitive solution to either. As they discussed, label switching can be handled in part by labelling components in ranked order. A detailed review of label switching regarded as a problem of identifiability under symmetric priors is given in Jasra et al. (2005), and we do not discuss this problem further here.

Multimodality is an unsatisfactory characteristic for two reasons. Firstly, it makes interpretation of the component fits difficult, and, secondly, it is the cause of the second effect where $\breve{\pi}(k|\mathbf{y})$ is biased high.

The following elementary numerical example demonstrates these two difficulties quite clearly. We apply the RJMCMC method to a sample of size $n = 100$ artificially generated from the two-component model

$$y \sim 0.5N(10, 0.5^2) + 0.5N(12, 0.5^2)$$

so that $k_0 = 2$. We analyse the sample by RJMCMC using NMix, carrying out 50,000 sweeps, and also by MAPIS, drawing 50,000 importance sampling observations.

Table 17.1 gives the estimated posterior distribution of k obtained from the two methods. With RJMCMC, the probabilities were non-negligible up to $k = 5$.

In our simple example, RJMCMC is clearly producing an estimate of the posterior probability of k, $\breve{\pi}(k|\mathbf{y})$ where the probability is high for $k = 3, 4, \ldots$, when we know that

Table 17.1 Estimated posterior distribution of k for artificial sample from a two-component normal mixture model

Method	k	2	3	4	5	6	7	8	9
RJMCMC	$\breve{\pi}(k\|\mathbf{y})$	0.35	0.27	0.17	0.10	0.05	0.03	0.02	0.01
MAPIS	$\hat{\pi}(k\|\mathbf{y})$	0.97	0.03	–	–	–	–	–	–

the sample comes from a two-component distribution. Richardson and Green (1997) suggest that multimodality in fitted posterior distributions of component parameters might genuinely be interpretable in terms of possible alternative fits to the data, but to take this viewpoint is not satisfactory in our example here and possibly in general.

To see what is actually happening, we need to consider the estimates of the posterior distributions of the parameters conditional on k. Figure 17.1 shows the parameter posterior distributions obtained by RJMCMC for the cases $k = 2$ and $k = 3$.

The case $k = 2$ is completely satisfactory, identifying the two components clearly and correctly. The distributions of the weights of the two components are located about $w_1 = 0.4$ and $w_2 = 0.6$. This imbalance in the weights, even though the true values from which the sample was generated were equal, can be seen in the data set, and reasonably reflects sampling variation.

However, in the overfitted $k = 3$ case, whereas the first and third components identify the two real components, the middle component is bimodal for the μ parameter, with the position of the two modes clearly matching those of the two genuine components. Recalling that these estimates are simply frequency histograms obtained for the observed component parameter values from each sweep in the Markov chain where k is the given value, the histograms for the middle component show that the μ, σ component pairs in each sweep must come either from a distribution located about $(\mu, \sigma) = (10, 0.5)$ or from a distribution located about $(\mu, \sigma) = (12, 0.5)$, with the weight being rather variable. Thus, such a 'middle' component pair should not be counted as arising from a genuine third component distinct from the two genuine components, but should, in some way, be counted as arising from one or other of these.

Though we have not done so, it is easy to give examples, where $k_0 > 2$, to show that the behaviour exhibited in our example occurs in general when using MCMC in any overfitted k-component model, with the excess $(k - k_0)$ fitted components not corresponding to genuine normal components that are distinct from the k_0 true normal components, but merely (still!) a mixture of them. Assessing how the multimodality has actually arisen is usually not too difficult. The difficulty is in quantifying its effect and in adjusting the posterior distributions to make them more meaningful.

What we can say is that multimodality seriously affects estimation of the posterior probability of k if we estimate this by $\breve{\pi}(k|\mathbf{y}) = m_k/m$, as in (17.9). If $k > k_0$, then any

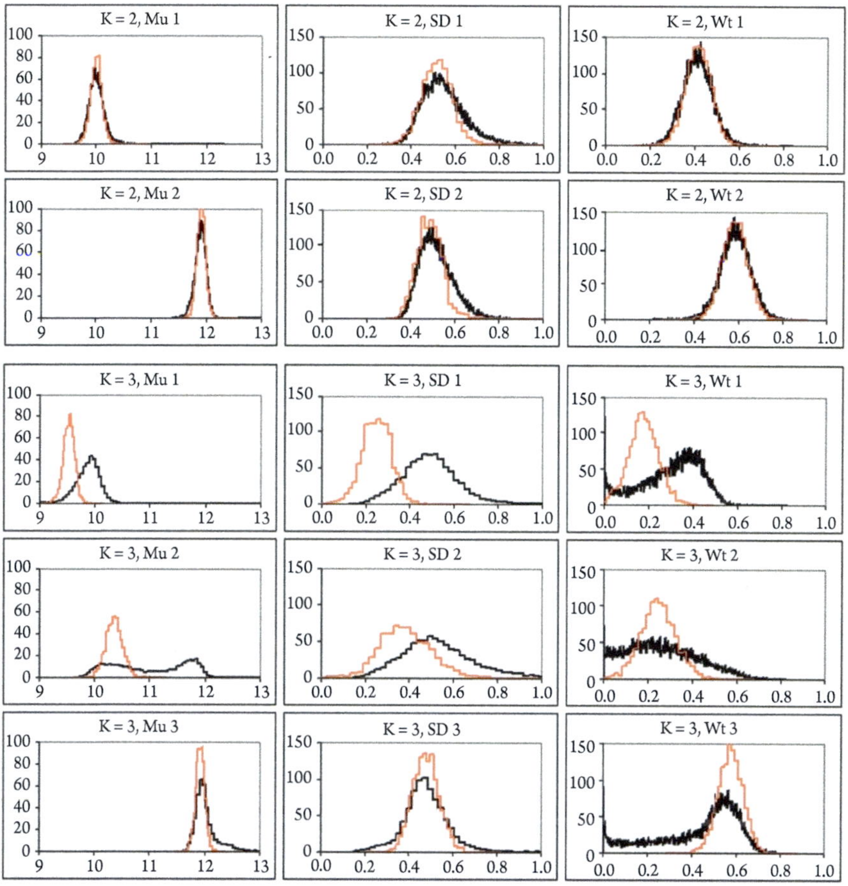

Figure 17.1 Posterior distributions of the parameter components and weights conditional on k for the cases $k = 2$ and $k = 3$ for the artificial two normal components data set as estimated using the RJMCMC (black graphs) and MAPIS (red graphs) methods.

fitted component whose parameters have posterior distributions that are multimodal actually only represents an aliased combination of genuine components. If it is counted as arising from a different genuine component, this will inflate m_k. This will then bias $\breve{\pi}(k|\mathbf{y})$ so that it is overly high for $k > k_0$.

By way of comparison, Table 17.1 shows that the MAPIS estimate of the posterior distribution of k has $\tilde{\pi}(2|\mathbf{y}) = 0.97$, $\tilde{\pi}(3|\mathbf{y}) = 0.03$, with $\tilde{\pi}(k|\mathbf{y})$ negligible for $k \geq 3$. A satisfactory result.

To see what has happened in this MAPIS case, Figure 17.1 also shows (plots in red) the estimates of the posterior distributions of the parameters conditional on k obtained by MAPIS for the cases $k = 2$ and $k = 3$. The distributions for $k = 2$ closely match those

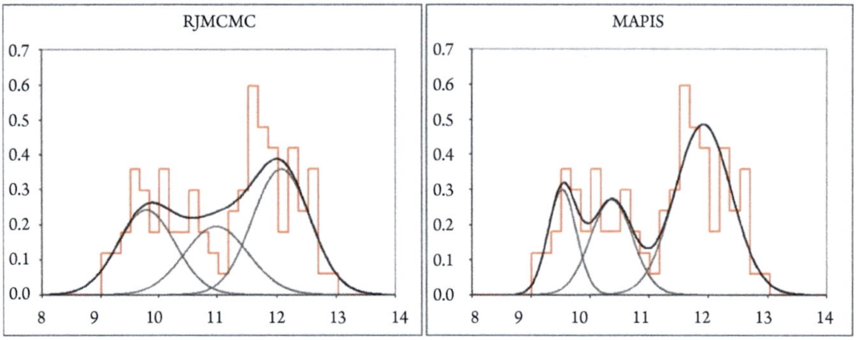

Figure 17.2 Fitted 3-component models for the artificial two normal components dataset using the RJMCMC (left-hand chart) and MAPIS (right-hand) methods. In both charts the black curve is the overall fit with the three components in grey. The middle component in the RJMCMC case does not correspond to a genuine component being placed where there are few data points. The resulting overall fit is oversmooth.

obtained by RJMCMC, identifying the two components clearly and correctly. However, for the case $k = 3$, unlike the RJMCMC case, the plots are *still unimodal*. From these plots, it is clear that MAPIS has now divided the first component, that was obtained when $k = 2$, into two distinct normal components, with similar weights $w_1 \simeq w_2 \simeq 0.2$. If we estimated the component parameters and weights of the three components from these plots using MAP, they would correspond quite closely to the actual MAP-estimated components for the case $k = 3$.

Figure 17.2 allows us to confirm this analysis by showing the predictive densities of the three-component models. The left-hand chart is the RJMCMC case, where the PDFs are drawn using 'plug-in' estimates of the parameter values, including weights, obtained by the *averaged MC parameter* method as given in equation 17.28. As mentioned, Richardson and Green (1997) point out that the method tends to give oversmooth fits, and there is some slight evidence of this in our example.

However, the main feature of note, as indicated in our discussion earlier, is that the third component obtained by RJMCMC in this example does not actually correspond to a genuine underlying normal component, and this is very evident by its location in the PDF plot where there are few observations.

The right-hand chart in Figure 17.2 shows the MAPIS three-component posterior predictive PDF obtained using the $k = 3$ MAP estimators as the parameter values. This displays a possibly meaningful three-component form for the sample, fitting two components to the underlying genuine left-hand component. However, the posterior probability of $\tilde{\pi}(3|\mathbf{y}) = 0.03$ correctly attaches little confidence to this fit compared with the two-parameter fit.

Summarizing, we see that the MAPIS method is much more satisfactory in this example, not being affected by either problem (B5) or (B6). We discuss the MAPIS approach more generally in the next section.

17.6.3 Overfitting with the MAPIS Method

In the MAPIS method, point estimates of component parameters and weights are obtained at each k, but this is done sequentially with $k = 1, 2, 3\ldots$ At each k, a component is added aimed at reducing the maximum discrepancy (17.19) between the previously fitted model for $k - 1$ and the EDF of the sample. Thus, the next component is placed so as to fit the *least well explained feature* of the current fit, and so will not in general simply replicate any previous component.

As already remarked in Section 17.3.1, if the shape parameter g in the prior (17.6) for the SD parameter of a component satisfies the condition $g > 1$, MAPIS will not suffer problem (D1) which can occur in ML, where an estimate of the SD parameter of some component, $\tilde{\sigma}_j \to 0$. In practice, g can usually be set much smaller, to reflect greater prior uncertainty concerning σ. In the example, having $g = 0.2$ is satisfactory, though components with small σ were present for fits corresponding to $k \geq 4$.

When fitting k components, we need to ensure we *do* actually fit k distinct components. To be sure of this, we need the estimates of all k weights to be positive. This is ensured by having $\delta > 1$. This might appear to run counter to R&M's recommendation that δ be chosen to satisfy $\delta < 1$. The reason given by R&M is that $\delta > 1$ can result in apparently different components having the same (μ, σ) estimates but different weights, with instability in the weight values. We have observed the effect when δ is made sufficiently large, but being an asymptotic result, this instability is not inevitable in a finite sample. We do not have a theoretical result to show fully what happens under MAPIS as δ is changed, but based on extensive experimentation, not reported here, we conjecture that for finite samples, there is an interval $1 < \delta < \delta^*$ where δ^* is dependent on the sample and the value of k, in which MAP will yield estimates of distinct components, all with positive weights.

In practice, setting $\delta_j = \delta < 1$ with MAPIS usually results in one or more weights $w_j \to 0$ for k sufficiently large. Consequently, our default is to set $\delta = 1$, as used by Richardson and Green (1997), but to increase δ in small steps if any $w_j \to 0$, with a check to stop if duplicated (μ, σ) values occur.

In the example of the previous section, choice of δ was not at all critical, with the duplicated (μ, σ) effect not occurring with either $\delta = 1.0$ or $\delta = 1.5$.

A more important concern with taking δ too large in MAPIS is that this makes the fit oversmooth. As a guard against this, the effect on the predictive density should be observed when changing δ. Estimation of the predictive density is discussed in the next section.

Overall, the MAPIS method will produce estimated posterior probabilities $\tilde{\pi}(k|\mathbf{y})$ for $k > k_0$ that tend to be much smaller than those obtained using RJMCMC or ADP. We can therefore regard MAPIS as a method that is parsimonious in yielding fits that are satisfactory, using far fewer componennts than MCMC methods.

We give detailed numerical examples in the next chapter to illustrate the methodological discussion of this chapter.

18 Finite Mixture Examples; MAPIS Details

In this chapter, we give numerical examples illustrating and comparing mainly the MAPIS and RJMCMC methods.

We also provide notes on supplementary aspects concerning the practical implementation of MAPIS regarding (i) mixtures where components have distributions other than the normal, (ii) elimination of hyperparameters, (iii) additional mathematical details on the form of the sampling distribution used in the MAPIS importance sampling, and (iv) the acceptance/rejection method used in the actual sampling.

18.1 Numerical Examples

We consider three data sets, comprising a related pair Galaxy and GalaxyB, and a separate set, the Hidalgo stamp issues, which we call HidalgoStamps for short. Galaxy and HidalgoStamps are celebrated samples, often used in testing fitting methods. The former is discussed in detail by Richardson and Green (1997), for example, and the latter by McLachlan and Peel (2000).

We shall continue to refer to Richardson and Green (1997) as R&G. We shall in a slight extension of the R&G terminology refer to the distribution corresponding to a predictive density as a *predictive distribution* so that we can refer to a *predictive CDF* as well as a *predictive PDF*.

In this section, we compare mainly the performance of MAPIS and RJMCMC, focusing on estimating the posterior distribution of k and the predictive distribution.

However, for the Galaxy and GalaxyB data sets, we include the ADP method in the comparison, and for the HidalgoStamps data set, we include ML.

For RJMCMC, we used the NMix simulation implementation, to which we added an Excel front-end interface. This allows estimation of $\pi(k|\mathbf{y})$ to be carried out using either the RJMCMC formula (17.9) or the reduced-k component estimator (17.29), this latter estimator being obtainable with some minor postprocessing of the NMix simulation

Non-Standard Parametric Statistical Inference. Russell Cheng.

results. The Excel front-end also allows implementation of the ADP method using NMix with a fixed $\delta = 1/N$ as proposed in Ishwaran et al. (2001). This allows $\breve{\pi}_+(k|\mathbf{y})$ to be calculated as in eqn (17.12) or $\breve{\pi}_{ADP}(k|\mathbf{y})$ to be calculated as in eqn (17.29), in this latter case, again with some minor postprocessing. The MAPIS implementation uses compiled bespoke C-code, again accessed by the Excel interface. This includes a simple switch to suppress the prior, which is all that is required to convert MAP estimation to ML estimation.

For the RJMCMC and ADP methods, we use point estimators calculated by the averaged MC parameters method, the averaging being over all sweeps i where $k^{(i)} = k$, and, when considering reduced-k distributions, recording $\underline{k}^{(i)}$ rather than $k^{(i)}$ as the k value visited. In MAPIS, we can just use the MAP component-parameter and weight estimators for the given k.

To make our comparisons as fair as possible, we use the Richardson and Green priors in all three methods. However, we have, in the MAPIS method, used the formulation where the basic mixture PDF of equation (17.1) is regarded, as mentioned in Richardson and Green (1997), simply as a parsimonious representation of a non-standard density. The allocation variables L_{ij} are therefore omitted from the hierarchical model (17.5). This is more in keeping with our focus on identifying the number of components and their parameter values and weights. This formulation is referred to in Jasra et al. (2005) as a *semiparametric* construction.

We used the default parameter settings for the MCMC computer runs with a burn-in of 5000, after which 100, 000 steps of the MC were recorded. The burn-in is somewhat smaller than the value used by Richardson and Green, but is quite sufficient to demonstrate the points we wish to make. In the MAPIS method, 100, 000 independent IS observations were made with no burn-in needed.

18.1.1 Galaxy and GalaxyB

We consider first the two data sets, Galaxy and GalaxyB.

The Galaxy data set was analysed in detail by Richardson and Green (1997), Ishwaran et al. (2001), and Ishwaran and James (2002).

To allow for the possibility that perhaps the Galaxy data set contains features which really are due to additional components that MCMC identifies but are missed by MAPIS, we also consider an artificially generated sample, GalaxyB, with the same sample size 82, but which we have generated from a model with just three components having parameter values $[\boldsymbol{\mu}_0, \boldsymbol{\sigma}_0, \mathbf{w}_0] = [(9.7, 0.4, 0.1), (21.0, 2.2, 0.85), (33.0, 0.8, 0.05)]$. This produces characteristics that are quite similar to those of the original galaxy data set, but of course comes from a normal mixture model whose components we know, with $k_0 = 3$. Methods that strongly support fitted models with $k > 3$ are therefore not very satisfactory for this sample.

Our main finding is that both the RJMCMC and the ADP method produce estimates of posterior distributions in which the problems (B5) and (B6) are quite pronounced. In particular, $\breve{\pi}(k|\mathbf{y})$, the estimated posterior distribution of k obtained using RJMCMC, is biased towards high values of k. This is not the case when MAPIS is used.

Table 18.1 Estimated posterior distribution of k for the GalaxyB and Galaxy samples using the MC method without and with k reduction, using $\delta = 1.0$ and 0.5; $g = 0.2$ throughout; $w_{\text{crit}} = 0.02$; the largest probability in each distribution indicated in bold

Sample	δ	k	2	3	4	5	6	7	8	9	10
GlxyB	1.5	$\breve{\pi}$		0.27	**0.33**	0.21	0.11	0.05	0.02	0.01	-
GlxyB	1.5	$\breve{\pi}_{RK}$	.06	0.32	**0.34**	0.18	0.07	0.02	-	-	-
GlxyB	2	$\breve{\pi}$		0.19	**0.33**	0.24	0.13	0.06	0.02	0.01	-
GlxyB	2	$\breve{\pi}_{RK}$		0.19	**0.37**	0.23	0.12	0.05	0.02	0.01	-
Glxy	1.5	$\breve{\pi}$		0.06	0.22	**0.25**	0.21	0.13	0.07	0.03	0.01
Glxy	1.5	$\breve{\pi}_{RK}$	0.04	0.18	**0.35**	0.27	0.12	0.04	0.01	-	-
Glxy	2	$\breve{\pi}$		0.04	0.18	**0.24**	0.21	0.15	0.09	0.05	0.02
Glxy	2	$\breve{\pi}_{RK}$	0.03	0.23	**0.35**	0.23	0.11	0.04	0.01	-	-

We stress that the problems we encountered in using RJMCMC and ADP for fitting finite mixture models to the the Galaxy and GalaxyB data set occurred quite generally, for instance, in all the examples appearing in Richardson and Green (1997) and in the papers by Ishwaran and his co-authors.

We consider first estimation of the posterior distribution of k.

Posterior Distribution of k Using RJMCMC

Consider first RJMCMC. The rows of Table 18.1 give two pairs of distributions for GalaxyB and two pairs for Galaxy, all calculated using RJMCMC. Each pair comprises $\breve{\pi}(k|\mathbf{y})$, the estimated posterior distribution of k and its corresponding reduced-k version, $\breve{\pi}_{RK}(k|\mathbf{y})$, this latter calculated from equation (17.29) with $w_{\text{crit}} = 0.05$. In all pairs, $g = 0.2$. The choice of δ is flexible in RJMCMC. The default value used by R&G is $\delta = 1$, but as we wish to compare results with MAPIS where $\delta > 1$ is required, we therefore used the values $\delta = 1.5$ and $\delta = 2$, these giving very similar results compared with $\delta = 1$.

Even though using the reduced-k approach leads to some reduction in the probability values for higher k, in all cases, high values are obtained for the posterior probabilities of $k > 3$, with $\pi(k|\mathbf{y})$ greater in all cases where $k = 4$ than for $k = 3$, and with all cases where $k = 5$ and many cases where $k = 6$ having posterior probabilities that are non-negligible.

For the GalaxyB data set, the true number of components is known to be $k_0 = 3$. The high posterior probability values $\pi(k|\mathbf{y})$ when $k = 4, 5$ seem somewhat unsatisfactory. For the Galaxy data set itself, $\breve{\pi}(k|\mathbf{y})$ is highest for $k = 5$, with $k = 4, 6$ also giving high $\breve{\pi}(k|\mathbf{y})$ values. The reduced-k probabilities are highest at $k = 4$, with $k = 3, 5$ also high. Given the similarity of the two data sets, these results would suggest that taking $k = 4$

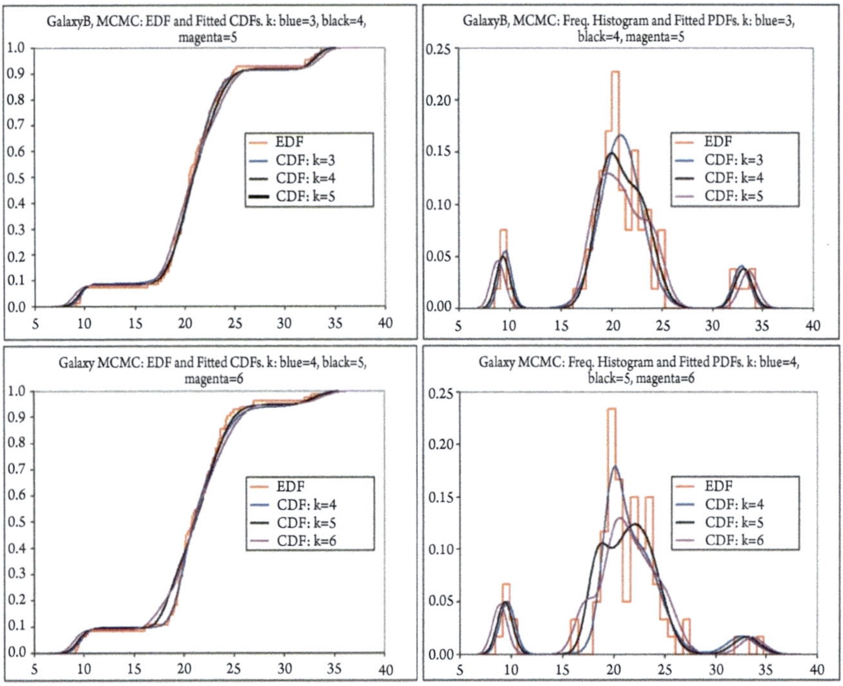

Figure 18.1 RJMCMC method. Predictive CDF and PDF conditional on k, the number of components. Upper charts: GalaxyB data. Lower charts: Galaxy data.

as the best k for the Galaxy data set would be preferable to $k = 5$ indicated by the basic RJMCMC analysis.

A high bias in the posterior distribution has repercussions when estimating the predictive distribution, which we consider next.

Predictive Distribution Estimation Using Averaged MC Parameters

Figure 18.1 shows, for both the GalaxyB and Galaxy data sets, the CDF and PDF of the predictive distribution conditional on k, calculated using the averaged MC parameters method. Only plots corresponding to those k with posterior probability $\breve{\pi}(k|\mathbf{y}) \geq 0.2$, as obtained by RJMCMC, are shown, with the plots corresponding to k with the highest $\breve{\pi}(k|\mathbf{y})$ highlighted in black.

Likewise, Figure 18.2 shows, for both data sets, the CDF and PDF of the predictive distributions, only this time corresponding to reduced-k with posterior probabilities $\breve{\pi}_{RK}(k|\mathbf{y}) \geq 0.2$, as obtained by RJMCMC; again, the plots corresponding to k with the highest $\breve{\pi}(k|\mathbf{y})$ are highlighted in black.

Both figures display the fits obtained using $\delta = 1.5$.

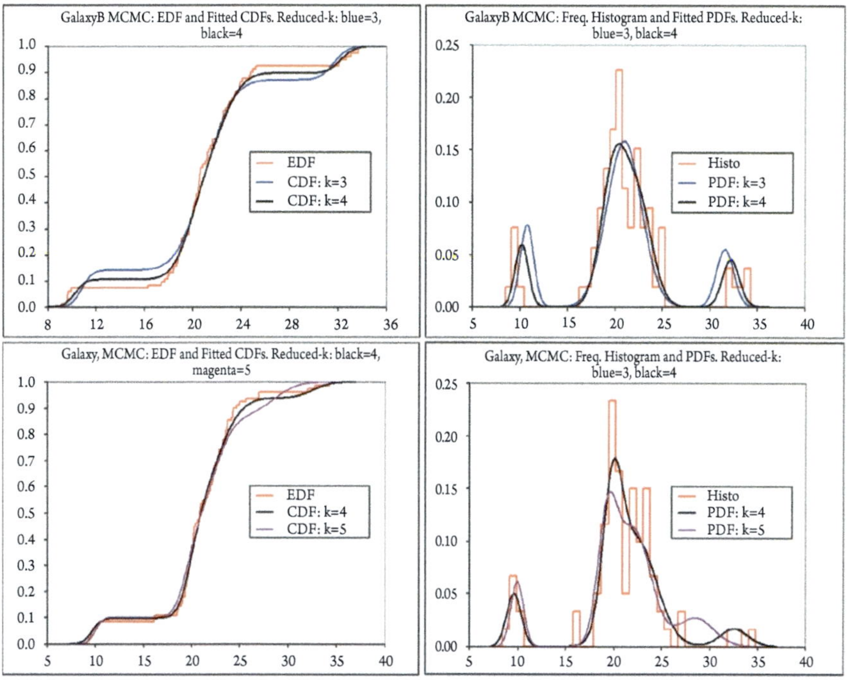

Figure 18.2 RJMCMC method using reduced-*k*. Predictive CDF and PDF conditional on reduced-*k*. Upper charts: GalaxyB data. Lower charts: Galaxy data.

The estimates obtained using RJMCMC, whether conditioning on k or reduced-k, are rather scattered, with inaccuracies that are not easily interpretable, particularly in the Galaxy k fit and in the GalaxyB reduced-k fit. Overall, the plots do not display any clear evidence that k should higher than $k = 4$.

Predictive Distribution Estimation Using Averaged Densities Method

In comparison, Figure 18.3 shows the predictive densities obtained using the averaged densities method.

The upper chart shows the $k = 3$, 4 and unconditional fits to GalaxyB. NMix output allows $k = 5$ and 6, but these are not depicted, being very similar to the $k = 4$, which is already very close to the unconditional case.

The lower chart shows the $k = 3$, 4, 5 and unconditional fits to Galaxy. It will be seen that both $k = 4$ and $k = 5$ are very similar to the unconditional fit.

The fits in both charts are more stable and closer than those obtained using the averaged MC parameters method to calculate the predictive distribution, and are visually much more satisfactory.

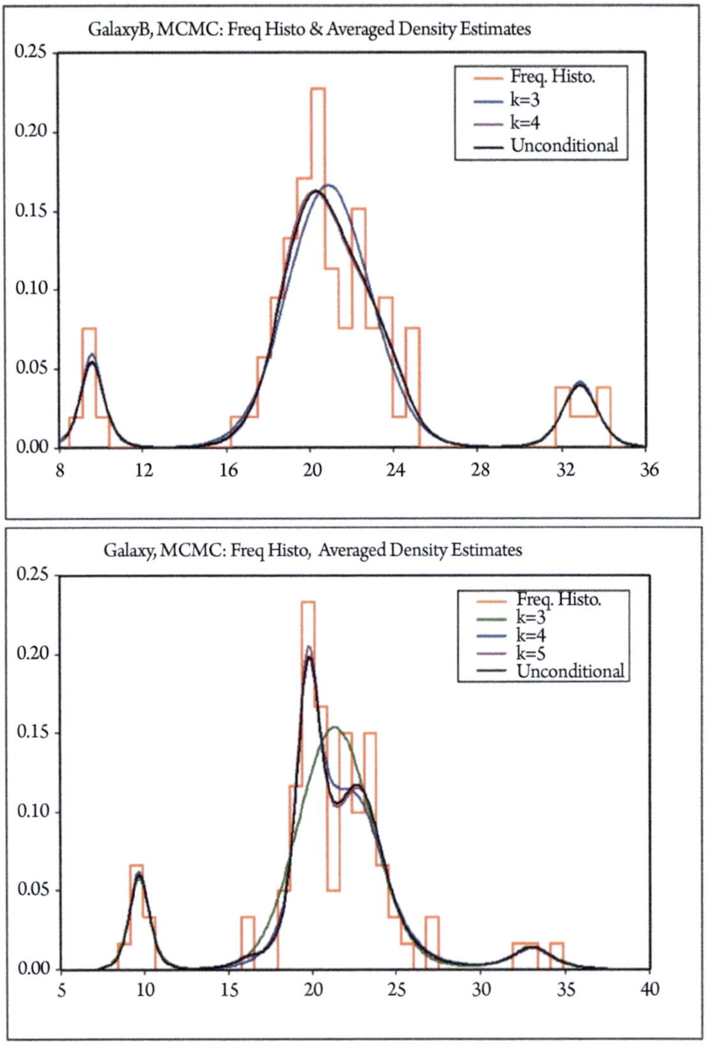

Figure 18.3 Predictive densities using averaged densities. Upper chart: $k = 3$, 4 and unconditional fits to GalaxyB data. Lower chart: $k = 3$, 4, 5 and unconditional fits to Galaxy data

The problem, as pointed out by Richardson and Green (1997), is that the averaged density method does not give a predictive distribution that is directly interpretable as a finite mixture model of the form (17.1). In particular, estimates of the individual component parameters and weights are not available, without careful postprocessing of the MCMC simulation results.

Table 18.2 Estimated posterior distribution of k for the GalaxyB and Galaxy samples using the MAPIS method; in all cases $g = 0.2$

Sample	δ	k	3	4	5	6
GalaxyB	1.5	$\tilde{\pi}$	**0.91**	0.01	-	-
GalaxyB	2.0	$\tilde{\pi}$	**0.88**	0.12	-	-
Galaxy	1.5	$\tilde{\pi}$	**0.68**	0.32	-	-
Galaxy	2.0	$\tilde{\pi}$	**0.58**	0.42	-	-

Posterior Distribution of k Using MAPIS

Consider now the MAPIS method. As remarked in Section 17.3.2, MAPIS method is sensitive to small δ, and, to a lesser extent, g being made too small as well. We set $g = 0.2$ in all cases to match the default value used in RJMCMC, but this is probably not best for MAPIS, making it a rather rigorous test for the method.

The top two rows of entries in Table 18.2 give the posterior distribution of k for the GalaxyB data set obtained using MAPIS. The value of $g = 0.2$ is near the lower limit below which fitted components with delta spike PDFs start to appear if, like g, δ is also made too small. Indeed, though a meaningful fit is obtainable when $\delta = 1$, a delta spike component is present when $k > 4$.

The results shown for $\delta = 1.5$ and 2 indicate the k-component fit with $k = 3$ is good.

The two lower rows of Table 18.2 give the posterior distribution of k for the Galaxy data using MAPIS. The choice of δ is again straightforward and the results are displayed for $\delta = 1.5$ and $\delta = 2.0$. In this case, though the maximum $\tilde{\pi}(k|\mathbf{y})$ is at $k = 3$, the probabilities at $k = 4$ are not negligible, with the probabilities increasing as δ increases.

Predictive Distribution Estimation Using MAPIS

Figure 18.4 shows for both data sets the CDF and PDF plots of the predictive distribution conditional on k. The only plots shown correspond to k with posterior probability $\tilde{\pi}(k|\mathbf{y}) \geq 0.2$, as obtained by MAPIS, with the plots corresponding to the k with the highest $\tilde{\pi}(k|\mathbf{y})$ denoted in black. In this case, there is only one plot in the GalaxyB case where $k = 3$ and two plots in the Galaxy case where $k = 3$ and 4.

It will be seen that the results give a more accurate predictive distribution fit than those obtained using the averaged densities method with RJMCMC.

Moreover, estimates are easily obtained for the component parameters and weights in the MAPIS case. In the GalaxyB case, they are

$$\tilde{\boldsymbol{\mu}} = (9.6, 20.9, 33.0), \tilde{\boldsymbol{\sigma}} = (0.24, 1.99, 0.64), \tilde{\mathbf{w}} = (0.08, 0.63, 0.08),$$

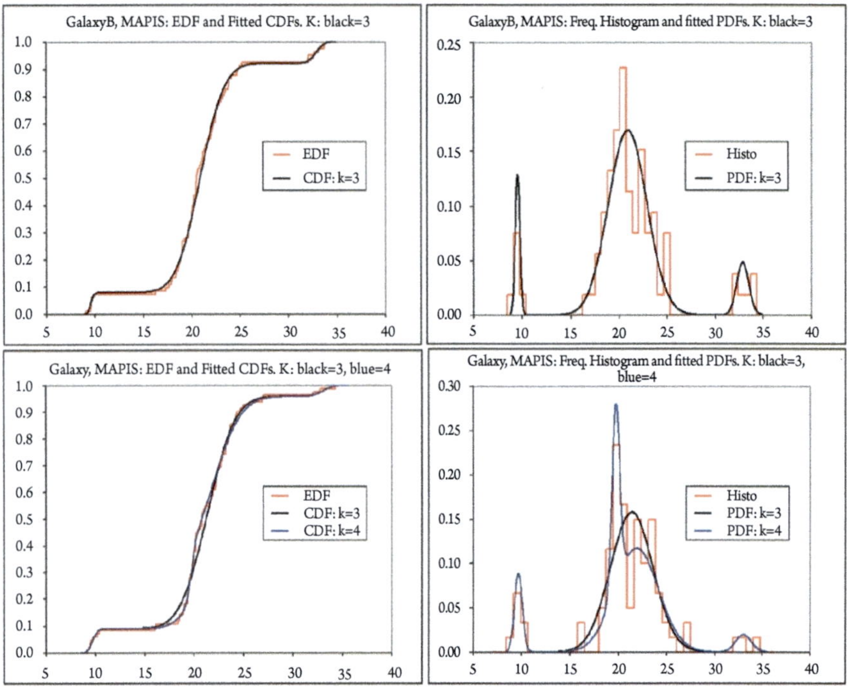

Figure 18.4 MAPIS method. Predictive CDF and PDF conditional on k. Upper charts: GalaxyB data. Lower charts: Galaxy data.

matching quite well the true values

$$\boldsymbol{\mu}_0 = (9.7, 21.0, 33.0), \boldsymbol{\sigma}_0 = (0.4, 2.2, 0.8), \mathbf{w}_0 = (0.1, 0.85, 0.05).$$

The estimated first and third weights are equal, which looks worryingly odd at first sight, but is actually an artifact of the small sample size. The two components are physically quite separate, so that their estimated weights depend totally on the number observations drawn from each corresponding component. These happen to be identical in the sample used, making the weight estimates identical.

ADP Results

The ADP came out the weakest of the three Bayesian approaches for both Galaxy data sets. We used $N = 15$ so that $\delta = N^{-1} \simeq 0.06667$. The predictive density estimators based on $\breve{\pi}_+$ of eqn (17.12) and $\breve{\pi}_{ADP}$ using eqn (17.29) are shown in Table 18.3. The $\breve{\pi}_+$ are particularly biased in favour of high k, indicating perhaps that n or N is not sufficiently large for the asymptotic theory to be applicable.

The predictive distribution estimators for the two data sets based on $\breve{\pi}_{ADP}$ are shown in Figure 18.5, where the CDF and PDF plots correspond only to those k with posterior

Table 18.3 Estimated posterior distribution of k for the GalaxyB and Galaxy samples using the ADP method; in all cases, $g = 0.2$

Sample	k	**2**	**3**	**4**	**5**	**6**	**7**	**8**	**9**	**10**
GalaxyB	$\breve{\pi}_+$	-	0.02	0.10	0.24	**0.30**	0.22	0.10	0.03	0.01
GalaxyB	$\breve{\pi}_{ADP}$	0.04	0.17	**0.32**	0.30	0.13	0.03	-	-	-
Galaxy	$\breve{\pi}_+$	-	0.01	0.07	0.22	**0.31**	0.24	0.12	0.04	0.01
Galaxy	$\breve{\pi}_{ADP}$	0.01	0.09	0.34	**0.37**	0.15	0.03			

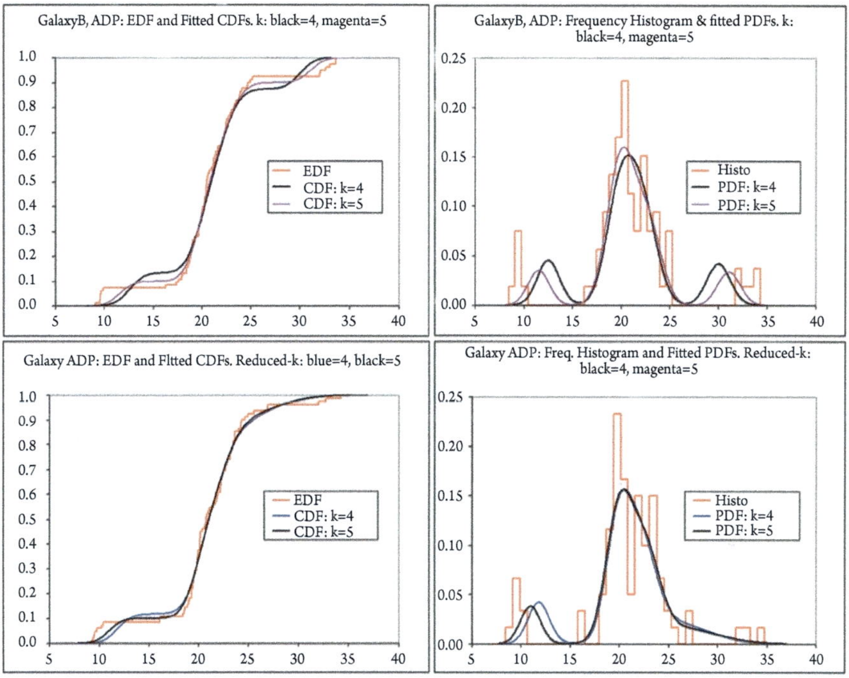

Figure 18.5 ADP method. Predictive CDF and PDF conditional on k. Upper charts: GalaxyB data. Lower charts: Galaxy data.

probability $\breve{\pi}(k|\mathbf{y}) \geq 0.2$, as obtained by ADP, with the plot corresponding to k with the highest $\breve{\pi}(k|\mathbf{y})$ denoted in black.

It will be seen that for both data sets, the predictive distribution is a poor fit, with components misaligned in the GalaxyB fits and the component with the largest mean not identified at all in the Galaxy fits.

Posterior Distribution of Parameters

Ideally, the posterior distributions of component parameters and weights should be compact, and preferably unimodal. The histograms of any method of simulation using repeated sampling should display these characteristics, or else would be hard to interpret.

To illustrate the problem of multimodality, we display in Figures 18.6, 18.7, and 18.8, the posterior distributions of the component parameters and weights for $k = 4$ for the GalaxyB data set obtained using RJMCMC, ADP, and MAPIS, respectively. It will be seen that the plots are unimodal in the MAPIS case, indicating that the main component of the underlying true f_0 distribution has been decomposed into two, thereby conditionally providing a fourth component. The plots in the RJMCMC display a significant degree of multimodality, with the third component clearly containing contributions from the MCMC output which really should be contributing to what is being taken as the second and fourth estimated components.

The ADP case is not very satisfactory in a different way. The ADP plots are based on reduced-k values, and this seems to have seriously affected the μ distribution plots which are highly non-smooth, though their general position is interpretable in terms of the known component positions. We have not investigated the matter further here.

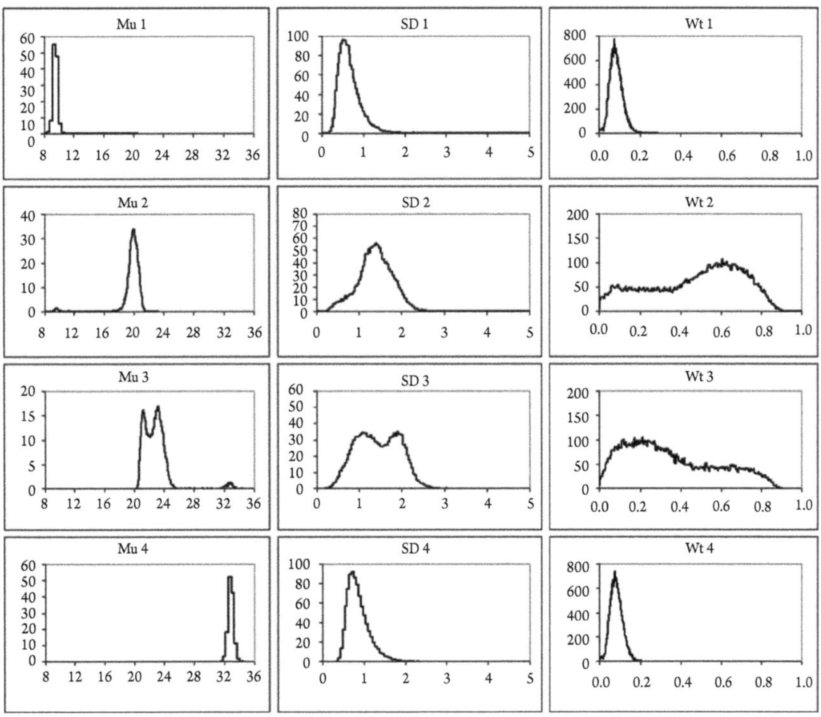

Figure 18.6 RJMCMC method. Posterior distributions of the component parameters and weights for $k = 4$ for the GalaxyB data.

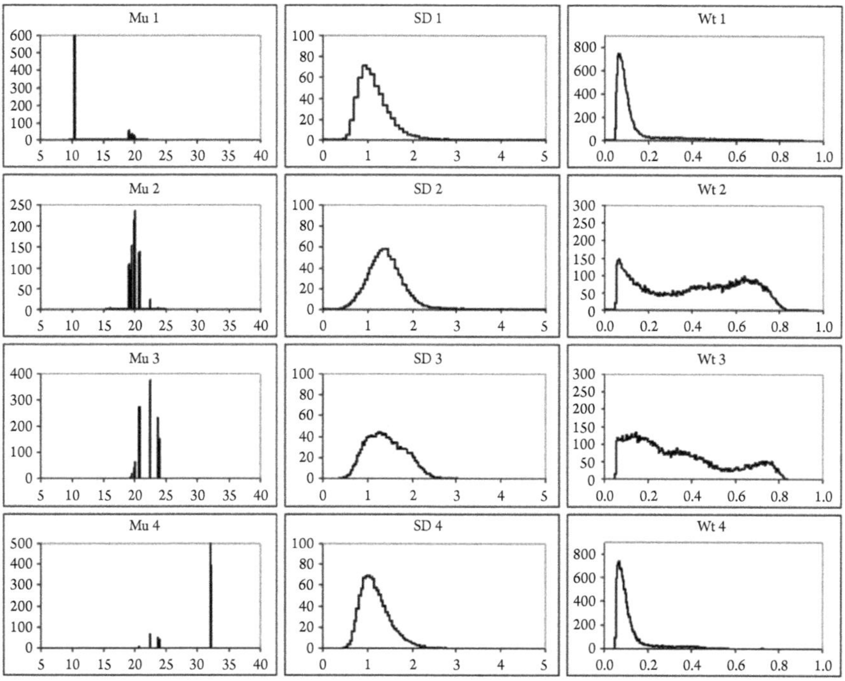

Figure 18.7 ADP method. Posterior distributions of the component parameters and weights for $k = 4$ for the GalaxyB data.

Conclusions

Our overall conclusion is that in the RJMCMC analysis of the artificial GalaxyB data set, the posterior distribution of k has favoured high k values. Given the similarity of GalaxyB to Galaxy, there are grounds for believing that the analysis of Galaxy has favoured high k values for the original data set as well.

In contrast, MAPIS has accurately identified $k_0 = 3$ in the GalaxyB data set. For the Galaxy data set, though $k_0 = 3$ is the more likely, the possibility that $k_0 = 4$ cannot not be discounted, but higher values of k_0 are unlikely.

The ADP approach was not competitive in these examples.

18.1.2 Hidalgo Stamp Issues

The Hidalgo stamp issues (HidalgoStamps) is an interesting data set discussed by Izenman and Sommer (1988) and reexamined by Basford et al. (1997). A summary of their findings is given by McLachlan and Peel (2000). The problem investigated was the modelling of the distribution of 485 stamp thicknesses using a finite mixture model.

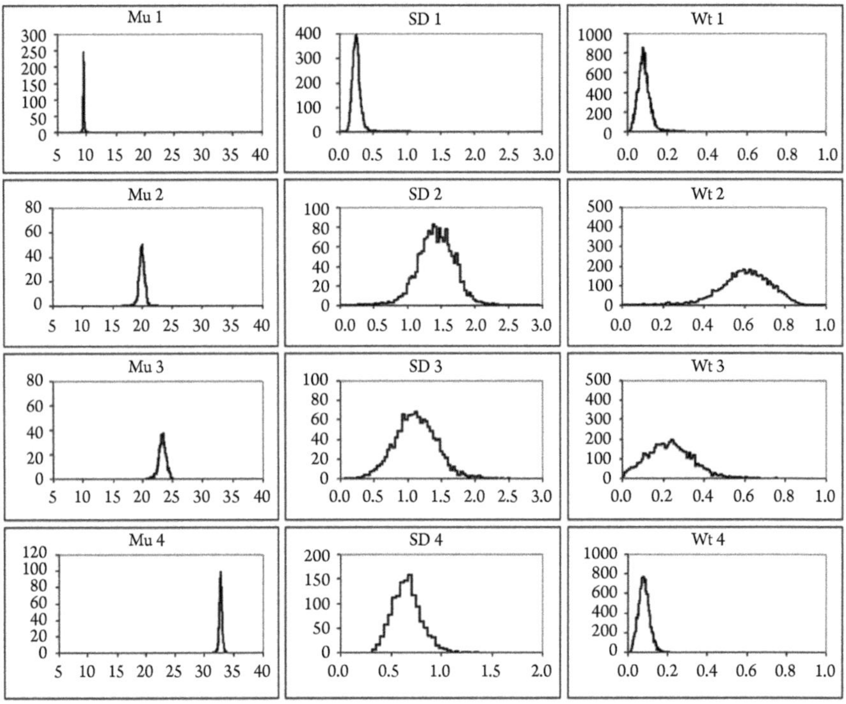

Figure 18.8 MAPIS method. Posterior distributions of the component parameters and weights for $k = 4$ for the GalaxyB data.

The frequency histogram appears genuinely multimodal, and the nonparametric technique considered by Izenman and Sommer (1988) gave seven components. Izenman and Sommer also fitted a normal finite mixture model with seven components, where the variances were unrestricted, which produced modes in almost the same locations as the nonparametric method. However, Izenman and Sommer also applied a likelihood ratio test sequentially in the normal model, which gave only three components. Basford et al. (1997) argued for some level of uniformity in variances in the normal model, and showed that when the variances are restricted to being equal, this gave seven or eight components. In all these three-, seven-, and eight-component fits, the fitted components were meaningful and interpretable. However, the selection of the number of components was dependent on the LR test, and, as we have discussed in Section 14.1.1 and in Chapter 14, in the case of overfitted models, the LR test is non-standard with a distribution that is not chi-squared, so it is not clear if the findings of Izenman and Sommer (1988) and Basford et al. (1997) are quantitatively reliable.

In this section, we fitted the ML, MAPIS, and RJMCMC fits without any restriction on variances. As the example is merely illustrative of how these methods work, we do not claim our findings are definitive. As remarked in Izenman and Sommer (1988), use of historical background information to augment the technical calculations is important in such an example before drawing serious conclusions.

Predictive Densities

Figure 18.9 gives eight charts depicting predictive densities in two columns.

The left-hand column shows the predictive densities conditional on $k = 3, 4, 7$, and 8 obtained by both ML and MAPIS for the normal mixture model. The fits are visually identical in the cases $k = 3, 4$, and 8. For the case $k = 7$, the fits obtained by ML and MAPIS are shown together, as they are almost identical, the only difference being in their choice of just one component at the left-hand end. It will be seen that the fitted components for all four k values are all readily interpretable.

The right-hand column shows the MAPIS fit in the case where the model is a finite mixture of *extreme value, EVMax,* distributions with component PDFs

$$f(y, \mu_j, \sigma_j) = \sigma_j^{-1} e^{(-w_j - e^{-w_j})}, w_j = (y - \mu_j)/\sigma_j, \ y \text{ unrestricted.}$$

Fitting this and other non-normal mixtures is discussed in Section 18.2.1. The extreme value mixture was fitted in the light of the comment made by Basford et al. (1997) that the $k = 3$ model found to be satisfactory by Izenman and Sommer (1988) had a component with a large variance, so that some of the thinnest stamps were being counted as coming from this component. This is unsatisfactory given the nature of the data. Basford et al. (1997) therefore suggested imposing the condition that all the component variances should be equal, showing that this gave satisfactory results, which, moreover, required $k = 7$ or even $k = 8$. Though we have not shown the PDFs of the individual components in the extreme value fits, none of them suffer from this problem, as EVMax components all have rapidly decreasing very small left-hand tails.

Figure 18.10 also gives eight charts depicting predictive PDFs in two columns, in this case all obtained by RJMCMC using the averaged MC parameters method.

The left-hand column shows the RJMCMC fits for $k = 3, 4, 7$, and 8 using $\delta = 1$. It will be seen that the only really satisfactory fit is where $k = 3$. The other fits all show oversmoothness and unsatisfactory irregularities.

For the HidalgoStamps sample, δ needs to be larger than unity when using RJMCMC. The asymptotic theory discussed earlier that suggests δ be made small is definitely unsatisfactory here, as it seems to make the MCMC simulation over-tolerant in allowing poor (μ, σ, w) parameter combinations. The right-hand column in Figure 18.10 gives the RJMCMC fits, only now using $\delta = 9$. (We did try δ as large as $\delta = 10$, but RJMCMC returned only observations with $k \leq 7$, and we wished to report results for $k = 8$.) The fits are generally satisfactory, except in the case $k = 7$, where what appears to be a fitted component does not correspond to any observed data cluster. The plots do not show individual fitted components, but, particularly when $k = 5$ and 6, there are fitted components with larger variances, not obvious in the plot of the full mixture PDF alone.

All four predictive densities conditional on $k = 3$ shown in Figures 18.9 and 18.10 are very similar.

This unanimity in the case $k = 3$ extends to the left-hand chart in Figure 18.11, which shows the predictive density conditional on $k = 3$, estimated using the averaged density method.

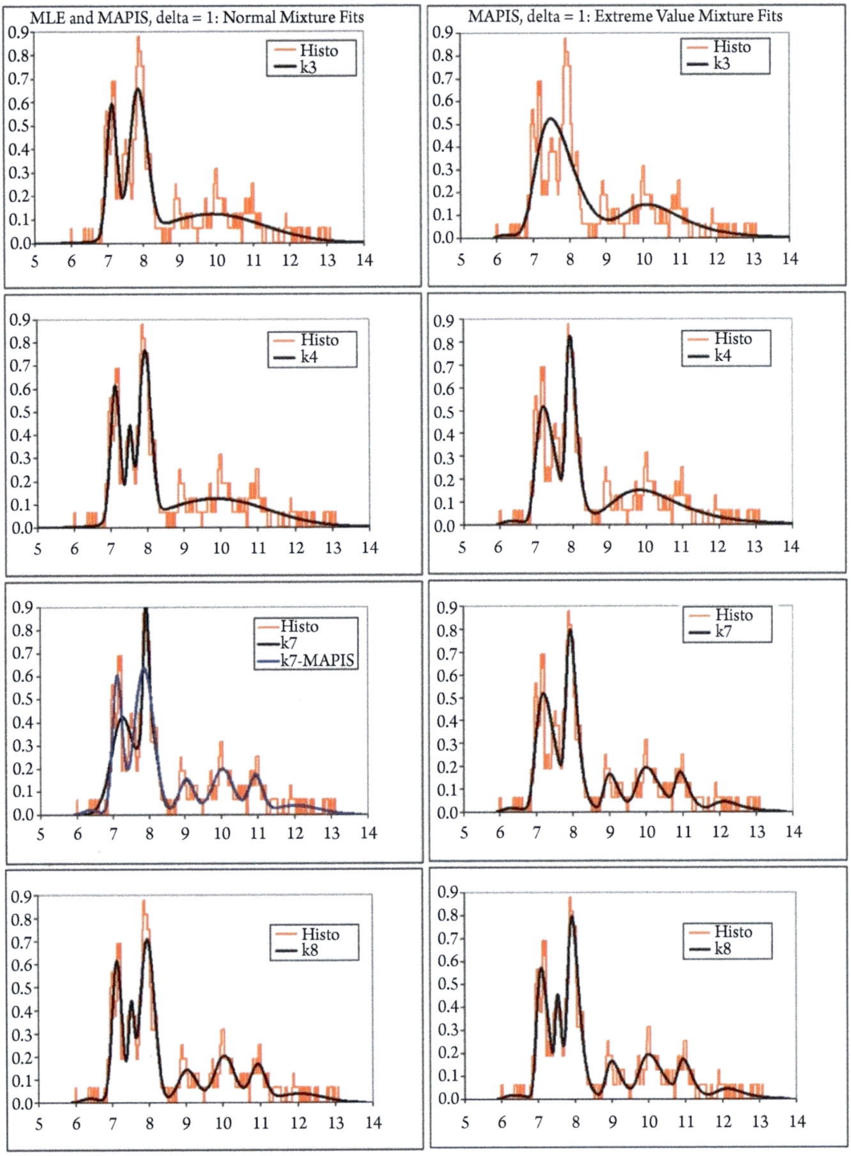

Figure 18.9 HidalgoStamps data. Left-hand column: the predictive densities conditional on $k = 3, 4, 7,$ and 8 obtained by both ML and MAPIS for the normal mixture model. Right-hand column: corresponding predictive densities for the MAPIS fit where the model is a finite mixture of extreme-value distributions.

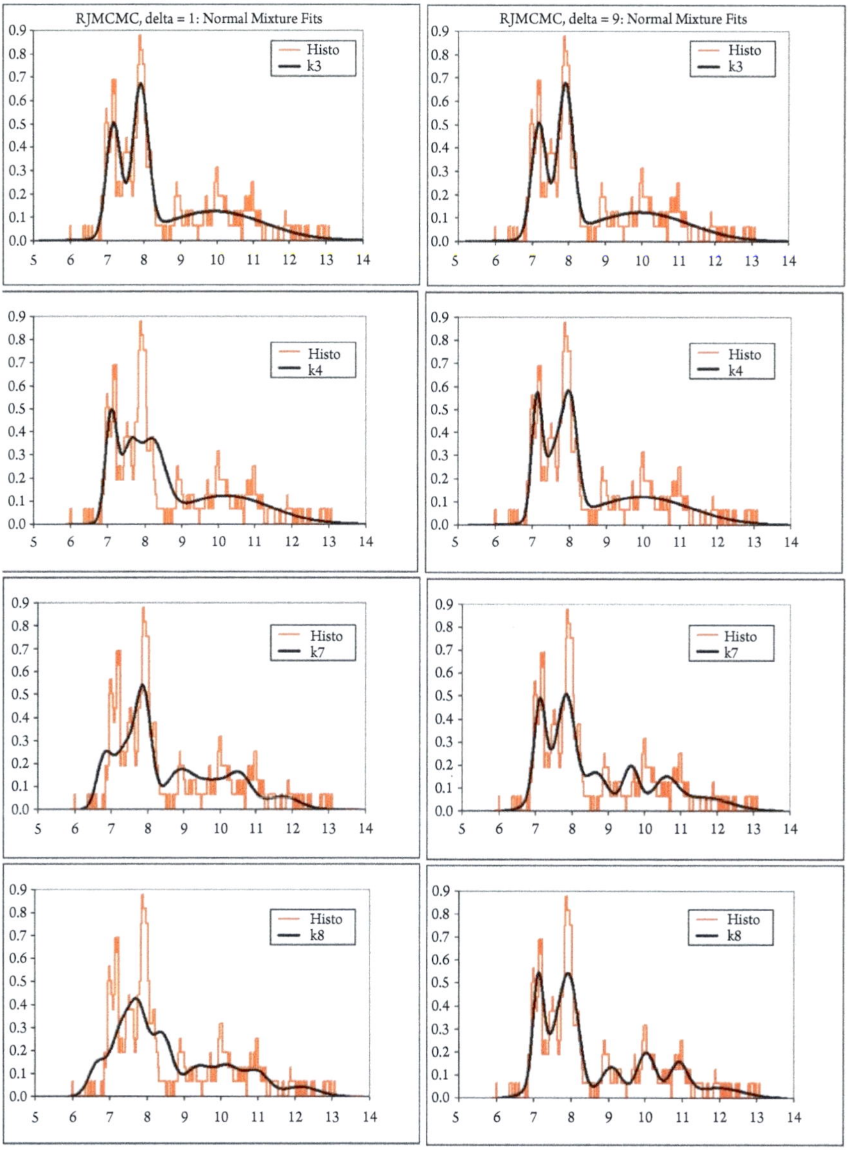

Figure 18.10 HidalgoStamps data. Left-hand column: the predictive densities conditional on $k = 3, 4, 7$, and 8 obtained by RJMCMC using the averaged MC parameters method for the normal mixture model with $\delta = 1$. Right-hand column: corresponding predictive densities obtained by RJMCMC, only now with $\delta = 9$.

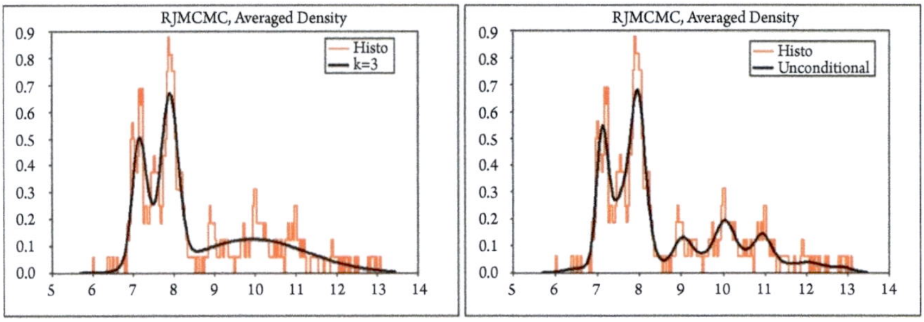

Figure 18.11 Predictive densities for the HidalgoStamps data set calculated by the averaged densities method. Left chart: conditional on $k = 3$. Right chart: unconditonal.

Thus, in this case where $k = 3$, all the methods discussed—ML, MAPIS, and RJMCMC, whether using the averaged MC parameters or density methods—give the same fit.

The right-hand chart in Figure 18.11 shows the *unconditional* predictive density calculated using the averaged density method. This gives a fit that is well matched to the data, and is very similar to that obtained by MAPIS conditional on $k = 7$ shown in the blue plot of the third chart in the left-hand column of Figure 18.9 and indeed in the $k = 7$ MAPIS fit of the extreme value mixture fit in the right-hand column.

It is unfortunate that the averaged density method of calculating the predictive distribution does not readily yield component distribution parameters.

We turn now to the estimation of k.

18.1.3 Estimation of k

Posterior Distribution of k

We consider first estimation of the posterior distribution of k, which can be carried out with either MAPIS or RJMCMC. We have already remarked that for the HidalgoStamps data set, the behaviour of the posterior distribution of k is sensitive to the choice of δ. Figure 18.12 shows the posterior distribution of k obtained by MAPIS using $\delta = 1, 2, 5, 9$.

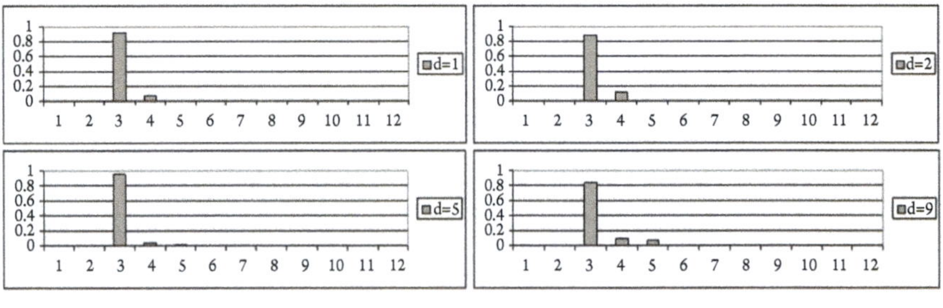

Figure 18.12 HidalgoStamps data. Posterior distribution of k using MAPIS with $\delta = 1, 2, 5, 9$.

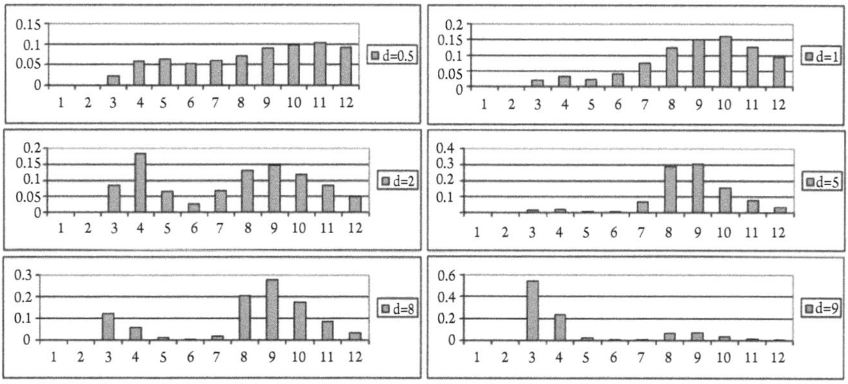

Figure 18.13 HidalgoStamps data. Posterior distribution of k using RJMCMC with $\delta = 0.5, 1, 2, 5, 8, 9$.

We have that $\tilde{\pi}(3|\mathbf{y}) > 0.89$ for $1 \leq \delta \leq 5$. Though not shown, the predictive density does not change very much if δ is varied in this range.

The situation is rather different for RJMCMC. Figure 18.13 shows that the posterior distribution of k varies significantly as δ varies in the range $1 \leq \delta \leq 9$, from being unimodal at $\delta = 1$ with the mode at $k = 10$, becoming bimodal as δ increases to $\delta = 2$ when there are modes at $k = 4$ and $k = 9$; then, if δ is further increased, switching to a single mode at $k = 9$ when δ increases to $\delta = 5$, before the mode moves back to $k = 3$ as δ increases to $\delta = 9$.

As already seen in the fitted predictive densities displayed in the previous section, arguably the most satisfactory choice of δ for RJMCMC with this particular data set is $\delta = 9$ when $\hat{\pi}(k|\mathbf{y})$ is maximum at $k = 3$.

Estimation of k under ML

In Section 17.2.2, we considered estimation of the best k under maximum likelihood, this being the most awkward aspect of fitting a finite mixture model when using ML. Our conclusion is that the Bayesian information criterion (BIC) method seems reliable and is easy to implement. Here, we illustrate its use to estimate the best k when fitting the normal finite mixture model to the HidalgoStamps data set, comparing it with the Akaike information criterion (AIC) and the GoF approach also discussed in Section 17.2.2.

We fitted the normal mixture model to the data set for a set of given k values, and calculated the corresponding BIC and AIC values. The upper two charts in Figure 18.14 show the BIC values BIC_k of eqn (17.4) for $k = 1, 2, \ldots, 10$. In the left chart, the values are obtained with the log-likelihood $\hat{L}_k$ evaluated at $(\hat{\boldsymbol{\psi}}(k), \hat{\mathbf{w}}(k))$, the component parameter and weight values estimated using ML. The maximum is located at $k = 3, 4$. In the right chart, the log-likelihood used is $\tilde{L}_k$ as evaluated at $(\tilde{\boldsymbol{\psi}}(k), \tilde{\mathbf{w}}(k))$, the estimated

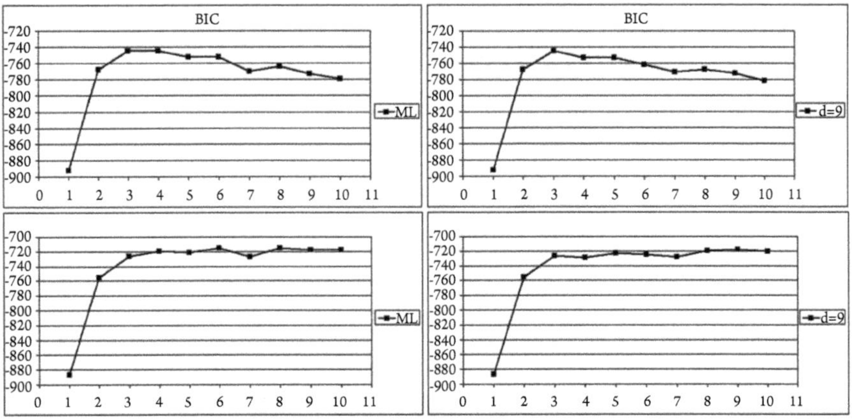

Figure 18.14 HidalgoStamps data. Upper charts: Fitted BIC_k values for $k = 1, 2, \ldots, 10$, using ML (left chart) and using MAPIS (right chart). Lower charts: Corresponding plots for AIC_k values.

component parameter and weight values using MAP with $\delta = 9$. The maximum is at $k = 3$. In fact, the latter plot changes very little for $1 \leq \delta \leq 9$.

Our estimate would therefore be $\hat{k} = 3$ or possibly $\hat{k} = 4$.

The AIC is not usually considered to be so satisfactory. The two lower plots in Figure 18.14 show the $AICC_k$ values as k varies, calculated using eqn (17.3). In this case, the values increase up to $k = 3$, but then level off. Though not so clear-cut, the value $\hat{k} = 3$ again appears to be an appropriate choice, but there is doubt whether k should not be larger.

For the GoF approach, we used the ML normal mixture fits to the original HidalgoStamps data given $k = 1, 2, \ldots, 10$, and obtained $B = 100$ BS samples for each k. We then tried three test statistics T: the Anderson-Darling statistic A^2 of eqn (4.12), the Cramér von Mises statistic of eqn (4.11), and the drift statistic $\Delta_0(m)$ of eqn (4.17), obtaining critical null test values for each k from the EDFs of the BS samples. Writing p_k for the resulting p-value of T_k obtained from the original sample, we found for $T = A^2 : p_1 = 1.0, p_2 = 0.91, p_3 = 0.00$, for $T = W^2 : p_1 = 1.0, p_2 = 0.88, p_3 = 0.00$, and for $T = \Delta_0(m = 5) : p_1 = 1.0, p_2 = 0.73, p_3 = 0.00$, all with $p_k = 0.00$ for $k \geq 4$, so that with all three GoF statistics, we would choose $k = 3$.

The main downside of bootstrapping is the computational intensiveness of the calculation. Having $B = 100$ is small by usual bootstrap implementation standards, but for the kind of assessment involved, was more than sufficiently large to give very clear results. Using $B = 50$ or even $B = 20$ is often sufficient for the choice of k to be evident.

Comments

The frequency histogram of the HidalgoStamps sample is strongly suggestive of multimodality, with $k_0 = 7$ or even 8. However, our analysis does not really offer confirming support. The posterior distribution of k obtained using RJMCMC seems to indicate $k_0 = 5$, with the possibility of k_0 ranging up to $k_0 = 8$ when δ is small. However, as shown when $\delta = 1$, the predictive density fits to the data for $k = 4, 7, 8$ are not very meaningful using the averaged MC parameters method, and only $k = 3$ is satisfactory. Meaningfully interpretable components from the predictive densities up to $k = 8$ are only achieved by setting a large $\delta = 9$ value, at which point the posterior distribution of k is concentrated at $k = 3$ and 4. Thus, the only reliable predictive density fit is $k = 3$, possibly $k = 4$. MAPIS, the BIC criterion, and GoF test approaches all strongly favour $k = 3$. Our analysis therefore tends to support the finding of Izenman and Sommer (1988) that the best parametric normal fit has $k = 3$. Evidence of more components would have to be based on the kind of contextual arguments suggested by Basford et al. (1997).

It may be that none of our tests are sufficiently sensitive for identifying small *regular* variation. However, in experiments not reported here, we have found MAPIS to be robust in identifying all the components in data generated to have specific multimodal characteristics, including components with large variance but 'hidden' by being mixed up with several components with smaller variances, and also where there are components with both small weight as well as small variance.

18.2 MAPIS Technical Details

18.2.1 Component Distributions

So far, in considering finite models, we have only considered the normal component distribution, so that MAPIS could be directly compared with RJMCMC and ADP, for which we could then use the NMix application.

We have implemented the MAPIS method as a compiled C application using an Excel front-end interface called FineMix. This allows not just the normal model but also a number of other component models, namely, the lognormal, extreme value (EVMax and EVMin), Weibull, gamma, and inverse Gaussian (IG) cases. We take all these in their two-parameter form. Table 18.4 lists the densities of all these cases except the EVMin, with the two parameters denoted by α and β, and appearing in the way that these distributions are conventionally defined.

When fitting the model (17.1), it is usually more intuitive to define components in terms of their mean μ and SD σ, as this allows the location and dispersion of fitted components to be easily identified, and allows the fits obtained using different component densities to be more easily compared.

For all six of the component distributions we consider, it is easy to express μ and σ in terms of the parameters α and β; however, if we are to use μ and σ as the parameters, we need to be able invert these relationships, as this then allows a numerical fitting procedure to be set out in terms of updating μ and σ, but where the final density and probability values can still be calculated and presented in terms of the conventional parametrization. Table 18.4 gives the α and β parameters as functions of μ and σ. The only difficult case is that of the Weibull.

Table 18.4 Conventional parametrizations of various component distributions, and the parameters considered as functions of the mean, μ, and standard deviation, σ, of the distribution; γ_E is Euler's constant, $\omega(\cdot)$ is as in eqn 18.1

Component	PDF	$\alpha(\mu,\sigma)$	$\beta(\mu,\sigma)$
Normal	$\frac{1}{\sqrt{2\pi\beta^2}} \times \exp\left[-(y-\alpha)^2/2\beta^2\right]$	μ	σ
Lognormal	$\frac{1}{\beta\sqrt{2\pi}y} \times \exp\left[-\frac{1}{2}(\frac{\ln y-\alpha}{\beta})^2\right]$	$\ln\mu - \frac{1}{2}\ln(1+(\frac{\sigma}{\mu})^2)$	$\sqrt{\ln(1+(\frac{\sigma}{\mu})^2)}$
EVMax	$\frac{1}{\beta}\exp\{-(\frac{y-\alpha}{\beta}) - \exp[-(\frac{y-\alpha}{\beta})\}$	$\mu - (\gamma_E\sqrt{6}/\pi)\sigma$	$(\sqrt{6}/\pi)\sigma$
Weibull	$\frac{\alpha}{\beta}(y/\beta)^{\alpha-1} \times \exp\left[-(y/\beta)^\alpha\right]$	$\omega(\sigma/\mu)$	$\frac{\mu}{\Gamma\left[1+\frac{1}{\omega(\sigma/\mu)}\right]}$
Gamma	$\frac{y^{\alpha-1}\beta^{-\alpha}\exp(-y/\beta)}{\Gamma(\alpha)}$	$(\mu/\sigma)^2$	σ^2/μ
IG	$\sqrt{\frac{\alpha}{2\pi y^3}} \times \exp\left[-\frac{\alpha(y/\beta-1)^2}{2y}\right]$	μ^3/σ^2	μ

For the Weibull case, α, the shape parameter, is an explicit function of the coefficient of variation $\gamma = \sigma/\mu$. We write this function as $\alpha = \omega(\gamma)$. A simple approximation for $\omega(\gamma)$ is

$$\alpha = \exp\left(0.5282 - 0.7565t - 0.3132\sqrt{6.179 - 0.5561t + 0.7057t^2}\right), \tag{18.1}$$

where $t = \ln(1+\gamma^2)$, which has a relative error of less than 1% in the range $0.0001 \le \gamma \le 1000$. This is derived in the next section. Using this approximation, we are thus able to express the usual parameters in terms of μ and σ over a reasonably practical range of values, so that in the Bayesian analysis, the Weibull distribution can be handled in exactly the same way as the other component distributions.

18.2.2 Approximation for $\omega(\cdot)$ function

Consider the Weibull distribution with PDF

$$f(y) = \frac{\alpha}{\beta}(y/\beta)^{\alpha-1}\exp[-(y/\beta]^\alpha).$$

This has mean $\mu = \beta\Gamma(\frac{1}{\alpha}+1)$ and variance $\sigma^2 = \beta^2\{\Gamma(\frac{2}{\alpha}+1) - [\Gamma(\frac{1}{\alpha}+1)]^2\}$. We therefore have

$$\begin{aligned}\ln\ln(1+(\sigma/\mu)^2) &= \ln\{\ln[\Gamma(2z+1)] - 2\ln[\Gamma(z+1)]\}\\ &= R(z),\ \text{say, where } z = 1/\alpha.\end{aligned}$$

Consider first the behaviour of $R(z)$ as $z \to 0$. Expanding $R(z)$ as a power series, we have

$$\ln(\ln(\Gamma(2z+1)) - 2\ln(\Gamma(z+1))) = \ln(\pi^2/6) + 2\ln z + O(z). \tag{18.2}$$

Now consider $R(z)$ as $z \to \infty$. Using a standard asymptotic formula, as given by Abramowitz and Stegun (1965, 6.1.41), we have

$$\begin{aligned}r_1 = \ln(\Gamma(2z+1)) &\sim (2z+1-\frac{1}{2})\ln(2z+1) - (2z+1) + \frac{1}{2}\ln(2\pi)\\ &+\frac{1}{12(2z+1)} - \frac{1}{360(2z+1)^3} + O(\frac{1}{z^4}),\end{aligned}$$

and

$$\begin{aligned}r_2 = 2\ln(\Gamma(z+1)) &\sim 2(z+1-\frac{1}{2})\ln(z+1) - 2(z+1) + \ln(2\pi)\\ &+\frac{1}{6(z+1)} - \frac{1}{180(z+1)^3} + O(\frac{1}{z^4}).\end{aligned}$$

The log factor in the first term in the expression for r_1 is

$$\begin{aligned}\ln(2z+1) = \ln 2z + \ln(1+\frac{1}{2z}) &= \ln 2z + \frac{1}{2z} - \frac{1}{2(2z)^2}\\ &+\frac{1}{3(2z)^3} - \frac{1}{4(2z)^4} + O(\frac{1}{z^5}).\end{aligned}$$

Therefore, the first term in r_1 is

$$\begin{aligned}(2z+1-\frac{1}{2})\ln(2z+1) &= (2z+\frac{1}{2})(\ln 2z + \frac{1}{2z} - \frac{1}{2(2z)^2}\\ &+\frac{1}{3(2z)^3} - \frac{1}{4(2z)^4} + O(\frac{1}{z^5}))\\ &= 2(\ln 2)z + 2z\ln z + 1 + \frac{1}{48z^2} - \frac{1}{96z^3}\\ &+\frac{1}{2}\ln 2 + \frac{1}{2}\ln z + +O(\frac{1}{z^4}).\end{aligned}$$

We also have that the first term in r_2 is

$$\begin{aligned}2(z+1-\frac{1}{2})\ln(z+1) &= 2(z+1-\frac{1}{2})(\ln z+\frac{1}{z}-\frac{1}{2(z)^2}+\frac{1}{3(z)^3}\\ &\quad -\frac{1}{4(z)^4}+O(\frac{1}{z^5}))\\ &= 2z\ln z+\ln z+2+\frac{1}{6z^2}-\frac{1}{6z^3}+O(\frac{1}{z^4}).\end{aligned}$$

The difference in these two first terms is therefore

$$\begin{aligned}&2\,(\ln 2)\,z+2z\ln z+1+\frac{1}{48z^2}-\frac{1}{96z^3}+\frac{1}{2}\ln 2+\\ &\frac{1}{2}\ln z-(2z\ln z+\ln z+2+\frac{1}{6z^2}-\frac{1}{6z^3})+O(\frac{1}{z^4})\\ &=2\,(\ln 2)\,z-\frac{1}{2}\ln z+\frac{1}{2}\ln 2-1-\frac{7}{48z^2}+\frac{5}{32z^3}+O(\frac{1}{z^4}).\end{aligned}$$

Thus,

$$\begin{aligned}r_1-r_2 &= \ln(\Gamma(2z+1))-2\ln(\Gamma(z+1))\\ &= 2\,(\ln 2)\,z-\frac{1}{2}\ln z+\frac{1}{2}\ln 2-1+O(\frac{1}{z^2}).\end{aligned} \tag{18.3}$$

Hence,

$$R(z)=\ln(r_1-r_2)=\ln(z(2\,(\ln 2)-\frac{1}{2z}\ln z+(\frac{1}{2}\ln 2-1)\frac{1}{z})+O(\frac{1}{z^3})))$$

$$=\ln z+\ln(2\,(\ln 2)-\frac{1}{2z}\ln z+(\frac{1}{2}\ln 2-1)\frac{1}{z}+O(\frac{1}{z^3})) \tag{18.4}$$

$$=\ln z+\ln(2(\ln 2))+O(\frac{\ln z}{z})\quad \text{as } z\to\infty. \tag{18.5}$$

If we write $x=-\ln z$, so that $\alpha=\exp(x)$, we have from (18.2) and (18.5) that

$$\begin{aligned}R(\exp(-x)) &= \begin{cases} g(x)+O(\frac{x}{\exp(-x)}) & \text{as } x\to-\infty\\ h(x)+O(\exp(-x)) & \text{as } x\to\infty\end{cases}\\ &= y(x),\ \text{say},\end{aligned}$$

where

$$\begin{aligned}g(x) &= 0.3266-x,\ (\text{using } \ln(2(\ln 2))\simeq 0.326\,63),\\ h(x) &= 0.4977-2x,\ (\text{using } \ln(\pi^2/6)\simeq 0.49770).\end{aligned}$$

The function $y(x)=R(\exp(-x))$ can be represented by one arm of the hyperbola

$$(y+x-a)(y+2x-b)=A,$$

for suitably chosen coefficients A, a, and b. Use of such a hyperbolic approximation allows inversion to express x in terms of y. The required solution is

$$x = \frac{1}{2}a + \frac{1}{4}b - \frac{3}{4}y - \frac{1}{4}\sqrt{4a^2 - 4ab + b^2 + 8A + (2b - 4a)y + y^2},$$

with a and b having values similar to $a = \ln(2(\ln 2))$, $b = \ln(\pi^2/6)$. The coefficients in (18.1) correspond to the approximation

$$x = 0.5282 - 0.7565y - 0.3132\sqrt{6.180 - 0.5561y + 0.7057y^2},$$

which gives an α relative accuracy within 1% over the coefficient of variation range $0.0001 \leq \sigma/\mu \leq 1000$.

18.2.3 Example of Hyperparameter Elimination

We consider the prior $\sigma_j^{-2} \sim \Gamma(\alpha, \beta)$ discussed in Section 17.3.1, where β is a gamma-distributed hyperparameter with PDF

$$\pi(\beta) = \frac{1}{\Gamma(g)}\beta^{g-1}h^g \exp(-h\beta),\ \beta > 0. \tag{18.6}$$

The precise way that the randomness of β affects σ is not very transparent. We can readily avoid having to deal with β directly as follows. Let X be a random variable which, conditional on the value of β, has the gamma distribution with PDF

$$\pi(x|\beta) = \frac{1}{\Gamma(\alpha)}x^{\alpha-1}\beta^{\alpha} \exp(-\beta x).$$

Assuming the prior distributions of X and β are independent, the density of the unconditional prior distribution of X is therefore

$$\begin{aligned}\pi(x) &= \int_\beta \pi(x|\beta)\pi(\beta)d\beta \\ &= \int_0^\infty \frac{1}{\Gamma(\alpha)}x^{\alpha-1}\beta^{\alpha} \exp(-\beta x)\frac{1}{\Gamma(g)}\beta^{g-1}h^g \exp(-h\beta)d\beta \\ &= \frac{\Gamma(\alpha+g)}{\Gamma(\alpha)\Gamma(g)}h^g (x+h)^{-\alpha-g} x^{\alpha-1},\end{aligned} \tag{18.7}$$

which is the PT VI distribution of eqn (9.2), that is, a beta distribution of the second kind.

In Richardson and Green (1997), this is the assumed prior distribution of $x = \sigma^{-2}$. The differentials of x and β satisfy $dx = -2\sigma^{-3}d\sigma$. This yields

$$\pi(\sigma) = 2\frac{\Gamma(\alpha+g)}{\Gamma(\alpha)\Gamma(g)}h^g \left(1 + h\sigma^2\right)^{-\alpha-g} \sigma^{2g-1}, \tag{18.8}$$

and this is equivalent to assuming $\sigma^{-2} \sim \Gamma(\alpha, \beta)$, with $\beta \sim \Gamma(g, h)$.

The form of this three-parameter prior is a generalization of the folded non-central-t prior considered by Gelman (2006), which is obtained from (18.8) by setting $g = 0.5$. Gelman's parametrization is

$$\pi(\sigma) = 2\frac{\Gamma\left(\frac{\nu}{2}+\frac{1}{2}\right)}{\Gamma\left(\frac{\nu}{2}\right)\Gamma\left(\frac{1}{2}\right)A\nu^{1/2}}\left(1+\frac{1}{\nu}\left(\frac{\sigma}{A}\right)^2\right)^{-(\nu+1)/2}, \tag{18.9}$$

where $\nu = 2\alpha$ and $A = R/(2\sqrt{10})$ when $h = 10/R^2$. Gelman observes that a special case of (18.9) is the half-Cauchy distribution obtained when $\nu = 1$. A is then a scale parameter, with large but finite values of A making (18.9) what Gelman terms a weakly informative prior.

The form of (18.8) shows that $\pi(\sigma) \to 0$, a constant, or ∞ as $\sigma \to 0$ according to whether g is greater than, equal to, or less than 0.5. It is well known that for any given j, there always exist finite combinations of the other parameter values for which the likelihood becomes infinite as $\sigma_j \to 0$. These correspond to mixture distributions with both a continuous component and discrete components located at one or more of the observations y_i. Such models can be eliminated from consideration, so that only purely continuous mixtures are allowed, by ensuring that the prior for each σ_j tends to zero sufficiently fast as $\sigma_j \to 0$. This is achieved by using (18.8) for the prior, provided we set $g > 0.5$. In practice, $g = 0.5$ will usually be satisfactory. R&G found that the precise value of g does not seem very critical, suggesting the default value of $g = 0.2$. This is our own experience; larger values do seem to dampen the effect of problem (B5), but not to a great degree. The choice of g is more delicate when (18.8) is used as the prior in the MAPIS method.

Concerning the value of the other parameter, α, in the PDF (18.8), R&G recommend the default value of $\alpha = 2$. The relatively simple form of $\pi(\sigma)$ in (18.8) allows an 'informed' choice that takes into account the sample being analysed. If S is a random variable with PDF $\pi(\sigma)$ as given in (18.8), we have

$$E(S^2) = \int_0^\infty \frac{\Gamma(\alpha+g)}{\Gamma(\alpha)\Gamma(g)}2h^g\left(1+h\sigma^2\right)^{-\alpha-g}\sigma^{2g+1}d\sigma = \frac{1}{h}\frac{g}{\alpha-1}. \tag{18.10}$$

If we select g, and use the R&G recommendation of $h = H/R^2$, where H is some user-chosen constant and R is the range of the data, then we can use (18.10) to fix α as

$$\alpha = 1 + \frac{gR^2}{HS^2},$$

where S in this formula is based on the sample. If we have $g = 0.5$ or slightly larger, $H = 10$, and $S \sim R/2$, then $1 < \alpha < 2$. This suggests that the precise choice of α is not too critical. In our numerical examples, we set $\alpha = 2$ to match the default RJMCMC setting.

18.2.4 MAPIS Method: Additional Details

Estimation of the Covariance Matrix

We give details of the candidate (IS) distribution used to carry out the importance sampling in the MAPIS method. We focus on the kth component, where k is given, continuing to use the notation $\boldsymbol{\theta}(k)$ to denote generic parameters of the finite mixture PDF of eqn (17.1), $\tilde{\boldsymbol{\theta}}(k)$ for the MAP estimator, and $\boldsymbol{\theta}^*(k)$ for a variate value generated by IS. When there is no confusion, we write L for $L(\boldsymbol{\psi}(k), \mathbf{w}(k), k)$, $\boldsymbol{\psi}$ for $\boldsymbol{\psi}(k)$, and $\mathbf{w}$ for $\mathbf{w}(k)$. The dimension of $\boldsymbol{\psi}(k)$ is $l = 2k$.

The IS distribution $\tilde{c}_k[\boldsymbol{\theta}(k)]$ is located at $\tilde{\boldsymbol{\theta}}(k)$, and we shall follow Geweke (1989) and set its variance equal to minus the inverse of the Hessian of $L(\boldsymbol{\psi}(k), \mathbf{w}(k), k) = \ln(p[\mathbf{y}|\boldsymbol{\theta}(k), k]$ evaluated at $\tilde{\boldsymbol{\theta}}(k)$. This is non-trivial, because the weights, w_i must sum to one.

Let the negative *unconstrained* Hessian of second partial derivatives be

$$\mathbf{H} = \begin{pmatrix} \mathbf{H}_{\boldsymbol{\psi},\boldsymbol{\psi}} & \mathbf{H}_{\boldsymbol{\psi},\mathbf{w}} \\ \mathbf{H}_{\boldsymbol{\psi},\mathbf{w}}^T & \mathbf{H}_{\mathbf{w},\mathbf{w}} \end{pmatrix}$$

with, in particular,

$$\mathbf{H}_{\mathbf{w},\mathbf{w}} = -\frac{\partial^2 L(\mathbf{w})}{\partial \mathbf{w}^2}.$$

These partial derivatives in $\mathbf{H}$ are unconstrained in that they are obtained ignoring the restriction that $\sum w_i = 1$. To include this restriction, we use an alternative parametrization $\boldsymbol{\lambda} = (\lambda_1, \lambda_2 \ldots, \lambda_k)^T$, and write w_i as

$$w_i = \lambda_i + k^{-1}\left(1 - \sum_{j=1}^{k} \lambda_j\right), \quad i = 1, \ldots, k.$$

This ensures that

$$\sum_{i=1}^{k} w_i = 1. \tag{18.11}$$

We do not place any restrictions on the λ_i, as the only requirement in addition to (18.11) is that $w_i \geq 0$ for $i = 1, 2, \ldots, k$, and this is handled separately later. The Jacobian matrix of the transformation is

$$\mathbf{J} = \frac{\partial \mathbf{w}}{\partial \boldsymbol{\lambda}} = (\mathbf{I}_k - k^{-1}\mathbf{1}_k\mathbf{1}_k^T), \tag{18.12}$$

where $\mathbf{I}_k$ is the k-component identity matrix and $\mathbf{1}_k=(1, 1, \ldots, 1)^T$ is the k-component vector with unit entries. The log posterior density, $L = L(\boldsymbol{\psi}(k), \boldsymbol{\lambda}(k), k)$, in terms of the $\boldsymbol{\lambda}$ parametrization, has Hessian

$$\mathbf{A}(\boldsymbol{\psi}, \boldsymbol{\lambda}) = \mathbf{A}(\boldsymbol{\psi}, \mathbf{w}) = \begin{pmatrix} \mathbf{H}_{\psi,\psi} & \mathbf{H}_{\psi,\mathbf{w}}\mathbf{J}^T \\ \mathbf{J}\mathbf{H}_{\psi,\mathbf{w}}^T & \mathbf{J}\mathbf{H}_{\mathbf{w},\mathbf{w}}\mathbf{J}^T \end{pmatrix}, \tag{18.13}$$

which we write as $\mathbf{A}$ from now on. Evaluated at $(\tilde{\boldsymbol{\psi}}, \tilde{\mathbf{w}})$, this is the required Hessian, because the inverse of $\tilde{\mathbf{A}}$ gives the covariance of $(\tilde{\boldsymbol{\psi}}, \tilde{\mathbf{w}})$ subject to $\sum_{i=1}^{k} \tilde{w}_i = 1$.

The matrix $\mathbf{A}$ must clearly be singular, and indeed the submatrix $\mathbf{J}\mathbf{H}_{\mathbf{w},\mathbf{w}}\mathbf{J}^T$ is singular, as $\det(\mathbf{J}) = 0$. Thus, $\mathbf{A}$ does not have a full inverse, but it does have a generalized inverse, $\mathbf{G}$, which by definition will satisfy

$$\mathbf{AGA}=\mathbf{A}. \tag{18.14}$$

To find a generalized inverse, we consider $\mathbf{P}$, the orthogonal matrix formed from the eigenvectors of $\mathbf{A}$. Being singular, $\mathbf{A}$ has at least one eigenvalue that is zero. Let the corresponding eigenvector be

$$\mathbf{p}_0 = \begin{pmatrix} \mathbf{0}_l \\ \mathbf{1}_k \end{pmatrix} \begin{matrix} \}l \\ \}k \end{matrix},$$

where $\mathbf{0}_l$ is the l-dimensional column vector of zeros. Let $\mathbf{P}_1 = (\mathbf{p}_1|\mathbf{p}_2|\dots|\mathbf{p}_\nu)$ be the matrix comprising the other eigenvectors

$$\mathbf{p}_j = \begin{pmatrix} \mathbf{p}_j^{\psi} \\ \mathbf{p}_j^{\mathbf{w}} \end{pmatrix} \begin{matrix} \}l \\ \}k \end{matrix}, \; j = 1, 2, \dots, \nu,$$

where $\nu = l + k - 1$, and write $\mathbf{P}$ as

$$\mathbf{P} = \left(\mathbf{P}_1 \; \mathbf{p}_0\right).$$

We have

$$\mathbf{P}^T\mathbf{AP} = \mathbf{D} = \begin{pmatrix} \boldsymbol{\Lambda} & 0 \\ 0 & 0 \end{pmatrix} \begin{matrix} \}\nu \\ \}1 \end{matrix}, \tag{18.15}$$

where $\mathbf{D}$ is the diagonal matrix of eigenvalues corresponding to the eigenvectors forming $\mathbf{P}$.

From now on, we assume that $\mathbf{A}$ is positive semidefinite, so that all its eigenvalues are non-negative. The main diagonal entries of $\mathbf{D}$ and $\boldsymbol{\Lambda}$ will therefore all be non-negative. In practice, it will usually be the case, in maximizing the logposterior density subject to $\sum w_i = 1$, that *all* the main diagonal entries in $\boldsymbol{\Lambda}$ will be strictly positive, but our construction of the generalized inverse does not require this. However, the last main diagonal entry of $\mathbf{D}$ is definitely zero by construction.

Let $\mathbf{S}$ be the $(l+k) \times (l+k)$ diagonal matrix

$$\mathbf{S} = \begin{pmatrix} \mathbf{R} & \mathbf{0}_\nu \\ \mathbf{0}_\nu^T & 0 \end{pmatrix} \begin{matrix} \}\nu \\ \}1 \end{matrix}, \tag{18.16}$$

where

$$\mathbf{R} = \operatorname{diag}(l_{ii} | \, l_{ii} = 1/\sqrt{\lambda_{ii}} \text{ if } \lambda_{ii} > 0, \; l_{ii} = 0 \text{ if } \lambda_{ii} = 0)$$

and λ_{ii} is the ith main diagonal entry of $\mathbf{\Lambda}$. Define $\mathbf{L}$ as

$$\mathbf{L} = \mathbf{PS}. \tag{18.17}$$

Define $\mathbf{G}$ as

$$\mathbf{G}{=}\mathbf{PSSP}^T{=}\mathbf{LL}^T. \tag{18.18}$$

For $\mathbf{S}$ as in (18.16), we find $\mathbf{DSSD}{=}\mathbf{D}$, and from (18.15) we have $\mathbf{A}{=}\mathbf{PDP}^T$. Using these two expressions we have

$$\mathbf{AGA}{=}\mathbf{PDP}^T\mathbf{PSSP}^T\mathbf{PDP}^T{=}\mathbf{PDSSDP}^T{=}\mathbf{PDP}^T{=}\mathbf{A}.$$

So $\mathbf{G}$ satisfies (18.14), and is thus a generalized inverse of $\mathbf{A}$.

Importance Sampling

We can now describe explicitly our proposed IS method of generating the parameters $\boldsymbol{\theta}(k)$ of Step IS2 from a modified multivariate t-distribution. Specifically, we generate this as

$$\boldsymbol{\theta}^* = \begin{pmatrix} \boldsymbol{\psi}^* \\ \mathbf{w}^* \end{pmatrix} = \begin{pmatrix} \hat{\boldsymbol{\psi}} \\ \hat{\mathbf{w}} \end{pmatrix} + \boldsymbol{\theta}_0^*, \tag{18.19}$$

where

$$\boldsymbol{\theta}_0^* \sim \text{StudentT}\left(\tilde{\mathbf{L}}\tilde{\mathbf{L}}^T\right),$$

and StudentT($\mathbf{V}$) is the multivariate t-distribution with mean $\mathbf{0}$ and variance $\mathbf{V}$. As a reminder, in this section, all the vectors are dependent on k, so strictly should have a k subscript, but for simplicity this has been omitted. A variate from this distribution, with $\mathbf{V}{=}\tilde{\mathbf{L}}\tilde{\mathbf{L}}^T$, can be generated using

$$\boldsymbol{\theta}_0^* = \tilde{\mathbf{P}}_1\tilde{\mathbf{R}}\mathbf{z}_\nu^*, \tag{18.20}$$

where $\mathbf{z}_\nu^*$ is a vector of independent Student-t variates, each normalized to have mean zero and variance unity. These can have arbitrary degrees of freedom, d, and are derived by rescaling non-standardized t-variates. The variance-covariance of $\boldsymbol{\theta}$ generated in this way is then

$$\mathrm{Var}(\boldsymbol{\theta}_0^*) = \mathrm{E}(\tilde{\mathbf{P}}_1\tilde{\mathbf{R}}\mathbf{z}_\nu\mathbf{z}_\nu^T\tilde{\mathbf{R}}^T\tilde{\mathbf{P}}_1^T) = \tilde{\mathbf{P}}_1\tilde{\mathbf{R}}\tilde{\mathbf{R}}\tilde{\mathbf{P}}_1^T.$$

But from (18.17) and (18.18) we find after some calculation that $\tilde{\mathbf{P}}_1\tilde{\mathbf{R}}\tilde{\mathbf{R}}\tilde{\mathbf{P}}_1^T = \tilde{\mathbf{P}}SS\tilde{\mathbf{P}}^T$. Therefore,

$$\mathrm{Var}(\boldsymbol{\theta}_0) = \tilde{\mathbf{P}}_1\tilde{\mathbf{R}}\tilde{\mathbf{R}}\tilde{\mathbf{P}}_1^T = \tilde{\mathbf{P}}\mathbf{SS}\tilde{\mathbf{P}}^T = \tilde{\mathbf{G}}.$$

Moreover, using the result that $(\mathbf{0}_l^T, \mathbf{1}_k^T)\tilde{\mathbf{P}}_1 = \mathbf{0}_\nu^T$, the sum of the component weights is given by

$$\begin{aligned}\sum_{i=1}^{k} w_i^* &= (\mathbf{0}_l^T, \mathbf{1}_k^T)\begin{pmatrix}\boldsymbol{\psi}^*\\ \mathbf{w}^*\end{pmatrix}\\ &= (\mathbf{0}_l^T, \mathbf{1}_k^T)\begin{pmatrix}\tilde{\boldsymbol{\psi}}\\ \tilde{\mathbf{w}}\end{pmatrix} + (\mathbf{0}_l^T, \mathbf{1}_k^T)\tilde{\mathbf{P}}_1\tilde{\mathbf{R}}\mathbf{z}_\nu^*\\ &= (\mathbf{0}_l^T, \mathbf{1}_k^T)\begin{pmatrix}\tilde{\boldsymbol{\psi}}\\ \tilde{\mathbf{w}}\end{pmatrix} + \mathbf{0}_\nu^T\tilde{\mathbf{R}}\mathbf{z}_\nu^*\\ &= \sum_{i=1}^{k}\tilde{w}_i = 1.\end{aligned}$$

Thus, under this sampling, we are restricted to the simplex $\sum_{i=1}^k w_i = 1$.

The vector of weights, $\mathbf{w} = (w_1, w_2, \ldots, w_k)$ clearly has a singular distribution. Let $\boldsymbol{\omega}$ be the $(k-1)$ dimensional vector the *reduced set of weights* formed from the first $(k-1)$ components of $\mathbf{w}$, and write

$$\boldsymbol{\phi} = \begin{pmatrix}\boldsymbol{\psi}\\ \boldsymbol{\omega}\end{pmatrix} \tag{18.21}$$

for the vector of component distribution parameters and this reduced set of weights. We have

$$\begin{pmatrix}\boldsymbol{\psi}^*\\ \boldsymbol{\omega}^*\end{pmatrix} = \begin{pmatrix}\tilde{\boldsymbol{\psi}}\\ \tilde{\boldsymbol{\omega}}\end{pmatrix} + \tilde{\mathbf{M}}\mathbf{z}_\nu^*, \tag{18.22}$$

where $\tilde{\mathbf{M}}$ is the matrix $\tilde{\mathbf{P}}_1\tilde{\mathbf{R}}$ but with the last row omitted. We have

$$\partial(\boldsymbol{\psi}, \boldsymbol{\omega})/\partial(\mathbf{z}_\nu) = \mathbf{M}$$

and the Jacobian of the transformation is $|\partial(\boldsymbol{\psi}, \boldsymbol{\omega})/\partial(\mathbf{z}_\nu)| = \det(\mathbf{M})$. This transformation is non-degenerate and invertible, with the PDF of $\boldsymbol{\phi}^*$, as given in eqn (17.22) by

$$\tilde{f}(\boldsymbol{\phi}^*) = \left|\frac{\partial \mathbf{z}_\nu}{\partial(\boldsymbol{\psi}, \boldsymbol{\omega})}\right|_{\boldsymbol{\phi}=\tilde{\boldsymbol{\phi}}} g_\nu(\mathbf{z}_\nu^*) = \left|\frac{\partial(\boldsymbol{\psi}, \boldsymbol{\omega})}{\partial(\mathbf{z}_\nu)}\right|_{\boldsymbol{\phi}=\tilde{\boldsymbol{\phi}}}^{-1} g_\nu(\mathbf{z}_\nu^*) = [\det(\tilde{\mathbf{M}})]^{-1} g_\nu(\mathbf{z}_\nu^*).$$

Finally, an acceptance/rejection procedure is needed to ensure parameters which should be positive necessarily are positive, and that all weights necessarily satisfy $0 < w_j < 1$. Thus a $\boldsymbol{\phi}^*$ sample is rejected if it breaks *any* such required constraint, so that given k, the IS distribution actually sampled is modified to

$$\tilde{c}_k^R[\boldsymbol{\psi}(k), \boldsymbol{\omega}(k)] = [\det(\tilde{\mathbf{M}}(k))]^{-1} g_\nu(\mathbf{z}_\nu)/\tilde{R}(k), \tag{18.23}$$

where $\tilde{R}(k)$ is the estimate of the probability $R(k)$ that a value sampled from (17.22) is accepted, which can be calculated as

$$\tilde{R}(k) = \frac{(\text{\# of replications sampled from (17.22) for the given } k \text{ and accepted})}{m_k},$$

where m_k = (# of replications sampled from (17.22) for the given k).

This gives $\boldsymbol{\phi}^*$. As $w_j = \omega_j, j = 1, 2, \ldots, k-1$, the final weight is $w_k^* = 1 - \sum_{j=1}^{k-1} w_j^*$.

Acceptance/rejection adds to the computational effort, but would only be problematic if $R(k)$ were ever to be small, which we have not encountered. Our proposed IS procedure appears acceptably fast in practice. We have tried more elaborate IS distributions which directly satisfy the required parameter constraints, but such distributions made the calculations significantly more complicated and less transparent, and actually slowed the IS procedure.

We summarize the IS as it applies to estimation of the posterior distribution of K.

In the previous section, we assumed, for ease of exposition, that the number of components k in each IS replication was sampled independently. However, we can remove the inherent variability in this sampling of k by using *stratified* sampling. We thus replace IS1 to IS4 by:

IS1′. Sample k *cyclically* with $k = 1, 2, \ldots, k_{\max}, 1, 2, \ldots, k_{\max}, 1, 2, \ldots$, and so on, so that if m replications are drawn, where for simplicity we assume that m is divisible by $k_{\max}$, we sample the *same* number of replications for each possible k, i.e. if m_k is the number of replications where $K = k$, then $m_k = m/k_{\max}$, $k = 1, 2, \ldots, k_{\max}$, so that all the m_k are equal. All the IS formulas derived in the previous section are unchanged by this.

IS2′. For each k_i obtained in IS1′, sample the mixture model parameters $\boldsymbol{\psi}(k_i)$, $\boldsymbol{\omega}(k_i)$ using (18.19), but applying acceptance/rejection so that each accepted $\boldsymbol{\psi}(k_i), \boldsymbol{\omega}(k_i)$ satisfies all parameter constraints for the given k_i component mixture model.

Record the $R(k)$ acceptance probabilities for each k.

IS3′. For each replication, calculate the IS ratio, ρ_i, as given in (17.25), with the divisor given by (18.23).

IS4′. Estimate $\tilde{\pi}(k|\mathbf{y})$, $k = 1, 2, .., k_{\max}$, the posterior distribution of the number of components from (17.17).

Other quantities of interest, such as the PDF of the parameters $\boldsymbol{\psi}(k)$, $\mathbf{w}(k)$ conditional on k, can then be estimated by appropriate weighted frequency histograms, using the IS ratios as the weights.

Bibliography

Aas, K. and Haff, I.H. (2006). The generalized hyperbolic skew Student's t distribution. *J. of Financial Econometrics*, **4**(**2**), 275–309.

Abramowitz, M. and Stegun, I.A. (1965). *Handbook of Mathematical Functions*. New York: Dover Publications Inc.

Akaike, H. (1973). Information theory and an extension of the maximum likelihood principle, in Petrov, B.N.; Cski, F., *2nd International Symposium on Information Theory, Tsahkadsor, Armenia, USSR, September 2–8, 1971*, Budapest: Akadmiai Kiad, 267–281.

Akaike, H. (1974). A new look at the statistical model identification, *IEEE Transactions on Automatic Control*, **19**, 716–723.

Álvarez, B.L. and Gamero, M.D.J. (2012). A note on bias reduction of maximum likelihood estimates for the scalar skew t distribution. *Journal of Statistical Planning and Inference*, **142**, 608–612.

Anatolyev, S. and Kosenok, G. (2005). An alternative to maximum likelihood based on spacings. *Econometric Theory*, **21**, 472–476.

Arcidiacono, P. and Bailey Jones, J. (2003). Finite mixture distributions, sequential likelihood and the EM algorithm. *Econometrica*, **71**, 933–946.

Arellano-Valle, R.B., and Azzalini, A. (2008). The centred parametrization for the multivariate skew-normal distribution, *J. Multivariate Anal.*, **99**, 1362–1382; *J. Multivariate Anal.*, **100**, (2009), 816 (Corrigendum).

Arellano-Valle, R.B. and Azzalini, A. (2013). The centred parameterization and related quantities of the skew-t distribution. *J. Multivariate Anal.*, **113**, 73–90.

Atkinson, A.C. (1970). A method for discriminating between models. *J. R. Statist. Soc. B*, **32**, 323–353.

Atkinson, A.C. (1985). *Plots, Transformations and Regression*. Oxford: Oxford University Press.

Atkinson, A.C., Pericchi, L.R. and Smith, R.L. (1991). Grouped likelihood for shifted power transformation. *J. R. Statist. Soc. B*, **53**, 473–482.

Azzalini, A. (1985). A class of distributions which includes the normal ones. *Scand. J. Statist.* **12**, 171–178.

Azzalini A. (2005). The skew-normal distribution and related multivariate families (with discussion). *Scand J. Statist.* **32**, 159–188 (C/R 189–200).

Azzalini, A. and Capitanio, A. (2003). Distributions generated by perturbation of symmetry with emphasis on a multivariate skew t distribution. *Journal of the Royal Statistical Society Series B*, **65**, 367–389.

Azzalini, A. and Dalla Valle, A. (1996). The multivariate skew-normal distribution. *Biometrika*, **83**, 715–726.

Azzalini, A. and Genton, M.G. (2008). Robust likelihood methods based on the skew-t and related distributions. *International Statistical Review*, **76**, 106–129.

Babu, G.J. and Rao, C.R. (2003). Confidence limits to the distance of the true distribution from a misspecified family by bootstrap. *J. of Statistical Planning and Inference*, **115**, 471–478.

Babu, G.J. and Rao, C.R. (2004). Goodness-of-fit tests when parameters are estimated. *Sankhya*, **66**, 63–74.

Bain, L.J. (1978). *Statistical Analysis of Reliability Data and Life-testing Models: Theory and Methods.* New York: Marcel Dekker.

Barnard, G.A. (1967). The use of the likelihood function in statistical practice. *Proc. 5th Berkeley Symp. Mathematical Statistics and Probability* (eds. L.M. LeCam and J. Neyman), vol. 1, 27–40. Berkeley: University of California Press.

Barndorff-Nielsen, O.E. (1978). *Information and Exponential Families in Statistical Theory.* Chichester: Wiley.

Barndorff-Nielsen, O.E. and Cox, D.R. (1994). *Inference and Asymptotics*. London: Chapman and Hall.

Bartlett, M.S. (1947). The use of transformations. *Biometrics*, **3**, 39–52.

Basford, K.E., McLachlan, G.J. and York, M.G. (1997). Modelling the distribution of stamp paper thickness via finite normal mixtures: The 1872 Hidalgo stamp issue of Mexico revisited. *Journal of Applied Statistics*, **24**, 169–180.

Bates, D.M. and Watts, D.G. (1988). *Nonlinear Regression Analysis and Its Applications*. New York: Wiley.

Becker, M. and Klöner, S. (2013). Package 'PearsonDS', Version 0.97, R Project. https://cran.r-project.org/web/packages/PearsonDS/PearsonDS.pdf Accessed: 22 July 2016.

Belov, I.A. (2005). On the computation of the probability density function of stable distributions. *Mathematical Modelling and Analysis. Proceedings of the 10th International Conference MMA2005&CMAM2* 333–341. Trakai: Technika.

Beran, R. (1987). Prepivoting to reduce level error of confidence sets. *Biometrika*, **74**, 457–468.

Berman, M. (1986). Some unusual examples of extrema associated with hypothesis tests when nuisance parameters are present only under the alternative. *Proc. Pacific Statistics Congr.* (eds. I.S. Francis *et al.*). Elsevier: North-Holland.

Bickel, P. and Chernoff, H. (1993). Asymptotic distribution of the likelihood ratio statistic in a prototypical nonregular problem. In: *Statistics and Probability: A Raghu Raj Bahadur Festschrift*, eds J.K. Ghosh, S.K. Mitra, K.R. Parthasarathy, and B.L.S. Prakasa Rao, New Delhi: Wiley Eastern Limited, pp. 83–96.

Biernacki, C., Celeux, G. and Govaert, G. (2000). Assessing a mixture model for clustering with the integrated completed likelihood. *IEEE Transactions on Pattern Analysis and Machine Intelligence*, **22**, 719–725.

Billingsley, P. (1986). *Probability and Measure.* Second Edition. New York: Wiley.

Bilmes, J. (1997). *A gentle tutorial of the EM algorithm and its applications to parameter estimation for Gaussian mixture and hidden Markov models.* TR-97-021, U.C. Berkeley.

Böhning, D. and Dietz, E. (1995). Contribution to discussion of Cheng and Traylor (1995). *J. Roy. Statist. Soc. Ser. B*, **57**, 33–34.

Böhning, D., Dietz, E., Schaub, R., Schlattmann, P. and Lindsay, B.G. (1994). The distribution of the likelihood ratio for mixtures of densities from the one-parameter exponential family *Ann. Inst. Statist. Math.*, **46**, 373–388.

Box, G.E.P. and Cox, D.R. (1964). An analysis of transformations, *J. R. Statist. Soc. B*, **26**, 211–243.

Brain, P. and Cousens, R. (1989). An equation to describe dose responses where there is stimulation of growth at low dose. *Weed Res.*, **29**, 93–96.

Branco, M.D. and Dey, D.K. (2001). A general class of multivariate skew-elliptical distributions. *Journal of Multivariate Analysis*, **79**, 99–113.

Branco, M.D. and Dey, D.K. (2002). Regression model under skew elliptical error distribution. *Journal of Mathematical Sciences*, **1**, 151–168.

Breiman, L. (1968). *Probability*. Reading, Mass.: Addison-Wesley.

Breiman, L. (1992). *Probability*. Philadelphia: SIAM. Reprint of original 1968 edition.

Brown, B.W. and Hollander, M. (1977). *Statistics: A Biomedical Introduction*. New York: Wiley.

Brown, L.D. (1986). *Fundamentals of statistical exponential families with applications in statistical decision theory*. Lecture Notes-Monograph Series, Volume 9. Hayward, CA: Institute of Mathematical Statistics, 284 pp. Acessed: https://projecteuclid.org/euclid.lnms/1215466757. 24 July 2016

Burley, D. (1974). *Studies in Optimization*. International Textbook Co. Ltd.

Burr, I.W. (1942). Cumulative frequency functions. *Ann. Math. Stat.*, **13**, 215–232.

Burr, I.W. and Cislak, P.J. (1968). On a general system of distributions, I. Its curve-shape characteristics, II. The sample median. *J. Amer. Statist, Assoc.*, **63**, 627–635.

Cabral, C.R.B., Lachos V.H. and Madruga, M.R. (2012). Bayesian analysis of skew-normal independent linear mixed models with heterogeneity in the random-effects population. *Journal of Statistical Planning and Inference*, **142**, 181–200.

Cabral, C.R.B., Lachos, V.H. and Prates, M.O. (2012). Multivariate mixture modeling using skew-normal independent distributions. *Computational Statistics and Data Analysis*, **56**, 126–142.

Chang, I.-S., Chen, C.-H. and Hsiung, C.A. (1994). Hayward, California: Institute of Mathematical Statistics. Estimation in change-point hazard rate models with random censorship. In *Change-Point Problems, Volume 23, IMS Lecture Notes Ű Monograph Series, eds* E. Carlstein, H.-G. Müller, and D. Siegmund, pp 78Ű-92. Hayward, California: Institute of Mathematical Statistics.

Chant, D. (1974). On asymptotic tests of composite hypotheses in nonstandard conditions. *Biometrika*, **61**, 291–298.

Chen, X., Ponomareva, M. and Tamer, E. (2014). Likelihood inference in some finite mixture models. *Journal of Econometrics*, **182**, 97–99.

Chen, Z. (2000). A new two-parameter lifetime distribution with bathtub shape or increasing failure rate function. *Statistics & Probability Letters*, **49**, 155–161.

Cheng, R.C.H. (1987). Confidence bands for two-stage design problems. *Technometrics*, **29**, 301–309.

Cheng, R.C.H. (2008). Selecting the best linear simulation metamodel. In *Proceedings of the 2008 Winter Simulation Conference*, eds S.J. Mason, R.R. Hill, L. Münch, O. Rose, T. Jefferson, and J.W. Fowler, 371–378. Piscataway, New Jersey: Institute of Electrical and Electronics Engineers. Available online via http://www.informs-sim.org/wsc08papers/043.pdfwww.informs-sim.org/wsc08papers/043.pdf [accessed 29 December 2008].

Cheng, R.C.H. (2009). Computer intensive statistical model building. In *Advancing the Frontiers of Simulation. A Festschrift in Honor of George Samuel Fishman*, Eds C. Alexopoulos, D. Goldsman and J.R. Wilson. Dordrecht: Springer, 43–63.

Cheng, R.C.H. (2011). Using Pearson Type IV and other Cinderella distributions. In *Proceedings of the 2011 Winter Simulation Conference*, S. Jain, R.R. Creasey, J. Himmelspach, K.P. White, and M. Fu, eds. IEEE, Piscataway, 457–468.

Cheng, R.C.H. (2014). Massively parallel programming in statistical optmization and simulation. In *Proceedings of the 2014 Winter Simulation Conference* eds A. Tolk, S. Y. Diallo, I. O. Ryzhov, L. Yilmaz, S. Buckley, and J. A. Miller, 3707–3717. Piscataway, New Jersey: Institute of Electrical and Electronics Engineers.

Cheng, R.C.H. and Amin, N.A.K. (1979). *Maximum product of spacings estimation with application to the lognormal distribution.* Mathematics Report 79–1, Department of Mathematics, University of Wales Institute of Science and Technology, Cardiff.

Cheng, R.C.H. and Amin, N.A.K. (1982). *Estimating parameters in continuous univariate distributions with a shifted origin.* Mathematics Report 82–1, Cardiff: University of Wales Institute of Science and Technology.

Cheng, R.C.H. and Amin, N.A.K. (1983). Estimating parameters in continuous univariate distributions with a shifted origin. *J. R. Statist. Soc. B*, **45**, 394–403.

Cheng, R.C.H., Evans, B.E. and Iles, T.C. (1992). Embedded models in non-linear regression. *J. R. Statist. Soc. B*, **54**, 877–888.

Cheng, R.C.H. and Iles, T.C. (1983). Confidence bands for cumulative distribution functions of continuous random variables. *Technometrics*, **25**, 77–86.

Cheng, R.C.H. and Iles, T.C. (1987). Corrected maximum likelihood in non-regular problems. *J. R. Statist. Soc. B*, **49**, 95–101.

Cheng, R.C.H. and Iles, T.C. (1990). Embedded models in three-parameter distributions and their estimation. *J. R. Statist. Soc. B*, **52**, 135–149.

Cheng, R.C.H. and Liu, W.B. (1995). Confidence intervals for threshold parameters. In *Statistical Modelling*, eds G.U.H. Seeber, B.J. Francis, R. Hatzinger, and G. Steckelberger., Lecture Notes in Statistics 104. New York: Springer-Verlag. 53–60.

Cheng, R.C.H. and Liu, W.B. (1997). A continuous representation of the family of stable law distributions. *Journal of the Royal Statistical Society, Series B*, **59**, 137–145.

Cheng, R.C.H. and Liu, W.B. (2001). The consistency of estimators in finite mixture models. *Scandinavian J. Statistics*, **28**, 603–616.

Cheng, R.C.H. and Stephens, M.A. (1989). A goodness-of-fit test using Moran's statistic with estimated parameters. *Biometrika*, **76**, 385–392.

Cheng, R.C.H. and Traylor, L. (1991). A hybrid estimator for distributions with a shifted origin. *Mathematics Report 91–2*, School of Mathematics, University of Wales, College of Cardiff.

Cheng, R.C.H. and Traylor, L. (1995). Non-regular maximum likelihood problems. With Discussion. *J. R. Statist. Soc. B*, **57**, 3–44.

Chernick, M.R. (2008). *Bootstrap Methods: A Guide for Practitioners and Researchers, 2nd Edition.* Hoboken, NJ: Wiley.

Chiogna, M. (2005). A note on the asymptotic distribution of the maximum likelihood estimator for the scalar skew-normal distribution. *Statistical Methods and Applications*, **14**, 331–341.

Choi, S.C. and Wette, R. (1969). Maximum likelihood estimation of the parameters of the gamma distribution and their bias. *Technometrics*, **11**, 683–690.

Clarkson, D.B. and Jennrich, R.I. (1991). Computing extended maximum likelihood estimates for linear parameter models. *J. R. Statist. Soc. B*, **53**, 417–426.

Cohen, A.C. (1969). A generalization of the Weibull distribution. *Marshall Space Flight Center*, NASA Contractor Report No. 61293 Cont. NAS 8-11175.

Cohen, A.C. and Whitten, B.J. (1988). *Parameter Estimation in Reliability and Life Span Models.* New York: Marcel Dekker.

Cooper, N.R. (1992). Statistical analysis of data from materials and components with two modes of failure. *Report 8/91*, Defence Research Agency, Sevenoaks.

Cordeiro, G.M. and de Castro, M. (2011). A new family of generalized distributions. *Journal of Statistical Computation and Simulation*, **81**, 883–898.

Cowan, R.J. (1975). Useful headway models. *Transportation Research*, **9**, 371–375.

Cox, D.R. (1961). Tests of separate families of hypotheses. *Proc. 4th Berkeley Symp. Mathematical Statistics and Probability*, vol.1, 105–123. Berkeley: University of California Press.

Cox, D.R. (1962). Further results on tests of separate families of hypotheses. *J. R. Statist. Soc. B*, **14**, 406–424.

Cox, D.R. and Hinkley, D.V. (1974). *Theoretical Statistics.* London: Chapman and Hall.

Cox, D.R. and Reid, N. (1987). Parameter orthogonality and approximate conditional inference. *J. Roy. Statist. Soc. B*, **49**, 1–39.

Cramér, H. (1946). *Mathematical Methods of Statistics.* Princeton, N.J.: Princeton University Press.

Cramér, H. and Leadbetter, M.R. (1967). *Stationary and Related Stochastic Processes.* New York: Wiley.

Crowder, M.J., Kimber, A.C., Smith, R.L. and Sweeting, T.J. (1991). *Statistical Analysis of Reliability Data.* London: Chapman and Hall.

Cuckle, H.S., Wald, N.J. and Thompson, S.G. (1987). Estimating a woman's risk of having a pregnancy associated with Down's Syndrome using her age and serum alpha-fetoprotein level. *British J. Obstetrics and Gynaecology*, **94**, 387–402.

D'Agostino, R.B. and Stephens, M.A. (1986). *Goodness-of-Fit Techniques.* New York: Dekker.

da Silva Ferreira, C., Bolfarine, H. and Lachos, V.H. (2011). Skew scale mixtures of normal distributions: Properties and estimation. *Statistical Methodology*, **8**, 154–171.

Darling, D.A. (1953). On a class of problems related to the random division of an interval. *Ann. Math. Statist.*, **24**, 239–253.

Dasgupta, A. and Raftery, A.E. (1998). Detecting features in spatial point processes with clutter via model-based clustering. *J. Amer. Statist. Assoc.*, **93**, 294–302.

Davidon, W. (1959). *Variable metric method for minimization.* Argonne Nat. Lab. report ANL-5990 Rev.

Davidson, R. and MacKinnon, J.G. (1981). Several tests for model specifications in the presence of alternative hypotheses. *Econometrika*, **49**, 781–793.

Davies, R.B. (1977). Hypothesis testing when a nuisance parameter is present only under the alternative. *Biometrika*, **64**, 247–254.

Davies, R.B. (1987). Hypothesis testing when a nuisance parameter is present only under the alternative. *Biometrika,* **74,** 33–43.

Davies, R.B. (2002). Hypothesis testing when a nuisance parameter is present only under the alternative: linear model case. *Biometrika,* **89,** 484–489.

Davison, A.C. and Hinkley, D.V. (1997). *Bootstrap Methods and their Application.* Cambridge: Cambridge University Press.

de Gusmão, F.R.S., Ortega, E.M.M. and Cordeiro, G.M. (2011). The generalized inverse Weibull distribution. *Stat. Papers,* **52,** 591–619.

Dempster, A., Laird, N. and Rubin, D. (1977). Maximum likelihood for incomplete data via the EM algorithm. *J. Roy. Statist. Soc. B,* **39,** 1–38.

Diebolt, J. and Ip, E. (1996). Stochastic EM: method and application. Chapter 15 in: *Markov Chain Monte Carlo in Practice,* W. Gilks, S. Richardson, and D. Spiegelhalter, Eds. London: Chapman and Hall.

Diebolt, J. and Robert, C. (1994). Estimation of finite mixture distributions through Bayesian sampling. *J. Roy. Statist. Soc. B,* **56,** 363–375.

Draper, J. (1952). Properties of distributions resulting from certain simple transformations of the normal distribution. *Biometrika,* **39,** 290–301.

Draper, N.R. and Cox, D.R. (1969). On distributions and their transformation to normality. *J. R. Statist. Soc. B,* **31,** 472–476.

Dubey, S.D. (1968). A compound Weibull distribution. *Naval Res. Logistics Quarterly,* **15,** 179–188.

Dudewicz, E.J. and Mishra, S.N. (1988). *Modern Mathematical Statistics.* New York: Wiley.

Dudzinski, M.L. and Mykytowycz, R. (1961). The eye lens as an indicator of age in the wild rabbit in Australia. *CSIRO Wildlf. Res.,* **6,** 156–159.

Dumouchel, W.H. (1975). Stable distributions in statistical inference: Information from stably distributed samples, *J. Amer. Statist. Assoc.,* **70,** 386–393.

Efron, B. (2010). *Large-scale inference.* Camridge: Cambridge University Press.

Efron, B. and Hinkley, D.V. (1978). Assessing the accuracy of the maximum likelihood estimator: observed versus expected Fisher information (with discussion). *Biometrika,* **65,** 457–487.

Ekström, M. (1996). Strong consistency of the maximum spacing estimate. *Theory Probab. Math. Statist.,* **55,** 55–72.

Ekström, M. (2008). Alternatives to maximum likelihood estimation based on spacings and the Kullback-Leibler divergence. *Journal of Statistical Planning and Inference,* **138,** 1778–1791.

Elandt-Johnson, R.C. (1976). A class of distributions generated from distributions of exponential type. *Naval Res. Logists. Quarterly,* **23,** 131–138.

Feng, Z.D. and McCulloch, C.E. (1996). Using bootstrap likelihood ratios in finite mixtures models. *J. Roy. Statist. Soc., B,* **58,** 609–617.

Fisher, R.A. (1930). Moments and product moments of sampling distributions. *Proc. London Math. Soc.,* **s2-30**(1), 199–238.

Fisher, R.A. and Tippett, L.H.C. (1928). Limiting forms of the frequency distribution of the largest or smallest member of a sample. *Mathematical Proceedings of the Cambridge Philosophical Society,* **24,** 180–190.

Fishman, G.S. (2006). *A first course in Monte Carlo.* Australia: Thomson Brooks/Cole.

Freedman, D.A. (1981). Bootstrapping regression models. *Ann. Math. Statist.*, **9**, 1218–1228.

Garel, B. (2005). Asymptotic theory of the likelihood ratio test for the identification of a mixture. *Journal of Statistical Planning and Inference*, **131**, 271–296.

Gelman, A. (2006). Prior distributions for variance parameters in hierarchical models. *Bayesian Analysis*, **1**, 515–533.

Geluk, J.L. and de Haan, L. (2000). Stable probability distributions and their domains of attraction: a direct approach. *Prob. and Math. Stat.*, **20** 169–188.

Geweke, J. (1989). Bayesian inference in econometric models using Monte Carlo integration. *Econometrica*, **57**, 1317–1339.

Ghosh, J.K. and Sen, P.K. (1985). On the asymptotic performance of the log-likelihood ratio statistic for the mixture model and related results. In: Le Cam, L.M., Olshen, R.A. (Eds.), *Proceedings of the Berkeley Conferences in Honor of Jerzy Neyman and Jack Kiefer, Vol. II*. Monterey: Wadsworth. 789–806.

Ghosh, K. and Jammalamadaka, S.R. (2001). A general estimation method using spacings. *Journal of Statistical Planning and Inference*, **93**, 71–82.

Giesbrecht, F. and Kempthorne, O. (1976). Maximum likelihood estimation in the three-parameter lognormal distribution. *J. R. Statist. Soc. B*, **38**, 257–264.

Griffiths, J.D. and Williams, J.E. (1984). Traffic studies on the Severn Bridge. *Traffic Engineering and Control*, **25**, 268–271, 274.

Godfrey, L.G. and Pesaran, M.H. (1983). Tests of non-nested regression models: small sample adjustments and Monte-Carlo evidence. *J. Econometr.*, **21**, 133–154.

Guerrero, V.M. and Johnson, R.A. (1982). Use of the Box-Cox transformation with binary response models. *Biometrika*, **69**, 309–314.

Gupta, A.K. (2003). Multivariate skew t-distribution. *Statistics*, **37**, 359–363.

Gurvich, M.R., Dibenedetto, A.T. and Rande, S.V. (1997). A new statistical distribution for characterizing the random strength of brittle materials. *Journal of Materials Science*, **32**, 2559–2564.

Hall, P. (1981). A comedy of errors: the canonical form for a stable characteristic function. *Bull. London Math. Soc.*, **13**, 23–27.

Hall, P. (1987). On the bootstrap and likelihood-based confidence regions. *Biometrika*, **74**, 481–493.

Hall, P. and Stewart, M. (2005). Theoretical analysis of power in a two-component normal mixture model. *Journal of Statistical Planning and Inference*, **134**, 158–179.

Harris, C.M. and Singpurwalla, N.D. (1968). Life distributions derived from stochastic hazard functions. *IEEE Trans. Reliab.*, **R-17**, 70–79.

Harter, L.H. and Moore, A.H. (1965). Maximum likelihood estimation of the parameters of gamma and Weibull populations from complete and censored samples. *Technometrics*, **7**, 639–643.

Harter, L.H. and Moore, A.H. (1966). Local maximum likelihood estimation of the three-parameter lognormal populations from complete and censored samples. *J. Am. Statist. Ass.*, **61**, 842–851.

Hartigan, J.A. (1985). A failure of likelihood asymptotics for the mixture model. *Proc. Berkeley Symp. in Honor of J. Neyman and J. Kiefer* (eds. L. LeCam and R.A. Olshen), vol. II, 807–810. New York: Wadsworth.

Heinrich, J. (2004). *A Guide to the Pearson type IV Distribution.* CDF/MEMO/STATISTICS/PUBLIC /6820. http://www-cdf.fnal.gov/physics/statistics/notes/cdf6820_pearson4.pdf Accessed 22 July 2016.

Henze, N. (1986). A probabilistic representation of the 'skew-normal' distribution. *Scandinavian Journal of Statistics,* **13,** 271–275.

Hernandez, F. and Johnson, R.A. (1980). The large sample behaviour of transformations to normality. *J. Am. Statist. Ass.,* **75,** 855–861.

Hill, B.M. (1963). The three-parameter lognormal distribution and Bayesian analysis of a point-source epidemic. *J. Am. Statist. Ass.,* **58,** 72–84.

Hill, B.M. (1995). Contribution to discussion of Cheng and Traylor (1995). *J. Roy. Statist. Soc. Ser. B,* **57,** 36.

Hill, I.D., Hill, R. and Holder, R.L. (1976). Algorithm AS 99: Fitting Johnson curves by moments. *J. Roy. Statist. Soc. C (Applied Statistics),* **25,** 180–189.

Hinkley, D.V. (1970). Inference about the change-point in a sequence of random variables. *Biometrika,* **57,** 1–17.

Hjorth, U. (1994). *Computer Intensive Statistical Methods.* London: Chapman & Hall.

Holt, D. and Crow, E. (1973). Tables and graphs of the stable probability density functions. *J. of Research of the National Bureau of Standards,* 77**B**, 143–198.

Hosking, J.R.M. (1984). Testing whether the shape parameter is zero in the generalized extreme-value distribution. *Biometrika,* **71,** 367–374.

Hougaard, P. (1984). Life table methods for heterogeneous populations: distributions describing the heterogeneity. *Biometrkia,* **71,** 75–83.

Hougaard, P. (1986). Survival models for heterogeneous populations derived from stable distributions. *Biometrika,* **73,** 387–396.

Howard, A. (1988). Degradation of aramid fibre under stress in dry, humid and hostile environments. *Proc. Composites 88IITT Int. Conf., Nice,* 229–241.

Huet, S., Bouvier, A., Poursat, M.-A. and Jolivet, E. (2004). *Statistical Tools for Nonlinear Regression: A Practical Guide with S-Plus and R Examples.* Second Edition. New York: Springer-Verlag.

Huzurbazar, V.S. (1948). The likelihood equation, consistency and the maxima of the likelihood function. *Ann. Eugen.,* **14,** 185–200.

Ibragimov, I.A. and Linnik, Y.V. (1971). *Independent and Stationary Sequences of Random Variables.* Groningen: Wolters-Nordho.

Ishwaran, H. and James, L. (2002). Approximate Dirichlet process computing in finite normal mixtures. *Journal of Computational and Graphical Statistics* **11,** 508–532.

Ishwaran, H., James, L. and Sun, J. (2001). Bayesian model selection in finite mixtures by marginal density decompositions. *J. Amer. Statist. Assoc.* **96,** 1316–1332.

Ishwaran, H. and Zarepour, M. (2000). Markov chain Monte Carlo in approximate Dirichlet and beta two-parameter process. *Biometrika,* **87,** 371–390.

Ishwaran, H. and Zarepour, M. (2002). Dirichlet prior sieves in finite normal mixtures. *Statistica Sinica,* **12,** 941–963.

Izenman, A.J. and Sommer, C.J. (1988). Philatelic mixtures and multimodal densities. *J. Amer. Statist. Assoc.,* **83,** 941–953.

Jain, K., Singla, N. and Sharma, S.K. (2014). The generalized inverse generalized Weibull distribution and its properties. *Journal of Probability*, **2014**, Article ID 736101, 11 pages.

Jasra, A., Holmes, C.C. and Stephens, D.A. (2005). Markov chain Monte Carlo methods and the label switching problem in Bayesian mixture modelling. *Statistical Science*, **20**, 50–67.

Johnson, N.L. (1949). Systems of frequency curves generated by methods of translation. *Biometrika*, **36**, 149–176.

Johnson, N.L., Kotz, S. and Balakrishnan, N. (1994). *Continuous Univariate Distributions Vol. 1*, Second Edition. New York: Wiley.

Johnson, N.L., Kotz, S. and Balakrishnan, N. (1995). *Continuous Univariate Distributions Vol. 2*, Second Edition. New York: Wiley.

Jones, M.C. and Faddy, M.J. (2003). A skew extension of the t distribution, with applications. *J. Roy. Statist. Soc. B*, **65**, 159–174.

Jones, M.C. (2009). Kumaraswamy's distribution: A beta-type distribution with some tractability advantages. *Statistical Methodology*, **6**, 70–81.

Jørgensen, B. (1982). Identifiability problems in Hadwiger fertility graduation. *Scand. Act. J.*, 103–109.

Jørgensen, B. and Labouriau, R. (2012). *Exponential families and Theoretical Inference*. http://www.impa.br/opencms/pt/biblioteca/mono/Mon_52.pdf 196 pp. Accessed 24 July 2016.

Kempthorne, O. (1966). Some aspects of experimental inference. *J. Am. Statist. Ass.*, **61**, 11–34.

Kingman, J.F.C. and Taylor, S.J. (1966). *Introduction to Measure and Probability*. Cambridge: Cambridge University Press.

Knight, K. (1997). Review of stable non-Gaussian random processes by Samorodnitsky, G. and Taqqu, M.S., Chapman and Hall 1994. *Econometric Theory*, **13**, 133–142.

Kumaraswamy, P. (1980). Generalized probability density-function for double-bounded random-processes. *Journal of Hydrology*, **46**, 79–88.

Lai, C.D., Xie, M. and Murthy, D.N.P. (2003). Modified Weibull model. *IEEE Transactions on Reliability*, **52**, 33–37.

Law, A.M. (2007). *Simulation Modeling and Analysis*. Fourth Edition. New York: McGraw-Hill.

Lawless, J.F. (1982). *Statistical Models and Methods for Lifetime Data*. New York: Wiley.

Le Cam, L. (1970). On the assumptions used to prove asymptotic normality of maximum likelihood estimates. *Ann. Math. Statist.*, **41**, 802–828.

Le Cam, L. (1986). *Asymptotic Methods in Statistical Decision Theory*. New York: Springer-Verlag.

Leroux, B.G. (1992). Consistent estimation of a mixing distribution. *The Annals of Statistics*, **20**, 1350–1360.

Lind, N. C. (1994). Information theory and maximum product of spacings estimation. *J. R. Statist. Soc. B.*, **56**, 341–343.

Liu, X. and Shao, Y. (2004). Asymptotics for the likelihood ratio test in a two-component normal mixture model. *J. Statist. Plann. Inference*, **123**, 61–81.

Loader, C.R. (1991). Inference for a hazard rate change point. *Biometrika*, **78**, 749–757.

Love, P.E.D., Sing, C.-P., Wang, X., Edwards, D.J. and Odeyinka, H. (2013). Probability distribution fitting of schedule overruns in construction projects. *J. of the Operational Research Society*, **64**, 1231–1247.

Lukacs, E. (1969). A characterization of stable processes. *Journal of Applied Probability*, **6**, 409–418.

MacKinnon, J.G. (1983). Model specification tests against non-nested alternatives. *Econometr. Rev.*, **2**, 85–110.

Mallows, C.L. (1973). Some comments on C_p. *Technometrics*, **15** 661–675.

Mallows, C.L. (1995). More comments on C_p. *Technometrics*, **37** 362–372.

Matthews, D.E. and Farewell, V.T. (1982). On testing for a constant hazard against a change-point alternative. *Biometrics*, **38**, 463–468.

Matthews, D.E. and Farewell, V.T. (1985). On a singularity in the likelihood for a change-point hazard rate model. *Biometrika*, **72**, 703–704.

McCullagh, P. (1980). A comparison of transformations of chimpanzee learning data. *GLIM Newsletter*, **2**, 14–18.

McCulloch, J.H. (1996). On the parametrization of the afocal stable distributions. *Bulletin of the London Mathematical Society*, **28** 651–655.

McCulloch, J.H. and Panton, D.B. (1997). Precise tabulation of the maximally-skewed stable distributions and densities *Computational Statistics & Data Analysis*, **23**, 307–320.

McKinnon, K.I.M. (1998). Convergence of the Nelder–Mead simplex method to a non-stationary point. *SIAM Journal on Optimization*, **9**, 148–158.

McLachlan, G. and Peel, D. (2000). *Finite Mixture Models*. New York: Wiley.

Meyer, R.R. and Roth, P.M. (1972). Modified damped least squares: an algorithm for non-linear estimation. *J. Inst. Math. Applns.*, **9**, 218–233.

Miller, R.G. Jr (1981). *Simultaneous Statistical Inference*. Second Edition. New York: Springer-Varlag.

Moran, P.A.P. (1951). The random division of an interval - Part II. *J. R. Statist. Soc. B*, **13**, 147–150.

Moran, P.A.P. (1971). Maximum likelihood estimation in non-standard conditions. *Proc. Camb. Phil. Soc.*, **70**, 441–450.

Mosteller, F. and Tukey, J.W. (1977). *Data Analysis and Regression*. Reading, Mass.: Addison-Wesley.

Nadarajah, S. (2008). On the distribution of Kumaraswamy. *Journal of Hydrology*, **348**, 568–569.

Nadarajah, S. and Kotz, S. (2005). On some recent modifications of Weibull distribution. *IEEE Transactions on Reliability*, **54**, 561–562.

Naik, P.A., Shi, P. and Tsai, C-L. (2007). Extending the Akaike information criterion to mixture regression models. *J. Amer. Statist. Assoc.*, **102**, 244–254.

Nakamura, T. (1991). Existence of maximum likelihood estimates for interval-censored data from some three-parameter models with a shifted origin. *J. R. Statist. Soc. B*, **53**, 211–220.

Nelder, J. and Mead, R. (1965). A simplex method for function minimization. *Computer Journal*, **7**, 308–313.

Nguyen, H.T., Rogers, G.S. and Walker, E.A. (1984). Estimation in change-point hazard rate models. *Biometrika*, **71**, 299–304.

Nigm, A.M. (1988). Prediction bounds for the Burr model. *Commun. Statist. -Theory Meth.*, **17**, 287–297.

Nishii, R. (1984). Asymptotic properties of criteria for selection of variables in multiple regression. *The Annals of Statistics*, **12**, 758–765.

Nolan, J.P. (1997). Numerical calculation of stable densities and distribution functions. *Commun. Statist. -Stochastic Models*, **13**, 759–774.

Nolan, J.P. (2005). Modeling financial data with stable distributions. Chapter 3 in *Handbook of Heavy Tailed Distributions in Finance: Handbooks in Finance* edited by S.T Rachev. Amsterdam: Elsevier Science, 102–130.

Nolan, J.P. (2015). *Stable Distributions - Models for Heavy Tailed Data,* Boston: Birkhauser. Note: In progress, Chapter 1 online at academic2.american.edu/~jpnolan

Owen, D.B. (1956). Tables for computing bivariate normal probabilities. *Ann. Math. Statist.,* **27,** 1075–1090.

Pace, L. and Salvan, A. (1990). Best conditional tests for separate families of hypotheses. *J. Roy. Statist. Soc. B,* **52,** 125–134.

Panton, D.B. (1993). Distribution function values for logstable distributions. *Computer Math. Applic.,* **25,** 9, 17–24.

Patefield, W.M. (1977). On the maximized likelihood function. *Sankhya B,* **39,** 92–96.

Pearson, E.S. (1963). Some problems arising in approximating to probability distributions, using moments. *Biometrika,* **50,** 95–112.

Pearson, K. (1895). Contributions to the mathematical theory of evolution. II. Skew variation in homogeneous material. *Philosophical Transactions of the Royal Society of London. A,* **186,** 343–414.

Pearson, K. (1916). Mathematical contributions to the theory of evolution. XIX. Second supplement to a memoir on skew variation. *Philosophical Transactions of the Royal Society of London. Series A,* **216,** 429–457.

Perlman, M.D. (1972). On the strong consistency of approximate maximum likelihood estimators. *Proc. of the Sixth Berkeley Symposium,* **1,** 263–282.

Pham, T.D. and Nguyen, H. T. (1990). Strong consistency of the maximum likelihood estimator in the change-point hazard rate model, *Statistics,* **21,** 203–216.

Pham, T.D. and Nguyen, H.T. (1993). Bootstrapping the change-point of a hazard rate. *Ann. Inst. Statist. Math.* **45,** 331–340.

Prentice, R.L. (1974). A log gamma model and its maximum likelihood estimation. *Biometrika,* **61,** 539–544.

Prentice, R.L. (1975). Discrimination among some parametric models. *Biometrika,* **62,** 607–614.

Prentice, R.L. (1976). A generalization of the probit and logit methods for dose response curves. *Biometrics,* **32,** 761–768.

Prescott, P. and Walden, A.T. (1980). Maximum likelihood estimation of the parameters of the generalized extreme-value distribution. *Biometrika,* **67,** 723–724.

Ranneby, B. (1984). The maximum spacing method: an estimation method related to the maximum likelihood method. *Scand. J. Statist.,* **11,** 93–112.

Ranneby, B. and Ekström, M. (1997). *Maximum spacing estimates based on different metrics.* Research Report No. 5, Department of Mathematical Statistics, Ume University.

Ranneby, B., Jammalamadaka, S.R. and Teterukovskiy, A. (2005). The maximum spacing estimation for multivariate observations. *J. Statist. Plann. Inference,* **129,** 427–446.

Ratkowsky, D.A. (1983). *Nonlinear Regression Modeling.* New York: Dekker.

Redner, R. (1981). Note on the consistency of the maximum likelihood estimate for nonidentifiable distributions. *Ann. Statist.,* **9,** 225–228.

Richardson, S. and Green, P. (1997). On Bayesian analysis of mixtures with an unknown number of components. *J. Roy, Statist. Soc. B,* **59,** 731–792.

Rinne, H. (2008). *The Weibull Distribution: A Handbook.* Boca Raton: CRC Press/Chapman and Hall.

Ritz, C. and Streibig, J.C. (2008). *Nonlinear Regression with R.* New York: Springer.

Rodriguez, R.N. (1977). A guide to the Burr Type XII distributions. *Biometrika,* **64,** 129–134.

Roeder, K. (1992). Semiparametric estimation of normal mixture densities. *Ann. Statist.,* **20,** 929–943.

Rotnitzky, A., Cox, D.R., Bottai, M. and Robins, J. (2000). Likelihood-based inference with singular information matrix. *Bernoulli,* **6,** 243–284.

Rousseau, J. and Mengersen, K. (2011). Asymptotic behaviour of the posterior distribution in overfitted mixture models. *J. Roy. Statist. Soc. B,* 73, 689–710.

Royston, P. (1993). Some useful three-parameter distributions arising from the shifted power transformation. Technical Report, Royal Postgraduate Medical School, Hammersmith Hospital, London.

Royston, P. and Thompson, S.G. (1995). Comparing non-nested regression models. *Biometrics,* **51,** 114–127.

Sakia, R.M. (1992). The Box-Cox transformation technique: a review. *The Statistician,* **41,** 169–178.

Salvan, A. (1986). Locally most powerful invariant test of normality (In Italian). *Atti XXXIII Riunione Societ Italiana di Statistica,* **2,** 173–179.

Samorodnitsky, G. and Taqqu, M. (1994). *Stable Non-Gaussian Random Processes.* New York: Chapman and Hall.

Sartori, N. (2006). Bias prevention of maximum likelihood estimates for scalar skew normal and skew t distributions. *Journal of Statistical Planning and Inference,* **136,** 4259–4275.

Schwarz, G. (1978). Estimating the dimension of a model. *Annals of Statistics,* **6,** 461–464.

Searle, S.R. (1971). *Linear Models.* New York: Wiley.

Seber, G.A.F. and Wild, C.J. (2003). *Nonlinear Regression.* Hoboken, New Jersey: Wiley.

Self, S.G. and Liang, K.-Y. (1987). Asymptotic properties of maximum likelihood estimators and likelihood ratio tests under nonstandard conditions. *J. Amer. Statist. Assoc.,* **82,** 605–610.

Shao, Q. (2002). A reparameterisation method for embedded models. *Communications in Statistics - Theory and Methods,* **31,** 683–697.

Shao, Q., Wong, H., Xia, J. and Ip, W.-C. (2004). Models for extremes using the extended three parameter Burr XII system with application to flood frequency analysis. *Hydrological Sciences Journal,* **49,** 685–702.

Shao, Y. (2001). Consistency of the maximum product of spacings method and estimation of a unimodal distribution. *Statistica Sinica,* **11,** 1125–1140.

Shao, Y. and Hahn, M. G. (1999). Strong consistency of the maximum product of spacings estimates with applications in nonparametrics and in estimation of unimodal densities. *Ann. Inst. Statist. Math.,* **51,** 31–49.

Shibata, R. (1981). An optimal selection of regression variables. *Biometrika,* **68,** 45–54.

Shuster, J. (1968). On the inverse Gaussian distribution function. *J. Amer. Statist. Assoc.,* **63,** 1514–1518.

Simonoff, J.S. and Tsai, C.-L. (1989). The use of guided reformulations when collinearities are present in nonlinear regression. *Appl. Statist.,* **38,** 115–126.

Smith, R.L. (1985). Maximum likelihood estimation in a class of non-regular cases. *Biometrika*, **72**, 67–92.

Smith, R.L. (1986). Maximum likelihood estimation for the NEAR(2) model. *J. R. Statist. Soc. B*, **48**, 251–257.

Smith, R.L. (1989). A survey of non-regular problems. *Proc. Inst. Statist. Inst. Conf. 47th Session, Paris*, 353–372.

Smith, R.L. and Naylor, J.C. (1987). A comparison of maximum likelihood and Bayesian estimators of the three-parameter Weibull distribution. *Appl. Statist.*, **36**, 358–369.

Stacy, E.W. (1962). A generalisation of the gamma distribution. *Ann. Math. Statist.*, **33**, 1187–1192.

Steen, P.J. and Stickler, D.J. (1976). *A sewage pollution study of beaches from Cardiff to Ogmore. Dept. Appl.* Biology Report. University of Wales Institute of Science and Technology, Cardiff.

Stephens, M.A. (1970). Use of the Kolmogorov-Smirnov, Cramér-von Mises and related statistics without extensive tables. *Journal of the Royal Statistical Society, Series B*. **32** 115–122.

Stephens, M.A. (1974). EDF statistics for goodness-of-fit and some comparisons. *J. Amer. Statist. Assoc.*, **69**, 730–737.

Stuart, A. and Ord, J.K. (1987). *Kendall's Advanced Theory of Statistics*, Fifth Edition vol. 1. London: Charles Griffin.

Stuart, A. and Ord, J.K. (1991). *Kendall's Advanced Theory of Statistics*, Fifth Edition vol. 2. London: Charles Griffin.

Stute, W., Mantega, W.G. and Quindimil, M.P. (1993). Bootstrap based goodness-of-fit-tests. *Metrika*. **40**, 243–256.

Tadesse, M.G., Sha, N. and Vannucci, M. (2005). Bayesian variable selection in clustering high-dimensional data. *J. Amer. Statis. Assoc.*, **100**, 602–617.

Tadikamalla, P.R. (1980). A look at the Burr and related distributions. *Inter. Statist. Rev.*, **48**, 337–344.

Taha, H.A. (2003). *Operations Research: an Introduction*. Seventh Edition. Upper Saddle River, NJ: Prentice Hall.

Thode, H.C., Finch, S.J. and Mendell, N.R. (1988). Simulated percentage points for the null distribution of the likelihood ratio test for a mixture of two normals. *Biometrics*, bf 44, 1195–1201.

Thornton, K.M. (1989). The use of sample spacings in parameter estimation with applications. PhD Thesis, University of Wales.

Tibshirani, R., Walther, G. and Hastie, T. (2001). Estimating the number of clusters in a data set via the Gap statistic. *J. Roy. Statist. Soc. B*, **63**, 411–423.

Titterington, D. M. (1985). Comment on 'Estimating parameters in continuous univariate distributions'. *J. R. Statist. Soc. B*, **47**, 115–116.

Titterington, D.M., Smith, A.F.M. and Makov, W.E. (1985). *Statistical Analysis of Finite Mixture Distributions*. New York: Wiley.

Traylor, L. (1994). Maximum likelihood methods applied to non-regular problems. PhD Thesis, University of Wales.

Tukey, J.W. (1949). One degree of freedom for non-additivity. *Biometrics*, **5**, 232–242.

Tukey, J.W. (1957). The comparative anatomy of transformations. *Annals of Mathematical Statistics*, **28**, 602–632.

Wald, A. (1943). Tests of statistical hypotheses concerning several parameters when the number of observations is large. *Transactions of the American Mathematical Society*, **54**, 426–482.

Wald, A. (1949). Note on the consistency of maximum likelihood estimate. *Ann. Math. Statist.*, **20**, 595–601.

Walker, A.J. (1977). An efficient method for generating discrete random variables with general distributions. *Assoc. Comput. Mach. Trans. Math. Software*, **3**, 253–256.

Wasserman, L. (2012). Mixture models: the twilight zone of statistics. https://normaldeviate.wordpress.com/2012/08/04/mixture-models-the-twilight-zone-of-statistics/ Accessed: 20 Sept. 2016.

Wei, J., Dong, F., Li, Y. and Zhang, Y. (2014). Relationship analysis between surface free energy and chemical composition of asphalt binder. *Construction and Building Materials*, **71**, 116–123.

Wilks, S.S. (1938). The large-sample distribution of the likelihood ratio for testing composite hypotheses. *Annal. Math. Statist.*, **9**, 60–62.

Wingo, D.R. (1975). The use of interior penalty functions to overcome lognormal distribution parameter estimation anomalies. *J. Statist. Computn. Simuln.*, **4**, 49–61.

Wingo, D.R. (1976). Moving truncating barrier-function method for estimation in three-parameter lognormal models. *Communs. Statist. Simuln. Computn.*, **5**, 65–80.

Wingo, D.R. (1983). Maximum likelihood methods for fitting the Burr Type XII distribution to life test data. *Biom. J.*, **25**, 77–84.

Wolfowitz, J. (1949). On Wald's proof of the consistency of the maximum likelihood estimate. *Ann. Math. Statist.*, **20**, 601–602.

Woodroofe, M. (1974). Maximum likelihood estimation of translation parameter of truncated distribution II. *Ann. Statist.*, **2**, 474–88.

Worsley, K.J. (1986). Confidence regions and tests for a change-point in a sequence of exponential family random variables. *Biometrika*, **73**, 91–104.

Wu, C.F.J. and Hamada, M. (2000). *Experiments: Planning, Analysis, and Parameter Design Optimization*. New York: Wiley.

Xie, M., Tang, Y. and Goh, T.N. (2002). A modified Weibull extension with bathtub-shaped failure rate function. *Reliability Engineering & System Safety* **76**, 279–285.

Yao, Y.C. (1986). Maximum likelihood estimation in hazard rate model with a change-point, *Comm. Statist. Theory and Methods*, **15**, 2455–2466.

Yao, Y.C. (1987). A note on testing for constant hazard against a change-point alternative, *Ann. Inst. Statist. Math.*, **39**, 377–383.

Young, G.A. and Smith, R.L. (2005). *Essentials of Statistical Inference*. Cambridge: Cambridge University Press.

Zhao, X., Wu, X. and Zhou, X. (2009). A change-point model for survival data with long-term survivors. *Statistica Sinica*, bf 19, 377–390.

Zhu, D. and Galbraith, J.W. (2009). *A generalized asymmetric Student-t distribution with application to financial econometrics*. Montreal: Cirano Scientific Series pp 1–37.

Zolotarev, V.M. (1964). On the representation of stable laws by integrals, *Trudy Mat. Inst. Steklov.*, **71**, 46–50.

Zolotarev, V.M. (1986). *One-dimensional stable distributions. Translations of Mathematical Monographs*, vol. 65. Providence: American Mathematical Society

Author Index

Subject Index

Notes: Tables and figures are indicated by an italic *t*, and *f* following the page number. *vs.* indicates a comparison